SOCIAL, ETHICAL, AND POLICY IMPLICATIONS OF ENGINEERING

IEEE Press
445 Hoes Lane, P.O. Box 1331
Piscataway, NJ 08855-1331

SOCIAL, ETHICAL, AND POLICY IMPLICATIONS OF ENGINEERING

Selected Readings

Edited by

Joseph R. Herkert
North Carolina State University

IEEE Society on Social Implications of Technology, *Sponsor*

A Selected Reprint Volume

The Institute of Electrical and Electronics Engineers, Inc., New York

This book may be purchased at a discount from the publisher
when ordered in bulk quantities. Contact:

IEEE Press Marketing
Attn: Special Sales
445 Hoes Lane, P.O. Box 1331
Piscataway, NJ 08855-1331
Fax: (732) 981-9334

For more information on the IEEE Press,
visit the IEEE Press Home Page: http://www.ieee.org/organizations/pubs/press

Printed in the United States of America

10 9 8 7 6 5 4 3 2 1

ISBN 0-7803-4712-9

IEEE Order Number: PP5397

Library of Congress Cataloging-in-Publication Data

Social, ethical, and policy implications of engineering : selected
readings / edited by Joseph R. Herkert.
p. cm.
"A selected reprint volume."
Includes bibliographical references and index.
ISBN 0-7803-4712-9
1. Engineering ethics. 2. Engineering—Government policy.
3. Engineering—Social aspects. I. Herkert, Joseph R., 1948–
TA 157. 5628 2000
174′.962—dc21 99-19982
CIP

Contents

Preface ix

Acknowledgments xi

PART I TECHNOLOGY, ENGINEERING, AND SOCIETY **1**

Chapter 1 Technology, Values, and Society **3**

Technology as "Big Magic and Other Myths"
Langdon Winner (*IEEE Technology and Society Magazine*, Vol. 17, No. 3, Fall 1998) (interview) **7**
Is Technology Neutral?
Robert J. Whelchel (*IEEE Technology and Society Magazine*, Vol. 5, No. 4, December 1986) **20**
Obscuring the Human Costs of Expert Systems
Todd D. Cherkasky (*IEEE Technology and Society Magazine*, Vol. 14, No. 1, Spring 1995) **25**
Information and Medical Ethics: Protecting Patient Privacy
Reid Cushman (*IEEE Technology and Society Magazine*, Vol. 15, No. 3, Fall 1996) **36**

Chapter 2 The Social Context of Engineering **45**

Engineers as Revolutionaries
Don E. Kash (*IEEE Technology and Society Magazine*, Vol. 9, No. 3, September/October 1990) **49**
The Humanist Engineer of Aleksandr Solzhenitsyn
Ingrid H. Soudek (*IEEE Technology and Society Magazine*, Vol. 5, No. 3, September 1986) **57**
Electricity and Socialism: The Career of Charles P. Steinmetz
Ronald Kline (*IEEE Technology and Society Magazine*, Vol. 6, No. 2, June 1987) **61**
Engineering and Social Equality: Herbert Hoover's Manifesto
Thomas A. Long (*IEEE Technology and Society Magazine*, Vol. 3, No. 3, September 1984) **70**

PART II SOCIAL AND ETHICAL RESPONSIBILITIES OF ENGINEERS **75**

Chapter 3 Moral Dilemmas in Engineering **77**

Collective and Individual Moral Responsibility in Engineering: Some Questions
John Ladd (*IEEE Technology and Society Magazine*, Vol. 1, No. 2, June 1982) **81**

Explaining Disasters: The Case for Preventive Ethics
Charles E. Harris, Jr. (*IEEE Technology and Society Magazine*, Vol. 14, No. 2, Summer 1995) **89**
"High-Reliability" Organizations and Technical Change: Some Ethical Problems and Dilemmas
Gene I. Rochlin (*IEEE Technology and Society Magazine*, Vol. 5, No. 3, September 1986) **95**
The Bureaucratic Ethos and Dissent
Robert Jackall (*IEEE Technology and Society Magazine*, Vol. 4, No. 2, June 1985) **102**

Chapter 4 Frameworks for Engineering Ethics 113

The Public Health, Safety, and Welfare: An Analysis of the Social Responsibilities of Engineers
Michael C. McFarland (*IEEE Technology and Society Magazine*, Vol. 5, No. 4, December 1986) **118**
Would Helping Ethical Professionals Get Professional Societies into Trouble?
Stephen H. Unger (*IEEE Technology and Society Magazine*, Vol. 6, No. 3, September 1987) **127**
Technological Hazards and the Engineer
Roland Schinzinger (*IEEE Technology and Society Magazine*, Vol. 5, No. 2, June 1986) **132**
The Role of the Law in Protecting Scientific and Technical Dissent
Alfred G. Feliu (*IEEE Technology and Society Magazine*, Vol. 4, No. 2, June 1985) **137**

PART III ENGINEERING ETHICS AND PUBLIC POLICY 145

Chapter 5 Technology Policy and Ethical Issues 147

Engineers and Social Responsibility: An Obligation to Do Good
Stephen Cohen and Damian Grace (*IEEE Technology and Society Magazine*, Vol. 13, No. 3, Fall 1994) **152**
When Expert Advice Works, and When It Does Not
E. J. Woodhouse and Dean Nieusma (*IEEE Technology and Society Magazine*, Vol. 16, No. 1, Spring 1997) **160**
Do We Need a Technology Policy?
S. J. Kline and D. E. Kash (*IEEE Technology and Society Magazine*, Vol. 11, No. 2, Summer 1992) **167**
Task for Engineers—Resuscitating the U.S. Rail System
John E. Ullmann (*IEEE Technology and Society Magazine*, Vol. 12, No. 4, Winter 1993) **176**

Chapter 6 Risk and Product Liability 185

The Conceptual Risks of Risk Assessment
K. S. Shrader-Frechette (*IEEE Technology and Society Magazine*, Vol. 5, No. 2, June 1986) **190**
Ethical Risk Assessment: Valuing Public Perceptions
Joseph R. Herkert (*IEEE Technology and Society Magazine*, Vol. 13, No. 1, Spring 1994) **198**
Health Risk Valuations Based on Public Consent
E. S. Cassedy (*IEEE Technology and Society Magazine*, Vol. 11, No. 4, Winter 1992/1993) **205**

Chapter 7 Engineering and Sustainable Development 215

Sustainability and the Design of Knowledge Tools
Alex Farrell (*IEEE Technology and Society Magazine*, Vol. 15, No. 4, Winter 1996/1997) **220**
The Role of Technology in Sustainable Development
Sharon Beder (*IEEE Technology and Society Magazine*, Vol. 13, No. 4, Winter 1994/1995) **230**

A Materials Life Cycle Framework for Preventive Engineering
Steven B. Young and Willem H. Vanderburg (*IEEE Technology and Society Magazine*, Vol. 11, No. 3, Fall 1992) **236**

Chapter 8 Engineering in a Global Context 243

The Ethics of Global Risk
Thomas Donaldson (*IEEE Technology and Society Magazine*, Vol. 5, No. 2, June 1986) **247**
Distributing Costs of Global Climate Change
John Byrne, Constantine Hadjilambrinos, and Subodh Wagle (*IEEE Technology and Society Magazine*, Vol. 13, No. 1, Spring 1994) **252**
Developing Telecommunications in Ghana
Godfred Frempong (*IEEE Technology and Society Magazine*, Vol. 15, No. 1, Spring 1996) **261**

Chapter 9 Technology and Health Care 271

Ethical Issues of Life-Sustaining Technology
Nicholas Cram, John Wheeler, and Charles S. Lessard (*IEEE Technology and Society Magazine*, Vol. 14, No. 1, Spring 1995) **274**
Genetic Tests: Evolving Policy Questions
David A. Asch and Michael T. Mennuti (*IEEE Technology and Society Magazine*, Vol. 15, No. 4, Winter 1996/1997) **282**
Defects and Deceptions—The Bjork-Shiley Heart Valve
John H. Fielder (*IEEE Technology and Society Magazine*, Vol. 14, No. 3, Fall 1995) **289**

Chapter 10 Information Technology 295

Humanizing the Information Superhighway
Michael C. McFarland (*IEEE Technology and Society Magazine*, Vol. 14, No. 4, Winter 1995/1996) **299**
Emerging Challenge: Security and Safety in Cyberspace
Richard O. Hundley and Robert H. Anderson (*IEEE Technology and Society Magazine*, Vol. 14, No. 4, Winter 1995/1996) **307**
Hidden Costs and Benefits of Government Card Technologies
Tom Hausken and Paula Bruening (*IEEE Technology and Society Magazine*, Vol. 13, No. 2, Summer 1994) **317**

Appendix 327

Author Index 331

Subject Index 333

About the Editor 339

Preface

THE Engineering Criteria 2000 of the Accreditation Board for Engineering and Technology (ABET) stipulate that graduates of engineering programs have:

1. an ability to apply knowledge of mathematics, science, and engineering
2. an ability to design and conduct experiments, as well as to analyze and interpret data
3. an ability to design a system, component, or process to meet desired needs
4. an ability to function on multidisciplinary teams
5. an ability to identify, formulate, and solve engineering problems
6. an understanding of professional and ethical responsibility
7. an ability to communicate effectively
8. the broad education necessary to understand the impact of engineering solutions in a global and societal context
9. a recognition of the need for, and an ability to engage in, life-long learning
10. a knowledge of contemporary issues
11. an ability to use the techniques, skills, and modern engineering tools necessary for engineering practice

These criteria are widely recognized as a mandate for professional training for engineers that goes beyond merely technical specialties. In recent decades, however, the overwhelming focus of engineering education has been on just that—creating engineering practitioners who are more specialized. As a result, few resources have been developed to address the breadth stipulated in the new ABET criteria.

The primary purpose of this book is to provide such a resource for engineering educators, students and practitioners, especially with respect to ABET criteria 6, 8, and 10: an understanding of professional and ethical responsibility, the broad education necessary to understand the impact of engineering solutions in a global and societal context, and a knowledge of contemporary issues. Professional and ethical responsibility is a thread that runs throughout this volume. Unlike most conventional treatments of engineering ethics, however, I have endeavored to demonstrate the theoretical and practical linkages between engineering ethics and issues that might more broadly be identified as social implications of technology. Hence, significant attention is given to the impact of engineering solutions in a global and societal context and contemporary issues, particularly policy and ethical issues relating to the development and use of technology.

A secondary purpose of this work is to anthologize some of the significant articles published in *IEEE Technology and Society Magazine* (*T&S*) over the past 17 years. All of the background readings and cases for discussion are reprinted from *T&S*, a publication of the Society on Social Implications of Technology (SSIT) of the Institute of Electrical and Electronics Engineers (IEEE). IEEE-SSIT recently celebrated its 25th anniversary (having begun as the Committee on Social Implications of Technology in the early 1970s). The reader will find that the vein of articles mined from *T&S* over the years is consistent with my approach to integrating conventional treatments of engineering ethics with social and policy analysis. Many of the articles taken from the 1980s focus explicitly on theoretical and applied aspects of engineering ethics, whereas more recent articles, though often grounded in ethics, are more oriented toward contemporary social issues and policy problems arising from the prominence of technology in society.

The book is organized in three parts, dealing respectively with the social, ethical, and policy contexts of engineering. Each part has two or more chapters with each chapter consisting of an extended introduction, one or two background readings reprinted from *T&S*, and one or two readings from *T&S* that contain specific case studies. The activities of SSIT and the scope of *T&S* are so wide ranging that the articles published in *T&S* and reprinted here should be of interest to all engineers, regardless of technical specialization, as well as others interested in the social, ethical, and policy implications of engineering.

Joseph R. Herkert
North Carolina State University

Acknowledgments

No reprint volume would be possible without the original contributions of the authors. Space limitations forced me to make difficult decisions, and a number of strong articles were left on the cutting-room floor. I was in the enviable position of having too much quality material to choose from due to the efforts of the authors themselves and the high standards set for *T&S* over the years by its editors: Norman Balabanian (twice), Robert Whelchel (twice), Leon Zelby, and Ronald Kline. Thanks are also due to the Managing Editor of *T&S*, Terri Bookman, and to the many guest editors and anonymous reviewers who have contributed to the excellence of the magazine. Special thanks to Ron Kline, who originally suggested the idea of a *T&S* reprint book and graciously handed the project off to me when IEEE Press expressed an interest, and to Steve Unger for supplying original copies of some of the reprinted articles.

I am grateful to all of the folks at IEEE Press for their encouragement, advice, and hard work on this project, especially Ken Moore (Director), Karen Hawkins (Executive Editor), Linda Matarazzo (Assistant Editor), and Mark Morrell (Assistant Production Editor). Karen was especially helpful in fine-tuning the scope of the volume.

I also owe a debt of gratitude to all of the individuals who have provided me with encouragement and mentoring in connection with the study of engineering ethics, especially Stephen Unger, Michael Pritchard, Vivian Weil, and Caroline Whitbeck. They bear no responsibility, however, for any errors or omissions in this work.

Portions of the text are based on material that first appeared elsewhere. I am grateful to the publishers who have kindly granted permission for its use, as follows: Part II (pp. 75–76) and portions of Chapters 2 (pp. 46–47), 3 (pp. 77–79), 4 (pp. 113–116), and 6 (p. 187) first appeared in, Joseph R. Herkert, "Ethics and Professional Responsibility," in John G. Webster, ed., *Encyclopedia of Electrical and Electronics Engineering*, Vol. 7: 173–182. Copyright © John Wiley & Sons, Inc., 1999. Reprinted by permission of John Wiley & Sons, Inc.; portions of Chapters 7 (pp. 215–217) and 8 (pp. 243–244) in, Joseph R. Herkert, "Sustainable Development, Engineering and Multinational Corporations: Ethical and Public Policy Implications." *Science and Engineering Ethics* **4**(3): 333–346. Copyright © 1998 by Opragen Publications; portions of Chapters 5 (pp. 148–149) and 10 (pp. 295, 296) in, Joseph R. Herkert, "Introduction: Special Issue on Computers and Society at a Time of Sweeping Change." *Social Science Computer Review* **16**(3): 237–239. Copyright © 1998 by Sage Publications, Inc. Reprinted by permission of Sage Publications, Inc.; portion of Chapter 5 (p. 149) in, Joseph R. Herkert (1995) "On Throwing Out the Baby" (President's Message). *IEEE Technology and Society Magazine* **14**(3): 2–3; Part I (pp. 1–2) and portion of Chapter 10 (p. 295) in, Joseph R. Herkert (1995) "Visualizing Grander Challenges" (President's Message). *IEEE Technology and Society Magazine* **14**(2): 2–5.

Last but not least, special thanks to my students and colleagues in Multidisciplinary Studies at North Carolina State University for providing a delightful and intellectually challenging work environment, and to Nell and "the guys" for their love and support.

Joseph R. Herkert
North Carolina State University

PART I

Technology, Engineering, and Society

VISUALIZING the future of technology and its social implications is an important activity within professional engineering societies and society at large. For example, the IEEE Technical Activities Board, the governing board for its 37 technical societies and councils, has a committee on New Technology Directions. This committee publishes periodic reports from the technical societies and councils on emerging technologies within their fields of interest, and a few years back compiled a list of "Grand Challenges" for electrotechnology (1). The Grand Challenges include the following items:

• Allow easy access to knowledge and information (e.g., to have instant access to all information)
• Simplify the transactions of daily life (e.g., the paperless office)
• Provide ready access to improved health (e.g., reliable, cost-effective medical diagnostics and prosthesis)

Although the committee should be commended for its work in identifying emerging technologies and in highlighting challenges for the electrical and computer engineering profession, confronting the social implications of technology entails considerable additional effort. For example, while the technical means to realize the paperless office and instant access to all information certainly exist, or will soon exist, serious questions need to be raised about the social implications of such developments. For example, do we really want or need a paperless office and instant information access? What ramifications would such developments have for individuals and for organizations? How would such changes affect the *quality* of communication and information? The social implications of technology pose questions relating to fundamental human values that individuals and social institutions have struggled with for centuries: individual freedom, dignity, and privacy; personal growth and fulfillment through meaningful education, work, and leisure; the making and remaking of the natural, built, and social environments; and decision making in a democratic society.

In this spirit, I suggest that grander challenges are embedded in the relationship between technology and society:

1. To insist on ethical behavior within the engineering professions, not merely because such a posture enhances the image of the profession, but because engineers have personal and professional obligations to act in an ethical manner.
2. To confront questions of risk assessment and other areas of sociotechnical decision making in a manner that incorporates and respects both the specialized knowledge, values, and experience of technical experts and the cultural knowledge, values, and experience of nonexperts.
3. To develop communication and information systems that improve the quality of information exchange, do not lend themselves to isolation or alienation of individuals and groups, and are sensitive to fundamental democratic notions, including personal freedom, dignity, and privacy.
4. To achieve equilibrium between ecological, economic, and social systems that provides for sustainable development for all peoples and nations of the world.
5. To provide health care that enhances quality of life without undermining human capabilities to deal with the value choices made possible or necessary by medical and biotechnologies.
6. To provide opportunities for life-long education, meaningful work, and rewarding leisure for all members of society.
7. To foster community, understanding, and peace throughout the world.

Some engineers might object to these challenges on the grounds that they come out of a critical tradition that is ultimately "antitechnology" in nature. Indeed, in recent years we have witnessed a considerable backlash against those who would identify with the Science, Technology, and Society or the Science and Technology Studies fields (collectively known as STS, although the emphases of these interrelated fields are somewhat different). I find this objection to be quite curious, since criticism is a fundamental tenet of all scientific and technological undertaking. But while we welcome critical review of our scientific theories and technical designs, we often look askance at someone who would question the purpose of a particular technology or wonder what the technology's unanticipated impacts might be. To draw an analogy from another field, perhaps the most well-known critics in the United States are the film critics Roger Ebert and the late Gene Siskel. Although they are often derided as "the fat guy" and "the balding guy," and their judgments on individual films sometimes run against the grain of box office popularity, two points are clear. By often focusing on films that are not products of the major Hollywood studios, they have helped lift some of the best films of our generation from relative obscurity into the national spotlight. And, most importantly, they are as devoted advocates of films and filmmaking as you will find anywhere.

Similarly, most, if not all, participants in STS and related fields, far from being antitechnology, are well aware of technology's past, present, and potential contributions to humanity. The Administrative Committee of the IEEE Society on Social Implications of Technology, for example, is comprised primarily of electrical and computer engineers from academia and industry. It includes a number of IEEE fellows, several former members of the IEEE Board of Directors, the editor of *Engineering in Medicine and Biology Magazine*, and current or past presidents of a number of technical societies and boards. Although SSIT often offers a critical view, we do so in the spirit of improving the responsiveness of electrical and computer engineers to personal and societal values. And, like Siskel and Ebert, we hope to shed light on technological developments that might not initially garner mainstream attention but that nonetheless have the potential to make significant contributions to humanity. For example, our 1991 International Symposium on Technology and Society, held in Toronto, focused on the theme of sustainable development, which has since become a topic of growing importance both within the IEEE and on the national and international political scenes (see Chapter 7).

Another objection to the challenges I have outlined might be that they are too idealistic—that it is not practical to assume that we could ever achieve a world free of unethical professional behavior, personal alienation, ecological damage, or military conflict. In other words, engineers should focus on solvable problems and not waste their time on unreachable goals. In replying to this objection, I will quote Eugene Ferguson (2), an engineer and noted historian of technology: "Social problems can have social solutions, which usually require discussion and compromise. Unless we insist on this, we can expect technological solutions to all problems because they are easiest to devise." (p. 31)

It might, indeed, be easier to design a paperless office or provide for instant access to all information than it is to deal with the social implications of these developments. This fact is all the more reason why engineers should not abandon the quest for social solutions. By clinging to the notion that we should only be concerned with technical matters, we disqualify ourselves and our profession from making vital contributions to the solution of social problems. Conversely, if we insist that social problems can be effectively addressed only by technological solutions, we risk building a civilization that accepts technological development without reflection on its human impacts.

My vision of the future is one in which individual engineers, our industries and universities, and our professional organizations continue to lead the way, not only in developing new and innovative technologies, but also in moving ethical concerns and awareness of the social implications of our work to the forefront of our efforts. This section lays the groundwork for that vision by exploring the relationships among technology, values, and society and the societal context of engineering.

References

[1] IEEE New Technology Directions Committee, *The Grand Challenges* (Available on the World Wide Web at http://www.ieee.org/newtech/challenges/.)

[2] Ferguson, E. 1979. The imperatives of engineering. In *Connections: technology and change,* ed. J. Barke et al., 29–31. San Francisco: Bond & Fraser.

Chapter 1

Technology, Values, and Society

TECHNOLOGY is pervasive in modern society. For example, virtually every section of a single issue of the *New York Times* (January 13, 1992) that I am fond of showing my students contains articles relating to some aspect of science or technology. This includes the front page, local section, national section, international section, business section, sports section, arts section, and even the celebrity column.

Despite technology's omnipresence in modern life, however, the average person tends to be characterized by what political philosopher Langdon Winner calls (1) "technological somnambulism"—that is, when it comes to technology, we sleepwalk through life taking its benefits for granted and pausing to consider its deeper character only when things go wrong. Although many nonengineers approach technology with a combination of awe and fear, engineers often tend to focus only on its technical character to the neglect of its societal context.

Defining Technology

In everyday usage technology often refers merely to machines and their use by human beings. However, the term can be defined in many richer ways. Following are some examples:

- The organization of knowledge for practical purposes—Emmanuel G. Mesthene (2)
- . . . the intelligent organization and manipulation of materials for useful purposes—Stephen Unger (3)
- Systems of rationalized control over large groups of men, events, and machines by small groups of technically skilled men operating through organizational hierarchy—John McDermott (4)
- . . . the knack of so arranging the world that we don't have to experience it—Max Frisch (5)

While many, if not most, definitions of technology seem to reflect the biases of the author, we often find common elements in such definitions. Technology is variously described as hardware, organized knowledge, applied science, or know-how. The purpose ascribed to technology often entails one or more of the following: practical purposes, production of goods and services, extension of human capabilities, or control of the environment. Some recent definitions of technology, however, attempt to incorporate living in harmony with the environment.

Many contemporary authors highlight the systems orientation of technology, for example, S. J. Kline (6), who focuses on the concept of sociotechnical systems of manufacture and use. These concepts are useful not only because they recognize the importance of technological systems, but also because they explicitly acknowledge the social context in which technology is developed and utilized.

Rather than defining technology, engineer Norman Balabanian (who served as the first editor of *Technology and Society Magazine*, with a second stint after an interlude of a few years) describes the dimensions or elements of technology (7):

1. *Physical objects*—hardware—tools, instruments, machines, weapons, appliances. This is the original, elementary conception of technology.
2. *Knowledge*—not abstract, scientific knowledge but *know-how*—methods, processes, *technique*. For [some people], this dimension *is* technology.
3. *Human beings*—not autonomous individuals but people who are largely interchangeable with one another—have the appropriate know-how to operate the hardware.
4. *Organization and system*—the organized structures, the

integrated networks within which the hardware is embedded and technique employed; the linkages that tie together the hardware and the technique with the social institutions.

5. *Political and economic power*—this dimension is implicit in the preceding one but it should be acknowledged explicitly.

Views of Technology

Building on such definitions as those cited above, various views of technology and its relationship with society emerge. A naively optimistic view described by Mesthene (2) sees "technology as an unalloyed blessing for man and society. Technology is . . . the motor of all progress," (p. 74).

According to Mesthene (2), an equally naive negative view is that "technology is an unmitigated curse. . . . It is seen as autonomous and uncontrollable" (p. 74). An extension of this view is the notion of technological determinism—that is, that technology marches to the beat of its own drum and people have no choice but to dance to the beat. (For more on technological determinism, see the interview with Langdon Winner in the background reading included at the end of this chapter.)

Mesthene himself sees technology more as a neutral tool (2): "What its effects will be and what ends it will serve are not inherent in the technology, but depend upon what man will do with technology" (p. 82). Moreover, he argues, one cannot separate technology's beneficial outcomes from its negative side effects: "(Technology) has both positive and negative effects, and it usually has the two *at the same time and in virtue of each other*" (p. 77; emphasis in the original). This attitude that technology is neutral (critiqued by Robert Whelchel in the background reading) is a popular belief, manifest, for example, through the statement often attributed to the National Rifle Association that "guns don't kill people, people kill people." A good test of such claims, however, is to substitute another term. For example, the expression "sofas don't kill people, people kill people" is liable to draw some snickers, even though sofas can certainly be used as weapons and can be silent killers of couch-potatoes. There is a difference between a gun and a sofa, however, that makes this appeal to the neutrality of technology suspicious.

A more sophisticated treatment is the notion of technology as a form of political power and influence, as noted in Balabanian's fifth element and as suggested by McDermott's definition of technology listed above. McDermott (4), like Winner and other critics of technology, believes that "technology creates its own politics."

The truth of this statement can be seen in the career of Robert Moses, the master planner of the mid-twentieth century responsible for much of New York City's public works and infrastructure. Though originally a popular figure, as Moses' power grew, critics began to object as his projects continued to displace more and more people and seemed to take on a life of their own. As historian Leo Marx has noted (8), the original Enlightenment view of technological progress was that it was a means to social progress—that is, the realization of democratic values. To Moses, however, technological progress became an end in itself rather than a means to an end. To him, democracy was an impediment to technological progress. While tenements were cleared as part of Moses' plans to build modern high-rise buildings, viable, albeit poor, communities were destroyed in the process. Other projects, such as the Cross-Bronx Expressway, destroyed vibrant, middle-class communities (9).

Moses' career thus reveals flaws in the popular notion of the technical or technological fix, a concept first used by physicist Alvin Weinberg in the 1960s (10). As originally coined by Weinberg, a technological fix was using technology to attempt to solve nontechnical or social problems. Rather than attempting to change how people behave, the technological fix changes the physical environment so that behavior change becomes unnecessary. Over the years, the concept has expanded to include technological solutions for the problems (technical and social) that technology itself creates.

The concept of a technological fix fails in two ways. Rarely can real-world problems be divided into those that are purely technical and those that are purely social. The common lament by some engineers that nuclear waste disposal is a political rather than a technical problem fails to appreciate that the two are inextricably linked. Moreover, as Weinberg himself admitted, technological fixes only attack symptoms and rarely address the underlying problems that result in those symptoms. Consequently, technological fixes usually amount to no more than band-aid solutions. The high-rise projects constructed by Moses, though intended as technical fixes for poverty, lacked the community values of the neighborhoods they replaced. Thus, drug abuse and crime plagued many of them. Similarly, most of the technological fixes predicted by Weinberg in his 1966 essay, most notably nuclear power, have failed to work as he envisioned.

Technological Controversies

Technological controversies have become an important part of our political landscape. Many of these controversies and their ethical and policy implications—risk, sustainable development, technology and health care, telecommunications and privacy—will be discussed in further detail in later chapters. Here it is important to note that a significant aspect of all technological controversies is the way in which the political process deals with the controversy. Learning to control technology in a manner consistent with democratic principles is a significant and ongoing problem.

Some people believe that technology should be controlled technocratically; that is, those trained as technical experts should be making all of the important decisions in technological controversies (and, some believe, on broader social issues as well). The belief in technocracy is based on the premise that only those with technical training can have a true understanding of complex issues and an effective ability to solve problems.

Others believe that technology should be controlled democratically. Although some observers think that normal democratic processes are sufficient to resolve technological controversies (11), others believe that public participation should be given a strong role in resolving these issues. Arguments for increased public participation are based on both ethical and practical grounds. In a democracy, people ought to have the right to self-determination and informed consent regarding risks they may be exposed to. Ultimately, a technology will fail in the absence of the consent of those most affected by it.

We have far to go in resolving technological controversies—at present, the outcomes are rarely satisfactory to all of the involved parties. Nevertheless, it is clear that controversy over technology serves many important functions, especially in focusing on important issues and filtering out unimportant ones (11).

The relationship between technology and society is a two-way street. The two are inseparable and have many mechanisms for feedback. The developers of technology often respond to society's needs and wants, but some technologies (for example, photocopiers and personal computers) seem to create their own demand. Technology and technologists, though a part of society, have a powerful influence in determining the possibilities for social change. Social change, on the other hand, can limit or nurture technological change. Setting reasonable and livable boundaries for technological change is one of our most significant societal challenges.

Background Reading

Noted political philosopher and technology critic *Langdon Winner* was featured in a 1998 interview in *Technology and Society Magazine* conducted by Managing Editor Terri Bookman. The far-ranging interview reprinted here offers a variety of insights into Winner's views on technology, engineering, and society.

Winner is heartened by the potential for grass-roots technological change, as evidenced by the antinuclear weapons and nuclear power movements of the 1970s, and more recently by attitudes toward tobacco and personal health. He even sees some cause for optimism concerning public response to institutionalized forms of power such as corporate control over mass media. Although he sees much growth in the scholarly understanding of the social context of technological development over the past two or three decades, he believes that too little of this thinking has made its way into the popular media, where most images of technology and its promise are still oversimplified.

Winner believes we should be more cautious about technological change (e.g., computers in education) and not take for granted that it will be automatically beneficial. He is also wary of the tradeoffs involved—in the education example, what other programs and teaching positions are being cut in order to pay for the computers and personnel to operate them? He believes we should seek positive alternatives for technological change, building on the intellectual criticism that has emerged over the past century. Winner argues that aberrations in this tradition, such as the Unabomber Manifesto, err in not seeing that hope for positive change. Movements such as community-supported agriculture are, he notes, seeking positive alternatives.

Winner asserts that other technological challenges require responses beyond the local level, in particular digital technology, flexible production, and genetic technology. Although he rejects technological determinism, he believes many powerful interests still appeal to it in order to persuade people that they have no choice other than adapt to inevitable technological change—change that focuses on individualism and economic growth, and may neglect community, equality, and environmental values.

Winner sees a key role for engineers in promoting socially responsible technological change, not just on the job, but also through dialogue within their professional societies and with the public in general. It is critical, Winner believes, for engineers to understand the ethical and sociopolitical contexts of their work, and to act on that understanding both through their career choices and their notions of engineering design. He challenges all engineers to ask, "What kind of world are we making here?"

Robert J. Whelchel, an electrical engineering professor currently serving for the second time as editor of *Technology and Society Magazine*, engages one of the myths of technology head-on in his provocative article "Is Technology Neutral?" first published in December 1986 and reprinted here. Stressing the reciprocal relationship between technology and society, Whelchel cautions that the growing importance of technology in society may result in a worldview dominated by technological values to the neglect of other viewpoints. Beginning with a discussion of neutrality, Whelchel concludes that since technology does affect the world, it cannot be causally neutral. Therefore, the key question is whether technology is morally neutral, or better yet, is technology value-free?

Whelchel responds to the latter question with a resounding "no," pointing to objectivity, quantification, and utilitarianism as primary technological values, and efficiency and a systems approach favoring function over device as secondary values. These values create a frame through which the technologist views the world. Whelchel argues, along the lines of the philosopher Heidegger, that there is a danger that the technological viewpoint, though correct in a number of important ways, will engender an attitude on the part of technologists and society alike that the technological frame captures all of reality.

Again noting the importance of society's impact on technology, Whelchel concludes by arguing for a broader framework for technology that provides for societal feedback. He believes this feedback is necessary for both society's well-being and the development of technology. He pleads for a reclamation of the "art" in technology and for its sense of wonder. To find these missing ingredients, however, we must look beyond the conventional technological frame to see technology as human experience.

Cases for Discussion

In a 1995 article *Todd D. Cherkasky* studies "Obscuring the Human Costs of Expert Systems," an example of the dual-edged relationship between technology and society. His focus is on the displacement and disaffection of technical professionals by such systems. He argues that advances in expert systems, artificial intelligence, and other knowledge-based systems create a conflict in national policy between research support for deployment of such systems and the purported solution for worker displacement by such systems of increased training and education.

Cherkasky rejects the technical fix provided by such systems on a number of grounds: the assumption of increased productivity upon downsizing of the workforce, including the technically trained sector; the growth in knowledge-based systems at the expense of maintaining human expertise; a mandate for technological progress in expert systems that outsiders may understandably perceive as autonomous technology; and a rejection of the notion that a "better, more fulfilling" world will emerge from technological developments that result in the wholesale disruption of the technical workforce. In addition, Cherkasky points to the likelihood that human users will be required to adapt to the imperatives of intelligent machines.

In conclusion, Cherkasky urges a reexamination of the goals of knowledge-based systems, which heretofore have focused primarily on their technical characteristics. He hopes that they might be redirected toward achievement of public policy goals such as improving job security, the quality of the workplace, and worker satisfaction. Such changes, he argues, will in the long run prove productive and flexible by taking advantage of the strengths of both human and machine experts.

In another case study of the interrelation between technology and social values, *Reid Cushman* discusses "Information and Medical Ethics: Protecting Patient Privacy," in an article published in 1996. According to Cushman, although the health care industry has lagged behind in moving from paper to electronic records, this situation has been changing rapidly in recent years owing to government pressure and the emergence of information-intensive managed care. While large databanks of patient information offer a number of benefits in administration, treatment, and research, they also raise legal and ethical questions regarding patient privacy and the access of medical records by unauthorized health care personnel, insurers, and employers. Obviously, these potential abuses could lead to discrimination against the individual. Indeed, Cushman argues, such use and misuse of information parallels that of databanks containing financial information on individuals.

Although some legal protection of patient rights is afforded by mechanisms such as the Americans with Disabilities Act and various state and federal laws governing use of personal information, Cushman is still concerned about misuse. Misuse could lead patients and physicians to withhold information from electronic records, resulting in incomplete or inaccurate records. In addition, it is too early to know if the benefits of information technologies will offset the risks to patient privacy. Although security of records is not a new issue and computer systems can incorporate security measures, Cushman believes that the advent of computer networks could result in widespread breaches of patient privacy, rendering meaningless the concept of informed consent to the use of one's medical records.

Cushman concludes by suggesting that this issue is tied to larger issues of health care reform—in particular, to equitable access to health care. Without such a reform, rather than narrowing inequities, information technology is likely to provide further bases for discrimination against individuals and classes of individuals.

References

[1] Winner, L. 1986. *The whale and the reactor.* Chicago: University of Chicago Press.

[2] Mesthene, E. G. 1969. The role of technology in society. *Technology and Culture* 10: 489–536. Reprinted in Teich, A. 1993. *Technology and the future.* 6th ed., New York: St. Martin's Press. 73–88.

[3] Unger, S. 1994. *Controlling technology: ethics and the responsible engineer.* 2nd ed. New York, Wiley.

[4] McDermott, J. 1969. Technology: The opiate of the intellectuals. *The New York Review of Books* (July 31). Reprinted in Teich, A. 1993. *Technology and the future.* 6th ed. New York: St. Martin's Press. 89–107.

[5] Frisch, M. 1987. *Homo Faber.* San Diego: Harcourt Brace Jovanovich.

[6] Kline, S. J. 1985. What is technology? *Bulletin of Science, Technology and Society* 1: 215–218.

[7] Balabanian, N. 1979. Technology, freedom and individual autonomy. *Science, Technology & Society,* No. 15 (December): 1–5.

[8] Marx, L. 1987. Does improved technology mean progress? *Technology Review* (January): 33–41, 71.

[9] PBS Video. 1988. *The world that Moses built.* Alexandria, Virginia.

[10] Weinberg, A. 1966. Can technology replace social engineering? *University of Chicago Magazine* 59 (October): 6–10.

[11] Mazur, A. C. 1981. Controlling technology. *Broadcasting Magazine.* reprinted in Teich, A. 1993. *Technology and the future.* 6th ed., New York: St. Martin's Press. 215–228.

Technology as "Big Magic" and Other Myths

Since the publication of his book Autonomous Technology *in 1977, political scientist Langdon Winner has been known for his provocative, beautifully-written critiques of technological development.*

*In the 1980s, Winner published a second much-read book (*The Whale and the Reactor, *1986) and held teaching positions at M.I.T., the University of California at Santa Cruz, and the University of Leiden in the Netherlands.*

Winner is now Professor of Political Science at Renselear Polytechnic Institute in Troy, NY. He recently completed a ten-year stint as a columnist at M.I.T.'s Technology Review. *A June 1997 profile in the* Wall Street Journal *called him "the leading academic on the politics of technology."*

In addition to his scholarly credits, Mr. Winner worked as a rock critic and contributing editor for Rolling Stone *magazine in the late 1960s and early 1970s, and has contributed articles on rock and roll to* The New Grove Dictionary of Music and Musicians *and* The Encyclopaedia Britannica. *He writes regular commentaries on technology and social issues for the on-line journal* NetFuture, *and he contributes regular commentaries on environmental issues for "Living on Earth," a National Public Radio program.*

Dr. Winner received his B.A., M.A., and Ph.D. degrees in political science from the University of California at Berkeley.

He spoke with T&S *Managing Editor Terri Bookman on Feb. 25, 1998, at IEEE offices in New York City.*

Prof. Langdon Winner can be reached at Rensselear Polytechnic Institute, 5118 Sage Hall, Troy, NY 12180; email: winner@rpi.edu.

History's Twists and Turns

Terri Bookman: *Dr. Winner, in a recent article entitled "Look out for the Luddite label," [*Technology Review, *Nov./Dec. 1997] you said: "Technical change ought to be guided by principles of social justice, psychological harmony, and personal dignity, rather than the untrammeled pursuit of efficiency and profit." I'm sure those inspiring concepts sometimes are carried out. But I wonder if the big picture, in terms of technological change, is that there really isn't a strong focus on social justice, psychological harmony, or personal dignity. Is there really hope for any of this? Hope that we can affect the process of technological development in beneficial ways?*

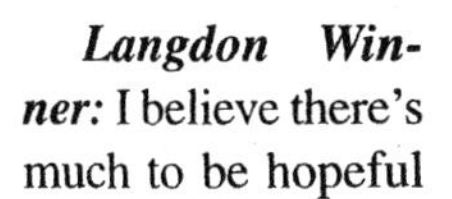

Langdon Winner: I believe there's much to be hopeful about. If people become aware of important choices about the interweaving of technology and society and speak up clearly about alternatives, history can take some surprising turns. The sudden shift in the nuclear arms race is a case in point. During the late 1970s and early 1980s the peace activists who called for an end to the balance of terror seemed foolish to most "responsible" political observers. Yet groups like Physicians for Social Responsibility and cantankerous souls like Helen Caldicott persisted in calling for the elimination of the instruments of mass destruction. I believe their voices were highly influential in ways not

Reprinted from *IEEE Technology and Society Magazine*, Vol. 17, No. 3, pp. 4–16, Fall 1998.

widely recognized. Eventually there were hundreds of thousands of people demonstrating in major cities around the globe. By October 1986 the idea of nuclear disarmament had become so powerful that the Reykjavik summit between Ronald Reagan and Mikhail Gorbachev hinged on proposals to eliminate the bombs and rockets altogether, much to the dismay of the policy advisors for both men. What seemed utter folly a few years earlier began to seem entirely feasible as the Cold War wound to a close.

Of course, there are still far too many nuclear weapons targeted and ready for use. But one thing that's changed significantly in the lives of my children and my students is that they don't live from moment to moment with the idea that they could be vaporized in a nuclear flash.

TB: *Another example might be anti-nuclear power efforts.*

LW: That's right. It's another case in which the unexpected twists and turns of history lent support to those who resisted what once seemed inevitable. Social movements of the 1970s that called attention to the hazards of nuclear power appeared rather futile in the beginning. But their message eventually intersected with growing economic problems within the industry. As the general public became aware of these combined troubles, what had been firm support for "the friendly atom" quickly vanished.

Another example of positive change in public attitudes and policies can be seen in the revolt against smoking and the tobacco industry. It turns out that when people finally have good information, a clear sense of who's at fault and an idea of what remedies are open to them, they often act quickly and resourcefully. This is especially true in matters over which people feel they have direct control; Americans today are willing to make substantial changes in diet, exercise, and other health related habits if they believe it will contribute to their well-being. But it's harder for them to get their hands on deeply entrenched, institutionalized forms of power that cause problems in our society and environment.

Challenging Institutionalized Power

TB: *What sorts of "institutionalized forms of power" are you talking about?*

LW: There are many of them. One can begin with the extraordinary concentration of power within the mass media. A few large corporations now control much of the content of radio, television, movies, book publishing, and other sources of information people encounter.

TB: *OK. But when it comes to issues like the mass media being concentrated in the hands of a few corporations or people's frustration with computerization, the problems don't seem so specific. Its not like organizing against nuclear power plants.*

LW: That's right.

TB: *How do we find ways to deal with matters as broad as, say, the power of mass media?*

LW: It requires a lot of imagination and ingenious strategy. One of my favorite political writers, Nicolo Macchiavelli, once observed that "Men make quite a number of mistakes about things in general, but not so many about particulars." It's always good policy to identify certain specifics that one cares about and act affirmatively to create new options and new possibilities. One interesting organization that's tackling the hold of the mass media is the Adbusters group which studies methods used to promote new products and to analyze the content of advertising campaigns. The group focuses upon ways in which people are manipulated by the media in subtle and not so subtle ways. In one of its "uncommercials" the group demonstrates how ads make women feel that their bodies are inadequate. In another one, the camera circles a a young man sitting watching TV. A voice softly proclaims, "The living room is the factory. The product being manufactured....is you."

The Adbusters have gone the full distance with these campaigns, even raising money to air their spots on national television. But the broadcasters have usually refused to run the "uncommercials" because they run counter to the whole drift of TV production and marketing. The name for this

> The motivation for trying to convince others that technological developments are inevitable is often that those doing the convincing have something to sell - or a position of power to protect.

response is, I suppose, "freedom of speech." But its interesting to see an organization push the power holders in the media to admit what the essence of their business is all about. I wish there were guerilla strategies like that at work in other industries where arrogant power mocks the public's trust.

Changing Perceptions of Technology

TB: *Do you think the understanding of the relationship between technology and society has improved over the past 20 years?*

LW: Yes, in some respects. When I started reading in this area, in the middle 1960s, the available literature on technology and society was simply deplorable. Its range of vision was limited to stories about heroic men and their wonderful inventions — the telegraph, telephone, radio, airplane, and the like. The general framework of explanation centered on notions of progress and modernization which today seem naive. Books, magazines, and news stories contained a very strong belief in the inevitability of technological change and the need for people to adapt without questioning anything. Most writers showed very little understanding of the ways in which inventions and technological innovations are the product of complicated social processes, of the many-centered human networks involved in all attempts to introduce new tools, instruments, techniques, and systems to the world. Of course, the contributions of women, ordinary workers, and non-Western people to our stock of technical abilities were totally ignored.

For the time being, the effect of all media — radio, television, and the computer — seems to be to reduce participation and amplify the power of the already powerful.

Since that time we've seen the rise of a variety of scholarly fields and the fields of journalism in which the notion of a socially shaped, socially created, socially constructed technologies has risen to prominence. Prominent myths formerly accepted as true have been replaced by some richer, more complex and, I believe, more accurate ways of understanding how technologies enter the world and how they interact with patterns of human activity. Scholarly interpretations here are far less deterministic than they were thirty years ago, far more inclined to understand changes in technology within a wide range of social, cultural, political, and intellectual contexts.

As fruitful as these developments have been for intellectuals, however, they have not done much to modify popular conceptions of technological change in the press and in general conversation. Most journalists, politicians, and ordinary folks are perfectly happy with the old, threadbare narratives. Here's the implement; here's the user; here are the wonderful possibilities it opens up; here are the "impacts" upon society. Even the "Nova" programs on PBS still peddle the same old stories, basically myths, about "those wonderful men and their flying machines" that seem to have come straight from the 1920s, only now its those wonderful men and their microchips and biotech projects.

For its own reasons, advertising still shows us technologies that arrive as if by miracle — the latest shiny automobile on a wide open highway free of congestion, for example. There are now even ads that show the highway lifting literally out into space; now we're cruising the planet! But there's still only one driver, and one automobile in the realm of ecstatic fulfillment. There's no sense of context, situation, connectedness, or of the social and ecological relations involved in any of these things. In that way our common notions about technology and human life are still based in tawdry illusions.

TB: *Are there more examples of old myths that have been replaced, at least in a scholarly context, by more complex understandings?*

LW: Take the rise of microelectronics. The best social histories talk about the matter in terms of the decades of government subsidized research and development, much of it directly connected to the military. Social scientists have described the formation of complex networks that link the universities to industry and, particularly in the early stages, to the military. To create a place like Silicon Valley you need decades of collective projects, publicly funded in large measure, producing not only specific products, but a vast social and material infrastructure that allows technical innovation to occur. Along with this there are all kinds of interesting, shared ideas and ideologies that emerge to help inform people's sense of what they are about, ideas about highly amorphous organizational arrangements and people shifting from firm to firm overnight. There are ways in which the complex social, economic, and political relationships involved here could be conveyed within the histories and policy discussions that most folks receive. But when it comes to writing magazine stories and TV programs about Silicon Valley, the publicists and journalists always gravitate back to the myth of the heroic inventor in his garage, coming up with the personal computer, modem, Web browser or whatever "it" happens to be that month.

TB: *Apart from myths about the origins of technological developments, what about myths about the results of a technology, such as electronics or the Internet? Do you think there's a difference between what the general public believes about the results of certain technologies, and the true impacts of those technologies? Also, while the average person may hold some mythological belief about a technology, in their daily lives they may feel frustrated by the same technology. Are there two different forces going — people feel frustrated, but also —*

LW: Yes, there's a real sense of excitement and promise.

TB: *Or, at the least, there's a tremendous media hype about information technology.*

LW: Exactly. And what the hype does is deflect people from asking questions they might otherwise raise if they were, say, buying a house — check out the market, check out the neighborhood, look for possible resale value. We are not especially clever or careful as we approach heavily promoted new technological innovations. A very good case is the rush to install computing in schools. People often don't ask some very basic, obvious questions. I occasionally sit down with members of school boards who are proposing to buy hundreds of thousands of dollars of PCs and network devices. Sometimes they're even preparing to take their districts into bonded indebtedness to accomplish what they see as a necessary technology upgrade. I ask them, "Do you realize the stuff you're proposing to buy on a 30-year bond issue will be obsolete in 2 years? Is that wise? And have you looked carefully at the software you'll be purchasing? That's really the heart of the pedagogical promise of these tools. If the software appeared in book form, would you be equally enthusiastic about buying it and requiring every student to use it?"

There are a good many very practical, no-nonsense questions that people ought to ask about all this apparatus. But often they set aside the most elementary common sense because they've come to regard technology as something like Big Magic. Behold: this will transform our lives for the better! Oh, don't ask how. Getting beyond these fantasies is a crucial task for any intelligent inquirer.

Technology and Education

TB: *In* Technology and Society Magazine *we have a special issue coming up on "computers in the classroom" this December. The Guest Editor of that issue, Kenneth Foster (who is also the current President of SSIT), asked me to ask you about it. He mentioned that there have been complaints published in some of the popular press about the "wired classroom" being a lot of hype. Do you agree with that criticism?*

LW: Again, I think you have to look at these proposals within the context as a whole. My approach to technology stems from questions in social and political theory. I begin by looking at practices, institutions, and patterns of human relationships. How do people associate? How do they live together? Which rules, roles, relations, and institutions are the good ones, which aren't so good, and how can we tell the difference? From that standpoint, I view technologies of various kinds as institutional components and varieties of practice that strongly affect who we are and what we do. What possibilities are afforded? Which constraints and conditions are imposed?

If you look at what is happening with the introduction of computer networks into the schools you have to ask what these devices replace or eliminate. And you need to identify what these instruments require of those who use them. Very often we see that as lots of money and resources are being given to computers in the classroom, at the same time arts education is being cut, physical education is being eliminated, foreign language teaching curtailed. The coming of the machine is the occasion for a whole set of institutional choices made invisible by the Big Magic of the computer.

TB: *So you're saying that computers are replacing these other areas of education.*

LW: Yes, in effect. You have to look at the whole situation and ask: What conception of quality education is being proposed? My own view is that in the best of circumstances every student ought to have access to the broadest array of opportunities, books, materials, equipment, and high quality teaching possible. In that light there's an educational center that I'm particularly fond of in Troy, NY, called The Ark.

TB: *You've written about that. ["Computers and Hope in an Urban Ark,"* Technology Review, *May-June 1997.]*

LW: Yes. One interesting feature of the place is that its material culture is fairly diverse. The Ark contains a good many musical instruments that people in the community have donated, along with a small library of books, and supplies and equipment for doing art — pottery wheels, kilns and the like. That wealth of equipment is made available to about 150 poverty-level, minority kids along with very good after-school tutoring and just plain human care. What captures the headlines, of course, are the state-of-the-art computers where the children make their own Web pages. But for the people who run the Ark and for the many volunteers who give their time, computing is just one among a rich set of possibilities made available to the boys and girls.

Alas, that sense of balance is often lacking in education these days. Computers are pushed into schools as if they were a source of redemption. They are treated as a one-dimensional solution to what is actually a very complex question: how does one inform and enliven young minds and souls?

TB: *So if you have a classroom where everyone has a computer on their desk that is hooked to the Internet, that is being used as the learning tool, getting information over the Internet, then that becomes what consumes everyone's attention almost all the time. Everything becomes Internet focused, and the students become Internet-philes —*

LW: Yes, it's amazing the devotion a box of silicon chips can inspire. What fascinates me is how reluctant people are to consider the evidence for productivity or lack of productivity from these innovations. There's a great deal of research available on the effects of technological applications in education. One prominent finding centers on the principal of "no significant difference." Study after study shows that information technologies help some students' learning, hinder others, but produce no significant difference for the great majority. [See T.L. Russell, "Technology Wars: Winners and Losers," *Educom Review*, vol. 32, no. 2, Mar./Apr. 1997.]

My own observations suggest this has probably been true for decades. American schools have been introducing the latest technological wonders into the classroom since World War I: motion picture machines, radio, tape strips, television, cable television, language laboratories and the like. I remember that in the 1970s when I first stated teaching, a number of my colleagues believed the camcorder was going to "revolutionize education."

The idea was that you could return knowledge production to the students. They would go out into the world and make their own television programs, thereby controlling the learning experience. It was a fabulous dream. But no one is calling for a television revolution or camcorder revolution in education today, despite the fact that the technology has gotten better and cheaper. Isn't that interesting? Now many people believe that the Internet will achieve the wonderful results that have eluded us in previous "revolutions." But if the research shows that the introduction of new technology makes no significant difference for the vast majority of students, why are we spending all this money on these boxes?

TB: *The answer that springs to my mind is that people are concerned that students, when they leave school, will not have total familiarity with the computer. To ensure that they are prepared to leave school and enter the working world, it needs to be introduced on their desks now.*

LW: You mean that exposure to computers acclimates students a particular work environment. That's probably true. The question is how well does this experience correspond to the goal of a getting a good education?

I was interested to see the results of the Third International Mathematics and Science Study released recently. The report compared test scores of students from the U.S. and fifteen other nations. Americans ranked consistently at the lower end of the scale. Even Cyprus scored higher than the U.S. in physics, for example, and there were no countries that scored significantly lower than the U.S. in advanced math. These were tests of twelfth graders taken in 1994-1995. This was, of course, exactly the generation of students bombarded with personal computers and widely heralded software like the Seymour Papert's Logo program during the grand and glorious computer revolution of the 1980s. How is it that when you test these students as they are leaving high school, they still do so poorly? The U.S. students lagged behind countries like Norway where there is very little computing in the classroom, just a strong emphasis on high standards in math, science, and other teaching. In my view, the American obsession with computers in education has begun to resemble a cargo cult.

TB: *What is a "cargo cult"?*

LW: Cargo cults were millenarian movements that arose sporadically in New Guinea and other islands in the Pacific during the past century or so. Some native people believed that the material goods they saw carried on ships and later on airplanes from distant places were actually meant for their benefit, but had been misappropriated by people from other lands. The cults conducted elaborate rituals, building bamboo ships or airplanes and tending them, in the conviction that the "cargo" would eventually arrive.

But, eventually, even members of a cargo cult have to ask: how well are the rituals working? Are our efforts bearing fruit? Many schools, for example, never figure in their planning the amount of staff support or repair expenses that computing equipment requires. They discover that, over time, more and more resources have to be devoted to making the software work, keeping the machines repaired. Around the country we see school systems hiring new computer support personnel at the same time they are laying off teachers.

Seeking Positive Alternatives

TB: *I want to get back to the big picture, to not being in control of technology. Do Unabomber Ted Kazinsky's writings have any relevance to this issue? What would cause someone to go "over the edge" like he did? What might that say about the relationships between technological development and people in our society today?*

LW: If the Unabomber Manifesto, "Industrial Society and Its Future," hadn't been written by a serial killer, no one would noticed it. It's fourth rate anarchist social theory of the most thick-headed kind. The essay describes the thoroughgoing domination of a system of technological economic arrangements that robs people of their autonomy, turning them into cogs in the machine. The same technological civilization, in the Unabomber's account, is rapidly destroying nature, producing an ecologically depleted planet. Ideas of that kind are common in a tradition of social thought that goes back to Thomas Carlisle and other early critics of industrialization. In the nineteenth century, writers like John Ruskin and William Morris in England and Henry Adams in the United States all offered strong critiques of a society in which machines prevail.

This tradition continues in the 20th century with writers like Lewis Mumford, Theadore Roszak, and other humanists who've warned about the spread of megatechnics.

What distinguishes that grand tradition of technology criticism from the writings of the Unabomber is worth noting. The best thinking in this vein draws a strong contrast between where our civilization is headed as compared to where it ought to go. These writers take care to offer positive, hopeful answers to the questions about technological and social alternatives. Lewis Mumford, for example, spent a lifetime describing the same kind of maladies that Kazinsky became fixated on. But his central purpose was to think about what human beings could become, what resources there were to help people lead rich, fulfilling, thoughtful lives and which social and material arrangements were compatible with the good life.

I think it's a sign of Kazinsky's madness that he thought himself into a corner from which he could not escape. He came to believe that we are so throughly overwhelmed by the power of modern technology that only the calamitous destruction of modern civilization would remove the ills. This is a totally unpalatable variant of technology critique which, unlike the teachings of Gandhi or Lewis Mumford or E.F. Schumacher, dwells upon fear, hatred and violence. At a certain point you've got to put aside analysis and explanation of what's wrong in the world and turn your attention to sources of cooperation, hope, love, and renewal, sources that are inexhaustible, if one knows where to look.

TB: *You gave some examples earlier of groups working towards or seeking positive alternatives to certain technological and social developments. Aren't groups of this kind exceptions — small "pockets" of activity in a larger picture?*

LW: I'm not so sure they are only small pockets. In most places that I'm aware of — cities, towns and neighborhoods — people are involved in trying fashion good places to live. Very often these days people have to confront technology-centered forces that make it more difficult to build livable, flourishing communities. Recent waves of downsizing, automation, offshore production, the dismantling of old workplaces, and the construction of new ones that don't have the wage rates or even the kinds of interesting work that the old places had — all of these are conditions that people have to contend with nowadays.

One "pocket" of renewal that I find quite interesting is the idea of community supported agriculture.

The idea that digital electronics necessarily allows greater access to information and therefore enhances the power of ordinary citizens is one of the recurring myths of the computer age.

There are about six hundred of these C.S.A.s in the U.S. now, using organic agricultural methods and contract marketing to bring farms back into production and revitalize farming as a way of life. When the Department of Agriculture recently tried to propose standards for labeling products "organic," there was a surprising flood of mail and email, much of it from ordinary folks, who objected to the inclusion of foods grown in sewage sludge and bioengineered plants on the list. Many people believe that the voluntary spread of truly organic agriculture would be good both for them personally and for the health of the planet. As a more general matter, I think there's lots of evidence at that people are taking a closer look at patterns in their lives — energy use, relationship to transportation, consumer habits, and so forth — looking for ways their actions can more closely match their deeply held beliefs.

TB: *You work at a university, you live in the academic world, the people you associate with see things from a similar perspective, so its easy to keep affirming that and keep going in that direction.*

LW: Right.

TB: *But the further you get from that academic environment, the harder a person has to work to keep that perspective.*

LW: I don't find that to be the case at all. I live in a small town — with ordinary people, people who run contracting businesses, plow the snow, go to the churches and run the PTA.

TB: *Do they go to the organic farm?*

LW: Some do, but I'm not talking only about that. I'm talking about the networks of social support, support for the elderly, or what's required to keep the local economy vibrant, keep public spaces interesting, or to attend to the needs of young people growing up.

TB: *You're talking about community.*

LW: Exactly. These are not people who've read great works of philosophy or social theory. But they understand a lot about how real communities operate. Around the country in recent years people have rallied, for example, when they thought the arrival of a Wal-Mart would pound a spike through the heart of the community as expressed in its downtown shops. There are some places where people say, oh fine, a Wal-Mart is coming in. But there have been lots of cases in New England, Washington State, all around the country, where people have correctly seen that the market power of a Wal-Mart, power based in electronic inventory systems, can be destructive to the economy of towns and neighborhoods. On occasion communities have stopped or at least modified the introduction of these stores.

I'm often surprised by the grasp of public issues exhibited by ordinary folks. In communities near my home, townspeople have battled environmentally and socially destructive developments proposed by the New York State Thruway Authority, McDonald's, and a large mining company. They've shown themselves to be a tough as nails in preparing environmental impact assessments, organizing campaigns aimed at the news media, and lobbying decision makers.

TB: *Are there types of technological change that are not so locally based, that are in a sense, truly beyond people's immediate control?*

LW: Yes, there are some kinds of technological change that demand much more than a local response.

TB: *What are some of the areas of technological change that you see coming up that are of concern, yet not so easily affected by, say local organizing against the introduction of a particular store.*

LW: First, in my view, is the overall influence of digital technology which affects every corner of social life. For better or worse, most social practices and social spaces are being rennovated to accommodate digital electronics.

Second are the various arrangements of flexible production. Some of these involve new hardware and software, while others are based on social policies, i.e., the use of temporary workers for many tasks. This new flexibility and ephemerality is changing how people experience their jobs, professions, family lives, and leisure.

A third area I would mention is the

rapid development of genetic technology in agriculture, industry, and health care which now presents a great challenge to how we understand all life forms and life processes.

All of these changes require widely shared, systematic responses that apply to every corner of the planet. That means that we'll have to find ways to connect specific local concerns to much broader patterns and to people in distant places who have similar problems.

Technological Determinism

TB: *So let's get back again to the idea of technology being "out of control." What does that term mean?*

LW: It's not really one idea, but a collection of related themes. For example, in fields of technical practice where things are changing rapidly, coming from so many sources, one sometimes hears even the most competent participants say that the technology has a "life of it's own." This means that no one controls or could possibly control what happens next.

Another version of the idea stems from a phenomenon that the ancient Greeks understood, that because the effects of any action are potentially limitless, we never fully control what we are doing. That's just as true of action through technology as any other kind. But because technology is seen as a distinctive domain of rationality and control, the unpredicted ramifications of seemingly insignificant acts are all the more astonishing. The Y2K problem, the millenium bug in computer software, is a perfect illustration.

TB: *What about "technological determinism?" Is that idea still important?*

LW: The basic thesis is that technological change has a necessary, linear trajectory and that technological development is the most forceful determinant of changes in society. As I mentioned earlier, this idea has gone out of fashion among most scholars. But it is still common in other quarters. Among cyber-libertarian thinkers, those who write for *Wired* magazine and right-wing foundations in Washington, DC, there is still a very pungent technological determinism. They argue that various technological waves are crashing over us and that people will simply have to adapt. They usually don't say: Here's a process of social creation with many wide open possibilities available to us, so let's make sure everybody is included in making the key choices. No, the message has to do with what is inevitable and necessary.

TB: *This is the idea we always hear, that the world is changing so fast, and you have to keep up, for example, that you must have a certain hardware or software.*

LW: Right. We are all advised to worship at the shrine of Our Lady of Perpetual Upgrade.

TB: *All right then, why do people want to convince other people that things are or must be a certain way in terms of the inevitability of technological development?*

LW: Quite often it's because they have something to sell or a position of power to protect. I think we had a very good example of this in how the so-called information superhighway was presented to the public. The images offered in news magazines and elsewhere had an extremely odd slant. At the top of the highway map were enormous information providers, the media networks, Time-Warner and Disney, for example, where the corporations control the content and pass it on to you. It was an extended model of cable TV. Many people still have this model in mind. Time-Warner spent an estimated hundred million dollars on creating the Full Service Network, one that tried to link shopping, news, entertainment, and information services all in one package. That particular system flopped. But rhetoric in these episodes is always the same: persuade people that a particular pattern is "inevitable" and they have no power to negotiate. Go along with Plan A or be rendered obsolete. Bow down! It cometh!

TB: *Are you saying that there is an underlying political basis, or agenda, for theories of technological determinism?*

LW: There is an ideology prevalent among today's economic, technological, and political elites that focuses on unfettered freedom in the marketplace. This idea of freedom turns out to be not only a justification for start-up entrepreneurs, but an ideology in defense of global capital. The underlying message is that the world is propelled by technological trends and global economic forces and that in such a world you're only responsible for yourself. That sounds fine until you begin to notice that you are no longer obligated to have any responsibility to community, to humanity, to the biosphere, to other people. Everyone's goal is

We are all advised to worship at the shrine of Our Lady of Perpetual Upgrade.

> With technological determinism, everyone's goal is assumed to be the sheer creation of wealth - if you run into people who aren't doing well, you say, well they're losers, they've just failed.

assumed to be the sheer creation of wealth. If you run into people who aren't doing well, you say, well they're losers, they've just failed. And certainly we don't need the action of government to provide public services other than the bare necessities like airports, roads, or bridges, and the like. Today the right-wing libertarians are not even sure public schools are necessary; perhaps private education and vouchers would be preferable. What we have here is an "I'm all right, Jack" ideology of a very comfortable elite, proclaiming freedom while they advise dismantling social programs that help ordinary folks and the poor (people without stock options and lavish mutual fund portfolios) get by from week to week.

Role of Engineers

TB: *What should be the role of the individual engineer in controlling technology, and what should be the role of the professional engineering society in controlling technology?*

LW: Engineers are often well situated to look at the changes that will affect the future of society and to make decisions that affect the broader public interest. They can, depending on how well they understand their place in the world, act to push a pattern of development in directions that would be safer, or more socially convivial, or more inclusive. I believe that a great many engineers already act responsibly in this respect. One reason that we don't hear about technology failures more frequently is that engineers do a good job quietly insisting on high standards in the domain of work in which they operate.

I especially admire the engineers who recognize that they have a role in communicating with the broader public and involving non-experts in decisions about technology. Computer Professionals for Social Responsibility does a very good job in its sphere of concern.

TB: *Or SSIT.*

LW: Yes. And very often these organizations take positions that shift the whole direction of public debate. Physicians for Social Responsibility started saying in the early 1980s, if the bombs start falling, don't call your doctor; your doctor can't help you. That message attracted considerable attention and helped redefine the arms race as an issue for public heath. Many people simply hadn't thought about nuclear war in that context before.

In my view, technical professionals — people who are involved in transportation, industrial production, electronic systems, and so forth — ought to be involved not just in preparing these systems, but in helping focus debate on the best and worst that developments in their fields might bring.

TB: *Do you mean focus debate within the professional organization?*

LW: Yes, there ought to be focused debate within professional societies, but also within public forums at large. One area of technological development that deserves more attention than it's getting right now is the development of electronic data collection and electronic surveillance. Much of what people do in their everyday comings and goings is being watched more closely than ever before. Electronic systems in the workplace monitor people's performance and their email. What people do in the realm of network computing is recorded and tracked. In addition, an increasing range of public and private gathering places — social spaces such as offices, shops, malls, streets — are covered by surveillance cameras and other devices that keep an eye on us. There is a new social world being created here in the name of safety and security. It contains some menacing possibilities.

Who builds these systems? Well-educated technical professionals employed by business firms and government agencies. As surveillance increasingly pervades society, most people receive the news almost as a kind of rumor. Oh, oh, the data banks and cameras have been installed and now we'll just have to put up with them. To impose conditions like that is a highly irresponsible form of professional practice. I'm surprised there hasn't been a greater outcry from engineering societies about these developments.

TB: *So are you saying that the engineers themselves who are designing these systems should be* *looking at the ethical implications of what they are doing on their own jobs, and that they should be talking to people about that?*

LW: Yes, definitely.

TB: *Or publicly saying, I have concerns about this — for example, writing an article in, or letter to, the local newspaper about their concerns about the work they are doing?*

LW: Right. In this case, technical professionals could write about the need to balance the concerns organizations have about security with public concerns for reasonable protection against surveillance. They could argue that truly good technologies ought to include these protections, but that for reasons of haste and cost-cutting, the best systems are not being installed. Presenting the issue in this way could create new bonds between engineeers and the general public and help re-open debate about important policy decisions.

TB: *Is that asking a lot of someone whose livelihood may be at stake? Would you ask someone to quit their job so they can talk about what they were doing there?*

LW: Sometimes that's necessary. As a teacher and writer I've quit jobs and been fired over matters of principle. It's part of the business. For engineers the decision to stay with an organization or leave is often a way of affirming deeply held values. These are very personal decisions, ones that outsiders can't easily judge. But I admire those who weigh concerns for public well-being heavily in their professional choices.

TB: *Explain more about how an engineer would incorporate such concerns into his or her professionl choices.*

LW: In addition to being knowledgeable, competent, creative workers they need to see how what they are doing has a moral and political component. The belief that technical skill somehow renders us immune from these matters is something the teaching of engineering ethics always struggles to overcome.

You've probably heard the joke about the engineer and the guillotine. During the French Revolution, three people have been condemned to death: a priest, a politician, and an engineer. The priest puts his head in the slot, down comes the blade, but it gets stuck. The executioner tells the priest, you can go, and the priest stands up and proclaims, "I've been saved by the will of God, glory hallelulia." The next victim is a political leader, but the blade gets stuck again. As he is released, he says, "I've been saved. It's the will of the people!" The next person to be guillotined is the engineer, who walks up to the apparatus, puts his head in the notch, sees where the blade is supposed to fall and says, "Hey, I think I see your problem..."

If a joke can have any significance, I suppose this one's is that you need to ask who you're working for — whose interests are served by this project, innovation, or piece of equipment. And what exactly is your role?

As a teacher during the 1970s and 1980s, I would ask my science and engineering students to question seriously their projected roles in the military industrial academic complex which for many of them was a virtual certainty. So I would say deliberately provocative things like "Why don't you move to Seattle and work for some small software company." (Some of them did, and are now multimillionaires.) Today my advice would be rather similar, although the context has changed. Think twice as a responsible technical professional before you sign on to the agendas

> Engineers need to ask who they're working for — whose interests are served by this project, innovation, or piece of equipment.

of the big transnational corporations and ongoing schemes of economic globalization. Perhaps your creativity would be better employed in a smaller organization, helping to rebuild economic vitality in a community closer to home or in the developing world. I don't think a message of this kind expects any great heroism. It asks people to ponder their roles and ask: What kind of world we are making here?

TB: *Steve Unger, a long-time member of the IEEE-SSIT AdCom, raised some of these points in a question he asked me to ask you, about how engineering talent is being utilized today. He says, "there has been some, though not nearly enough, reduction in the percentage of engineers employed on war-related projects. But this reduction may have been more than compensated for by a channeling of engineers (particularly in the computer software area) into what I would label casino work, i.e., work for stock brokerage firms. It seems that an amazing number of our best students, at all levels, are being snapped up by such firms as Bears Sterns. They* *are being paid very high salaries. Just as in the case of military work, I don't blame the engineers for doing work that is, in my view, nonproductive, but rather our system, which seems to focus on work that is least beneficial to people." Steve says he would be interested in what you would have to say about this.*

LW: It's interesting that Steve would call working for brokerage firms non-productive. I see it that way as well. Very often the highest salaries are paid to the movers and shakers of the financial world, by the people who are interested in moving capital around the globe. Students who are thinking of joining such institutions might well ask themselves how comfortable they are in participating in corporations that are moving jobs overseas and trying to dominate the economic future of people on other continents.

TB: *I wonder if some of them might feel pretty comfortable, for a first job, especially if it seems like what the company is doing is not "all that bad," and you're 22 years old and it looks like you're being offered a lot of money.*

LW: That's right. But it's always been that way, if you're only in it for the money.

TB: *Well, maybe they're not only in it for the money. But maybe it's hard to see — particularly given the wider cultural acceptance —how what the corporation is doing is really so bad. And that kind of job might be a hard thing to turn down.*

LW: That's true. But if people seriously look at the whole fabric of which they are a part, the kind of economic development that they are sponsoring, they may think differently about the matter. Today's corporate economy concentrates the worlds' wealth in the hands of fewer and fewer hands, creating millionaires and billionaires while average wage levels in the U.S. and the rest of the world are standing still or falling. I ask my students how comfortable they are with that.

The experience of David Korten, a businessman who taught at the Harvard Business School, is worth noting here. For many years he worked for organizations he thought were contributing to the economic development that would lift the fortunes of the poor and downtrodden. His book *When Corporations Rule the World* describes his realization that something quite different was happening. The actual policies of global corporations, he argues, degrade working conditions, living conditions, and environmental quality as they seek to increase profits. Eventually Korten recognized that his work was not compatible with his sense of what it meant to be a decent human being. He's now involved in efforts to educate people about how to reclaim their futures from this kind of maldevelopment.

One area of technological development that deserves more attention than it's getting is the development of electronic data collection and electronic surveillance.

The Internet

TB: *I am interested in any other comments you might have about the Internet.*

LW: That's an enormous topic. In many ways I'm hopeful about what the Internet can do to enable people to express their ideas, eliminate barriers to communication and find information useful for their own purposes. But I have long criticized the idea that the spread of information technology is inherently democratizing. The idea that digital electronics necessarily allows greater access to information and therefore enhances the power of ordinary citizens is one of the recurring myths of the computer age — "mythinformation," I like to call it. That canard has received a new boost with publications like the "Magna Carta for the Knowledge Age" written by Esther Dyson and other so-called "conservatives" at the Progress and Freedom Foundation. They predict that the spread of new technologies like the Internet will liberate us, causing hierarchical structures to collapse and freedom to flourish. Research like that being done by political scientist Bruce Bimber gives a more balanced view. It turns out that those who already have power and have resources — those able to make use of newspapers, mass mailings, and other ways of mobilizing public opinion — are using the Internet to do this as well.

TB: *You don't see it changing the balance of power, but as continuing the existing balance.*

LW: Yes, it seems to reproduce existing forms of politics with a few new wrinkles. Bimber seems to be finding that involvement on the Internet by women and minorities is somewhat greater than one would expect. But predictions of a revolution that will produce substantial leveling in political society are probably mistaken. For the time being the effect of all media — radio, television, and the computer — seems to be to reduce participation and amplify the power of the already powerful. I hope the Internet and World Wide Web will provide new channels for citizen involvement. But to realize that promise will take a great deal of work.

TB: You have a nice Web page.

LW: (*laughs*) Well, I don't know about that — it's rather clunky actually. Like many Web pages, it does contain advertising for new products. My latest invention, the Automatic Professor Machine, marketed by Educational Smart Hardware Alma Marter, Inc., is on display there: www.rpi.edu/~winner.

It's interesting that during the popular spread of the Internet during the 1990s, the corporate broadcast model is not the one that seems to have taken hold. There's an amazingly diverse range of voices and interests, a lot of people who find it exhilarating to reach out to people with similar needs and problems. People who have certain kinds of illnesses, who need the support of others who understand what they are going through, find support on the Internet. Minorities of various kinds, gays and lesbians for example, discover ways of making contact that were more difficult previously.

TB: Does this mean that the Internet provides community of a certain kind?

LW: I'm skeptical of the idea that community is enhanced greatly by the Internet and the World Wide Web. That has to do with my understanding of what community is about. These days people use the term to encompass people with very similar traits. But in a historical and sociological sense, living communities have been composed of people from different backgrounds who found some way to come together in face-to-face interaction and work things out. I worry that the kinds of homogenization and social isolation that we see in other parts of society — the creation of "gated communities," for example — might be reinforced by Internet communication.

Design and Ethics

TB: Could you talk a little about engineering design, including its relationship to engineering ethics.

LW: I see design as a crucial boundary between ends and means. Designs for architectural spaces, for example, have often been occasions for implementing important social policies. And designs for technological systems often express deliberate or unconscious choices about social and political values. The design of the federal highway system in the 1950s, for example, was actually a blueprint for the future of American society.

I spent some time recently finding out what engineers mean by design. Very often for them design is the point at which they narrow their focus and try to solve solve internal, technical problems — to get exactly the right size bolt or the proper configuration of electronic components. For me, the challenge of design is something different, even as it involves technology creation. Engineering design is the effort to relate decisions about technology to their context, to connect them ultimately to our ideas of the good life. Engineering design involves making a technical framework for what human activity ought to be. Both the ends and means of this "ought" should be matters for widespread study and debate.

Much of my thinking and writing these days involves design. It is a topic around which professionals and students can come together to discuss creativity, problem solving, moral and political theory, and aesthetics in settings of practical significance. But design is an ambiguous notion; no one quite knows how to define it. Many people like to say, yes, we are doing design. Students often come to engineering school with idea that they are going to be making new things which will make life better. Often, when they get there, they find they are involved with something quite different than that — namely, the application of engineering science to narrowly defined tasks. In teaching, I see design as an arena for inquiry that stretches across many disciplines. It provides good opportunities to talk about ethics. What should we make? What should we do?

TB: Can you comment on your critique of the way engineering ethics is traditionally taught. I have a quote from your article "Engineering Ethics and Political Imagination" [in Broad and Narrow Interpretations of Philosophy of Technology, *P.T. Durbin, Ed. The*

Netherlands: Kluwar, 1990] that I am told summarizes your views: "Ethical responsibility...involves more than leading a decent, honest, truthful life, as important as such lives certainly remain. And it involves something much more than making wise choices when such choices suddenly, unexpectedly present themselves. Our moral obligations must...include a willingness to engage others in the difficult work of defining the crucial choices that confront technological society and how to confront them intelligently."

In addition to being knowledgeable, competent, creative workers, engineers need to see how what they are doing has a moral and political component.

LW: My objection is that engineering ethics is sometimes presented only in the crisis mode. What would you do if someone asked you to lie, or if you saw some obvious danger? Knowing how to respond to these crisis points is obviously important. But we need more from technical professionals than that. Somehow they must understand that their everyday work is profoundly connected to the moral life of society. They need to ask: Whose interests are embodied in the project I'm engaged in? How does this match what I want my life's work to produce? Many engineers do ask these questions all the time.

It's the teaching of engineering ethics that concerns me. Sometimes classes dwell upon little cartoon problems and narrow case studies. Students end up thinking that ethics is pretty silly stuff. I believe that our teaching here ought to convey issues — ones about distributive justice in engineering decisions, for example — in ways that show why they are important and how they affect day-to-day engineering practice.

Towards the Future

***TB:** I was wondering if there are any specific ways your thinking has evolved since your two books,* Autonomous Technology, *and* Whale and the Reactor, *were published - ideas that have changed or solidified. Of course a lot has changed in 10 or 20 years — some of what you wrote might seem more strongly confirmed, or then again, are there ways you have shifted in your thinking?*

LW: Since the time I wrote *Autonomous Technology*, I've become more aware of the fragility of large sociotechnical systems. What appears to be a juggernaut or unstoppable colossus usually turns out to be something people hold together, or allow to fall apart, depending on how enough of them feel about it. Under the right set of circumstances it's possible for there to be rapid change in ideas, policies, and structures. In that way the Cold War dissolved and much of its supporting apparatus collapsed overnight. That's why I'm somewhat more hopeful than I was when I wrote *Autonomous Technology*. And that's why I believe all the more strongly that people should persist in advancing even their most improbable ideas about human well-being.

***TB:** What part of the relationship between technology and society would you most like to see changed now?*

LW: What would make me most hopeful is if I saw many more people stand up more frequently to announce their own agendas and needs for projected paths of technical and social change, rather than take somebody else's story as the one that defines the possibilities.

***TB:** Do you have a vision of the future?*

LW: I believe I can see some of the questions that will be central in the coming century, although the answers remain unknown.

- Will a global, high-tech economy continue to foster radical inequalities of wealth and living conditions? Or will forces emerge to alter that destructive trend?
- Will regional conflicts spawn a new era of competition in nuclear, chemical, and biological weapons? Or will the arts of peace and reconciliation lead us in more promising directions?
- Will the world's people own up to pending environmental disasters, including climate change, in time to make a difference? Or will ecological decline become a dreary fact of life?
- Will the genomes of the world's species come to be regarded as an elaborate Lego set, subject to manipulation for fun and profit? Or will caution about these god-like powers prevail?

***TB:** Thank you very much, Dr. Winner, for sharing your knowledge and insights with our readers.*

FEATURE ARTICLE

Is Technology Neutral?

ROBERT J. WHELCHEL

Abstract—Since technology has a value system it is not neutral. One consequence of technology having values is that it has a viewpoint, *i.e.*, it "frames" reality according to these values. Because of technology's growing influence in society, the danger exists that this restricted manner of viewing reality could become accepted as the "proper" way to look at reality—with the effect that other viewpoints would be attenuated or eliminated. Because of the bilateral nature of the interaction between society and technology, the occurrence of the dominance of the technological ethos would have repercussions for technology itself. In order for technology to prosper, it needs to be imbedded in a diverse cultural matrix. Thus, it is to the advantage of technology, as well as society, to be sure that the technological ethos maintains its proper sphere of influence.

THE MEANING OF NEUTRALITY

The question "Is technology neutral?" appears to be a polarizing element in many discussions of the social implications of technology. Before outlining any answer to this question, it seems necessary to clearly understand what is being posed by the question itself. To achieve this, it is desirable to clarify the substantive terms in the question—"technology" and "neutral." Lelt us begin with "neutral."

If one takes "neutral" as meaning "having no effect" (as in a chemical reaction), then obviously technology—whatever it is— is not neutral; technology profoundly affects our world. But clearly this is not what is meant when this question is posed by others. For example, Langdon Winner contends that one of the tenets of the "stultifying technical orthodoxy" is "that technologies are neutral: they are simply tools that can be used one way or another; the benefit or harm they bring depends on how men use them." [1] David Sarnoff is an advocate of this view: "The products of modern science are not in themselves good or bad; it is the way they are used that determines their *value*" (emphasis added). [2] This view appears to be broadly held by the technical community, so it seems fruitful to scrutinize this outlook for the implication it has relative to the meaning of "neutral."

The implication seems to be that technology is morally neutral, *i.e.*, innocent. Thus, our original question can be rephrased as: Given that technology is not causally neutral (it does affect our world), can it be morally neutral? This question is akin to the perplexing philosophical personal question: If one influences the world in which one lives, can one be truly innocent? (a question with Dostoevskian overtones). Melvin Kranzberg seems to be saying that, yes, it is possible to be morally neutral while causally non-neutral when he formulates what he refers to as Kranzberg's First Law: "Technology is neither good nor bad, nor is it neutral." By this he means that technology does indeed interact with society to produce "consequences which go far beyond the immediate purpose of the technical devices and practices themselves." Yet, for Kranzberg, "the problem is not technology itself," a position that seems to exonerate technology from moral culpability. [3]

This postulated moral neutrality of technology is a two-edged sword from the standpoint of salving the consciences of practicing technologists. Initially it seems an enviable position: There is an aura of childlike innocence about our activities—like a child we are not responsible for what we do. There is, however, an obverse side which is not nearly so attractive. This moral neutrality is based upon viewing technology purely as a means (providing tools for society to use) with the ends (the actual usage of technology) lying beyond and outside the realm of engineering; this position also assumes that available means have no causal influence on the ends chosen. If technology truly is only a means, then engineering is a second-class profession since we are the mere pawns of the real power brokers. We buy our innocence at a tremendous cost: To be innocent, we must be powerless.

Yet technology is a power. It would seem at a gut level that technological innocence it too good to be true. Contrast the following remarks of Arnold Pacey to Kranzberg's First Law discussed above.

> People often say that the fault is not with technology but with the way it is applied. But the two are not separable. Technology is about application. It is not an abstract, unbiased and morally neutral collection of useful skills and knowledge. It also includes ideas about how knowledge may be used, based on presuppositions about nature, and how one should study nature and use it, and presuppositions about society, and how technical change and social change are linked. If the practice of technology leads to frequent or dangerous dislocations in the natural environment or in society, then it is perfectly right to suspect that there is something wrong with technology itself, as a discipline . . . . [4]

Pacey admits the Dostoevskian late night thoughts that Kranzberg carefully avoids.

Let us back up yet one more time to see if we cannot finally pin down this term "neutral." Sarnoff's remark is a good reference point. Technology is neither good nor evil; this has a good ring to it. Engineers often respond as if the critics of technology are claiming that technology is an unmitigated evil; this is clearly an untenable position and one which, in fact, few critics actually take. It

The author is with the Department of Electrical and Computer Engineering, Tri-State University, Angola, Indiana 46703.

Reprinted from *IEEE Technology and Society Magazine,* Vol. 5, No. 4, pp. 3–8, December 1986.

is just as untenable to claim that technology is totally good. However, granting these two positions is not the same as assigning moral neutrality to technology. Some aspects of technology may be good, some may be bad, some ambiguous, and some (morally) neutral. Thus, although technology, as a whole, is neither good nor evil, it cannot claim innocence.

A better way to phrase our original question can be gained by returning to the emphasized word in the Sarnoff quotation—"value." In place of "Is technology neutral?" let us put "Is technology value free?" This is clearly a better way to state the question since it implies neither a wholesale condemnation of nor a total acquiescence in the activities of technology; the question itself is more nearly neutral. If technology does have a set of values, some of these may be good, some bad, some neutral, some debatable, etc.

THE MEANING OF TECHNOLOGY

Having refined the question "Is technology neutral?" down to "Is technology value-free?", we are now in a position to clarify the term "technology." A good place to begin is the etymology of the word "technology." It is a combination of two Greek roots: *techne,* art (skill, craft), and *logos,* reason (word, reason manifested through speech). Thus, technology is a combination of art and reason, a truly venerable union. The word "technology" does not simply refer to the artifacts but to the entire body of methods and materials used in combining science and art to produce items and concepts to safisfy industrial, commercial, and social objectives. Thus, I am using the word "technology" in a very broad sense (as contrasted to a narrow usage as, e.g., in "the MOS technology").

Although the above is certainly not a rigorous description of the concept of "technology," it should be adequate to prevent misunderstanding.

THE VALUES OF TECHNOLOGY

Is technology value-free? Sarnoff clearly thinks so in his statement, but I feel this response is almost an automatic defensive reply that has a longer pedigree than most engineers realize. Although this reply is partially a response to current attacks on technology, it has historically deeper roots that can be traced to the fact/value dichotomy that has engaged philosophers basically since the origins of modern science. Although interesting and relevant, there seems no need to trace the history of the fact/value controversy for our purposes here. A simple way to establish that technology does possess values is to list some of them that I think everyone would agree to.

A representative list of major value components present in any technical enterprise is: objectivity, quantification, and utilitarianism. From these follow some lesser but more specific value terms: efficiency and a systems approach which emphasizes function over device (this last a somewhat paradoxical turn for technology). All of these are nicely combined in what I consider one of the major technological concepts of our age—the cost/benefit ratio, or simply the cost function.

Although perhaps not everyone would agree with all the items listed, most would agree that some of these concepts underpin modern technology. It is equally clear that these terms are not value-free.

What, then, is the significance of technology having a value system? The consequence pertinent to our investigation is that technology creates a "frame" through which the functioning technologist views his world. This imagery of a frame is a quite potent one and it has been used by a number of authors. Alan Watts observes that "Every framework sets up a restricted field of relationships . . .," i.e., if we "frame" reality by a technological viewpoint, we are leaving out a lot of other considerations that clutter our picture, just as a professional photographer (or perhaps better, painter) excludes clutter from his pictures by "proper" framing. As Watts concludes: " . . . as soon as we introduce a frame anything does *not* go." [5]

It should be obvious that this technological framing is beneficial to the practice of technology; it is basically the Cartesian method for solving complex problems. But let's carry the photographic metaphor just a little further to introduce what Heidegger has called the "danger" of enframing. The danger arises when someone starts thinking that an Ansel Adams photograph is reality, i.e., when we think that the "clutter" that has been eliminated by "framing" is truly clutter—insignificant, valueless, perhaps even improper.

To illustrate this restricting properly of technological framing, let's consider the previously listed value components: objectivity, quantification, and utilitarianism. The first, objectivity, excludes individual experience and feelings (the individual is subordinate to the universal, possibly contributing to the linkage in some people's minds between technology and authoritarian political structures). The second, quantification, excludes qualitative thought and discourse (a woman is no longer radiant, matronly, stunning, horse-faced, etc., but ranked on a scale of one to ten). The last, utilitarianism, excludes the frivolous (or forces us to justify the frivolous by false utilitarianism: a walk in the woods is no longer just enjoyable but contributes to our health). To emphasize the point once more, framing and the exclusions which accompany that act are not necessarily bad. What is dangerous is to consider the framed view as reality, i.e., as a complete picture of what is. This last trap is especially easy to fall into when one becomes so accustomed to framing in one specific way that one no longer realizes that one is framing.

HEIDEGGER'S PERSPECTIVE

These last several paragraphs serve as an introduction to the admittedly difficult treatment of technological framing given by Martin Heidegger in his essay "The Question Concerning Technology." [6]: see also [7], [8]. According to Heidegger, one major consequence of the anthropocentric utilitarian viewpoint of technological framing is that it treats the world as if it were a "standing-reserve" (Heidegger's term), something at our beck and call, just standing by waiting for us to decide what to do with it. This in turn implies a predominance of function over device, or basically a systems viewpoint. We can (within constraints) shuffle the objects of reality around, independent of their present physical manifestations, and substitute one object for another to perform whatever function we currently wish it to do. An illustration of this point is the beer can shim in Pirsig's *ZAMM*. [9] The narrator proposes cutting up a beer can to use as a shim to repair the handlebar on John's motorcycle. John is aghast; for John, a shim is a shim and a beer can is a beer can. A shim is something you buy at a motorcycle parts dealership, not something you've just finished drinking beer out of. Pirsig claims that John is trapped in the world of "immediate appearance" and doesn't recognize the world of "underlying form."

Exploring this abstract systems viewpoint a little further, consider that one can analyze the stability of a given block diagram without knowing whether it is the model of a steel mill, the national economy, or the central nervour system (see, e.g., the homework problems in a popular undergraduate textbook in control systems by Dorf [10]). Thus, curiously, technological framing leads to a certain degree of abstraction in viewing the world. Objects are not simply seen as what they are but in terms of functions they can perform. As Heidegger states, "the object disappears into the objectlessness of standing-reverse." [11] Given Heidegger's viewpoint, those concerned that technology will eventually make us treat man as an object are chasing a pseudo-problem since technology no longer treats objects like objects.

Heidegger contends that this technological framing of the world into a "standing-reserve" is the "supreme danger" to our society. It is important to understand his reasoning on this point since the arguments themselves give new insight into the problems being discussed in this essay. Summarizing Heidegger's line of reasoning:

1) The technical framing viewpoint is correct. If the viewpoint were incorrect then there would be no crucial danger since we would eventually discover its invalidity. Being (reality) is so structured that it can in fact be treated as "standing-reserve." Thus, man alone is not solely responsible for the creation and dominance of the technological world-view; man has been aided and abetted by Being (the structure of reality).

2) Although correct, the technological world-view is not complete, i.e., it does not exhaustively describe our world nor is it the only way to reveal reality. For example, art (in its broadest sense) is another way to reveal (disclose, expose) being.

3) The danger is, then, that the correct is mistaken for the complete, and technological framing is not recognized as being merely one paradigm of reality but is mistakenly claimed to be the sole structure of reality. Restating, the danger is that *a* mode of revealing pretends that it is *the* mode of revealing and thus excludes other ways of exploring and understanding the world around us.

It is important to understand that Heidegger is not ascribing this technological framing viewpoint solely to technologists; it is being assimilated by society and becoming a commonplace attitude. The danger of this is that our society will become so ensconced in the technological framework that we will not realize how much it pervades our thought and how frequently we regret non-technical viewpoints as frivolous.

In an attempt to sharply focus on the point being made, consider this one-sentence summary: The technological "standing-reserve" frame is highly successful because it works and is correct; but this very success engenders an undue emphasis on this mode of looking at the world which results in a falsified picture of reality due to the exclusion (or at least attenuation) of other modes of apprehending the world about us. Heidegger's conclusion from this chain of reasoning is in an almost poetic or even prophetic style:

> As soon as what is unconcealed no longer concerns man even as object, but does so, rather, exclusively as standing-reserve, and man in the midst of objectlessness is nothing but the orderer of the standing-reserve, then he comes to the very brink of a precipitous fall; that is, he comes to the point where he himself will have to be taken as standing reserve. [12].

This, as Heidegger sees it, is the "supreme danger" of the technological framing viewpoint. Yet (for Heidegger) within this very danger lies our "salvation" since man can now recognize technical enframing as merely a mode of revealing and not *the* sole mode and can thus transcend its limits by embracing other modes of revealing in addition to the "standing-reserve" mode.

Heidegger's prescription is thus deceptively simple: one merely changes the way one looks at the world "Here and now and in little things. . . ." [13] To most people I'm sure this sounds a meager and impotent suggestion. Yet other thinkers have come to essentially the same conclusion. Robert Pirsig states:

> The place to improve the world is first in one's own heart and head and hands, and then work outward from three. Other people can talk about how to expand the destiny of mankind. I just want to talk about how to fix a motorcycle. I think that what I have to say has more lasting value. [14]

What Heidegger, Pirsig and others are calling for is basically a cultural transformation by peaceful means which is centered on each individual interacting with other individuals. They eschew legislation, promotional hype, and violent revolution as not only counter-productive but simply not applicable.

Many engineers are either bored or repelled by these "philosophical" considerations and view them as extraneous to the practice of technology. Basically due to the nature of their proclivities, engineers feel more comfortable talking about specific manifestations of technology than discussing philosophical concepts that lie at the core of technology. But understanding these concepts seems vital to meaningful discourse about the implications of technology. It serves no purpose to answer critiques such as Heidegger's with a ritual listing of specific devices which have contributed to a better world; yet this is precisely what many engineers do. It is crucial for engineers to understand that the views of Heidegger, Pirsig, *et al.*, would still pertain in a world without nuclear weapons, pollution, and artificial mind-destroying drugs. Their critiques are at a level which transcend specific devices and are independent of the relative merits of any specific device. it is also important to recognize that these people are not rejecting technology but are struggling to find a rewarding way to live and interact with technology. Pirsig's work is especially interesting because he directly addresses this transformation from within the framework of technology rather than approaching it as an outsider. Further, these considerations are not truly extraneous to technology, as the next section argues.

TECHNOLOGY NEEDS A BROADER FRAMEWORK

Up to this point, I have dealt with the flow of influence from technology to society. But the relationship between technology and society is bilateral rather than unilateral. Social influence on technology is not merely a fact but also a desirable situation in most cases. Technology needs outside forces driving and shaping it; we need independent inputs to spur creativity, guide our direction, and provide inspiration. The type of society a technology exists in does directly affect that technology.

Combining this conclusion with the preceding analysis of technological framing, we can see that Heidegger's "danger" to society is also a danger to technology. When the technological ethos dominates society, these "outside" forces that shape technology are no longer independent inputs. Metaphorically, technology is like a feedback system with no input. The system is driving itself; no external "reference" signal is given and the system response is determined solely by its internal structure. We are like an artist who embraces one "school" of thought to the exclusion of all others and whose creativity thus becomes stifled. In a nutshell, the enterprise of technology needs a framework broader than that provided by technology itself in order for technology to prosper and grow.

This topic has been explored extensively by the Spanish philosopher Jose Ortega y Gasset. Ortega comments that modern man (including the technologist) is largely "unaware of the artificial, almost incredible, character of civilization, and does not extend his enthusiasm for the instruments to the principles which make them possible." [15] It must be clearly understood that Ortega is not using the word "artificial" in a pejorative sense. Civilization is not a natural occurrence; it is man-made and thus artificial. Ortega's point is that it thus requires constant effort on our part to prevent this artificial civilization from degenerating back into a natural state. Ortega continues by stating that our present-day "super abundance" of technology and the ease with which it is used may allow man to "lose sight of technology and of the conditions—the moral conditions, for example—under which it is produced and return to the primitive attitude of taking it for a gift of nature which is simply there;" or, as Ortega states metaphorically, man may come to believe "that aspirin and automobiles grow on trees like apples." [16]

Similar conclusions were reached almost a century earlier by that astute observer of civilization Alexis de Tocqueville. Quoting from his chapter in *Democracy in America,* "That excessive care of worldly welfare may impair that welfare."

> There is a closer tie than is commonly supposed between the improvement of the soul and the amelioration of what belongs to the body. The soul must remain great and strong, though it were only to devote its strength and greatness from time to time to the service of the body. If men were ever to content themselves with material objects, it is probable that they would lose by degrees the art of producing them; and they would enjoy them in the end, like the brutes, without discernment and without improvement. [17]

Thus, my concern with technological framing is not limited to the broad non-technical implications of this view (obviously very important), but it is also much more parochial in nature. I am greatly concerned about its effect upon technology itself.

One way to phrase my point is that we need to regain certain art (craft) aspects of our *techne-logos* combination. I am not advocating a return to the mediaeval guild system, not am I simply pushing a restoration of skill and quality (though there is nothing wrong with that). But I am referring to some of the attitudes traditionally (perhaps sometimes erroneously) associated with craftsmen, *e.g.*, enthusiasm and excitement and personal involvement.

As Samuel Florman has stated, people used to exhibit a sense of wonder and awe before great engineering feats, yet "now it seems that no sense of wonder can be summoned at all." [18] It seems to be that there is a real danger of this sense of wonder being lost by engineers themselves. Modern science and technology are basically awe-destroying in at least two ways. First, modern science is predicated on destroying mystery from our universe by providing a rational explanation of its workings. Secondly, as A. N. Whitehead noted, the true breakthrough of the technological revolution was the "invention of the method of invention." Although technology has been with us for centuries, it is this aspect of modern technology which distinguishes us from previous eras. As a consequence of this methodological invention, technological change has become, as Whitehead said, "quick, conscious, and *expected*" (emphasis added). [19] Basically, awe follows from the unexpected. In our modern world of technological expectations, we are more awe-struck by the failure of a specific technology to develop than by continued technical progress. It seems to me that a certain amount of wonder is advantageous to the practice of technology; it also seems necessary to look outside the technological frame to find a source for this wonder.

TECHNOLOGY AS A HUMAN ENTERPRISE

Many engineers consider concepts such as awe and mystery as having no relevance to technology, and, strictly speaking, they are correct. Technology *qua* technology does have nothing to do with these things (i.e., we employ the technological frame which does exclude these concerns and which is indeed very useful). But the practice of technology (by human beings) does depend on these factors. I contend that, given two people of equal talent and knowledge, the person excited about what he is doing will usually produce better results than the person indifferent to his task. Even in the cold, impersonal world of technology, the subjective attitudes of its practioners are significant to success or failure.

The continued progress of technology depends upon the attitudes of its practitioners. Technology *qua* technology cannot provide these proper attitudes. This is where we need those independent societal inputs to drive us. Although technology itself cannot generate these attitudes, it can stifle them, through the processes discussed in the preceding paragraphs. Thus, as practitioners of technology, we need to ensure that the technological viewpoint does not become the sole ethos of our society. It is our duty, to ourselves as well as to society, to abet cultural diversity.

Furthermore, the appreciation of technology by those outside its realm should be furthered if its artistic side is recognized as well as its scientific side. As Arthur Drexler comments:

> Engineering is among the most rewarding of the arts not only because it produces individual masterpieces but because it is an art grounded in social responsibility. Today we lack the political and economic apparatus that would facilitate a truly responsible use of our technology. But it may be that a more skillful and humane use of engineering depends on a more knowledgeable response to its poetry. [20]

Drexler's concluding comment applies equally to technical practitioners and laymen. The danger of the dominance of the technological framing viewpont is not simply that society will disregard the poetry in engineering but that engineers themselves will fail to address the poetic aspect of their art and thus stifle the creative basis of their profession. Engineering needs poetic enrichment as much as society does.

CONCLUSION

Technologists need to become articulate about technology. We need a more mature grasp of the implications of technology than is suggested by the "love it or leave it" response all too frequently heard. Technology is a potent force in our world, and we should strive to understand the serious critiques of it. Only after this can we respond in a creative and meaningful fashion.

No one would contend that every individual practicing engineer needs to study and contemplate the type of questions broached in this essay. But the profession as a whole does need to address these issues. If our profession chooses to confine itself to narrow technical specialties, then we will be ignoring precisely those aspects of technology which are most significant to society. Such a stance is not only embarrassing but irresponsible.

ACKNOWLEDGMENTS

The original version of this paper was presented at the Carnahan Conference on Harmonizing Technology with Society, Lexington, Kentucky, August, 1985. I should like to thank editor Norman Balabanian for making suggestions that improved the form of this essay.

REFERENCES

[1] Winner, L., "The political philosophy of alternative technology," rpt. in A. H. Teich (ed.), *Technology and Man's Future,* 3rd ed. New York: St. Martin's, 1981, p. 370.

[2] Quoted in Marshall McLuhan, Understanding Media: The Extensions of Man. New York: Signet, 1964, p. 26.

[3] Kranzberg, M., "Technology: the half-full cup," *Alternative Futures,* vol. 3, Spring 1980, p. 8.

[4] Pacey, A., *The Maze of Inqenuity: Ideas and Idealism in the Development of Technology.* Cambridge, MA: MIT, 1976, pp. 317–318.

[5] Watts, A., "Beat zen, square zen, and zen," rpt. in N. Wilson Ross (ed.), *The World of Zen.* New York: Vintage, 1960, p. 335.

[6] Heidegger, M., *The Question Concerning Technology and Other Essays.* New York: Harper, 1977, pp. 3–35.

[7] Biemel, W., *Martin Heidegger: An Illustrated Study.* New York: Harvest (Harcourt Brace Jovanovich), 1976,ch. 8.

[8] Hood, W. F., "The Aristotelian versus the Heideggerian approach to the problem of technology," in C. Mitcham and R. Mackey (eds.), *Philosophy and Technology: Readings in the Philosophical Problems of Technology.* New York: Free Press, 1972, pp. 347–363.

[9] Pirsig, R. M., *Zen and the Art of Motorcycle Maintenance.* New York: Morrow Quill, 1979, pp. 57 ff.

[10] Dorf, R. C., *Modern Control Systems,* 3rd ed. Reading, MA: Addison-Wesley, 1980.

[11] Heidegger, *op. cit.,* p. 19.

[12] ___ p. 27.

[13] ___ p. 33.

[14] Pirsig, *op. cit.,* p. 297.

[15] Ortega y Gasset, J., *The Revolt of the Masses.* New York: Norton, 1957, p. 82.

[16] Ortega y Gasset, J., "Man the technician," in *History as a System and Other Essays Toward a Philosophy of History.* New York: Norton, 1962, p. 153.

[17] de Tocqueville, A., *Democracy in America,* vol. II. New York: Colonial Press (Co-operative Publication Society), 1900, p. 157 (Second Part, Second Book, Chapter XVI).

[18] Florman, S. C., *Blaming Technology: The Irrational Search for Scapegoats.* New York: St. Martin's, 1981, p. 160.

[19] Whitehead, A. N., *Science and the Modern World.* New York: Free Press, 1967, p. 96.

[20] Quoted in Florman, *op. cit.,* p. 161.

TODD D. CHERKASKY

Obscuring the Human Costs of Expert Systems

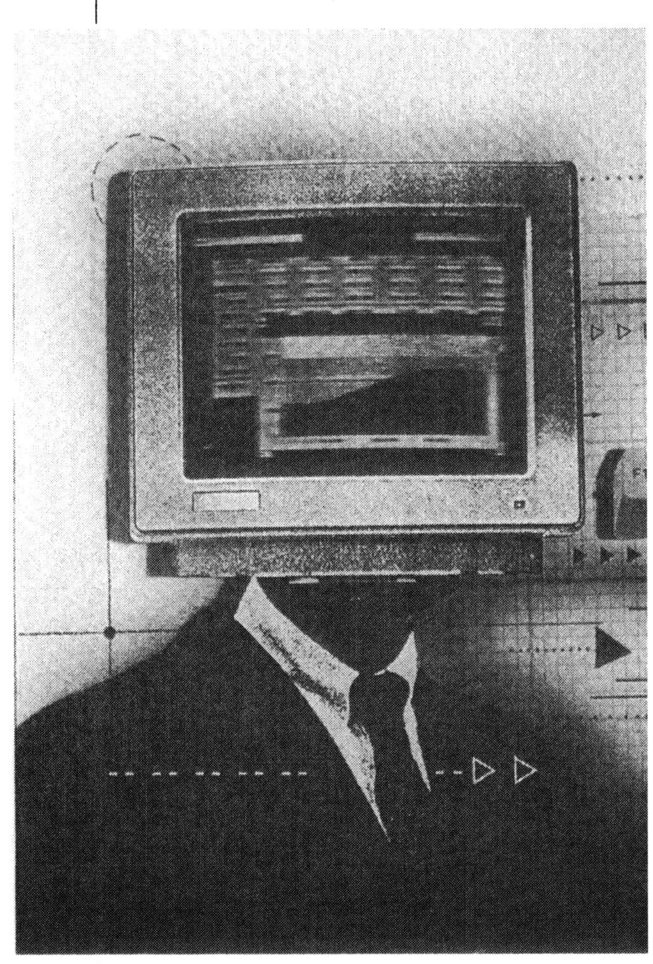

Credit: The Image Bank/©Richard Schneider/1999

Technical, specialized knowledge is generally seen as the key to personal, corporate, and national success. Top-level policy makers from the federal government encourage workers to master knowledge-intensive skills via continual training and education, as a way to help them maintain secure jobs in the global marketplace. Business leaders, threatened by intense competition and driven to continuously expand markets, require their organizations to use new technologies in sales, marketing, administration, inventory, design, manufacturing, etc. A quick review of these policy and business prescriptions evokes images of competitive industries maintaining a constant demand for sophisticated, well-trained knowledge workers. Contrary to these optimistic views, however, a close look at current trends in manufacturing — a significant segment of American business — reveals that technical professionals are being increasingly displaced and disaffected.

Decision makers from both the policy and corporate arenas must understand the implications of these trends in manufacturing to engage technology policy problems constructively. Applications of expert and knowledge-based systems deserve special attention because recently they have been used extensively in new industrial applications.[1] In particular, the rhetoric of the applied re-

The author is with Rensselaer Polytechnic Institute, Department of Science and Technology Studies, Troy, NY 12180-3590.

[1]Although well known expert systems (e.g., MYCIN for diagnosing infectious blood diseases) have been available since the 1970s, the 1980s saw the real birth in expert system utility as a few start-up companies began to commercialize them. Since that time "thousands of knowledge systems have been developed and deployed in industrial and commercial settings." [1, p. 27]

Reprinted from *IEEE Technology and Society Magazine*, Vol. 14, No. 1, pp. 10–20, Spring 1995.

Tom Toles/The Buffalo News

searchers[2] who develop these systems is problematic. Their research projects provide the first opportunity for managers and users of industrial systems to see what may become product offerings of the future. These applied scientists and engineers are particularly well connected to both innovative industrial production and advances in artificial-intelligence (AI) theory and existing technology. They write of their experiences with applications they helped install and then present visions of future possibilities.

By emphasizing enhanced working conditions and increased productivity, the researchers' language obscures the countervailing effects of pervasive expert-system development. Certainly, expert systems are useful. In fact, as applied researchers currently understand and apply them, expert systems compete against white-collar experts, devaluing the technical professionals' assets. And the researchers have powerful, if misguided, constituents to support the development of their products. However, the growing number of artificial-intelligence applications in industry creates a tension between the hopeful perspective that the white-collar workforce will fill business' increasing need for knowledge, and the reality that experts will increasingly be displaced by expert systems.

Conflicting Prescriptions

Policy makers propose education and training as one solution to industrial "downsizing." At the same time, they plan to increase funding for new industrial technology to increase industrial competitiveness. Secretary of Labor Robert Reich, for example, sees a promising future for "symbolic analysts," knowledge workers who leverage their value to the globally competitive firm based on problem-solving skills and data analysis.[3] In *Technology for America's Economic Growth*, the Clinton administration recognizes that

"the new growth industries are knowledge-based. They depend on the continuous generation of new technological innovations and the rapid transformation of these innovations into commercial products the world wants to buy.

[2] "Applied," as the term is used here, is intended to distinguish these computer science projects from theoretical and epistemological AI research projects in the same field.

[3] See [2], especially chapter 14, "The Three Jobs of the Future."

That requires a talented and adaptive work force capable of using the latest technologies and reaching ever-higher levels of productivity." [3, p. 7]

President Clinton's new technology initiative proposes not only increased training and education as a means for job growth, but also "new investments in technology that will help the private sector create high-wage, high-skill jobs." [4] Specifically, in manufacturing, President Clinton and Vice President Gore point to the automobile industry and its "new construction technologies [and] intelligent control and sensor technologies" as representative of the technologies that they wish to fund. [3] These new technologies are being tightly linked to industry by government-supported extension centers, national labs, and educational research institutions.[4] President Clinton plans to measure the success of this plan by "our ability to make a difference in the lives of the American people, to harness technology so that it improves the quality of their lives and the economic strength of our nation." [3, p. 2] Although President Clinton sees new manufacturing technology creating "stable, rewarding jobs for large numbers of workers," the implementation of this technology, including "intelligent control and sensor technologies," may actually undermine social stability. [3, p. 2]

The reality is that experts will increasingly be displaced by expert systems.

Leaders from corporate boardrooms and management schools also describe the imperatives of global competition and contemplate employment patterns affected by the demands of agile, flexible manufacturing. Lean, dynamic organizations well directed by corporate management and well linked to strategic partners have learned "the principles of CIM (Computer Integrated Manufacturing) warfare." These principles include the conviction that "fewer first-line supervisors and middle managers are going to be required."[5]

Strategic theorists from the corporate world, however, are not unaware of potential social costs of the trend towards an increasingly contingent workforce.[6] Tension has been building in the corporate community to address the social instability induced by the economic imperative. Recently, for example, the *Journal of Business Strategy* ran a cover story on the corporate addiction to downsizing and warned business leaders of possible worker disaffection. [9] In considering business's "social compact," Bernard Avishai of the *Harvard Business Review* offers a proposal for efficient education and better matching between the needs of industry and the skills of the workforce that would, he believes, attenuate the disturbing trend towards joblessness. [10] Like the policy makers, business leaders understand technology as a force to which technical professionals must react, to which school administrators must adapt their educational agendas, around which all workers must weave their personal lives and social commitments. In response to this ostensibly inevitable technological development, both business leaders and policy makers promise that more education, tailored to business and manufacturing needs, creates jobs.

Plans for education and training, however, are in a race against new knowledge-embedded technologies. Artificial-intelligence researchers and engineers writing for the trade presses understand that practical and strategic management of a firm's resources depend upon detailed knowledge about pertinent manufacturing processes, (e.g., the nature of chemical reactions, the procedure for proper part assembly). Their solution, however, is to embed symbolic analysis in the machine since manufacturing and service processes are becoming increasingly complex, while available, specialized expertise is becoming increasingly rare. The tension builds: as industry prepares to meet its particular needs, sometimes with the aid of government research-and-development funding and often with knowledge gained from government-supported academic research, roles for blue and white collar personnel are regularly excluded. Although both policy makers and applied researchers seem

[4]The issue of worker involvement in directing work organization by increased training is taken up by the U.S. Senate. [5] The relationship between small businesses and federally funded research centers as a part of technology policy is discussed in a hearing before the U.S. House. [6]

[5]Guy Potok, Senior Vice President of Fanuc Robotics North America, Inc. [7]

[6]*Fortune* magazine recently described this group as "people hired by companies to cope with unexpected or temporary challenges — part-timers, freelancers, subcontractors, and independent professionals. As the name suggests, such workers typically lead far riskier and more uncertain lives than permanent employees; they're also usually paid less and almost never receive benefits." The article also refers to them as "just-in-time employees, throwaway execs." [8]

to be working towards the same goals (e.g., efficient information processing, higher productivity, better working conditions, economic prosperity, etc.), incongruity grows between the social solution of training and educating workers and the technical solution of developing AI systems. It may not be possible to educate and train technical professionals before technological systems have absorbed expert functions that previously required well-trained, well-educated, white-collar workers. Feeling the practical strains of technological momentum, managers, trade editors, scholars, and applied researchers often argue that, beyond a certain point of technological development, too much is invested in expert and knowledge-based systems to remove them from industrial applications.

In the normal sequence of events, applied researchers receive government grants. AI solutions diffuse throughout the industrial community via the trade press, expositions, and word-of-mouth. High-technology companies combine this new AI research with their own advanced studies. They develop products and supply them to distributors. Entire systems are sold by these distributors to system integrators and to manufacturers. In all, the new technology generated by government and privately funded research is incorporated by the supply and distribution channel of original-equipment manufacturers. At this point, the technological infrastructure is complete. A complex system of interdependencies develops where research institutions and private firms supply goods demanded by the distribution channel and the manufacturing companies. The alternative use of human experts in manufacturing processes is dismissed because existing AI technology can handle the same tasks that the person was trained to perform. Therefore, it is crucial to understand the pervasive development of technologies that are incongruous with current policy agendas and hostile to the long term social prospects of knowledge workers.[7]

Justifying KBS Implementation

The rhetoric of those who promote knowledge-based systems obscures the tension between the social programs of training and education, on the one hand, and technical fixes, on the other. To make this tension explicit, let us look more closely at the criteria that applied researchers, sales engineers, marketing managers and corporate staff use to justify knowledge-based system development. Examining their rationale emphasizes their attempts to advance, in whatever way they can, the technology that they identify so thoroughly with continuing research funding and growing sales orders.

▾ Productivity Increased

As might be expected, cost reduction and increased productivity are usually cited in the trade press and journal articles as reasons for developing artificial intelligence-based manufacturing technologies. Although all articles discussing expert system technology will highlight the economic advantages of increased productivity, Frederick Hayes-Roth and Neil Jacobstein of Teknowledge Corporation, a knowledge engineering company, effectively summarize them in bullet form: "order-of-magnitude increases in speed of complex-task accomplishment, increased quality, reduced errors, reduced cost, decreased personnel required..." [1, p. 31]

In an industry-wide expert-system application used in paper manufacturing, these "decreased personnel" are thoroughly experienced, highly specialized experts. To make clean paper and to ensure reliable machinery, process engineers have had to prevent sticky wood resin, called pitch, from depositing on the machinery or ending up in the final pulp mixture. Lawrence Allen, an expert with more than twenty years of experience in diagnosing pitch problems in the batch process of paper manufacturing, may have contributed to the eventual extinction of his kind. His insight served as the foundation for the central knowledge domain of a relatively standard rule-based expert system called Pitch Expert.[8] Extensive training of more experts in this area was discarded in favor of having applied researchers develop and install a knowledge-based system.

Over time, mill management (via knowledge engineers) will attempt to augment the system's knowledge domain with any additional knowledge coming from research or operation. Seemingly unaware of cautions from critics of artificial intelligence technologies, managers

---

[7]In my research, I selected application stories from a broad spectrum of industries within manufacturing that illustrate these new technologies. The applications discussed in this article are representative of the many published in AI journals, trade magazines from the machine control industry, and recent engineering textbooks. I did not look especially for authorities in the area; instead, I attempted to gather many application examples throughout manufacturing and from the various perspectives of suppliers, customers, system integrators, journal editors, and researchers.

[8]From Kurt Vonnegut's *Player Piano* comes a pertinent excerpt about Bud Calhoun, the engineer who designed himself out of a job [11, p. 63]:

"Ah haven't got a job any more," said Bud. "Canned."

Paul was amazed. "Really? What on earth for? Moral turpitude? What about the gadget you invented for—"

"Thet's it," said Bud with an eerie mixture of pride and remorse. "Works. Does a fine job." He smiled sheepishly. "Does it a whole lot better than Ah did it." ... "Ouah [*sic*] job classification has been eliminated. Poof." He snapped his fingers.

continue the trend towards implementing expert systems because their sights are currently on increasing productivity and reducing costs to achieve the truly "lean" firm. Aware of the potential difficulties that would result if manufacturers were to eliminate significant numbers of human experts, critics suggest that firms will lose their ability to adapt their manufacturing processes over time because adaptation requires human input. If managers were to engage in this debate, they might respond that expert systems now being developed to create expert systems (meta-expert systems) will enable manufacturing processes to adapt to changing circumstances.[9] Mill managers seem confident that Pitch Expert will successfully overcome these difficulties, despite the reduced number of human pitch experts in the field.

There are social costs of the trend towards an increasingly contingent workforce.

Pitch Expert now permeates an entire industry; fewer experts of Allen's stature, people who have years of experience in the field, will be needed. Pitch Expert was designed at a research center in Montréal and transferred to the Canadian pulp and paper consortium, Papricon. Once there, all mills in the industry were given access to the knowledge base via modem. While emerging from a Canadian effort, the authors write that Pitch Expert's "possible use by mills in other countries is now being considered; some companies outside Canada have already expressed interest." [12, p. 97] It is important to note that Pitch Expert is a technical and social reality, not a theoretical case study about the feasibility of machine intelligence. "Pitch Expert serves to highlight the fact that large and sophisticated expert systems can and do provide distributed and up-to-date expertise in a readily available and accessible fashion, which translates into improved productivity and a more competitive industry." [12, p. 97]

In realizing benefits to the firm, we must not lose sight that these efforts to improve productivity come at the cost of decreased personnel. Ironically, for the society as a whole, retraining these experienced professionals is not an efficient or productive solution. In *Things That Make Us Smart*, Donald Norman estimates that over 5000 hours of intensive training are required to achieve expert performance. [13] Furthermore, frequent retraining of technical professionals (as prescribed by Reich, Clinton, and Avishai, in the expectation that training and re-education will solve the human problems) requires not only expensive training and education, but also costly social lifestyle transitions for the professionals and their uprooted families. Social costs manifest themselves in the constant anxiety of unemployment and increased work hours exhausting the workforce and removing them from other social activities. As contingency-employment firms sprout up, distributing corporate executives on a temporary basis, and as "rightsizing" continues to trim the slivers of fat from already lean firms, observers of work conditions and employment suggest the nature of work has permanently changed:

"[The] striking effect of these cuts is that they are not just the belt-tightening response of many companies to cyclical hard times. These layoffs are a sign of fundamental and permanent structural changes in the way high-tech companies will conduct business in the 21st century. Thus, many of those former employees — including the engineers — may never get their jobs back even when recovery is fully under way, because those jobs will no longer exist." [14, pp. 6-7]

Workers will be pitting themselves against intelligent machines and competing against each other over a decreasing pool of jobs. Retraining will not ameliorate this public policy nightmare.

▾ *Expertise Needed*

Generally, trade engineers and applied researchers believe that fewer workers with the needed expertise are available to do the jobs that advanced manufacturing requires, for what manufacturers are willing to pay. Contrary to the hopes of policy experts in the current administration, these researchers do not see education and training as a viable solution to the problems the manufacturing industry faces. In *A Practical Guide to Expert Systems*, William Bechtold,

[9] Piyush Jain, of the Department of Mechanical and Aeronautical Engineering at Clarkson University, and Charles Mosier, of the Department of Management at Clarkson, describe several "intelligence" levels of AI technologies in manufacturing. The lower levels involve simple operation and optimization of processes. "A cell of the third level of intelligence can adapt its operation to changing economic and technical circumstances, and is made possible by adaptive control and some elementary learning ability. A cell of the fourth and highest level of intelligence can perform more wide-ranging functions like high-level planning, learning, and self-optimization." [16, p. 379]

Vice President of Expert Edge, a vendor of expert system software for personal computers, explains that

"the costs of failing to understand modern manufacturing processes ... have grown dramatically. It seems as if today's manufacturers need an army of high-technology workers who are more efficient than the old-fashioned worker with the screwdriver and oil can. Yet, well-educated workers are scarce and expensive, and will probably continue to be so on into the next century. Enter the expert system..." [15, p. 41]

In a summary paper titled *Artificial Intelligence in Flexible Manufacturing Systems*, Jain and Mosier emphasize this need for computer systems to tackle contemporary design and control problems. They argue that "the increasing complexity associated with the control of these functions [scheduling, tool management, maintenance, fault diagnosis, process planning] and the relative dearth of true manufacturing system expertise has forced the application of artificial intelligence (AI) techniques to decision making in this environment." [16, p. 379] AI technology is increasingly being applied to manufacturing systems because of increased technical demands and the reduced availability of experts. To their way of thinking, it is an unavoidable necessity, with no degrees of choice preserved.

In the Fall 1993 issue of *AI Magazine*, knowledge engineers Adam Cunningham and Robert Smart describe the implementation of a computer-aided part estimator (CAPE), which is the "first large-scale production expert system to be deployed within Ford of Europe." The rationale for implementing CAPE is based upon scarcity of human experts in cost estimation and increasing competition:

"Because skill shortages and economic pressures prevent the replacement of expertise lost through retirement, fewer estimators with less knowledge must produce more estimates faster. Thus, there is less time to investigate alternatives to sufficient depth, resulting in sometimes shallow comparisons to previously purchased part prices and possible propagation of previous errors going unrecognized." [17, p. 40]

If it is true that there are shortages of skilled workers, one solution would be to educate and train workers to become skilled in the area of cost-estimating. However, education and training of skilled engineers has been dismissed by Ford management in favor of CAPE implementation.

Called a "knowledge-based estimator assistant," CAPE increasingly reduces the need for human experts over time. Once cost estimating is understood as a rigid job function — as contrasted to an ingrained skill of production engineers who have detailed knowledge of automobile components and the manufacturing production process — this function can be compartmentalized and programmed. In other words, a knowledge domain pertinent to cost estimation can be developed to "capture knowledge of estimating experts in all areas of automotive manufacture"; an inference engine can be applied, and an expert system employed to reduce estimating response time and costs. Because "the majority of estimator time is spent on detailed estimating, analyzing, and reporting," and because these functions can be done by expert systems, "each of these functions is now performed entirely using CAPE as the standard day-to-day estimating platform and interface to other financial systems." [17, p. 48] The human estimators, no longer needed, are eliminated from the process because their entire repertoire of analysis and reporting has been mastered by the expert system (and maintained by a few knowledge engineers). Although the primary motive for implementing CAPE involved a lack of available expertise, a closer examination of this study suggests that firms will increasingly forego training of additional engineers to replace those who left through retirement by the less costly task of implementing a knowledge-based system — in this case, an "estimator assistant" that is more accurately named an "estimator replacement."

While President Clinton sees expert systems creating stable, rewarding jobs, implementing the technology may actually undermine social stability.

▼ *Leveraging "Technological Progress*

"Once the plant floor manager is convinced that expert systems will increase productivity and compensate for the lack of available human expertise, the company's senior management must be persuaded to support the new technology's adoption. In terms familiar to management scholars, expert systems need a product champion. For this purpose, high-technology company presidents and applied researchers write scripts in trade journals that outline the procedure to follow for successful implementation of expert system technology. Savvy, knowledgeable writers instruct journal readers on how to best leverage the first expert system project so that many more can follow. Bechtold describes the process well:

Plans for education and training are in a race against new knowledge-based technologies.

"It will be easier to justify the use of an expert system if you pick an application in which the system can provide real leverage—for example, where a large monetary penalty is incurred if problems aren't dealt with quickly and accurately, where experts on the application are in short supply, or where the system can respond online to a process problem without operator intervention. Finally, to make it easier to assess the effectiveness of the expert system, select an application in which the payback can be quantified in terms of lower cost, higher product quality, and improved uptime and production efficiency." [18, p. 75]

Positioning product approval in this manner, the product champion is a political actor hoping to convince powerful managers that the increased use of AI applications assures the technological progress of their firms.

It is important to understand this process so as not to dismiss the trends towards implementing expert systems as inevitable, self-propelling technological development. Decisions are being made by managers within the organization based upon their understanding of resources and constraints. Plant floor managers, for example, actually do not have well-trained experts available who can quantify the cost savings they would impart to the company. But they do have expert-system sales engineers who have well-structured plans that substantiate the advantages of expert system implementation.

The scale of some expert-system projects reinforces the tendency for technical professionals to submit helplessly to the predominant vision of another ostensibly inevitable technological innovation. Recalling the pulp mill example, Pitch Expert's implementation and its impact on an entire industry raise the requirements for each firm in the industry. Either one must become a part of the current technological infrastructure or else pit one's human expertise against the intelligent machine. What is important here is the evolution of intelligent AI-based technology and its increasing integration into the industry. Such developments appear as autonomous technological changes when seen from the perspective of the outlying mills not involved in the Pitch Expert project development. Mill managers distant from Papricon, the brain-center of Pitch Expert, "see the technology coming." At this point, the choice has already been made—embedded in the efficiency criterion years ago. To the outlying mill managers, the technology arrives as if through its own volition. Even if technology does not actually proceed autonomously, the complex nature of technological development leads to the perception that it does. The pervasive influence of expert systems like Pitch Expert that are implemented on an industry-wide scale is combined with mill managers' helpless perspective of technology out of their control. The resulting danger is that technological development actually is out of control of the expert personnel and managers in the outlying mills. As their judgement is subordinated to a technological Pitch Expert (potentially located hundreds of miles away), anxiety increases and autonomy decreases for technical professionals.

Prior to the application of KBS technology, the expert was protected by immature technology that could not reason effectively or explain clearly its decisions. In the Pitch Expert application, however, KBS technology helps to determine pitch problems from information gathered from complex, chemical processes:

"All this information must be considered when attempting to diagnose a pitch problem in a kraft pulp mill. Furthermore, the information and the conclusions drawn from it carry various degrees of certainty, reliability, and subjectivity. Missing, incomplete, and even inaccurate data are also a fact of life in pulp mills, as is the use of ambiguous and synonymous terminology. The expert must consider all these factors when determining further questions and making recommendations. A conventional programming language would clearly be incapable of describing this situation adequately." [12, pp. 82-83]

Knowledge-based systems like Pitch Expert, in contrast, have improved on this earlier technological deficiency and have become useful for practical applications. The appearance of an application similar to Pitch Expert in *Instrumentation & Control Systems*, a trade magazine that dedicates its space to successful application stories, strengthens the point. [18, p. 78] The additional technological capability of knowledge-based systems has been effective and clearly threatens the continued survival of extremely experienced technical professionals in the industry.

It is useful, then, to re-examine how the rhetoric used by industrial manufacturers obscures the threat to technical professionals' job security. In the trade press and book series centered on new manufacturing technology, intelligence is depicted as something thoroughly disembodied. The 1993 International Programmable Controls Conference and Exposition, a leading forum for plant-floor automation and control technologies, followed the theme of "intelligent manufacturing" where "all the factors of production—including the human element" are to be "integrated" into the manufacturing process. [19, preface] A significant portion of the conference, however, was devoted to applications that included "problem-solving technologies, such as model-based reasoning, neural nets, expert systems and case-based reasoning" and that "seamlessly integrate these approaches into a single, useful problem solving methodology." [20] Embedded in the machine, intelligence has been severed from the human expert. *Human-Intelligence-Based Manufacturing* proposes "thought model-based manufacturing" as the solution to contemporary manufacturing productivity problems. [21] Although the title seems to indicate the centrality of human expertise, the "flair and thought processes of an experienced process engineer, and the decision-making processes of a mature designer" are to be extracted from the expert. "The objective is to transfer such uncertainty-related information processing to the computer environment." [21, pp. 23-25] In the same text, T. Ihara discusses this process in the wider context of a knowledge-value society:

"The industrialized nations are moving towards increasing manufacture of human-sensitivity-oriented products and aesthetic objects. This form of manufacture may be interpreted as an application of anthropocentric intelligence-based manufacturing. Its core technology is the *use of computer representation of the knowledge of experienced engineers to control processes.* [22, p. 151, emphasis added]

Based upon the commodification of "thought objects" and their integration into computer models, thought-based manufacturing "strives to make the machines appear 'human' by analysing ... the thought patterns of an engineer and how that engineer conducts the decision-making process." [22, p. 152] This model of human-intelligence-based manufacturing fully intends to replace the real human completely with human-like machinery.

The tension grows between the social solution of training and educating workers, and the technical solutions of developing AI systems.

▾ *A Better, More Fulfilling World*

Situated in social contexts where resources are negotiated, plant floor managers, product champions, expert system vendors, and applied researchers are motivated to provide an optimistic vision of a more productive, more humane world. These actors, then, project a view where jobs become more attractive for the worker. Application stories use language that makes the original task performed by human experts sound unappealing; implying that the central purpose of expert systems is actually to save technical professionals from doing tedious work. According to Spur and Specht, for example, "the solution of problems in these knowledge areas is often *relegated* to a human specialist or expert in the factory who acts on the basis of acquired experience." [23, p. 303, emphasis added] To relieve the burden, the authors recommend that "expert systems ... imitate the approach of a specialist."[10]

Much more grand visions come from those who see AI development paralleling the rapid evolution of home computer technology and the many opportunities that radical technological change affords. In an introduction to a journal

[10] I do not mean to imply that expert systems cannot aid technical professionals and relieve the drudgery that analysis of complex processes demands. However, it would be naive to understand expert systems as simply tools to relieve technical professionals of undesired tasks.

summarizing current artificial-intelligence applications in industry, Stuart Rubin states that "AI is intrinsic to the future of a better, fairer, more productive society. The attainment of these goals requires the large-scale investment and cooperation of world governments." [24, p. 8]

From the technical professionals' perspective, however, the view of fierce global competition and the need for increased productivity omits the "better" and "fairer" components of the latter, grander vision. The two views are contradictory; yet the tension between productivity and human benefits often appears in the same article. For example, although Hayes-Roth and Jacobstein argue that "quantitative changes [in KBS technology] constitute a qualitative difference in economics and human possibilities" [1, p. 31], they also emphasize that

The rhetoric of those who promote knowledge-based systems obscures the tension between the social programs of training and education, on one hand, and technical fixes, on the other.

"knowledge processing has permeated nearly every area of industry and government. *The basic logic of these systems has not changed for some time: automated knowledge processing enables organizations to make better-quality decisions, with reduced variance, using fewer people.* ... This is the basic "physics" of why knowledge processing is becoming a building block in the new economy." [1, p. 33, original emphasis]

It is unclear how a better world will evolve for technical professionals while as many workers as possible are being eliminated from their jobs. Only if all of the displaced technical professionals were to find jobs in knowledge engineering, an unlikely prospect, would these visions be compatible, and even then only for a time.

Most of the preceding argument has been centered on the harm that results from technical professionals being displaced by machines that perform their job function. Another consideration is that as the abilities of expert systems continue to advance, human contributions to the production process are structured by what the machine is currently capable of handling. A strange inversion takes place when applied researchers describe how their expert systems will be used. Typical of the applied researchers, D. Spur and G. Specht, of the Institute for Machine Tools and Manufacturing Technology, offer a view of the human worker being adapted to the machine:

"A knowledge-based system must be predictable for the system user. The human being should be able to estimate the computer's area of competency. Only when this prerequisite is met can an intelligent machine be a tool of man or an intelligent assistant in its normal task areas. This is related to the expectation that the machine serves as a tool for compensating man's weaknesses. ... The concept of the computer as an assistant to man allocates to the computer above all the function of *restructuring human tasks* on the basis of knowledge-based reasoning processes," [23, p. 305, emphasis added]

Ironically, the human becomes the tool of intelligent machines.

In this sense, the term "tool" is used to indicate that a person is subordinated to the intelligent machine. The technical professional's tasks depend upon what the machine is currently capable of doing.[11] In *Towards Intelligent PID Control,* K.J. Åström *et al.*, engineers specializing in computer control technologies, describe the design of a "knowledge-based feedback controller" that may be used to automate proportional integral derivative (PID) controllers. [25] PID controllers are often used in industrial applications that sense temperature, pressure, or position changes. Highlighting the trends in automating design tasks, they write that

"autotuners for PID controllers have been commercially available since 1981. These controllers automate some tasks normally performed by an instrument engineer. The autotuners include methods for extracting process dynamics from experiments and control design methods. ... To make systems with a higher degree of automation *it is desirable to also auto-*

[11]For an insightful fictional account written forty years ago, see Vonnegut (1952) — particularly references to the Reconstruction and Reclamation Corps.

mate tasks normally performed by process engineers." [25, p. 1, emphasis added]

Research "towards intelligent PID control" highlights the increasing capabilities of machine decision making and the development of technologies that depend less and less upon human interaction. This research assumes that "it is desirable to incorporate the expert knowledge of design engineers so that the controller can make decisions on the choice of control algorithm and provide diagnostics on the effectiveness of the control system. ... [T]he controller can interact with the operator and advise him on the choice of control algorithms. If desired, it can also make the choice automatically and explain its reasoning." [25, p. 1] It seems plausible, then, that intelligent machines might serve as just another tool for engineers specializing in narrow areas of expertise. Conservatively speaking, the use of such intelligent instruments as PID controllers can be seen as merely a tool to aid instrument and process engineers by performing some of their tasks. But heavily funded research in areas like the one described here projects a world in which there is a vast increase in the capabilities and influence of machine decision making that increasingly impinges on the tasks performed by technical professionals. Thus, a machine-centered view of manufacturing-system development "provides a mind-set that artificially elevates some aspects of life [e.g., efficiency, productivity] and ignores others [e.g., quality working conditions, challenging tasks]..." [13, p. 15] In the development and application of expert systems in manufacturing, managers' concern for the effectiveness of intelligent machines diverts their attention from issues central to workers' well-being, like job security and the cultivation of satisfying work.

Intelligent Manufacturing?

Developers of expert systems too easily impart the machine-centered perspective that humans are the weak link. In doing so, they limit options for the design of systems using intelligent machines, resulting not only in harm to workers, but also to their firms. In 1988 Shoshana Zuboff published *In the Age of the Smart Machine*, a classic study that highlighted how workers could take advantage of computer-based control technology to contribute productively to the manufacturing process. [26] Zuboff's contribution to the study of work-life quality was to recognize the "informating" capabilities of computer-based programmable logic controllers.[12] She noted correctly that shop-floor technologies of the 1980s, while intensively automating manufacturing processes, actually generated information about the processes they controlled. So, Zuboff suggested that instead of de-skilling workers, managers ought to encourage workers to develop intellective skills — the ability to reason and analyze information.

Although the capabilities of new manufacturing technology continue to increase — even intellective skills can now be handled by intelligent machines — their implementation is not determined by the force of autonomous technological development, but by decisions made throughout the complex product development process. Therefore, Zuboff's lesson for managers and designers is relevant today as well. She outlined the tendency for managers to design programmable logic control technology to reinforce traditional relationships of power and control. The managers, therefore, sacrificed opportunities for alternative arrangements that were more flexible, productive and able to synthesize the strengths of both humans and machines. In the same vein, the machine-centered emphasis of expert systems development suggests that human needs and contributions to the production process may be neglected, threatening future manufacturing productivity and flexibility.[13]

A superficial review of "new intelligent control technologies," one that does not closely examine the discourse of the various business managers, applied researchers, and engineers, produces a misleading vision of stable, secure technical professionals working for efficient, productive firms. On the other hand, examining the wide range of application stories, as told by actors intimately involved with KBS development, provides a view of the tension between what is being proposed (better jobs, more security, more productivity, etc.) with what is actually being implemented (devaluation and displacement of technical professionals). Looking at the designers' discourse in this way reveals their assumptions and expectations. Making these explicit and open to challenge increases the

[12]By the time Zuboff carried out her study, advanced manufacturing technology had already greatly displaced labor in the manufacturing process and radically altered the nature of product assembly. Her concern with work life quality did not focus on saving the vast majority of jobs. Instead, she hoped to ensure that the remaining jobs helped workers to leverage increased skill.

[13]Similar lessons can be learned from Scandinavian applications of expert systems ("systems for experts") to production processes. A seminal example that takes into account the knowledge and experience of sophisticated workers involves the Maintenance Control System (MCS) of SAS, the Scandinavian Airline. The creative design of MCS allowed for increased productivity and efficiency and better quality jobs for airline personnel. See *Computers in Context*, a video tape available from California Newsreel, 149 Ninth Street, No. 420, San Francisco, CA 94103, (415) 621-6193.

possibility that technological development can be better directed towards ends consistent with publicly deliberated goals.

Acknowledgment

The author would like to thank Deborah G. Johnson, Langdon Winner, Chuck Huff, P. Thomas Carroll, David Levinger, Anthony Robbi, Brian Rappert, and anonymous referees for helpful comments and criticism.

References

[1] Frederick Hayes-Roth and Neil Jacobstein, "The state of knowledge-based systems," *Commun. ACM*, vol. 37, pp. 27-39, Mar. 1994.
[2] Robert B. Reich, *The Work of Nations*. New York: Vintage Books, 1992.
[3] Bill Clinton and Albert Gore, *Technology for America's Economic Growth: A New Direction to Build Economic Strength*. Washington, DC: U. S. Government Printing Office, 1993.
[4] "President outlines comprehensive new technology initiative: Technology to create jobs, protect the environment, improve government, bold changes proposed to redirect, focus U.S. efforts," press release dated February 22, 1993, from the White House, Office of the Press Secretary. Available via ftp from cpsr.org; INTERNET.
[5] U.S. Senate, Committee on Labor and Human Resources, *Making the Future Work: Technology, Workers and the Workplace: Hearing before the Committee on Labor and Human Resources on S. 1020*. 103rd Cong., 1st sess., July 1, 1993.
[6] U.S. House, Committee on Science, Space, and Technology, *High Technology Small Business*. 102nd Cong., 2nd sess., Aug. 3, 1992.
[7] Guy Potok, "The principles of CIM warfare," in *IPC'93 Intelligent Manufacturing: Proc. ESD Int. Programmable Controllers Conf. & Expo*. Ann Arbor, MI, 1993, pp. 371-379.
[8] Jaclyn Fierman, "The contingency work force," *Fortune*, vol. 129, pp. 30-36, Jan. 24, 1994.
[9] Laurel Touby, "The business of America is jobs," *J. Bus. Strategy*, vol. 14, pp. 21-31, Nov.-Dec. 1993.
[10] Bernard Avishai, "What is business's social compact?" *Harvard Bus. Rev.*, vol. 72, pp. 38-48, Jan.-Feb. 1994.
[11] Kurt Vonnegut, *Player Piano*. New York: Dell, 1954.
[12] Allan Kowalski, Diana Bouchard, Lawrence Allen, Yves Larin, and Oliver Vadas, "Pitch expert: A problem-solving system for kraft mills," *AI Mag.*, vol. 14, pp. 81-98, Fall 1993.
[13] Donald A. Norman, *Things That Make Us Smart: Defending Human Attributes in the Age of the Machine*. Reading, MA: Addison-Wesley, 1994.
[14] "Manpower comments," *Jobs at Risk*, vol. 30, p. 6, Mar. 1993.
[15] William R. Bechtold, "A practical guide to expert systems," *Instrumentation & Control Syst.*, vol. 66, pp. 41-43, Dec. 1993.
[16] Piyush K. Jain and Charles T. Mosier, "Artificial intelligence in flexible manufacturing systems," *Int. J. Comput. Integrated Manufact.*, vol. 5, pp. 378-384, Nov.-Dec. 1992.
[17] Adam Cunningham and Robert Smart, "Computer-aided parts estimation," *AI Mag.*, vol. 14, pp. 39-49, Fall 1993.
[18] William R. Bechtold, "A practical guide to expert systems," *Instrument. & Control Syst.*, vol. 67, pp. 75-78, Feb. 1994.
[19] The Engineering Society, *IPC'93 Intelligent Manufacturing: Proceedings of the ESD International Programmable Controllers Conf. & Expo.*, Ann Arbor, MI, 1993.
[20] David M. Kennedy, Using an integration of expert systems, neural nets, predictive maintenance and control technologies to reduce equipment downtime," in *IPC'93 Intelligent Manufacturing: Proc. ESD Int. Programmable Controllers Conf. & Expo.*, Ann Arbor, MI, 1993, pp. 357-369.
[21] Y. Ito, Ed., *Human Intelligence-Based Manufacturing*. London: Springer-Verlag, 1993.
[22] T. Ihara, "Anthropocentric intelligence-based manufacturing," in *Human-Intelligence-Based Manufacturing*, Y. Ito, Ed. London: Springer-Verlag, 1993, pp. 151-170.
[23] G. Spur and D. Specht, "Knowledge engineering in manufacturing," *Robot. & Computer-Integrated Manufact.*, vol. 9, pp. 303-309, Aug.-Oct. 1992.
[24] Stuart H. Rubin, "Artificial intelligence for engineering" (Rev. of Editorial), *ISA Trans.* vol. 31, pp. 5-8, 1992.
[25] K.J. Åström, C.C. Hang, P. Persson, and W.K. Ho., "Towards intelligent PID control," *Automatica*, vol. 28, pp. 1-9, Jan. 1992.
[26] Shoshana Zuboff, *In the Age of the Smart Machine: The Future of Work and Power*. New York: Basic, 1988. T&S

Information and Medical Ethics: Protecting Patient Privacy

One day, at a university medical center of my acquaintance, a famous patient arrived. Television crews and trucks with satellite transmitters took up their posts, clogging the parking garages and corridors, ready to broadcast each detail of the patient's condition to a concerned nation. Finally, after a week or so, the patient left, and the media moved on. The excitement was over.

What occurred in the wake of this event, however, was of greater concern. During the patient's stay, the medical center's computers recorded several hundred attempts by separate individuals to access the famous patient's on-line clinical records. Under normal circumstances, the number of doctors, nurses, and technicians who would access an individual's records would number only several score.

Management then sent out a memorandum to the "browsers," whose computer identifications had been recorded, reminding them of the institution's privacy and confidentiality policy.

The result? Not a few expressed shock, *shock* that their behavior was being questioned, or even checked up on in the first place!

Health Care (Information) Reform

It is well known that health care has made astonishing gains in effectiveness in recent decades through the use of new medical technologies. It is less well known that the sector has lagged on the central technologies of the late 20th century: computers and telecommunications. Traditionally paper-intensive information

---

Reid Cushman is a Robert Wood Johnson Foundation Scholar at Yale University, New Haven, CT. Email: <cushman@yale.edu>

Reprinted from *IEEE Technology and Society Magazine,* Vol. 15, No. 3, pp. 32–39, Fall 1996.

processes at the core of operations have remained largely paper-intensive, long after sectors like banking and finance have converted to electronic operation [1]. Billions of reimbursement claims and other administrative documents still float around on paper. Hundreds of millions of patient files sag shelves in hospitals and clinics. One should not be misled by the fancy "telemedicine" demos that appear on the news now and again, in which doctors beam their services over video and computer links. The familiar file wall at the doctor's office is, for the moment, just as representative of the state of the art.

U.S. health care institutions have moved more aggressively to automate administrative and clinical systems in this decade. Policy makers have tried to spur the effort. Several reform bills, aimed particularly at the reduction of administrative costs, were introduced in the 102nd Congress. President Clinton's health care initiative, along with numerous competitor proposals, came in the 103rd. Most of these bills embraced a heavy reliance on information technology to promote "efficiency" — including the generation of lifetime, cradle-to-grave electronic health records on all system participants [2], [3]. Comprehensive governmental reform did not pass, of course. But private-sector and state-level restructurings under the umbrella of "managed care" have proceeded apace. More than any previous organizational scheme, managed care relies on an enhanced flow of information to organize care delivery and control costs. Information technology advances over the last two decades — in processing, storage and telecommunications capacities — make such large-scale reliance possible.

Individual medical records, carrying ever more sensitive personal information, are already being gathered and stored by the tens of thousands in databanks maintained by hospital networks, health maintenance organizations, and drug companies. As health care becomes more fully electronic, pieces of everyone's health care history will reside in public and private computer repositories. Strong legal, administrative, and technical controls will be required to structure uses and disclosures of such data. Unlike many industrialized countries, unfortunately, the U.S. has no standing national privacy body, nor privacy legislation reaching effectively to both the public and private sector [4], [5]. Narrow legislation, covering some government activity and a very limited range of private transactions, was passed in the 1970s and 1980s. Despite continual attention from Congress, manifested in myriad hearings, reports, and legislative proposals in the intervening decades, nothing comprehensive has emerged. Nor has federal legislation specific to medical privacy yet passed in the U.S. (though a proposal is now before the 104th Congress). State-level statutory protections remain an inconsistent patchwork. And the courts, to date, have found only a limited constitutional right to privacy [6].

Information technologies designed to improve efficiency thus also bring personal risks. Most of us don't need to worry that our health care histories will titillate the staff at our next hospital stay, much less be sought by television programs like "Inside Edition" or the nightly news. But our personal data could well find an interested readership in more mundane venues — at the desktops of our insurers, employers, or bankers, for example. The adverse consequences of such "leaks" can potentially be quite severe, depending on the reader and what happens to be in the patient's file. What people know about an individual can compromise not just a reputation, but the ability to obtain employment, housing, education — and, by its effects on insurability, to access affordable health care. For better or worse, the U.S. retains a system of health services financing and delivery that is predominantly private. Strong incentives for risk segregation (as the insurers like to call it), cost shifting, and discrimination are an inevitable part of this regime.

Computer Systems, Old and New

Hospitals and large clinics have long used computers for individual tasks, such as accounting or resource management. Such systems were information islands, however, typically unable to communicate with those of other departments, much less computers at other institutions. Anyone who has worked in large organizations with the old machinery knows the consequences: costly multiple entry of common data, redundant storage (with inevitable inconsistencies), periodic conversions to and from paper, and highly variable capacities for rapid information retrieval in times of need. Administrative functions are better integrated today, and more effectively computerized; electronic billing is increasingly common. Yet paper remains the dominant storage medium for patient clinical data in most settings. Administrative, evaluative, and research data is recorded separately, or "mined" post hoc from the paper histories.

The goal for the next century is an ambitious one: systems of fully computerized clinical records, "paperlessly" generated at every point of care, complete for all health episodes and fully linked to administrative and research databanks. Such a conversion is predicted to reduce overhead expense, given the logistics of moving and storing voluminous paper files. More importantly, linked electronic records will be potentially accessible immediately, whatever the time, location, or situation, and contain a complete health history regardless of when or where care was previously rendered. And unlike often-voluminous paper files, electronic records can be organized in ways that highlight important details immediately, promoting better diagnoses and treatments. Indeed, computer data management and decision support tools may eventually be used to examine information in such records, supplementing the evaluations of human caregivers. (For more information on electonic medical records, see [7].)

As medical records become electronic, pieces of everyone's health history will reside in public and private computer repositories.

Longer term benefits flow from the hope that electronic clinical records will be more easily and cheaply translatable into useful research data. Old textbook models of medical technology development follow a standard formula. Somewhere in a lab or clinic or garage there is an innovation. Controlled trials are used to establish safety, efficacy, and cost-effectiveness. Regulatory approval is gained, and routine use begins. The reality is messier [8]. Practitioners continually find new indications for use of a drug, device, or procedure after its release, for which no previous testing is performed. Development and refinement occur in day-to-day practice. Rare side effects and contraindications appear in large populations that go undetected in small-scale trials. Long-term effects emerge, but only with the passing of many years. New "outcomes" databases, generated by aggregating individual clinical and administrative records, might provide an effectively universal regime of ongoing monitoring to detect such problems. Public health disease surveillance could employ such data as well.

Managed Care, Old and New

Over the past two decades, broad geographic and interinstitutional variations in practice patterns have been detected by many medical researchers (see, e.g., [9]). (Rates of procedures such as hysterectomy and prostatectomy, for example, vary beyond anything predicted by epidemiology.) Narrowing of such disparities has been considered desirable on efficiency grounds. Yet greater standardization has been stymied both by insufficient data on the "correct" rates of application and the natural resistance of practitioners to reduced autonomy. Expanded outcomes research would potentially ameliorate the former problem — ideally telling us what works and what doesn't. The latter is already under attack by managed care's characteristically intensive review of practice patterns — and its aggressive provider "behavior modification," affected by a range of explicit "practice guidelines" and implicit financial incentives [10], [11].

Managed care depends on an intensive information foundation, which is facilitated considerably by electronic recordkeeping. Each patient's health care record is an input to cost-effectiveness analyses of the providers and institutions that constitute the "practice plan." In addition to the traditional historical data, extensive justifications are typically recorded at stages in each individual's course of care. Intimate medical information — such as with mental health services — becomes the fodder for examination by a host of persons beyond the practitioner and patient at these points. These reviews can serve to weed out inappropriate or unnecessary care. They may also, to a cynical eye, verge on a form of rationing by inconvenience. Regardless of the intent, practitioners in managed care environments can rarely assure the patient about confidentiality, given the range of reviewing parties.

In addition to these "internal" viewings, health care information has for many years been examined, transferred, collected, and aggregated by a broad range of private and public entities [12]. Much has been written about the particular propensities of private employers to seek access to medical histories. Approximately half of U.S. firms (with 50 or more employees) offering health insurance benefits self-insure; that is, they pay claims out of their own pockets. Most of the remainder, large and small, are "experience-rated" by insurers (i.e., their premiums depend on their specific claims history) [13]. Even those that do not provide health insurance must worry about other economic consequences, such as workers' compensation rates or leave utilization. The few surveys available suggest widespread structuring of employment decisions on the basis of medical histories and preexisting conditions as a consequence [14], [15]. It is unclear how much the available antidiscrimination laws, such as the Americans with Disabilities Act (ADA), have limited such behavior (see [16], [17]).

Private-sector health data collection is expanding all the time, facilitated by a long-standing culture of corporate "data swapping" [18], [19]. (Anyone who has made a large credit-card purchase, and then received related product catalogs in the mail within a few weeks, knows the culture first hand.) Some health care providers, such as physician group practices, turn over information from their patients' records to companies that provide computer software and hardware at a discount in exchange for such access [3], [20]. The data, which can contain diagnoses of illness as well as prescription information, is usually used for marketing of medical products and services. Companies such as IMS America buy large volumes of patient data from state governments, medical clinics, and drug store chains, and resell them to pharmaceutical corporations. Prescription information is also routinely sold by large drugstore chains. Companies such as Equifax, a pioneer in financial records databases, are expanding into the medical information business [21]. The insurance industry's data clearing house, the Medical Information Bureau, already keeps health-related data on 15 million persons for use in rating life, disability, and individual health policies [22]; (for other examples, see [12]).

In the public sector, at least 28 states are now expanding their own health data collection efforts, primarily motivated by cost control for publicly funded programs like Medicaid, but also as an aid to consumer choice [23]. For example, a new Maryland law requires every contact between a patient and a doctor to be recorded in a state databank. (Exemplifying the security problems with such efforts, Maryland Medicaid clerks were recently found to have sold records containing patients' names, addresses, incomes, and medical records to HMO recruiters [21], [24].) Neighboring Virginia has established databanks covering all inpatient episodes (hospital and nursing home stays) and is now constructing a repository for outpatient encounters. Taken together, these efforts amount to a broad regime of public and private health care data collection, akin to, and in many ways inspired by, databanks for individual financial information.

Privacy, Old and New

It is easy to understand the desire to limit access to one's health records, which contain or allude to some of the most intimate information conceivable about an individual. Medical files routinely itemize current and past physical status, diseases, disabilities, treatments, and medications — not just of the patient but of his or her family. Records can include material on mental health and psychological stability (including details of therapy), dietary and drug use (legitimate and illegitimate), recreational and exercise habits, and sexual practices (disclosed or supposed). Demographic information such as education, employment, marital status, and family composition is often included as well [25]. Sophisticated diagnostic tests, such as afforded by emerging genetic technologies, provide the ability to predict future disease and disability with increasing specificity. These results will increasingly be "part of the file" as well. The U.S. Equal Employment Opportunity Commission in 1995 ruled that genetic susceptibility to a disease constitutes a "disability" and therefore qualifies for protection under the ADA; bills now pending before the Congress would limit use of genetic test results for health insurance rating.

Such provisions cannot alone make the *incentives* to discriminate go away, or dislodge discriminatory dispositions. An atmosphere of distrust about the confidentiality of computer-resident information, and about behavior in response to it, accordingly breeds fears of personal humiliation, loss of reputation, and risks to financial status. In medical settings, it may cause persons increasingly to withhold sensitive information from their health care providers. Such nondisclosure presents obvious risks for the patient, since it could materially affect the course of care. Equally, physicians may elect to keep some types of information out of patient records because of confidentiality concerns (or keep duplicate, private records of sensitive information). Incomplete or inaccurate records have the potential to contaminate the knowledge base for health outcomes research. Sorting out the privacy problem is thus not just an ethical nicety, but a matter that conditions the abilities of the clinical and research apparati of health care to perform appropriately.

How much patient and practitioner "hiding" actually goes on is unknown, as is the actual magnitude of discrimination. We have for now mostly anecdote and speculation. Recent public opinion polling does, however, confirm popular concern about privacy generally. Almost 80% of respondents to the latest Equifax-Harris "consumer privacy survey" [26] felt they had lost control of personal information on them gathered by computerized systems. Some 43% were concerned about private-sector data collectors. Even more (51%) were worried about the activities of government. The same survey confirmed a strong faith in the benefits of electronic medical information, however, implying some willingness to trade off risks to privacy.

Unfortunately, we cannot now specify the terms of exchange between privacy and health benefits. Productivity and investment payoff are notoriously difficult to measure for information technologies, particularly in service industries like health care [1]. Rapid changes in price/performance ratios of electronic devices continually shift "engineering" boundaries. Predictions for unprecedentedly large, ambitious system designs are often wide of the mark. (Consider efforts to replace information technologies underlying the U.S. air traffic control system [27].) Nor are health system benefits easy to specify for any level of engineering. Improved clinical decision making logically flows from faster access to richer patient-specific data. The fraction of patients for whom the improvement will be substantial, particularly enough to justify the associated cost, is not yet known. The ultimate usability and cost of outcomes research data, derived from large-scale records mining, are also unknown. Hopes that such efforts could be a cheap, high-quality substitute for rigorous controlled trials have to date not been realized [28].

Security, Old and New

Paper documents have always been vulnerable to unauthorized access, for a variety of low-tech reasons. Access validation and logging procedures by file clerks, to control and trace which records are sent where, are rarely adequate. Physical security for central repositories, and for individual records moving within and among institutions, is often defective. And now the photocopier and fax are everywhere available to reproduce documents. An example of paper's (in)security was recently provided in the U.K., where the *Sunday Times* obtained "information black market" copies of the National Health Service records of more than a dozen politicians and celebrities. (The "victims" gave prior permission for the exercise.) All it took was a telephone order to one of a number of firms listed in the London yellow pages, and a few hours' wait [29].

At least in theory, access can be controlled and recorded in electronic systems, on a user by user basis. Indeed, access can be granted to only parts of a record, given "need to know," rather than transferring an individual's entire paper file. Once security is breached, however, electronic environments can make records vulnerable on a large scale. Software allows high-volume data searches and retrievals on criteria of interest. High-capacity telecommunications and portable storage media permit the transfer and creation of voluminous new databases quickly. Such mining can be conducted surreptitiously and anonymously — as can addition, alteration, or deletion of information in the original database — when protective measures fail completely. (Maryland's Medicaid clerks allegedly sold thousands of records over a five-year period; earlier this year, U.S. government employees were indicted for allegedly selling at least 11 000 records from the Social Security Administration's databanks to support a credit card fraud operation [24], [30].)

Networked computer systems, in the name of convenience for multiple users at multiple sites, provide numerous routes of access to information. Concomitantly, they also render up numerous routes of attack and a large, knowledgeable constituency capable of intrusions — both inadvertent and deliberate. Evidence on the actual level of security breaches in current systems is sketchy, but strong enough to raise concerns. Nearly half of the 1300 corporate respondents to a 1995 Ernst and Young survey [31] reported significant information losses in the preceding two years. Nearly 70% indicated their security risks had worsened in the last five years. While the outside "hacker" is the paradigmatic threat in the popular imagination, surveys indicate the vast majority of detected violations are still due to internal users [32]. Insiders are harder to defend against.

When computers were safely behind glass and cut off from each other, information security could be effected on an institution-by-institution basis. But networked computer systems are as vulnerable as their weakest interconnection. Computer and telecommunications protections must obtain across all participating institutions, as must effective administrative measures to complement technical safeguards [33]. As envisioned by the most aggressive promoters of electronic health records, the network of health data users potentially embraces every hospital, nursing home, physician's office, pharmacy, medical lab, and third-party payer, as well as the spectrum of public and private institutions conducting research. It will be a monumental task (and an expensive one) to manage the risks of such a regime. Few projects on the scale of it have ever been attempted.

Security uncertainties aside, it is clear that the capabilities for data extraction and manipulation dramatically expand the number of interested data users — legitimate and otherwise — for new uses that would have been prohibitively expensive in a paper environment. While it is easy to become preoccupied with the arcana of information systems protection, the reality is that a vast array of users are — and will in future likely still be — granted access to such personal data. They will have no need to "break in."

Informed consent is now a core foundation of both medical ethics and information ethics. Yet consent for others to access one's personal information becomes increasingly abstract — and, accordingly, ethically unmeaningful — as unknowable downstream users and uses proliferate.

Legal Protections, Old and New

Health providers' confidentiality obligations are routinely traced back to the Hippocratic Oath, which enjoins that what is seen or heard in the course of treatment be kept to oneself, and not "spread abroad." The current AMA Code of Medical Ethics requires that patient disclosures be safeguarded "to the greatest possible degree"; confidential communications or information are not to be revealed "without the express consent of the patient" unless required by law [34]. (Nurses and other allied health professionals have similar codes.) The American Hospital Association Patient Bill of Rights states that patients may expect "all communications and records" to be treated as confidential by the hospital "and any other parties entitled to review" information [35]. Yet in a world where information exchange is common, and providers have little control over data uses, these fine sentiments have limited effect. Armies of data users are by law, custom, and contractual arrangement "entitled to review."

Legal protections at the state level present a variable and inconsistent patchwork, and tend to set few limits on reviewing parties. Almost every state makes some statutory provision for medical privacy, but coverage and sanctions vary widely. State protections are customarily spread across medical and other professional practice acts, or embedded in hospital and other institutional licensure laws. Consequently, requirements not uncommonly conflict even within the same state. Statutes may specify different levels of obligation and protection for information held by different classes of health provider, vary according to the institution or setting of treatment, and haphazardly mix disease- and condition-specific protections with mandatory reporting requirements. Ironically, given all the disclosures permitted, only 28 states mandate patient access to their own medical records, with varying provisions about what may be withheld [36].

State statutes usually provide minimal regulation of the information practices of insurers and other payers. The evolving information strategies of managed care are thus not likely to be addressed. Only 13 states have general statutes that prescribe "fair" information practices, and these mostly reach to government activity only [6]. Since the 1970s more than a dozen states have adopted constitutional amendments designed to protect privacy interests generally, but these too are aimed predominantly at government behavior [36]. Most health care facilities, like most health care payers, are private. Even with better protections, reaching to the private sector, regulation of health care information practices *solely* at the state level accords with neither the current structure of health care delivery and finance, nor the capabilities of information storage and access technologies. Data flows don't respect such borders.

Current law at the national level does provide some privacy protections. The Privacy Act of 1974, and the Computer Matching and Privacy Protection Act of 1988 amending it, place (weak) general restrictions on federal agencies' behavior. Data collection is formally limited to "relevant and necessary" materials, linking and matching of personal information from multiple databanks is circumscribed, and disclosures to third parties are restricted. Individuals are required to have some means to learn what information is kept on them, and to request corrections of erroneous data. Other federal acts, reflecting a piecemeal approach, protect privacy category by category: e.g., the Fair Credit Reporting Act of 1970 (credit records), the Family Educational Rights and Privacy Act of 1974 (educational records), Right to Financial Privacy Act of 1978 (bank records), the Employee Polygraph Protection Act of 1988 (polygraph tests), and the Video Privacy Protection Act of 1988 (video rental records) [37].

Hospitals operated by the federal government, and private facilities operating or maintaining records under federal contract, are covered by these laws. No federal statute currently defines an individual's rights regarding personally identifiable health information held by state and local governments (although there are some restrictions on the use of social security numbers). As with the state privacy protections, private health care entities are also largely beyond the reach of these laws. Thus only a small fraction of the health databanks in the U.S. are covered [38].

Federal health care privacy legislation has been introduced many times, and is currently pending before the 104th Congress. This session's "Medical Records Confidentiality Act" (104 S.1360) as currently rendered affords some standardized protections — such as the ability to access one's medical records and submit corrections. It also legitimizes the continuing collection and inspection of medical data by a host of parties. Although individuals must formally give "informed consent" to collection and use of their health care data, the actual voluntariness is debatable. One is really free to "opt out" only if able to self-pay health care bills, or willing to forgo treatment. Large categories of uses and users are exempted from consent requirements in any case (e.g., law enforcement, medical researchers). Medical industry groups have generally been supportive, viewing this as an opportunity to move electronic recordkeeping forward. Many privacy groups oppose the legislation because of its limited protections, and because it preempts the ability of states to set higher standards. (As of April 15, 1996, subsequent revisions to the bill have tightened disclosure requirements, but not nearly to everyone's satisfaction.)

Privacy and Discrimination

It is a commonplace of data security that risks increase with the value of the information at issue, and in proportion to the number of persons who potentially have access to it. The current expansion in computer networking, and in sophisticated linking/matching capabilities for data on those networks, raise the stakes in both categories. The U.S. health sector can potentially gain much from improvements in its computer and telecommunications infrastructure, especially given its relatively lackluster utilization of such technology to date. Individual patients are likely to receive better care, given more complete, accurate, and timely medical records, and with diagnostic systems making use of that information. Systemic research using computer-aggregated data may yet point to improved modes of care, and certainly should be able to point out "problem" areas for further investigation. Paperwork should be less of a burden too.

Yet information is not, as discussed, always benign. Increasing availability of personal data raises the spectre not only of compromised privacy per se, but of a range of substantive discrimination. The U.S. health system retains a substantial degree of private financing and delivery, with attendantly strong incentives to shift risks and costs. In the name of efficiency, and private freedom of action, such shifting is allowed to a considerable degree. Despite public program "safety nets," substantial disparities in access to health care result [39]. Managed care may well have brought new efficiencies, but it has probably exacerbated the inequalities too. Ironically, if electronic health records yield the systemic data its most ardent promoters project, they will tell us with much greater specificity about such disparities. It is sometimes popular to pretend that substantive inequalities don't exist, and that rationing of care, rather than mandated by the inherent scarcity of economic resources, is a feature only of other countries' health systems.

When and if it comes in the U.S., substantive health care reform must decide the extent to which current allocation arrangements will yield to more egalitarian visions of payment and access [40]. Technology will not obviate this problem. Indeed, technological improvements — in information science and in medicine — will continually push forward the kinds of sorting, categorizing, and segregating of our fellow citizens that are possible. As a society, citizens of the U.S. have yet to agree on the limits of what government, corporations, or individuals should legitimately know about each other, or the limits of differential treatment based on that knowledge. The ethical territory is daunting. Yet in few areas of public policy are the stakes of finding appropriate balance greater.

References

[1] National Research Council, *Information Technology in a Service Society.* Washington, DC: National Acad. Press, 1994, pp. 52ff.
[2] L.O. Gostin *et al.*, "Privacy and security of personal information in a new health care system," *JAMA*, vol. 270, Nov. 23, 1993, pp. 2487-2493.
[3] S. Alpert, "Smart cards, smarter policy," Hastings Center Rep. 23, Nov./Dec. 1993, pp. 13-23.
[4] D.H. Flaherty, *Protecting Privacy in Surveillance Societies.* Chapel Hill, NC: Univ. of North Carolina Press, 1989.
[5] C.J. Bennett, *Regulating Privacy.* Ithaca, NY: Cornell Univ. Press, 1992.
[6] P.M. Schwartz, "Privacy and participation," *Iowa Law Rev.*, vol. 80, pp. 553-618, 1995.
[7] Institute of Medicine, *The Computer-Based Patient Record.* Washington, DC: National Acad. Press, 1991.
[8] Institute of Medicine, *Modern Methods of Clinical Investigation.* Washington, DC: National Acad. Press, 1990.
[9] J. Wennberg, "Variations in medical care among small areas," *Scientific Amer.*, Apr. 1982.
[10] S.R. Eastaugh, *Health Care Finance.* New York, NY: Auburn House, 1992.
[11] M. Angell, "The doctor as double agent," *Kennedy Inst. Ethics J.*, vol. 3, pp. 279-286, 1993.
[12] J. Rothfeder, *Privacy for Sale.* New York, NY: Simon and Schuster, 1992.
[13] J.C. Cantor *et al.*, "Private employment-based insurance in ten states," *Health Aff.*, vol. 14, pp. 199-211, Summer 1995.
[14] D.F. Linowes, *Privacy in America.* Urbana, IL: Univ. of Illinois Press, 1989.
[15] Office of Technology Assessment, *Medical Monitoring and Screening in the Workplace.* Washington, DC: U.S. Government Printing Office, 1991.
[16] P.T. Kilborn, "Backlog of cases is overwhelming jobs-bias agency," *New York Times*, Nov. 26, 1994.
[17] J. Mathews, "Disabilities act failing to achieve workplace goals," *Washington Post*, Apr. 16, 1995.
[18] N. Cobb, "The end of privacy," *Boston Globe*, Apr. 26, 1992.
[19] "House of cards," *Consumer Rep.*, Jan. 1996.
[20] "Who's reading your medical records?," *Consumer Rep.*, Oct. 1994.
[21] G. Kolata, "When patients' records are commodities for sale," *New York Times*, Nov. 15, 1995.
[22] J. Birnbaum, "Worried about access to insurance?," *New York Times*, July 9, 1995.
[23] C.A. Mowll, "The availability of health care information for consumer use," *J. Health Care Finance*, vol. 21, pp. 31-41, June 22, 1995.
[24] P.W. Valentine, "Medicaid bribery is alleged," *Washington Post*, June 14, 1995.
[25] Institute of Medicine, *Health Data in the Information Age.* Washington, DC: National Acad. Press, 1994.
[26] *Equifax-Harris Mid-Decade Consumer Privacy Survey 1995.* New York, NY: Louis Harris and Assoc., 1995.
[27] Matthew Wald, "Flight to nowhere," *New York Times*, Jan. 29, 1996.

[28] Office of Technology Assessment, *Identifying Health Technologies That Work*. Washington, DC: U.S. Government Printing Office, 1994.
[29] L. Rogers and D. Leppard, "For sale: Your secret medical records," *Sunday Times* [U.K.], Nov. 26, 1995.
[30] S. Hansell, "U.S. Workers stole data on 11,000, agency says," *New York Times*, Apr. 6, 1996.
[31] *Third Annual Ernst & Young/Information Week Security Survey*. New York, NY: Ernst and Young, 1995.
[32] "Computer crime greater than expected," *Newsbytes News Network*, Oct. 25, 1995.
[33] R.J. Anderson, "Security in clinical information systems," British Medical Assoc. Rep., 1996.
[34] *Code of Medical Ethics Current Opinions with Annotations*. Chicago, IL: American Medical Assoc., 1994.
[35] *Patient Bill of Rights*. Chicago, IL: American Hospital Association, 1992.
[36] L.R. Gostin *et al.*, "Privacy and security of health information in the emerging health care system," *Health Matrix*, vol. 5, pp. 1-36, 1995.
[37] P.M. Regan, *Legislating Privacy*. Chapel Hill, NC: Univ. of North Carolina Press, 1995.
[38] Office of Technology Assessment, *Protecting Privacy in Computerized Medical Information*. Washington, DC: U.S. Government Printing Office, 1993.
[39] U. Reinhardt, "Coverage and access in health care reform," *New England J. Medicine*, vol. 330, pp. 1452-1453, May 18, 1994.
[40] V.R. Fuchs, "Economics, values and health care reform," *Amer. Economic Rev.*, Mar. 1996.

T&S

Chapter 2

The Social Context of Engineering

Defining Engineering

LIKE technology, engineering resists precise definition. At the beginning of each semester in my course on "Ethics in Engineering," the class is required to come to a consensus on a definition of engineering. Ironically, most engineering students are never asked to do this in any of their engineering classes.

The class-generated definitions that have emerged from this process over the past several years are as follows:

- Engineering is the practical application, integration, and development of the ideas of applied science and math to satisfy a demand or create a new one.
- Engineering is the application of mathematical and scientific knowledge to manipulate resources to solve a problem.
- Engineering is using scientific and mathematical knowledge to develop and implement innovative solutions to technical problems presented by social wants and needs.
- Engineering is a field that uses problem-solving skills to design, build, and/or perfect biological, chemical, physical, or systematic artifacts, with the goal to enhance the quality of life.
- Engineering is a discipline creatively applying pure sciences, past experiences, and problem-solving skills fueled by curiosity and the intent to have a positive effect on society.

One insight that emerges from such attempts to define engineering is a distinction between *normative* and *descriptive* definitions. A normative definition defines something the way it ought to be, whereas a descriptive definition purports to define something objectively and value-free. Many, if not most, definitions of engineering are normative in the sense that they ascribe some beneficial outcome for engineering activity. Such definitions become problematic, however, when considering engineering products designed for evil intent (e.g., the gas chambers designed for Nazi Germany), designed to do harm to others even though the legitimacy of the intent may be arguable (e.g., weapons systems), or even everyday technologies that may have a negative impact on particular individuals or groups (e.g., assembly-line robots that displace workers). Descriptive definitions, on the other hand, are often couched in terms such as "efficiency" or "innovation." Such terms may sound value-free, but they come with their own set of implicit and explicit meanings (see, for example, Whelchel's background article in Chapter 1).

Many contemporary definitions of engineering also stress the reliance of engineering on science and mathematics. Although this is certainly true of modern engineering education, many of the great engineering marvels of the past were created with only limited scientific knowledge and mathematical tools. Even today, many real-world engineering decisions are based to some degree on "engineering judgment," acquired as much through experience as through formal training in science and mathematics. Most definitions of engineering incorporate the notion that engineers are problem solvers. While this often entails quantitative solutions to problems, it is usually acknowledged that there is a creativity, an art, if you will, associated with engineering.

Finally, it should be recognized that definitions of engineering are subject to change over time, as broader cultural values change. Many traditional definitions of engineering, for example, have highlighted engineering's role in human manipulation of the environment, in contrast to emerging paradigms that emphasize cooperation with nature. Concepts that have appeared more recently in definitions of engineering include the view of engineering as a process and the notion of technological systems.

Although most engineers are engaged in design work of some degree or another, they can be found in all facets of technical work, including basic research (sometimes referred to as "pure science"), applied research, development, design, construction, operation, and maintenance. Whereas others with scientific or technical backgrounds are members of the "technical team," engineers comprise the largest single group, outnumbering even scientists in the area of applied research and development (sometimes referred to as applied science) (1).

Engineers though of unquestioned importance to the technical success of projects, often overlook the social context of engineering. This is quite ironic, given the tendency of most engineers to define their work normatively. More subtle than the technical context, the social context of engineering can be as important as the technical with respect to engineering's ultimate contributions to humankind.

A complete understanding of the impact of engineering requires an introduction to its societal role. This can be done in many ways, ranging from a historical treatment of engineering to a social constructionist's view of technology (2). However, here I propose to consider only three aspects of the engineer's role in society: the way the engineer views the world, societal perceptions of engineers and engineering, and the relationship between engineering and business.

The Engineering View

A number of authors have attempted to describe the characteristic "engineering view," some much more favorably than others. Samuel Florman, a civil engineer and author of several books that sing the praises of technology, summarizes the engineering view as follows (3):

> a commitment to science and to the values that science demands—independence and originality, dissent and freedom and tolerance; a comfortable familiarity with the forces that prevail in the physical universe; a belief in hard work, not for its own sake, but in the quest for knowledge and understanding and in the pursuit of excellence; a willingness to forgo perfection, recognizing that we have to get real and useful products "out the door": a willingness to accept responsibility and risk failure; a resolve to be dependable; a commitment to social order, along with a strong affinity for democracy; a seriousness that we hope will not become glumness; a passion for creativity, a compulsion to tinker; and a zest for change. (pp. 76–77)

In a more critical tone, Eugene Ferguson, a noted historian of technology who also studied engineering, decries what he calls the "imperatives of engineering" (4):

> An engineer (1) strives for efficiency, (2) designs labor-saving systems, (3) tries to design the control of a system into it, so the user will have limited choices.
>
> The engineer is also fascinated by his or her ability to disregard human scale, so he (4) favors the very large, the very powerful, and (in the electronic revolution) the very small.
>
> Finally, because an engineering problem is inherently interesting, (5) it becomes an end in itself, rather than a means to satisfy a human need. (p. 30)

A more descriptive view than any of these can perhaps be found in B. D. Lichter's "core principles" of engineering (5):

> (1) A concern for the efficiency of practical means;
> (2) A commitment to concrete problem-solving, constrained to some necessary degree by time and available resources;
> (3) The pursuit of optimal technological solutions based on scientific principles and/or tested technical norms and standards;
> (4) The pursuit of creative and innovative designs;
> (5) And the development of new tools for the accomplishment of each of these.

All three of these views, regardless of ideological slant, characterize the engineering view to one extent or the other as being focused mainly on technical solutions to problems. This characteristic of the engineering view may account for the reluctance of some engineers to stray into the uncharted waters of the social and ethical dimensions of engineering. As I note in my article reprinted in Chapter 6:

> The prevailing engineering culture is readily recognized from both inside and out. Engineers are no-nonsense problem solvers, guided by scientific rationality and an eye for invention. Efficiency and practicality are the buzzwords. Emotional bias and ungrounded action are anathemas. Give them a problem to solve, specify the boundary conditions, and let them go at it free of external influence (and responsibility). If problems should arise beyond the work bench or factory floor, these are better left to management or (heaven forbid) to politicians.

The Engineering Image

Unfortunately, the limitations of the characteristic engineering view play into the popular image of the engineer as a one-dimensional person submerged in technical detail, a stereotype most engineers are quick to disown. Engineers rarely appear as characters in popular entertainment vehicles, and when they do, they are either confused with scientists or portrayed in a one-dimensional fashion (6). For example, in the 1991 feature film *Homo Faber*, based on the book by Max Frisch (7) and originally released in the United States with the title, *Voyager,* the protagonist is a globe-trotting civil engineer who is readily absorbed in gadgetry and technical discussions of risk, but who is adrift when it comes to discussing the arts or dealing with his own emotions, chance social encounters, and personal moral dilemmas.

Like all stereotypes, this image of the engineer is formed

by a small element of the truth surrounded by shallow generalities. Consequently, engineers are often pigeonholed when they have to participate in decisions that have social, policy, or ethical implications. Consider, for example, the infamous instruction to an engineering manager during the Challenger incident (discussed in Chapter 3) to take off his engineering hat and put on his management hat, a reflection of the notion that the engineering view is too narrow when it is necessary to consider "the big picture."

The Engineer as Professional

A third characteristic of the social role of engineering—and perhaps the one with the most significant implications for the social and ethical responsibilities of engineers—is the relationship between engineering and business eloquently described by E. T. Layton (8). Layton depicts the engineer as part scientist and part businessperson, yet not really either; that is to say, the engineer is marginal in both cases. This situation, which resulted from the co-evolution of engineering as a profession and technology-driven corporations, sets up inevitable conflicts between engineers' professional values and their employers' business values. Roughly three quarters of all engineers work in the corporate world. This statistic stands in contrast to other professions such as law and medicine where the model has been, at least historically, for professionals to work in private practice, serving clients or patients as opposed to employers. Professionals value autonomy, collegial control, and social responsibility, whereas businesses value loyalty, conformity, and ultimately, the pursuit of profit as the principal goal. This tension is exacerbated by the fact that the career path of engineers often leads them into management. Consequently, engineers who hope to advance in the corporate hierarchy are expected to embrace business values throughout their careers. A further drawback of this situation, discussed in more detail in Chapter 4, is the extent to which business interests exert control over the professional engineering societies.

The work of an engineer is not an isolated technical activity. It takes place in a particular institutional context—typically a corporation—as well as a broader societal context. Engineers bring distinctive viewpoints to this social arena, where others often perceive them in stereotypical ways. The social context of engineering is both a challenge and an opportunity for the profession. To rise to this challenge, engineers need to push beyond mere technical training and overturn the one-dimensional view of their profession. In so doing, they will be in a position to help identify and address broader social problems and issues.

Background Reading

In his 1990 article, "Engineers as Revolutionaries," *Don E. Kash* places a high premium on the role of engineers in society. He argues for the importance of technological innovation in the transition from an industrial to a postindustrial, information-based society. Ironically, most engineers are unaware of their role as what Kash terms "revolutionaries."

Contrary to traditional revolutions, where enough individuals adopt an ideology that so far outstrips the reality of their society that social disruption occurs, Kash argues that the present revolution consists of realities outstripping current ideology as manifested by the government's growing inability to effectively solve problems of public policy. Most such failures are revealed through unanticipated consequences of technological development and innovations that are either deficient or characterized by an inertial preference for sophisticated or large-scale systems.

The cause of the current revolution, according to Kash, is the growth in complexity of technological systems in terms of both substance specialties and the range of basic and applied skills needed to produce them. Only complex organizational systems or networks, such as those that exist in the defense and medical sectors, are capable of marshalling and synthesizing the resources necessary to cope with this complexity. This reality is in stark contrast to the prevailing ideology of individualism as the seat of technical and managerial success.

Engineers, Kash argues, are key players in today's revolutionary cadre. Evidence shows that engineers are more successful at running organizational systems than those with more traditional managerial training. Kash encourages engineers to acknowledge their revolutionary role and contribute to formulating an ideology consonant with the reality of how today's society works. Though somewhat technocratic in tone, Kash's suggestion that engineers become more aware of and responsive to their central role in organizational and social change is an important message for all engineering professionals.

Ingrid H. Soudek, in her 1986 article "The Humanist Engineer of Aleksandr Solzhenitsyn," focuses on the individual character of engineers as described in the fiction and nonfiction writings of the noted Russian author who began his career as a mathematician. Solzhenitsyn's work reveals the human side of engineers when exposed to the ordeal of being political prisoners during the Stalinist era. The values expressed by Solzhenitsyn's humanist engineer, Soudek argues, are consistent with those called for in modern codes of engineering ethics.

As reported by Soudek, Solzhenitsyn distinguished between the engineers trained before and after the Russian Revolution. The humanist engineers of the earlier era, who valued social and artistic concerns along with technology and placed the public welfare ahead of personal interests, were either eliminated or exploited for their technical skills, eventually giving way to newly trained engineers who were lacking in both the technical skills and breadth of character of their predecessors. Even when faced with the brutality and deadliness of prison, however, the old guard maintained their basic humanity, often refusing to work on projects that would benefit the police state.

Soudek suggests that the humanist engineers described by Solzhenitsyn, who preserved their values without the benefit of a formal code of ethics and very little in the way of a support system, should serve as a model for contemporary engineers faced with ethical dilemmas and professional engineering societies that have the means to provide support for ethical engineers.

Cases for Discussion

In our first case study *Ronald Kline*, an historian of technology and a former editor of *Technology and Society Magazine*, discusses "Electricity and Socialism: The Career of Charles P. Steinmetz," published in 1987. Steinmetz, the chief consulting engineer of the General Electric (GE) Company in the early years of the twentieth century and a member of the American Socialist party, was intrigued by Vladimir Lenin's plans to build a communist state in Russia based on electric power development. Steinmetz saw similar benefits of electrification in creating a socialist state in the United States.

Steinmetz became a socialist in his college days in Germany, eventually fleeing his home country to escape political reprisals and commencing his engineering studies in Zurich. Upon immigrating to the United States in 1889, Steinmetz initially withdrew from political activity as he built a highly successful engineering reputation and career and enjoyed the rewards of capitalism. In 1910, he returned to politics in Schenectady, New York, the home of GE, and participated in state and national socialist activities.

At the heart of Steinmetz's political activism was the notion that electrification, by promoting "industrial cooperation" through the creation of large-scale integrated networks, would eventually lead to the realization of a socialist state in the United States in the form of a technocratically managed planned economy. A proponent of evolutionary change as opposed to revolutionary social change, Steinmetz eventually embraced cooperation between labor and capital, rather than socialism per se, as the means of achieving a productive form of industrial democracy.

Philosopher *Thomas A. Long* offers some interesting insights on the role of engineering in public policy in his 1985 essay "Engineering and Social Equality: Herbert Hoover's Manifesto." The focus of Long's article is Hoover's book, *American Individualism*, published in 1922.

Hoover's thesis rested on the notion that equality of opportunity was the most important moral virtue; it distinguished the United States from its European roots. But Hoover, himself an engineer, believed that equality of opportunity could not be separated from engineering efficiency, which he also believed to be a moral virtue. By providing the basis of a growing material culture, engineers make equality of opportunity more realizable, offering everyone in American society more leisure and hope for a better life. Hoover therefore viewed engineers not merely as technicians but also as public servants imbued with a spirit of volunteerism. Long hastens to add, however, that Hoover's view was not technocratic in nature. Rather, Hoover pictured engineers as participating in voluntary coalitions with others, including government, as a means of furthering the goal of equality of opportunity.

Long asserts that Hoover's challenge, though simplistic in many ways, still poses a relevant challenge to professional groups regarding their social responsibilities. Engineers of today, having lost the essence of Hoover's moral ideal, see themselves merely as solvers of problems posed by others, with improvements in technical efficiency inevitably leading to equality of opportunity. Sorely lacking in the profession, Long argues, is a desire to participate in the political process of deciding what problems need to be solved. Long disputes the notion that engineers have neither the right nor the ability to participate in matters of public and social policy.

Long believes that engineers should develop a collective sense of purpose, at least with respect to the principle of equality of opportunity. He bases this argument on the fact that engineers have an extraordinary degree of power over the amount and distribution of material resources, and therefore a responsibility to see that technology is applied in the best interests of all of society. Although he admits that equality of opportunity is a vague concept, Long believes it is a starting point for recognition by engineering societies that the work of their profession is a political activity.

References

[1] Hartman, J. P. Undated. *What is an engineer, engineering technologist, and engineering technician.* College of Engineering, University of Central Florida, Orlando, Florida.

[2] Sladovich, H., ed. 1991. *Engineering as a social enterprise.* Washington, D.C.: Academy Press.

[3] Florman, S. 1987. *The civilized engineer.* New York: St. Martin's Press.

[4] Ferguson, E. 1979. The imperatives of engineering. In *Connections: technology and change,* ed. J. Burke et al., San Francisco: Bond & Fraser 29–31.

[5] Lichter, B. D. 1989. Safety and the culture of engineering. In *Ethics and risk management in engineering*, Lanham, Md.: University Press of America, 211–221.

[6] Bell, T., and Janowski, P. 1988. The image benders. *IEEE Spectrum* (25th Anniversary): 132–136.

[7] Frisch, M. 1987. *Homo Faber.* San Diego: Harcourt Brace Jovanovich.

[8] Layton, E. T. 1986. *The revolt of the engineers.* Baltimore, Md.: Johns Hopkins University Press.

DON E. KASH

Revolutionary War - tearing down the Statue of George III in New York City.

Corbis/Bettmann Archive

Engineers as Revolutionaries

A growing number of authors assert that the United States and a large portion of the world are experiencing a fundamental historical transition. It is a transition from an industrial to what has been labeled a new or post industrial, technological, services, information, or synthetic society. The transition is of proportions that appear revolutionary; that is, it involves a basic restructuring of the socioeconomic situation.

Don E. Kash is with the Science and Public Policy Program, University of Oklahoma, Norman, OK 73019. He is the author of Perpetual Innovation : The New World of Competition (Basic Books, 1989).

Technical innovation is a key element, if not the primary cause of the transition. Engineers represent a major component of the cadre which is making this revolution. Engineers are seldom seen, and seldom see themselves, as social revolutionaries. On the contrary, most evidence indicates that engineers fall on the conservative side of a liberal-conservative scale [1]. Assuming such a scale can measure one's taste for social-economic-political change, most engineers apparently have a less than ravenous appetite for such change.

What challenges does the technology driven transition pose for society? How is it that engineers are part of the revolutionary cadre and do not recognize it? Answers to these questions require an understanding of: 1) the dominant

Reprinted from *IEEE Technology and Society Magazine*, Vol. 9, No. 3, pp. 11–18, September/October 1990.

characteristics of the contemporary revolution; and 2) the causes of the revolution.

The Revolution

Revolutions occur when ideas and/or expectations about how societies should work are seriously at odds with how they actually work. Ideas and expectations about how a society should work are derived from ideologies, which are sets of systematic concepts about how human life and culture work, or should work. Successful societies, those that provide their members with acceptable levels of satisfaction and thus enjoy stability without excessive coercion, have ideologies with two characteristics. First, the ideologies are subscribed to by the vast majority of the members of society; and second, they provide a reasonably accurate picture of reality, of how society works [2, p. 7].

Traditional revolutions evolve through a period of increasing social stress climaxed by a sudden disruptive effort by a revolutionary cadre to take control of society for the purpose of making reality fit with expectations. In traditional revolutions, the underlying cause of radical change has been the formulation, by an intellectual elite and the acceptance by substantial portion of the members of society, of a new ideology which posits that the way society

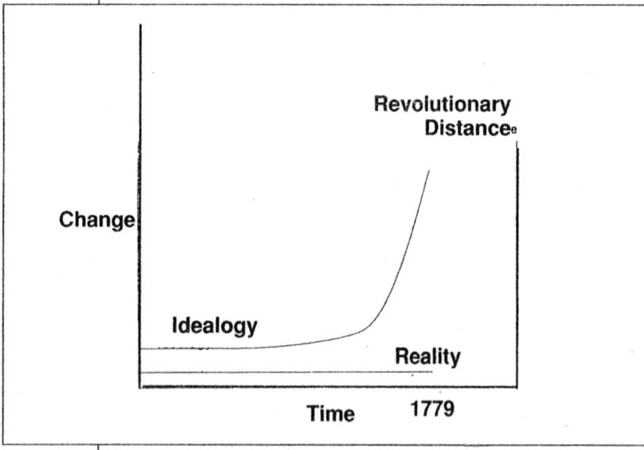

Fig. 1. Traditional revolution.

presently works is undesirable. It was this experience with new ideologies which brought about the radical change which led to the assertion that "the pen is mightier than the sword."

Alexis de Tocqueville's description of the period leading to the French Revolution is illustrative. The revolutionary cadre "built up in men's minds an imaginary ideal society in which all was simple, uniform, coherent, equitable, and rational in the full sense of the term. It was this vision of the perfect State that fired the imagination of the masses and little by little estranged them from the here-and-now" [3].

The development of traditional revolutions is illustrated in Figure 1. The horizontal axis represents time and the vertical axis represents some hypothetical unit of change. The bottom line represents reality, the way society works. The top line represents ideology. In traditional

Destruction of the Bastille, July 14, 1789. Corbis/Bettmann Archive

revolutionary periods, the ideology line reflects an exponential change as increasing numbers of people reject the old ideology and subscribe to the new. When the distance between the ideology line and the reality line becomes too large, social disorder occurs. At that point, in successful revolutions, society is fundamentally restructured.

The causes and patterns of development of the contemporary transition are just the opposite of the traditional revolution. Michael Harrington has aptly characterized the contemporary situation: "Where these conscious revolutionists of the past proposed visions which outstripped reality, the unconscious revolutionists of the present (e.g., engineers) create realities which outstrip their vision" [4].

In sum, the traditional revolution was driven by a new ideology which was exponentially subscribed to by members of the society. The contemporary transition is characterized by an exponentially changing reality. Rapid and fundamental changes are occurring in the way society works and those changes are coming into conflict with existing ideology. Some of the consequences of this conflict are heightening individual and group dissonances and alienation. A major factor in this alienation is the growing inability of public policy to solve public problems. The reason is that the conceptual framework of our ideology results in the formulation of ineffective governmental programs and actions.

The development of the contemporary situation can be shown on a graph with the same ideology and reality lines used to illustrate the traditional revolution. Figure 2 indicates an exponential change in the reality line. The way society works has been changing at an accelerating rate since World War II, with the rate of change and the pervasiveness of its impact on Americans becoming very large in the 1980s. The distance between our ideology and reality may be reaching a point where a socially disruptive reaction can occur.

Understanding the technically driven revolution requires appreciation of the complex nature of contemporary innovation.

Consequences of the New Reality

What categories of problems and issues resulting from the new reality cause social stress? A perusal of the front page of the New York Times over a three month period during 1988 indicated that a large number of the issues and problems reported fall into one or another of three technology caused or related categories. They are: 1) unanticipated consequences; 2) inertial innovation; and 3) deficient commercial innovation. The problems which fall in these categories have become both increasingly important and increasingly intractable during the 1980s.

Technical innovations generate negative unanticipated consequences so frequently that the notion that there is "no free lunch" has become an integral part of our expectations from advancing technologies. Such unanticipated consequences as toxic wastes from the chemical industry, or illnesses caused by the Dalkon Shield, are widely recognized examples of a pervasive phenomenon. Expectations of negative unanticipated consequences now complicate the development and use of large numbers of technologies. Even the cleanup of the last generation's unanticipated consequences (e.g., toxic wastes) is frequently resisted not only because of its high cost but also because there is a suspicion that the cleanup will create even more serious unanticipated problems. The record of negative unanticipated consequences generated by ever more complex technologies has contributed to an environment of social distrust. In the absence of a public understanding of why unanticipated negative consequences occur, there is tendency to explain them as the consequence of business greed and of government incompetence.

The second set of technology driven contemporary problems flows from the fact that certain sectors continuously deliver ever more sophisticated, higher cost technologies. In the United States, this is especially the case in defense and in medicine where continuous innovation has a seemingly uncontrollable momentum: it has become an inertial process. In both the defense and medical sectors, there are organizational systems which continuously drive the technological frontiers forward. These organizational systems have a powerful preference for the most sophisticated technologies. Thus, we have an abundance of nuclear aircraft carriers that are too expensive to risk in the escort of tankers in the Persian Gulf, whereas on the other hand we have no modern minesweepers. The United States leads the world in sophisticated medical technologies, yet ranks near the bottom among industrial nations in the crude death rate because of the failure to address mundane maladies. Million-dollar organ transplants are provided to a selected few, prenatal care to indigent mothers is inadequate. Such conundra are frequently explained the same way as the unanticipated consequences: they are the result of government incompetence and business greed.

The third category of technology related problems, deficient or lagging commercial innovation, results from the growing economic importance of international trade to the United States and the role of manufactured goods in that trade.

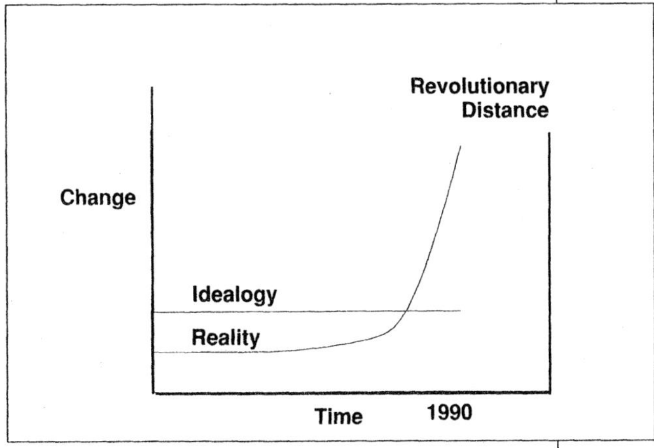

Fig. 2 Contemporary revolution.

Roughly 80 percent of international merchandise trade involves manufactured goods, and the innovation of both goods and the processes used to manufacture them is the key to competitive success. Our capacity to innovate commercial products and processes is clearly deficient when compared to some of our competitors, especially to the Japanese. Americans are bombarded daily with information on trade deficits and foreign debt. How can the country which invented the microchip, the computer, and the video cassette recorder, as well as the mass produced auto, run a manufacturing trade deficit? Like the explanations for the problems associated with unanticipated consequences and inertial innovation, the explanation for deficient commercial innovation — all too frequently — is business greed and government incompetence with the added explanation that our foreign competitors are not playing fairly.

If our changing reality is building toward a point of social disorder, it seems likely that the trigger will be economic. The continued escalation of foreign debt, accumulated to support consumption as opposed to investment, will doubtless lead to a decline in the nation's living standard. At some point, the pressure generated by that decline, when added to the existing public unease, will likely lead to popular demands for radical change.

Causes

Will the revolutionary cadre of the contemporary period (e.g., engineers) be able to offer effective answers to our problems? Will this cadre be able to formulate an ideology which provides direction for the future? It seems unlikely that a revolutionary cadre which does not know that it is playing a revolutionary role will be able to offer a better social design. *If engineers are to offer insight and direction, they must appreciate the causes of this very different revolution and understand their role in it.*

Understanding the technically driven revolution requires appreciation of the complex nature of contemporary innovation and of the organizational arrangements necessary to carry it out continuously. It is part of the conventional wisdom that modern technology is complex and is becoming more complex. Complexity refers to the fact that technologies require the synthesis of increasingly large and diverse bodies of knowledge, information, and skill. Innovation is defined as the first successful use of new ideas or methods. Thus, innovation involves manufacture and marketing as well as research, design, and development [5].

Some insight into the nature of contemporary technological complexity can be gained by referring to Figure 3, the Expertise Matrix. It is roughly correct to characterize expertise as running along two spectra. One, shown along the vertical axis of the Expertise Matrix, is organized around the substantive focus of the expertise, for example the disciplines of physics, mathematics, and chemistry. The other, shown along the horizontal axis, is organized around the degree to which the character of the expertise is fundamental versus applied. The degree of complexity is a reflection of the number of cells of expertise needed to deliver a useful technology.

As technology becomes more complex, it increasingly becomes the product of complex organizational systems which have the capacity to tap, integrate, and synthesize the needed

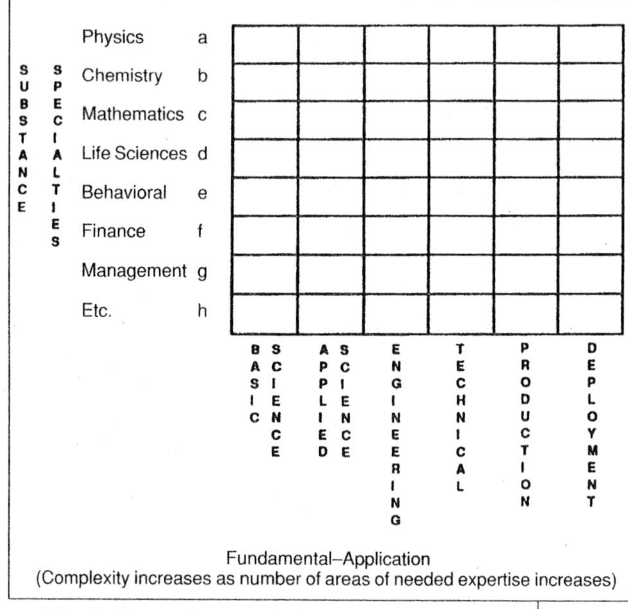

Fig. 3. Expertise matrix (source: [2,p.45])

diverse expertise. Producing complex technologies demands much greater capabilities than those available to any individual, even a genius. Complex organizations know how to do more than any individual; but more importantly, they have capabilities which are greater than the sum of the individuals who make up the organizations [6].

What distinguishes the contemporary period is the existence of organizational systems which are continuously producing complex technologies with capabilities which never existed before. These systems continuously innovate technologies. The label "high tech" connotes technologies with capabilities not available yesterday; it also connotes, however, that these new technologies will be obsolete tomorrow because other, newer technologies will be produced with capabilities superior to the present ones.

It is precisely this phenomenon which drives the exponential growth in the contemporary reality line shown in Figure 2. It is the capacity to continuously innovate complex technologies which causes changes seen by some as leading to a revolution.

Our ideology assumes that individuals have capabilities which they do not actually possess.

One point appears indisputable: never before in human history has the capacity existed to rapidly and continuously create capabilities to manipulate or exploit nature. Continuous technological innovation reflects a capacity to create which is historically unique, a capacity which not many decades ago was thought the sole province of a supreme being. That such creativity flows from organizational systems populated by ordinary people distinguished only by their in-depth but narrow expertise is quite remarkable [7]. Recognition that technological innovation is now the continuous and routine output of organizational systems full of ordinary people working in groups that routinely produce creativity is the first step to the understanding of the source of change in the contemporary period [2, pp. 38-49].

It is this characteristic of contemporary reality which is most inconsistent with existing ideology. A central postulate of the existing ideology is that creativity and innovation are the product of distinctive individual effort (e.g., great engineers, scientists and entrepreneurs). Contemporary American ideology rests on the assumption that everything can be understood through the use of the scientific method, that is, that exceptional individuals are capable of gaining understanding and thus capable of manipulating the physical world. Individuals become then the basis for explaining both successes and failures. That is, when things go wrong the cause must be bad motives or incompetence on the part of individuals. Alternatively, we commonly explain success as the result of exceptional engineers, scientists, innovators, entrepreneurs, or managers. Succinctly stated, our ideology assumes that individuals have capabilities which they do not actually possess, and it then incorrectly attributes both successes and failures to those individuals.

Organizational Systems

To deal effectively with the challenges posed by the transition in reality, it is essential that we appreciate the lack of understanding of the way organizational systems deliver continuous innovation of complex technologies. The genius of the contemporary period lies in that we know how to build and operate organizational systems with the capability for continuous innovation yet without understanding how it is done. This assertion requires some explication because it offers insight both into the source of misunderstanding of the many of our problems and into the lack of understanding by engineers of their revolutionary role.

As noted previously, the central characteristic of contemporary technological innovation is synthesis. Synthesis involves the combination of information, knowledge, experience, and materials in ways they have not been previously combined in order to produce capabilities that did not exist previously. Continuous synthetic innovation requires the ability to combine in a fashion which makes synergism routine, that is the combination produces more than the sum of the parts.

Continuous innovation requires continuous creativity. That is what has occurred in the United States in defense and medicine since World War II and it is what now characterizes the production of many commercial products in Japan but not so frequently in the United States [2, pp. 38-49].

It is now useful to investigate the assertion that "we know how to do synthetic innovation but we do not understand how to do it." Understanding requires a theory, or at least a conceptual system, which identifies the fundamental variables of a system and their mutual interactions; it identifies causes and effects. No such conceptual system currently exists for synthetic innovation. What we do know is that if one builds and supports an organizational system, and provides it with incentives which make continuous experimentation in pursuit of superior products and processes attractive, it will learn how to deliver continuous synthetic innovation. The system will accumulate the organizational learning necessary for continuous synthetic innovation of technologies; it will develop capabilities no one understands.

In discussing organizational systems which learn how to accomplish continuous synthetic innovation, it is useful to reemphasize that innovation involves use. Continuous innovation requires the continuous input of information and expertise which emanate from various points along an innovation spectrum which includes conceptualization, design, development, production, marketing, and use. Successful synthetic innovation of commercial products and of processes requires organizational systems which value market expertise as much as they value the expertise of theoretical scientists and of design engineers. Successful commercial innovations require organizational systems which emphasize manufacturing processes as much as they do product design. Innovation is not linear [8]. That we do not understand the workings of synthetic innovation is a given.

In this paper it has been repeatedly asserted that "organizational systems" are the source of continuous synthetic innovation. Organizational systems must be distinguished from "organizations." Because complex synthetic technologies require the combining and gaining of synergism from large and diverse sets of expertise, organizational networks must be built. The networks are essential because not even the largest and most complex organizations contain all of the expertise needed.

It is especially true in the United States that the needed expertise is dispersed among industry, government, universities and non-profit organizations. It is only when such networks exist which make it possible for expertise to flow horizontally, as needed among these organizations, that continuous innovation can be achieved.

How are such organizational networks built and maintained? How do they operate? For answers, we need to look at the two sectors in United States with the most consistent record of continuous synthetic innovation: defense and medicine. Building and sustaining the innovative organizational systems in defense and medicine involved three kinds of governmental action: First, the government absorbed a substantial portion of the high-risk front-end costs through R&D funding. Second, the government acted as the marriage broker in linking public and private sector organizations into systems which exist to produce continuous innovation. Third, government created a demand for new products and processes by manipulating the market [2, pp. 104-184].

Getting people and organizations with very different motives (e . g., industry pursues profit, universities pursue reputation) to work together in search of non-existent capabilities requires a distinct set of incentives. Reduced to its essentials, continuous synthetic innovation requires the ability to get: 1) groups made up of people with differing expertise located in different organizations to think creative thoughts; and 2) organizations to try those thoughts in trial and error tests even though it is recognized that most will fail. In sum, continuous innovation requires the capacity to turn on its head the conventional wisdom about how groups and organizations behave.

Government does this in defense and medicine by absorbing a high portion of the experimental risk with R&D contracts and by assuring a market. In the case of defense, government has provided the market by buying the innovative weapons systems, and in medicine it has provided a market through medical insurance coverage. It must be remembered that an R&D contract pays the cost of experimentation whatever the outcome; and where the goal is a particular performance capability, the R&D contract will cover cost plus terms which assures that the government will continue to pay until the desired capability is developed and produced.

Needed expertise is dispersed among industry, government, universities, and non-profit organizations.

Appreciation of the operation of synthetic organizational systems requires an appreciation that these systems are populated by ordinary people who behave in a way which routinely produces creativity. This occurs in a context where no one understands the actual process of continuous synthetic innovation. That is, even those who are a part of the process do not understand their role because they are continuously interacting as a parts of ever changing groups. It is roughly accurate to say that the organizational systems which produce continuous innovation require decision making and management processes which are perfectly designed to blur accurate perception of what is occurring, because decision making and management takes place by ever changing groups making consensus recommendations to other groups. To the extent that individual decision making authority is exercised, it results from narrow substantive expertise. That is, the expert defines the boundaries of what is possible in his or her specialty at a particular point in time. Only group expertise which is combined and synthesized by organizational systems can produce complex technological innovations.

The Role of Engineers

Now to the role of engineers as revolutionaries. Continuous synthetic innovation requires the integration and synthesis of many kinds of expertise: that which comes from theoretical scientists, engineers, skilled craftsmen, salesmen, and the like. Thus, engineers frequently represent only one source of the expertise which drives the contemporary revolution; they are only a part, *albeit an important one*, of the revolutionary cadre of our time. This fact, when added to the general lack of appreciation of the historically unique nature of the contemporary revolution, explains in part why engineers are not seen and do not see themselves as revolutionaries.

Engineers must help formulate an ideology for contemporary society which provides an accurate picture of reality, of how society works

The understanding of the role of engineers is further blurred because they are overwhelmingly "organization" people. Engineering is, for the most part, a group enterprise and engineers generally work for organizations, frequently large ones. Given the propensity of the American culture to attribute disproportionate credit for success to those who hold high, or visible, organizational positions, and the parallel propensity for these positions to be held by managers, the role of engineers has not been a major focus of attention.

In sum, the complex nature of contemporary technological innovation, the fact that innovation is a product of organizational systems which work in ways no one really understands, and the tendency to give credit to managers of organizations explains why there little inclination by either the engineers themselves or by the society in general to see engineers as revolutionaries.

Although hard evidence is elusive, there is a growing belief that *organizational systems in which engineers play a major or dominant role tend to be more innovative than other systems.* Increasingly, observers of the American manufacturing system suggest that the control of top management positions by financial, legal, or management people has contributed to the declining innovation and competitiveness of their organizations. Alternatively, it is common to attribute some of the competitive success of the Japanese to the dominant role which engineers play at every level in their manufacturing companies. This perception is said to have resulted in a trend back to engineers as top executives in U.S. companies, obvious examples being the recent CEOs of General Motors and Ford.

There is good reason to argue that among those experts necessary for continuous technological innovations, engineers are if not the most important then at least among the more important. They are, by any measure, a key component of the revolutionary cadre of the contemporary period.

It is compellingly important that engineers take a much more comprehensive view of the process of synthetic innovation and how they relate to it. It is important because they must help to formulate an ideology for contemporary society which provides an accurate picture of reality, of how society works. Only such an ideology will provide the society with the means to solve its problems and to provide for the well being of Americans. We need to develop an ideology which is both accepted by the vast majority of our citizens and provides an accurate picture of reality. Nothing is more central to this changed ideology than the recognition that innovation is the product of organizational systems which use processes whose key players are ever-changing groups of experts.

Coming to Grips

This paper began with the thesis that a very large number of our most serious and intractable problems resulted from continuous technological innovation (i. e., unanticipated consequences, inertial innovation, deficient commercial innovation). It concludes by asserting that the solution to most of these problems will have to come from technological innovation. Until we come to grips with how our new reality operates, we will have little chance of dealing effectively with these problems. Until we openly and consciously integrate into our ideology the recognition that continuous synthetic innovation is a product of organizational systems which are not understood, we will only add to the pressures which lead to social stress and disorder. Engineers are the best located professional group to provide ideological clarity. Unfortunately, they are among the least socially inclined to play this role, especially when it requires rejecting the central role of the exceptional individual.

References

[1] E. C. Ladd, Jr. and S. M. Lipset, "Politics of academic natural scientists and engineers," *Science*, VOL. 176, pp. 1091-1100, June 9, 1972.

[2] D. E. Kash, *Perpetual Innovation: The New World of Competition*. New York: Basic Books, 1989.

[3] Alexis de Tocqueville, *The Old Regime and the French Revolution*. Garden City, NY: Doubleday Anchor Books, 1964.

[4] M. Harrington, *The Accidental Century*. New York: Penguin Books, 1966, p. 16.

[5] C. Hill, "Technological innovation: Agent of growth and change," in *Technological Innovation for a Dynamic Economy*, C. Hill and J. Utterback, Eds. New York: Pergamon Press, 1979, p. 3.

[6] R. B. Reich, *Tales of a New America*. New York: Times Books, 1987, pp. 116-129.

[7] J. K. Galbraith, *The New Industrial State*, 3rd ed. Boston, MA: Houghton Mifflin, 1978, p. 64.

[8] S. J. Kline, "Innovation styles in the United States: Cultural bases; implications for competitiveness," from *The 1989 Thurston Lecture, Typescript, Values, Technology. Science and Society Program*. Stanford University, Stanford, CA, 1989. **T&S**

SPECIAL FEATURE

The Humanist Engineer of Aleksandr Solzhenitsyn

INGRID H. SOUDEK

In *The First Circle* and *The Gulag Archipelago I, II,* Aleksandr Solzhenitsyn depicts engineers as professionals and as human beings under the most stressful and negative circumstances—as political prisoners of the state. Their humanity in highly adverse conditions of life seems to be compatible with the codes of ethics which most professional societies espouse—to uphold the integrity of the profession by showing concern for human welfare and by serving the public, as well as clients, honestly and faithfully.* Although many American engineers are not aware of the formal codes of ethics of their engineering societies and thus do not feel compelled to subscribe to them, the codes are available as a support structure when the need arises. The profession in general still shows a disconcerting lack of interest in such professional codes of ethics, although some professional societies are now taking a greater interest in promoting such ethics.** These codes imply more than just following a set of rules; they imply a commitment to be a caring and responsible member of society—an ethical human being. This aspect of professional engineering life is seldom discussed by Western writers. For a literary exposition of it, we can turn to Aleksandr Solzhenitsyn, a mathematician turned novelist who had much contact with engineers, and who describes engineers who live up to those ethical ideas both in the non-fictional work *The Gulag Archipelago I, II* and in the novel *The First Circle.* By examining Solzhenitsyn's humanist engineers, we see how a code of ethics, whether explicit or implicit, becomes manifest in the pursuit of professional engineering.

The Gulag Archipelago I, II portrays conditions and situations in the Soviet Union prison system from 1918-1956; Solzhenitsyn experienced them for part of that time and was told of various other experiences by other prisoners. Many of these prisoners, both older and younger, were engineers, and, in describing them, Solzhenitsyn draws a picture not only of what kind of people they were, but also of his view of "the humanistic engineer."

While in *The Gulag Archipelago* Solzhenitsyn describes events and people as he saw and heard them, in the *The First Circle* he uses the form of a novel to describe some of the same situations and to express the same sentiments. The artistic vehicle of the novel allows the author to develop his characters more fully, and he can shape events, though based on real life, to underline his messages to the reader. Accordingly, the material found in *The Gulag Archipelago I, II* provides the factual bases for the more substantial fictive character development and portrayal of the engineers found in *The First Circle.* Also, the main characters in *The First Circle,* according to Solzhenitsyn, are based on real people, and the events are as he experienced them when he was imprisoned and worked with engineers. Thus *The First Circle* is largely autobiographical.

THE GULAG ARCHIPELAGO I, II

Particularly in the first two books of the four volumes of *The Gulag Archipelago,* Solzhenitsyn distinguishes between the engineers trained before the revolution and those trained afterwards. He discusses in detail why the earlier generation of engineers was eliminated, what distinguished them as "humanistic" engineers in *his* mind, and how they differed from the "new" engineers.

He explains in *The Gulag Archipelago I,* for example, that in the 1920's in Russia

> . . . The time had long since arrived to crush the technical intelligentsia, which had come to regard itself as too irreplaceable and had not gotten used to catching instructions on the wing. (P. 43, *G.I.)*

In order to create a "new society," all established members of the old society were held highly suspect, and engineers, especially, had been an important force in building up the old society. The new government wanted people in important positions who would serve only its own interests, who would not think and reason for themselves or influence others to view conditions critically; the older generation of engineers obviously would not do so. Thus, Solzhenitsyn points out:

> . . . If any pre-revolutionary engineer was not yet exposed as a traitor, then he could certainly be suspected of being one. (p. 44)
>
> . . . The only thing which at times delayed the arrest of the old engineers was the absence of a new batch to take their place. (p. 45)
>
> . . . In other words, we never did trust the engineers and from the very first years of the Revolution we saw to it that those lackeys and servants of former capitalist bosses were kept in line by healthy suspicion and surveillance by the workers. However, during the reconstruction period, we did permit them to work in our industries, while the whole force of

**See for example Code of Ethics of ASCE as accepted by the Engineers' Council for Professional Development (ECPD).*

***The Influence of Engineering Societies on Professionalism and Ethics." William H. Wisely, Executive Director Emeritus, ASCE. ASCE, Ethics, Professionalism and Maintaining Competence, pp. 51-62. As cited in R. J. Baum, Ed.* ETHICAL PROBLEMS IN ENGINEERING, VOL 2: CASES *(2nd edition). Troy, NY: Center for the Study of Human Dimensions of Science and Technology. 1980. 215.*

The author is Associate Professor of Humanities, Engineering of Applied Science, University of Virginia, Charlottesville, VA 22901.

Reprinted from *IEEE Technology and Society Magazine,* Vol. 5, No. 3, pp. 19–22, September 1986.

> the class assault was directed against the rest of the intelligentsia. (p. 43)
>
> . . . Thus in the course of a few years they broke the back of the Old Russian engineers who had constituted the glory of the country, who were the beloved heroes of such writers as Garin-Mikhailovsky, Chekhov, and Zamyatin. (p. 46)

Also, the new Russian government needed a scapegoat on whom to blame most of the country's economic woes. Thus "all the shortages in the country, including famine, cold, lack of clothing, chaos, and obvious stupidities, were blamed on the engineer-wreckers." (p. 398, *G.I.*)

While the government needed the engineers to rebuild the country, it made sure that the old engineers' work situations were intolerable and that they could be gotten rid of when convenient:

> The engineers were looked on as a socially suspicious element that did not even have the right to provide an education for its own children. Engineers were paid immeasurably low salaries in proportion to their contribution to production. But while their superiors demanded successes in production from them, and discipline, they were deprived of the authority to impose this discipline. Any worker could not only refuse to carry out the instruction of an engineer, but could insult and even strike him and go unpunished—and as a representative of the ruling class the worker was *always right* in such a case. (p. 391, *G.I*)

Solzhenitsyn writes in *The Gulag Archipelago I* that he had grown up among engineers and that he remembered that generation of "old" engineers of the twenties. This is how he describes them:

> . . . Their open, shining intellect, their free and gentle humor, their agility and breadth of thought, the ease with which they shifted from one engineering field to another, and, for that matter, from technology to social concerns and art. Then, too, they personified good manners and delicacy of taste; well-bred speech that flowed evenly and was free of uncultured words; one of them might play a musical instrument, another dabble in painting; and their faces always bore a spiritual imprint. (p. 197, *G.I*)

While this may seem quite an idealized picture of engineers to a modern reader, Solzhenitsyn tries to prove that this portrait of the older engineer was accurate by describing both in *The Gulag Archipelago I, II* and in *The First Circle* individuals he knew personally, and by telling how heroically they conducted themselves under adverse circumstances. To Solzhenitsyn, these engineers do not just represent the engineers of the twenties, but portray his ideal of the "humanistic engineer" of any era. His engineer is an intellectual human being, who is at home with technology, cares about and involves himself in social concerns, and appreciates some form of art; he is a vital part of his society and will put the human interests of that society before himself.

Because an entire generation of engineers had been swept away by the Bolsheviks, it became very urgent in the 1930's for the Soviet government to recruit its own engineers, people who would not be influenced by the past, who would be completely loyal to the government, and who would not try to think for themselves. Solzhenitsyn met one of these "new" engineers in prison; the man had been denounced by a jealous rival, and some political mistakes made in the past came back to haunt him and ended his prosperous life as a free citizen. Solzhenitsyn describes this man:

> . . . And here before me stood—an engineer, one of those who had replaced *those* destroyed.
>
> No one could deny him one point of superiority. He was much stronger, more visceral, than *those others* had been. His shoulders and hands retained their strength even though they had not needed it for a long time. Freed from the restraints of courtesy, he stared sternly and spoke impersonally, as if he didn't even consider the possibility of a dissenting view. He had grown up differently from *those others* too, and he worked differently. . . . If there had been no revolution, he would have plowed the land, and he would have become well-to-do because he was energetic and active, and he might have raised himself into the merchant class.
>
> It being the Soviet period, however . . . shot him like a rocket through the Workers' School right into the Industrial Academy. He arrived there in 1929 when *those other* engineers were being driven in whole herds into Gulag. (p. 197, *G.I*)

According to Solzhenitsyn, by meeting some of these "new" engineers in prison, he was able "to perceive through them a whole generation" of "new" engineers. He found them to be sadly lacking when compared to the older generation, both "in the breadth of their technical education" and "in their artistic sensitivity and love for their work." (p. 289, *G.II*) Yet these "new" engineers also seem to have some positive qualities: they are strong and ambitious, driven by a desire to succeed on the terms laid out for them by their government. In their own way, they view themselves as serving their country and supporting its ideology, no matter how wrong the latter may seem to Solzhenitsyn and to us.

Solzhenitsyn would have the engineer move within the system, but not hesitate to step outside of it and even oppose it when some fundamental human right is in jeopardy. In order to recognize and be sympathetic to the various needs of society, Solzhenitsyn touts the "educated" engineer, who has learned to see and appreciate different points of view, who is "humanistic" in his orientation. The new generation of engineers, according to Solzhenitsyn, lacked a well-rounded education and exposure to different ideas, which lack gave them "tunnel-vision" and allowed them to justify or ignore infractions against basic human rights. Thus *The Gulag Archipelago I, II* provides a factual basic for portraying the "humanistic" engineers who appear in *The First Circle,* even though the latter predates the publication of *Gulag.*

THE FIRST CIRCLE

With such emphasis on class background and ideology, the quality of engineering education under the Soviets had to decline. Some of the "free workers," "new" engineers with a technical education described in *The First Circle,* for example, are sadly lacking in technical skills because their teachers allowed them to get by with very little knowledge. As Solzhenitsyn explains in the novel, because there was such a need for technically trained people and quotas had to be met, schools during the Stalin era put the burden on the teachers and pressured them to graduate large numbers of technically trained students, no matter how little the students had learned. If a student failed, it was the teacher's fault and not the student's. Thus, according to Solzhenitsyn, failing students were given passing grades, accounting for a new generation of technical "experts" who could not compare in skill and knowledge to the previous generation, which had been made to study very hard and demonstrate true expertise. The young people trained during Stalin's regime were often directed into technical areas even though they might have been more suitable for another area of study, whereas the older generation of engineers had studied engineering because of aptitude and inclination. An example of this "new" engineer in *The First Circle* is Simochka, who learned so little in school that she really cannot do any higher level technical work and is little more than a watchdog over the prisoners who work alongside her in the prison laboratory (pp. 27-29).

The older generation of engineers Solzhenitsyn so admires were able to carve out some sort of niche for themselves in the vast prison system. Even as human beings, these engineers distinguished themselves in the camps:

> . . . such clearheaded, smart fellows who managed to get around the arbitrary camp rules, who helped to organize the common life so not all would die, and so as to deceive both the trust and the camp. Those heroes of the Archipelago, who understood their duty not in terms of feeding their own persons but as a burden and an obligation to the whole prison herd. . . And most of them came from the engineers. And . . . glory to them! (p. 264, *G.II*).

The government realized that these engineers were wasting away in prison doing unskilled work and decided to use those skills and talents for government purposes—for free, of course. Thus, the engineers who were able to work in their profession within the prison system found themselves in a place like the *sharashka* Solzhenitsyn describes in *The First Circle.* In 1949, the time of the novel, the term *sharashka* meant a special scientific or technical institute staffed with prisoners, or, in Soviet slang, *zeks* (p. ix). The *sharashka* was a very special kind of prison, and the prisoners had at least a few vital privileges, such as enough food—even good food—not to be had in the ordinary prison camps. This is why Solzhenitsyn calls his book *The First Circle;* the *sharashka* is like Dante's first circle of hell, where the wise men of ancient times were put. As a Christian, Dante had to put these enlightened pagans in hell, but as a humanist he could not put them with the ordinary sinners; so, like Solzhenitsyn's engineers, these wise men were put into a special place in hell where there was no actual physical torture.

In this "hell," the prisoners struggled not only to stay alive, but also to hold on to their humanity in the face of a very brutal and repressive regime. Under Stalin, according to Solzhenitsyn, people were punished whether they were guilty or not, and the government and the prison system set prisoner upon prisoner and even induced paranoia in the free citizenry. In the prison "jungle," the individual was soon reduced to a creature struggling for naked survival; to hold on to humane ideals became a luxury only special people were able to do. Among these special people were many engineers.

Solzhenitsyn devlops the characters of the engineers and other specialists in his novel carefully, and many of them emerge in the mind of the reader as real people. Yet these "real" people are also representative of many others Solzhenitsyn knew when he was himself a prisoner in camps and in a *sharashka.* Interestingly, almost all prisoner engineers in *The First Circle* were humanistic.

The pictures Solzhenitsyn draws of the engineers Potapov and Pryanchikov, for example, are complex. They are wedded to their work and may at times even be called "robots"; they are both talented, profoundly honest, sincere, and men of principle. While both were professionally successful when free and when in prison, the hardship of war and of prison revealed them as men of character who would never exploit others and compromise their principles for personal gain. Neither was politically oriented before prison, when they simply accepted what the government dictated; it was prison that caused them to question the injustice of their situation and that of their fellow prisoners and the dynamics of the Soviet society of the time. Solzhenitsyn describes this potential awakening and the rage they felt as another positive attribute of these engineers: they grew in prison; they cared not only about work, but also about justice and freedom for themselves and their fellow man. After some time in prison, neither Potapov nor Pryanchikov were robots any longer.

An engineer who understands the prison system, who knows how to deal with officials and make the most of his situation without losing his integrity is Sologdin. He is a survivor of the infamous northern camps and a very energetic, intelligent man who enjoys intellectual games and calls himself a "creator, meaning in the Language of Maximum Clarity an engineer" (p. 208).* Solzhenitsyn portrays Sologdin as a man with "inviolable peace in his soul" (p. 151), perhaps because his mind is ever fascinated by intellectual and even physical challenges. He is also a very egocentric man who says of his work, "You see, so far I have done this only to try out my own strength. For myself" (p. 204). His ultimate aim is to win his freedom, but he is not willing to sacrifice everything for that freedom. He does not try to ingratiate himself

**From now on all the references will be from* The First Circle.

with the prison administration. On the contrary: "By every possible means Sologdin avoided the attention of the authorities and spurned their favors" (p. 156). Among his fellow prisoners he is known as a "character" and is always a welcome participant in their discussions. So even though he is egocentric, it is not to the detriment of others; rather it is an indication of his faith in himself and his mental powers. He finds much pleasure in the challenge of finding a solution, and when he finds it, his pleasure consists in planning how best to use his discovery to get him freedom. He is like a chess master, thoroughly enjoying planning strategy and carefully anticipating his opponent's every move. He is successful and will probably be the first prisoner in the *sharashka* to be set free, and that freedom will be won without loss of integrity or hurting anyone else.

While Sologdin carefully plans for his freedom, others knowingly throw away such chances, even though this means leaving the *sharashka* and going back to the dreadful and very often fatal camps.

Gerasimovich is such a man; he is a physicist specialist in optics who is offered freedom if only he will work on a device that would photograph unsuspecting people who enter certain doorways at night, perhaps causing their arrest. Gerasimovich, "who . . . wore a pince-nez for his nearsightedness, which made him look exactly like the intellectual depicted in spy posters during the thirties," (p. 260), absolutely refuses. He could have agreed to work on it and then never quite completed a successful device, which is what many of the prisoners would have done. He had been pressured by his wife at her last visit to invent something so that he would be freed and she would no longer be an outcast of society, having lost her job and been ostracized for being the wife of an "enemy of the people." But Gerasimovich stands by his principles:

> "No! That's *not* my field!" he said in a clear high voice. "Putting people in prison is not my field! I don't set traps for human beings! It's bad enough that they put *us* in prison . . ." (p. 583).

Of course, this means that he will be sent to the deadly camps. Solzhenitsyn claims in *Gulag* that he knew many such engineers and technicians, who knowingly chose almost certain death rather than compromise their principles and their humanity.

Thus, by describing primarily the "old" and briefly comparing him to the "new" engineer, Solzhenitsyn develops his concept of the humanistic engineer. The American engineering societies' professional codes of ethics cited in the introduction place the highest values on the welfare of humanity, on honesty and integrity; Solzhenitsyn's engineer as seen in *The First Circle* and in *The Gulag Archipelago I, II* adheres to such a code of ethics even when his life and the life of his family is threatened. Solzhenitsyn does not cite a particular "official" code of ethics in either *The Gulag Archipelago I, II* or *The First Circle.* He uses examples from real life, as he and others experienced them, to portray his code of ethics. American societies have official codes of ethics for engineers, which can provide social support in some instances for engineers who are faced with unreasonable and perhaps unethical demands from their employers and clients. These codes are drawn up and sometimes supported by ethics panels who have no vested interests in particular projects or work. In this way, professional societies offer a support structure to their members to uphold the integrity of engineering and to show concern for human welfare. Not all American engineers have knowledge of such support systems or use them, but at least the structure of support is in place and available. The monthly magazine *Civil Engineering* lists telephone numbers for an Ethics Advisory Service. This service makes available expert advice on "potential problems about ethics, from extortion to conflict of interest" to any member of ASCE.* Such hotlines, however, do not appear to be widespread; the largest societies, IEEE and ASME, do not offer this kind of service.

Solzhenitsyn's engineers do not have this formal, social support provided by an official code of ethics. They are living in an environment that puts politics before human welfare, and thus they must try to maintain an ethical point of view in the face of enormous, sometimes life-threatening obstacles. By giving examples of the old, educated engineers in particular, Solzhenitsyn hopes to create a model of the humanistic engineer for the profession that would transcend even time and place. His "humanistic engineer" has solid technical skills which he applies competently and sometimes brilliantly to the work at hand. Although devoted to his work, sometimes to the exclusion of most other things, he is well read and culturally aware. He may also be talented in nontechnical areas, such as the arts or even cooking; but most important to Solzhenitsyn is that the "humanistic engineer" sets ethical considerations above all else. He is always aware of a responsibility to his fellow man and to himself; he never just "does a job," but strives to be a professional and a person of integrity in all aspects of life. By painting such a detailed and complete picture of his ideal engineer, Solzhenitsyn is appealing to the conscience of the reader and to other engineers to question what we must be as people and as professionals. He uses his art to make a statement about an important profession that, in his view, must always take a humanistic approach to work and life. To Solzhenitsyn, the engineer becomes a symbol for the creative, caring person who helps make society better for the good of all.

BIBLIOGRAPHY

[1] Solzhenitsyn, A., *The Gulag Archipelago I, II, 1918-1956, An Experiment in Literary Investigation.* Thomas P. Whitney, trans. New York: Harper and Row.

[2] Solzhenitsyn, A., *The First Circle.* Thomas P. Whitney, trans. Toronto: Bantam Books, 1968.

[3] Baum, R. J. (ed.), *Ethical Problems in Engineering: Vol. Two: Cases* (2nd edition). Troy, NY: Center for the Study of Human Dimensions of Science and Technology, 1980.

[4] *ASCE News,* Vol. 2, No. 12, Dec. 1977.

[5] Flores, A. (ed.), *Ethical Problems in Engineering: Vol. One: Readings* (2nd edition). Troy, NY: Center for the Study of Human Dimensions of Science and Technology, 1980.

*ASCE News, *Dec. 1977, Vol. 2, No. 12.*

FEATURE ARTICLE

Electricity and Socialism: The Career of Charles P. Steinmetz

RONALD KLINE

In 1920, during the throes of the economic crisis facing Soviet Russia, Vladimir Lenin proclamed that "Communism is socialist power plus the electrification of the whole country." "Only when the country has been electrified, and industry, agriculture, and transport have been placed on the technical basis of modern large-scale industry, only then shall we be fully victorious" over capitalism within and outside Russia.

Electrification had many virtues. It would "accelerate the transformation of dirty, repulsive workshops into clean bright laboratories worthy of human beings." Electrifying the home would ease the burdens of millions of "domestic slaves." And electricity would "make it possible to raise the level of culture in the countryside and to overcome, even in the most remote corners of the land, backwardness, ignorance, poverty, diseases and barbarism." [1]

To achieve these results, Lenin had Russian engineers draw up a ten year plan to build regional power networks covering the entire country. Such systems were not without precedent in 1920. Regional networks, larger and more advanced than those planned by Lenin, had evolved in the United States, Britain, and Germany prior to World War I. And still larger systems had been proposed in all three countries shortly after the war, including the socialization of a unified electric supply system in Germany in 1919. Although Lenin's project owed much to these precedents, both in technology and ideas, it stood out for its comprehensiveness, degree of state control, and ideological imperative. All of Russia must be electrified to bring about the Communist state. [2]

Lenin's program appealed to many socialists and perhaps to a few engineers. It certainly fired the imagination of Charles Proteus Steinmetz, socialist office holder and Chief Consulting Engineer of the General Electric Company. In 1922, Steinmetz wrote to Lenin, offering his engineering assistance in the great work of electrifying Russia. As an engineer, he admired the efficiency and rationalization of Lenin's plan. As a socialist, he favored the project because a decade before Lenin pronounced his famous dictum equating electricity and communism, Steinmetz had worked out a similar relationship between electricity and socialism.

Of course , the idea of electricity as a social force was not new with Steinmetz nor Lenin. After the introduction of electric light and power in the 1880s, several writers in the United States and Europe thought electricity would decentralize industry, clean up the cities, and improve life in rural areas. Politicians who sought state-controlled electric supply, such as Gifford Pinchot in his giant power scheme for Pennsylvania, also attributed many of these virtues to widespread electrification, including the decentralization of industry. [3] But in Steinmetz's view, electricity was not decentralizing industry. It was centralizing it in the inevitable evolution of society from a capitalist to a socialist, technocratic state.

SOCIALIST BACKGROUND, 1884-1913

The roots of Steinmetz's socialism go back to his student days at the University of Breslau where he joined a socialist group in 1884. The 1880s were an exciting, if not dangerous, time to be a socialist in Germany because of Bismarck's anti-socialist laws. It was legal for a Social Democrat to hold a seat in the German parliament. But it was illegal for others to belong to the Party or to engage in socialist work, a situation which enlivened Steinmetz's university life considerably. [4]

Steinmetz became a socialist through his friendship with Heinrich Lux, a fellow student of mathematics and a leader of the Breslau socialists. Impatient with the distant socialist society predicted by Marxist theory, Lux and several comrades formed an Association, the Gesellschaft Pacific, to establish a utopian community in the United States along the lines of that outlined in Etienne Cabet's celebrated *Voyage to Icaria.* They were even naive enough to register the Association with the Breslau police, who took no action against them. In preparation for settling their community, the young idealists dispatched their leader, Alfred Ploetz, to study the Icarian colonies in America in the spring of 1884. Late that year, he returned with bad news: The few remaining colonists in Iowa lived in wretched conditions and would unlikely make it through another year. With the end of Gesellschaft Pacific, Ploetz left for Zurich and Lux turned away from utopian socialism to Marxism and joined the Social Democratic Party (SDP).

It was at this time, in the winter of 1884-1885, that Lux invited Steinmetz to join the socialist group—beginning the "most exciting" time of his life. Intellectually, there were heated discussions about the failure of the Icarian dream that dominated his first socialist meetings. There were also close brushes with the police, particularly after the group became affiliated with the local branch of the SDP in the summer of 1885.

Things were going well until early 1887 when the Breslau police arrested Lux and 37 others. He and two comrades were charged with belonging to the Gesellschaft Pacific, which was claimed to have ties to the SDP, while

The author is with the IEEE Center for the History of Electrical Engineering, 345 East 47th Street, New York, NY 10017-2394.

Reprinted from *IEEE Technology and Society Magazine,* Vol. 6, No. 2, pp. 9–17, June 1987.

the others were charged with forming an umbrella secret organization in violation of the anti-socialist laws. Steinmetz recalled being present when the police apprehended Lux at home, but he was not arrested, probably because he was not a member of the Gesellschaft Pacific. While Lux and his friends sat almost nine months in prison awaiting trial, Steinmetz edited the *Breslauer Volkstimme* (The People's Voice), the local socialist paper. [5]

Editing this paper proved to be a turning point in Steinmetz's life. He wrote an inflammatory article which got the paper banned in November 1887 and brought him under surveillance as the suspected author. Then after the trial, he, Fritz Trappe, and another comrade helped found a new party paper, the *Schlesische Nachrichten* (Silesia News). As the police were building up evidence for a case against them for working on this paper, Steinmetz and Trappe decided to flee. Steinmetz woke his father on the morning of March 29, 1888, and told him he was going on a short trip to Austria. Instead, he fled with Trappe to Zurich, thus forfeiting his doctorate and a possible career teaching mathematics in Germany. [6] Like other political refugees from Bismarck, Steinmetz found a haven in the flourishing socialist life of Zurich. A regular in the salon of Carl Hauptmann, brother of Gerhardt, the noted playwright, Steinmetz retained his interest in socialism while studying engineering at the Zurich Polytechnic. [7]

He most likely chose engineering because of his solid background in mathematics, but there were other considerations as well. At Breslau, Steinmetz took courses in electrophysics chiefly because of his friendship for Lux, who became a successful electrical engineer and socialist writer. [8] At Zurich, Steinmetz studied electrical engineering on his own to supplement his courses in mechanical engineering. Within a year, he published two important papers on the theory of the transformer and electric circuits in the major electrical engineering journal in Germany. [9] There is no link between his socialism and the content of these articles. But his decision to become an electrical engineer bears the imprint of Lux, who showed him that engineering and socialism were far from incompatible. In fact, Steinmetz later argued that engineers, especially electricals, had a vital role to play in bringing about the Socialist Commonwealth.

Steinmetz began his new career sooner than he expected. In March of 1889, his Zurich residence permit expired, one year before his schooling was up. He had little hope of getting another one since the Breslau authorities had finally issued a warrant for his arrest the month before. Faced with another uprooting, Steinmetz decided to accept the offer of his roommate and fellow socialist, Oscar Asmussen, to emigrate with him to the United States. [10]

After landing at Castle Garden in June of 1889, Steinmetz quickly found work with the Eickemeyer & Osterheld Company in Yonkers, a manufacturer of electric streetcar equipment. Rudolph Eickemeyer, the principal partner of the firm, was a German refugee from the revolution of 1848. Since both he and Steinmetz came to America to escape prosecution for their socialism, they had much in common and often discussed socialist theory. But Eickemeyer, who was a successful inventor, had given up the political struggle to enjoy the economic fruits of capitalism. [11]

Steinmetz came under the same spell. Although he gave a lecture on electricity to a socialist group shortly after his arrival, he soon became enamored with the profit system. In a letter to his father in 1890, Steinmetz predicted that a motor he and Eickemeyer had invented would make them wealthy if they could obtain a patent monopoly on it. [12] And in October 1893, eight months after joining General Electric, Steinmetz wrote Lux that he had "withdrawn almost completely from politics." Much of the reason lay with the personal character of the socialists he had met in America. Most of them were, like Steinmetz, German immigrants. But unlike Steinmetz, most refused to learn English or to become citizens, a trait he deplored. He also disliked their personal denunciations in the party press, which was then controlled by the vituperative Daniel DeLeon. Consequently, Steinmetz "had nothing to do with the party except to send in my contribution when they send a subscription list, something they never forget to do." [13] In 1918, Steinmetz alluded to another reason he was not politically active in the 1890s: American bias against socialism would impede his rise in the engineering world. Steinmetz said that the America of the 90s was unlike Germany in that "It was not illegal to be a socialist, but no man could be one and be considered respectable or admitted to positions of trust." [14]

By 1911, after Steinmetz had made an international reputation as an authority on alternating current theory and practice, [15] socialism had become much more respectable in America. No longer was it associated with the German-dominated Socialist Labor Party, which Steinmetz had objected to in the 1890s. The movement was now largely in the hands of the Socialist Party of America, which embraced a wide political spectrum, from revolutionary syndicalism on the left, to evolutionary, step-at-a-time reformism on the right. Formed in 1901 from the merger of Eugene Deb's Social Democratic Party and a splinter group from the SLP, the Socialist Party gained a substantial following in every section of the country except the southeast. It was particularly strong in Milwaukee, New York City, Oklahoma, and Pennsylvania. [16] In 1911, near the peak of its power (Debs polled six percent of the presidential vote in 1912), the party counted over 80,000 members, and socialists won mayorships of 74 American cities, including Schenectady, NY, headquarters of General Electric. [17]

Responsible for the socialist victory in Schenectady was George R. Lunn, the candidate for mayor. A Christian Socialist minister, Lunn campaigned more on a municipal reform ticket than on a doctrinaire Marxist platform. Once in office, he made good his pledge to eliminate graft and corruption in street contracting and enacted other reforms. The major socialist action Lunn took was to establish municipally owned ice and coal distributorships, which soon closed because of opposition from private firms. General Electric hardly opposed the election, however. President Charles Coffin praised Lunn's reform measures; Ward 11, home of many G.E. executives, voted heavily for Lunn; and the company employed several socialists, including the future Nobel-prize winner, Irving Langmuir, at the Research Laboratory. [18]

It was Lunn's activism that brought Steinmetz back into socialist politics. In June 1910, he began subscribing to the *Citizen,* a new paper founded by Lunn that turned socialist when Lunn joined the party that December. In the spring of 1911, the paper reported a speech by Steinmetz on needed reforms in the city. And after the election, Mayor Lunn appointed him to the school board, most likely because of his experience teaching in G.E.'s educational programs and at nearby Union College. Elected president of the board of education, Steinmetz waged a successful campaign to improve the condition of the public schools during his first term. [19]

Seeing the possibilities of reform under socialism, Steinmetz began a decade of active socialism. In 1913, he ran for President of the Schenectady Common Council on the socialist ticket. He and Lunn lost that year to a "fusion" ticket of the Democratic and Republican parties. But when the citizens of Schenectady returned Lunn to office in 1915, they also elected Steinmetz president of the common council, a post he held for one term. The term was a stormy one. Lunn was forced out of the Socialist Party in the sping of 1916 for appointing non-socialists to office. Steinmetz supported his friend Lunn, but then broke with him when Lunn ran for Congress as a Democrat that fall. The socialist *New York Call* hailed Steinmetz's decision with the headline "Steinmetz Shows the Way" and quoted Steinmetz as saying that he had "never been out of the Socialist party." When Lunn was elected to congress, Steinmetz was in line to be mayor since he was president of the common council. Steinmetz said he would not be able to accept the position because he could not devote enough time to the job. But Lunn apparently did not trust this statement and returned to Schenectady from Washington often enough to retain his mayorship. Lunn said he did so in order to avoid turning the city over to Steinmetz, whom Lunn now objected to because of Steinmetz's pro-German stance during the war. [20]

Steinmetz was also active in national and international politics during this period. He joined the advisory council of the socialist journal, the *New Review,* as a contributing editor in 1914 and actively wrote for it in 1914 and 1915. [21] Such political involvement was not without its price, however. At least one newspaper argued that Steinmetz was unsuited for the wartime Naval Consulting Board, for which he was being considered in 1915, because of his socialism and his public predictions of a German victory in the War. [22] Steinmetz did not get the post and resumed his political work after the war. In 1921, he became the first Vice President of the League for Industrial Democracy, the successor to the Intercollegiate Socialist Society. He ran for New York State Engineer as a socialist in 1922. And he served on the advisory committee of the Soviet Kuzbas Colony and on the board of the *Soviet Russia Pictorial* in 1923. [23]

ELECTRICITY AND SOCIALISM, 1913-1922

A major theme in Steinmetz's decade of socialist politics was the relationship between electricity and socialism. Steinmetz may have considered such a connection prior to 1911, but his private and published papers bear no evidence of this. Before 1911, he was busy establishing a reputation as an internationally known electrical engineer. After Lunn's election, he began writing about the political consequences of his engineering work. He then saw electricity as a force to transform society into the "cooperative commonwealth," a practical fulfillment of his student utopian dreams.

Steinmetz made his first public statements on the relationship between electricity and socialism at a conference sponsored by the Society for Electrical Development in 1913. Held at the rustic setting of Association Island in Lake Ontario, the conference brought together representatives of electrical manufacturers, utility companies, government, finance, and the press for two days of speeches and camaraderie. The society hoped that "Camp Co-operation," as it was called, would convince these often antagonistic groups of the value of working together to promote the electrical industry. Specifically, the society solicited contributions from all sectors of the industry to finance an advertising campaign based on the slogan "Do It Electrically," [24]

Among the invited guests were such notables as the utility magnate Samuel Insull, the banker Frank Vanderlip, and the writer Elbert Hubbard. Many speakers proclaimed the virtues of private enterprise and questioned such means of "cooperation" as government regulation. Steinmetz, however, took the opportunity to address a subject he had apparently been mulling over for some time—electricity's ability to usher in a socialist state. Steinmetz did not preach the inevitable triumph of socialism to the gathering of capitalists and friends of capital. Rather, he argued that electrification would bring about "industrial cooperation," a phrase he later identified with socialism.

Steinmetz argued as follows. The electrification of industry will become universal because electricity is the most efficient means of transmitting and distributing energy. Most electrical production is by steam power, which is more economical in large units. But electricity cannot be stored efficiently and has to be consumed as it is produced. Therefore, the production and consumption of electricity are most economical by means of large interconnected systems with even loads throughout the day and night, that is, with a high "load factor." To develop such a system requires cooperation on two fronts. On the production side, cooperation between utilities is essential to create a nationwide power network. On the consumption side, industries must cooperate in regard to their energy usage to ensure a high load factor. [25]

Implied in the lecture was a planned economy, run by technocrats, who would engineer this "cooperation," for example, by deciding which utilities to interconnect and when industries should consume electricity. Also implied in the lecture was that electrification would usher in socialism, as seen in the following statement:

> The relation between the steam engine as a source of power and the electric motor thus is about the same as the relation between the individualist [capitalist] and the socialist, using the terms in their broadest sense; the one is independent of everything else, is self-contained, the other, the electric motor, is dependent on every other user in the system. That means, to get the best economy from the electric power, co-ordination of all the industries is necessry, and the electric power is probably today the most powerful force tending towards co-ordination, that is co-operation [socialism]. [26]

In response to a letter asking him about the national electrical grid implied by this speech, Steinmetz replied that he had only dealt with the engineering side of it in his address. He hoped "sometime to express my views also on the political and social side of the problem of the transportation of materials and energy, as necessities of life, and thus entities which society must control . . " [27] Steinmetz began to do this three years later, in 1916, when he clearly identified "industrial cooperation" with socialism. In an article on commission government, Steinmetz concluded that "industrial socialism is the final outcome of the present development of industrial cooperation." [28] And in *America and the New Epoch,* his major work on technocracy, Steinmetz predicted that "industrial co-operation would be so near socialism, would so imperceptibly merge into it, that nobody would ever be able to see where 'capitalist society' ended and the 'socialist commonwealth' began." [29]

The last quotation capsulizes Steinmetz's theory of socialism. Like many American socialists of that period, Steinmetz believed in evolutionary, rather than revolutionary, socialism. That is, he thought that industrial changes were occurring within capitalistic society that were preparing the framework for socialism. Steinmetz joined such leading socialists as Morris Hillquit, John Spargo, and Charles H. Vail in viewing the consolidation of industry by trusts and corporations as an important, inevitable step toward a socialist society because these organizations "socialized" production. Steinmetz, however, went much further than his colleagues and predicted that such a society would be firmly based on the organizational model of the modern corporation. [30]

Steinmetz laid out this evolution in some detail in *America and the New Epoch.* Writing during the turmoil of World War I, Steinmetz saw the war as a watershed in civilization, comparable to past catastrophes that brought about the epochs of slavery, feudalism, and capitalism. Extending this variation of Marxist theory to current conditions, Steinmetz viewed the war as an economic contest between nations representing the individualistic era of capitalism and the approaching cooperative era of socialism. Germany had learned how to cooperate economically through the state control of cartels. The individualistic countries, including the United States, would have to learn this lesson or "fall by the wayside." Integral to achieving industrial cooperation was the corporation. Steinmetz predicted that American industry would continue consolidating until it formed one large corporation, one large technocratic state that would be governed by an "industrial senate," composed of the corporate leaders of industry, and a democratically elected "tribunal," which only had veto power over the "senate." [31]

In *America and the New Epoch,* as in previous writings, Steinmetz prescirbed how corporations had to develop in order for this evolution to occur. Currently, the corporation was very efficient economically becasuse it performed three of its functions superbly: the administrative, financial, and technical functions. But corporations failed utterly in their fourth function: human welfare, that is, programs to protect workers against the vagaries of unemployment, accident, sickness, and old age. A few enlightened corporations, such as General Electric, had taken modest steps in this direction. But their minimal efforts were the cause of much of the social clamor against the corporation. By developing a substantial human welfare function, the corporation would not only solve the labor question, it would become the model for the future socialist state. [32]

As pointed out by historian James Gilbert, the major influences on this theory came from Steinmetz's experiences with General Electric, national engineering societies, and the National Association of Corporation Schools, of which he was president in 1914. The NACS, for example, helped corporations improve their human welfare function by furthering industrial education. [33] Also important was Steinmetz's experience with municipal government in Schenectady. In 1913, he said that the socialists' "watchwork when they got in office was: copy the methods of private corporate management." [34] Contemporary intellectual influences on Steinmetz's theory likely came from Lunn, who viewed the consolidation of capital as a step toward socialism; Hillquit, who was an early advisor to Lunn; and Spargo, whose *Applied Socialism* Steinmetz ordered a few months after joining Lunn's administration in 1912. [35] The utopian bent of his student socialism and a likely reading of Edward Bellamy's *Looking Backward* (1887) probably contributed to his particular brand of evolutionary corporate technocracy, as well.

Electricity's role in this evolution was to bring about industrial cooperation through large scale, interconnected networks. Steinmetz's role was to educate managers and electrical engineers in their responsibility to assist this evolution, rather than to hinder it, for example, by opposing interconnection. For although Steinmetz believed that "economic laws" made such an evolution inevitable, he wanted technocrats to make the transition as orderly as possible. [36]

In 1919, Steinmetz pursued this aim by joining the organizing committee of a technocratic group, based in New York City, called the Technical Alliance. Headed by Howard Scott, who also added Thorstein Veblen's name to the masthead, the Alliance intended to conduct a survey of American industry, past and present, in order to gather data for designing a more efficient system. Although the survey was never made, Scott resurrected the idea in the 1930s when he established the technocracy movement. Then, he claimed Veblen and Steinmetz as its founders because of their role in the Technical Alliance. [37] While there is no evidence that Steinmetz did more than lend his name to the organization, the Alliance's program coincided with his socialist views.

The Alliance also stood for the type of technocracy inherent in Lenin's state-planned electrification project, which Steinmetz so eagerly supported in 1922. But he did not embrace the plan without reservation. In fact, one year before Lenin unveiled his electrification program, Steinmetz spoke out against the Russian revolution. During the Red Scare of 1919, *American Magazine* published an interview with him, entitled "The Bolshevists Won't Get You, But You've Got to Watch Out!" As an evolutionary socialist, Steinmetz abhorred Lenin's revolutionary tactics and similar methods by the Industrial Workers' of the World in the United States. Steinmetz thought Bolshevism was a symptom of an industrial disease, which, in America, the corporations could cure by developing their human welfare function. [38]

But after Lenin announced his technocratic plan to electrify Russia in 1920, Steinmetz became an ardent supporter. For example, in 1921, he praised Einstein and Lenin as the "two greatest minds of our time." Lenin was great because he was bringing order out of chaos in Russia. [39] Despite his distaste for Lenin's political methods, Steinmetz saw that Lenin intended to accomplish for Russia the dream Steinmetz held for the United States: state-planned electrification—and Steinmetz wanted to take part.

In February 1922, two months after the project was formally approved, Steinmetz wrote to Lenin, via a courier, and praised his

> wonderful work of social and industrial regeneration [which] must not be allowed to fail. If in technical and more particularly in electrical engineering matters I can assist Russia in any manner with advice, suggestions or consultation, I shall always be very pleased to do so as far as I am able. [40]

Lenin wrote back that he was grateful for the offer, but the absence of diplomatic relations between the United States and Russia made it impossible for him to accept it. [41]

Steinmetz was personally disappointed, but that did not dampen his enthusiasm for Russian electrification. Harold Ware, the person who hand-delivered Lenin's reply to Steinmetz, recalled how excited Steinmetz became when he discussed Lenin's plan:

> Young man, do you realize what Russia has been doing? In this short time they have developed a standardized, planned electrification scheme for the whole country. There's nothing like it anywhere. It's wonderful what they have done. I would give anything to go over there myself and work with them. [42]

Since that was out of the question, Steinmetz agreed to publicize the project in the United States. He received information from Russian engineers and officials, one of whom called him "our beloved comrade and leader in electrification as the liberator of labor." [43] In two articles published in the *Electrical World* in 1922, Steinmetz informed American engineers of the Soviet plan for electrifying the country through eight regional power networks and described the opening of one plant near Moscow. Although plants were to be interconnected only within these regions, Russian engineers designed the separate networks as a system, bearing in mind the natural resources of the country and the transportation of fuel between regions. At the end of the first article, Steinmetz stated that the cost of the first part of the project would be $570 million and that its completion in the planned time of ten years "will depend on the assistance which can be enlisted from America." [44]

The reaction of the *Electrical World* to pouring Amreican money into Soviet Russia was predictable. The journal's editor doubted "whether this or any material amount of American capital can be coaxed into Russia until both its ideals and its practice of government are radically changed." [45] Despite Steinmetz's technical reputation, the editor also questioned the second article's accuracy because it was prepared "at [the] instance of the putative Russian government." [46]

But the *Electrical World* also begrudgingly admired the comprehensiveness of the Russian program. A month prior to publishing Steinmetz's articles, the journal characterized Lenin's plan as "sane and far-reaching in its effect." The editorial concluded with the provocative thought that "Good and lasting things grew out of the destruction wrought by the French Revolution, and who knows but that an electrified Russia may arise in the future due in part to the visions of the Moscow dictator." [47]

Unknown to the *Electrical World,* Steinmetz's employer was also taking an interest in the electrification of Russia. In late March 1922, about six weeks after Steinmetz wrote to Lenin, the president of International General Electric informed the U.S. State Department that he believed the time was ripe to resume business with Russia and that G.E. and the German Allgemeine Elektricitats-Gesellschaft (AEG) were discussing an agreement to this effect. In early May, one week after Lenin released Steinmetz's letter to the press, the Soviet Electro-Technical Trust invited International G.E. to participate in a joint-mixed capital company to help develop the electrical industry in Russia. But all these negotiations fell through and G.E. did not become a large supplier of electrical equipment to Russia until after 1930. [48] It is unknown what role Steinmetz's letter to Lenin and his articles in the *Electrical World* played in this matter. It seems clear, though, that G.E. had good reason not to discourage Steinmetz's overtures to Russia.

Besides supporting Lenin's program in the technical press, Steinmetz carried the issue of state-planned electrification into the political arena by running for New York State Engineer. Steinmetz was an excellent choice as a candidate because of his national popularity as an "electrical wizard," the "Oracle of Schenectady." Over the past decade, newspapers and magazines had touted the humped-backed, General Electric socialist, who purportedly earned $100,000 a year for performing scientific miracles in a home laboratory equipped with test tubes, cacti, and gila monsters. After America entered the war, Steinmetz fell from grace because of his earlier predictions of a German victory. But he rebounded to new heights in March 1922, when *The New York Times* enthroned him as "Modern Jove" for creating artificial lightning. The front-page headline read "Modern Jove Hurls Lightning at Will—Million-Horse-Power Forked Tongues Crackle and Flash in Laboratory." [49] At the peak of Steinmetz's popularity, Lenin released their correspondence to the press. In April, Steinmetz made front-page news again, this time with the headline "Steinmetz Offers Russia Help in Electric Projects." [50]

Such a figure was bound to draw attention when the Socialist and Farmer-Labor parties nominated him as their candidate for New York State Engineer. The New York papers took much interest in a Modern Jove, Socialist Chief Engineer of General Electric, and Friend of Lenin running for the top engineering office in the state. [51]

But Steinmetz ran an unusal campaign in that he made no public appearances. He was slated to address two rallies in Manhattan and Brookly, and Westinghouse was to broadcast his speeches by radio. Both times, Steinmetz canceled by telegram at the last moment. [52] One suspects that the attendant publicity of G.E.'s most famous engineer addressing these rallies, and greater New York via radio, prompted the company to persuade Steinmetz to forgo his plans. In any event, reporters went north to sit at the feet of the Oracle of Schenectady to hear what he would do if elected.

What they heard was his dream of a technocratic state. First, he would investigate the personnel in the State Engineers department, remove political appointees, and bring in "socially minded" engineers like his friend Vladimir Karapetoff at Cornell. Next he would conduct a survey of the state's natural resources, much like the survey planned by the Technical Alliance. He would then submit a plan for the "co-ordinated development" of these resources. Transportation would be gradually socialized, beginning with the state owning and operating a fleet of barges, which would force down rail and canal shipping rates. The centerpiece of his program, however, was the development of the state's waterpower. One proposal was to harness more fully the hydroelectric power of Niagara Falls by diverting the Niagara River, which would dry up the Falls. To overcome resistance from the tourist industry, he would turn the falls back on for Sundays and holidays. [53] Steinmetz was serious about this; he had proposed a similar scheme during the power shortage of World War I, which brought no response from the government nor the utilities. [54] With this experience in mind, he now told the voters that such technocratic plans were impossible under capitalism and required a "Socialist world."

Much more practical was his plan to develop the scattered water power of New York state. Since 1918, Steinmetz had discussed technical means of tapping the vast amount of hydroelectric power of the state's many rivers and streams. He then proposed to build small, automatic generating plants at these sites and electrically connect them to large central stations. [55] Now Steinmetz made this plan the major theme of his campaign. Although the scheme for turning Niagara Falls on and off like a faucet made better newspaper copy, Steinmetz succeeded in making the state's scattered water power an issue in the election, on which even the Republican and Democratic gubernatorial candidates took a stand. [56]

It was an issue uncomfortable for the electrical industry. The editor of the *Electrical World* commented on the election by noting that Steinmetz was well qualified technically, but pointed out that he was also a socialist. The editor feared that the "general public will not differentiate between Steinmetz the engineer and Steinmetz the politician, and hence unwittingly Steinmetz will bring hurt to his industry" by advocating a program that required the public ownership of utilities for it to be economically feasible. [57]

The election was not much of a contest. The United Socialist and Farmer-Labor candidates ran about a million votes behind their Republican and Democratic counterparts. But Steinmetz ran far ahead of his ticket, polling nearly 300,000 votes, about two and one-half times that of his fellow socialists. [58] *The New York Times,* at least, was happy about Steinmetz's defeat. After noting that the socialists had put Steinmetz on the ballot to give them "much-needed prestige," the *Times* editor said that a Steinmetz victory "would have been an appreciable loss to science" because his "proper place" was the G.E. laboratory. [59]

Steinmetz had other ideas about his proper place and continued to speak out for his view of a just society. Following the November elections, he made a speech at the Unitarian Church in Schenectady, where he spelled out his political beliefs in detail, something he had not done since writing *America and the New Epoch* six years before. Steinmetz's thinking had undergone some changes since the war and his recent political defeat. Again, he distinguished between three economic systems: capitalism, socialism, and cooperation. He identified socialism with cooperation in 1916. But he now observed that socialism was out of consideration, "regardless of whether it is justified or not," because "only a small percentage of Americans accept this viewpoint today." What was needed was cooperation between capital and labor, by which

Steinmetz meant a form of industrial democracy, whereby labor could "participate in the profits of the company through dividends, and in the management of the company through the Board of Directors." [60]

Steinmetz delivered a similar message in a speech before the Babson Institute in February 1923, in which he declared that both capitalists and socialists were working toward a corporate technocracy. Lenin had already made great progress toward this end:

> You may not admire Lenine [sic] and the soviet organization of Russia. His motive may be right or wrong, but this much is clear. He has organized and maintained a government in Russia through one of the most trying periods that any nation has ever experienced and he has been able to do it because he is using the same system of control that is being used by the best managed corporations in this country. He considers all his followers as stockholders and appoints and promotes the officials of his government on the basis of their efficiency and fitness to perform the task required of the office. [61]

He concluded the speech by describing a future America that was similar to that outlined in *America and the New Epoch,* a society he thought Lenin was achieving in Russia through the help of state-planned electrification. Lenin's main method was political revolution, while industrial evolution would be the path of the United States toward a technocratic, corporate state.

There is some indication that Steinmetz discussed his theories of corporate technocracy with G.E. president Gerard Swope in 1922 and early 1923. [62] But he did not get to counsel Swope on the development of G.E.'s welfare capitalism in the mid-1920s. On the morning of October 26, 1923, Steinmetz died peacefully after a short illness. General Electric, the engineering community, and the national press paid homage to the Oracle of Schenectady, the Modern Jove who hurled thunderbolts about the G.E. laboratory. [63] Missing, except in the socialist press, were references to his letter to Lenin and his campaign for New York State engineer on the Socialist and Farmer-Labor ticket.

John Winthrop Hammond, a publicity man for G.E. did mention these matters in his biography of Steinmetz, published in 1924. In his hands, socialism became an integral part of the Horatio Alger story of a penniless, deformed immigrant fleeing the political intolerance of Bismarck's Germany for the Promised Land, where he rose to fame and fortune on his abilities alone. Evidence of this was in clear view. G.E. permitted Steinmetz to retain his socialism in exchange for engineering marvels. But in Hammond's eyes, it was a mild, idealistic type of socialism, one that could be easily overlooked. [64]

A good example of how easily it could be forgotten occurred in 1926. When Governor Pinchot proposed his Giant Power scheme for Pennsylvania, critics charged it would lead to socialism. One opponent, James Francis Burke, went so far as to bring up the spectre of Lenin: "Is Pennsylvania to lead America in following Russia into the dismal swamp of commercial chaos and financial disaster? Are we to forget our Franklins, our Westinghouses, or Edisons, our Steinmetzes, and all the geniuses whose names light up the horizon of industrial progress?" [65] Either Burke was not aware that Steinmetz offered his help to Lenin four years before, or he dismissed it as a forgivable foible of an ecentric genius, a necessary, yet disposable, part of the Steinmetz legend.

CONCLUSION

Anyone who worked among cacti and gila monsters in his home laboratory was bound to be viewed as eccentric. But Steinmetz was serious about his socialism. Socialism was the passion of his youth, the reason he fled his homeland for the United States. After he made his technical reputation, it became associated with his other passion—electrification. Together, they provided the means to create a more just society in his adopted land.

The particular type of socialism advocated by Steinmetz grew out of his technical interests and experience. Going to G.E. after working for small, family-owned firm showed him the economic virtues of a large corporation in an era of vast industrial consolidation. Corporations seemed much more efficient than small firms, a virtue highly regarded by Steinmetz the engineer. They were also beginning to address the needs of their employees somewhat through educational and other "welfare" programs. And they were beginning to "cooperate" with other corporate-like entities—national engineering societies, the NACS, and municipal governments, for example—to engineer a less competitive economic order. Observing such a system naturally suggested that industry would be more efficient if professional engineers and managers ran it cooperatively, instead of democratically-elected politicians outlawing the trusts. It is little wonder, then, that Steinmetz thought future society would be based on the corporation and that the centralizing power of electricity was a major force in the evolution to technocractic socialism.

ACKNOWLEDGMENTS

I would like to thank James Brittain, Jonathan Coopersmith, Robert Frost, and George Wise for reviewing an earlier version of this paper that was presented at the history of technology colloquia at Rutgers University and the University of Delaware in 1985. I would also like to thank Elsa Church at the Schenectady County Historical Society; Pauline Wood at the History Center, Schenectady City Hall; and Ellen Fladger at Schaffer Library, Union College, Schenectady for their assistance and cooperation in using the Steinmetz archives at their institutions.

NOTES

[1] Quotations cited in Jonathan Coopersmith, "The Electrification of Russia, 1880 to 1925," Ph.D. disst., Oxford Univ., 1985, pp. 144, 168–169.

[2] ——, chs. VI-VII; and Thomas P. Hughes, *Networks of Power: Electrification in Western Society,* 1880–1930. Baltimore: Johns Hopkins Press, 1983, chs. 12-14.

[3] ——, pp. 297-312.

[4] Guenther Roth, *The Social Democrats in Imperial Germany: A Study in Working-Class Isolation and Rational Integration.* Totowa, NJ: Bedminister Press, 1963, p. 73.

[5] Heinz Lux, "Der Breslauer Sozialistenprozess," in Walter Heynen, ed., *Mit Gerhardt Hauptmann* (Berlin, 1922), pp. 69-82; *Der Sozialdemokrat* (Zurich), November 25, 1887; Theodor Muller, ed., *Die sozialdemokratische Presse Schlesiens* (Breslau, 1925), pp. 36-37, 43-44; and John W. Hammond, *Charles Proteus Steinmetz; A Biography.* New York: Century, 1924, pp. 67-72, 82-111.
[6] C. H. Steinmetz (his father) to C. P. Steinmetz, May 28, 1889, Steinmetz Papers, Schenectady County Historical Society, Schenectady, NY; and Hammond, *Steinmetz,* pp. 107-114. Unless otherwise noted, all archival material is from this source.
[7] Gerhardt Hauptmann, *Das Abenteuer meiner Jugend.* Berlin: S. Fischer, 1937, pp. 445-446.
[8] "Autobiography of Charles P. Steinmetz," *Electrical World,* vol. 41, 1903, pp. 524-525. On Lux's engineering career, see *Elektrotechnische Zeitschrift,* vol. 66, 1945, p. 45. His socialist writings include *Sozialpolitisches Handbuch* (Berlin, 1892) and *Etienne Cabet und der ikarische kommunismus* (Stuggart, 1894).
[9] Ronald R. Kline, "Charles P. Steinmetz and the Development of Electrical Engineering Science," Ph.D. disst, University of Wisconsin, 1983, pp. 52-55.
[10] C. H. Steinmetz to C. P. Steinmetz, March 7, May 28, and June 17, 1889; Eidgenossisches Polytechnikum, "Entlassungs-Zeugniss, April 30, 1889; and "Autobiography of Steinmetz," p. 524.
[11] C. P. Steinmetz, "Rudolph Eickemeyer," *Electrical World,* vol. 25, 1895, pp. 331-332; and Hammond, *Steinmetz,* ch. 10.
[12] C. P. Steinmetz to C. H. Steinmetz, June 7, 1890. On his lecture on electricity, see Hammond, *Steinmetz,* p. 140.
[13] C. P. Steinmetz to Mr. & Mrs. Hinz [Heinrich Lux], October 25, 1893, in Justin G. Turner, "Steinmetz," *Manuscripts,* vol. 15 (Fall 1965), pp. 27-34. On the German control of the New York Socialist Labor Party, see Ira Kipnis, *The American Socialist Movement, 1897-1912* (New York: Columbia Univ. Press, 1952), p. 20. On LeLeon's domination of the Party, see Howard H. Quint, *The Forging of American Socialism,* 2nd ed. (New York: Bobs-Merrill, 1963), ch. 5.
[14] C. P. Steinmetz, "The World Belongs to the Dissatisfied," *American Magazine,* vol. 85, May 1918, pp. 38-40, 76, 79-80.
[15] Kline, "Steinmetz," chs. 2-6.
[16] Kipnis, *American Socialist Movement,* ch. 16.
[17] James Weinstein, *The Decline of Socialism in America, 1912-1925* (New York: Monthly Review Press, 1967), Table 2, p. 116. The membership figure is from W. E. Walling, *et al* (ed)., *The Socialism of Today* (New York, 1916), p. 191.
[18] Kenneth E. Hendrickson, "Tribune of the People: George R. Lunn and the Rise and Fall of Christian Socialism in Schenectady," in Bruce M. Stave, ed., *Socialilsm and the Cities* (Port Washington, NY: Kennikat Pr., 1975), pp. 72-98; Chad Gaffield, "Big Business, the Working-Class, and Socialism in Schenectady, 1911-1916," *Labor History,* vol. 19, 1978, pp. 350-372; and George Wise, "Schenectady, Strikes, and Socialists, 1886-1923," unpublished manuscript, May 1985.
[19] Steinmetz to Lunn, June 16, 1910; Hendrickson, "Lunn," pp. 77-80; Wise, "Schenectady," p. 20; and H. W. Bibber, *Union Worthies No. 11: Charles Proteus Steinmetz* (Schenectady, NY: Union College, 1965), pp. 13-14.
[20] Hendrickson, "Lunn," pp. 90-94; *Knickerbocker Press,* Sept. 29, 1916; *The New York Call,* Oct. 13, 1916; *Knickerbocker Press,* Oct. 14, 1916; *Schenectady Gazette,* Apr. 19, 1917; and *Knickerbocker Press,* Nov. 4, 1917. All newspaper citations, except for *The New York Times, The New York Call,* and *Citizen,* come from the Steinmetz clipping collection at the History Center, City Hall, Schenectady, NY.
[21] William English Walling to Steinmetz, May 5, 1914, and H. C. Senior to Walling, May 13, 1914.
[22] C. D. Wagoner to Steinmetz, Aug. 12, 1915; and *The New York Press,* Aug. 17, Aug. 20, and Sept. 13, 1915. For an example of Steinmetz's prediction of a German victory, see *The New York Times,* Sept. 13, 1914.
[23] Harry Fleischmann, *Norman Thomas: A Biography, 1884-1968* (New York: W. W. Norton, 1969), p. 90; and Sender Garlin, *Charles P. Steinmetz; Scientist and Socialist (1865-1923)* (New York: American Institute for Marxist Studies, 1977), pp. 18-19.
[24] Society for Electrical Development, *Camp Co-operation: Book of Proceedings* (Association Island, 1913), pp. 151-166.
[25] ——, C. P. Steinmetz, "The Future Development of the Electrical Business," pp. 53-72.
[26] ——, p. 70. This was a theme of much of Steinmetz's writing. See, for example, his "Effect of Electrical Engineering on Modern Industry," *Journal of the Franklin Institute,* vol. 177, 1914, pp. 115-124, quote on pp. 122-124, and "Electrifying America," *Collier's,* vol. 56 (Nov. 27, 1915), pp. 12-13, 34-35, quote p. 13.
[27] Steinmetz to Harry N. Slattery, Dec. 24, 1913.
[28] C. P. Steinmetz, "Commission Control," *Colliers,* vol. 57 (Apr. 8, 1916), pp. 17, 27, quote on p. 22.
[29] C. P. Steinmetz, *America and the New Epoch.* New York: Harper, 1916, p. 222.
[30] Charles H. Vail, *Principles of Scientific Socialism* (New York, 1899), pp. 22-34; Morris Hillquit, *Socialism in Theory and Practice* (New York, 1909), pp. 111-113; and John Spargo, *Applied Socialism* (New York, 1912), pp. 118-119. See also James Gilbert, *Designing the Industrial State: The Intellectual Pursuit of Collectivism in America, 1880-1940* (Chicago: Quadrangle Books, 1972), pp. 70-71, ch. 7; and Donald Stabile, *Prophets of Order: The Rise of the New Class, Technocracy and Socialism in America* (Boston: South End Press, 1984), chs. 5-6.
[31] Steinmetz, *America and the New Epoch, passim,* especially, chs. 11-17. The quote is on p. 229. Steinmetz stated an early version of this theory in 1913, when he predicted that there "will be a Corporation of the United States, owning everything, running everything." *The New York Times,* Nov. 2, 1913.
[32] *America and the New Epoch,* ch. 16; and C. P. Steinmetz, "Opening Address," National Association of Corporation Schools, *Proceedings,* vol. 2, 1914, pp. 55-60.
[33] Gilbert, *Designing the Industrial State,* pp. 195-198. On the role of the engineering societies, see Ronald Kline, "Professionalism and the Corporate Engineer: Charles P. Steinmetz and the American Institute of Electrical Engineers," *IEEE Transactions on Education,* vol. E-23, pp. 144-150, 1980.
[34] *The New York Times,* Nov. 2, 1913.
[35] See Gaffield, "Socialism in Schenectady," pp. 357, 360; and Steinmetz to Rand School, July 3, 1912.
[36] See, for example, Steinmetz, "The Future Development of the Electrical Business."
[37] Joseph Dorfman, *Thorstein Veblen and His America* (New York: Viking, 1934), pp. 459-460, 462; and William E. Akin, *Technocracy and the American Dream: The Technocratic Movement, 1900-1941* (Berkeley: Univ. of Cal. Press, 1977), pp. 33-37.
[38] *American Magazine,* vol. 87, Apr. 1919, pp. 9-11, 132, 134-135.
[39] *Citizen,* Apr. 29, 1921.
[40] Steinmetz to Lenin, Feb. 16, 1922, H. W. Bibber Papers, Schaffer Library, Union College, Schenectady, NY. The original letter is in the New York State Library in Albany. The text of the correspondence appeared in *The Nation,* vol. 115 (1922), p. 78.
[41] Lenin to Steinmetz, April 12, 1922.
[42] Quoted by Jessica Smith, "Some Memoirs of Russia in Lenin's Time," in Daniel Mason, *et al,* ed. *Lenin's Impact on the U.S.,* (New York: NWR Pr., 1970), pp. 95-104, quote on p. 103.
[43] Telegram, Peter Bogdanoff to Steinmetz, June 11, 1922, GE Main Library, Schenectady, NY.
[44] C. P. Steinmetz, "The Soviet Plan to Electrify Russia," *Electrical World,* vol. 80 (Sept. 30, 1922), pp. 715-719; and "Russia's First Regional Power Station," *ibid.,* Nov. 25, 1922, 1155-1158. The request for American capital is on p. 719.
[45] [Editor], "Comprehensive Electrical Projects in Russia," *ibid.,* p. 701.
[46] [Editor], "Superpower in Russia Now Under Way," *ibid.,* p. 1141.
[47] [Editor], "The Electrification of Soviet Russia," *ibid.,* p. 412. In an introduction for A. A. Heller's *The Industrial Revival in Soviet Russia,* (New York: Thomas Seltzer, 1922), pp. xi-xv, Steinmetz responded to these criticisms of Lenin by stating that Americans did not have to agree with Lenin's revolutionary methods, but that they should learn from the industrialization occurring in Russia.
[48] Anthony Sutton, *Western Technology and Society Economic Development, 1917 to 1930.* Stanford: Hoover Institution, 1968, pp. 186-187, 198.
[49] Kline, "Steinmetz," pp. 401-412; and David E. Nye, *Image Worlds: Corporate Identities at General Electric* (Cambridge: MIT Press, 1985), pp. 29, 42-45, 107.
[50] *The New York Times,* April 22, 1922, p. 1.
[51] Steinmetz was named a candidate of the Socialist Party at its state convention in July 1922. The Farmer-Labor Party (1919-1923), which was considering a fusion with the Socialist Party for the campaign, said they would probably accept Steinmetz. See *The New York Times,* July 3, 1922, p. 5. In October, the New York

American Labor Party was formed as a "political partnership of the Socialist Party, the Farmer-Labor Party, and a large number of progressive organizations" for the 1922 campaign. See *The New York Times*, October 16, 1922, p. 2. On the history of the Farmer-Labor Party, see James Weinstein, *The Decline of Socialism in America, 1912-1925* (New York: Random House, 1967), pp. 222-272.

[52] "Dr. Steinmetz Won't Conduct a Campaign," *The New York Times*, July 8, 1922, p. 2; and *The New York Call*, Oct. 13, 1922, p. 9, Oct. 16, 1922, p. 2; Oct. 19, 1922, p. 7: and Oct. 20, 1922, p. 4.

[53] William M. Feigenbaum, "Steinmetz Would Chain Roaring Niagara Torrents," *The New York Call*, July 16, 1922, pp. 1, 3.

[54] C. P. Steinmetz, "Mobilizing Niagara to Aid Civilization," *Electrical World*, vol. 71, Feb. 23, 1918, p. 399.

[55] C. P. Steinmetz, "America's Energy Supply," American Institute of Electrical Engineers, *Transactions*, vol. 37, 1918, pp. 985-1009. See also, his "Hydro-electric Power Collection," *General Electric Review*, vol. 22, Nov. 1919, pp. 960-963.

[56] *The New York Times*, Oct. 29, 1922, VIII, p. 2.

[57] *Electrical World*, vol. 80, Oct. 28, 1922, pp. 919-920.

[58] *The New York Times*, Dec. 8, 1922, p. 10.

[59] *The New York Times*, Nov. 9, 1922, p. 18.

[60] *The New York Times*, Dec. 9, 1922, p. 14. Also, see C. P. Steinmetz, "Industrial Cooperation," March 1923, Steinmetz Papers, Schaffer Library, Union College, Schenectady, NY.

[61] New Brittain, CT, *Herald*, Feb. 3, 1923.

[62] Wise, "Schenectady," p. 37.

[63] Kline, "Steinmetz," pp. 409-410.

[64] Hammond, *Steinmetz*, chs. 5-7, 17, 19.

[65] Hughes, *Networks of Power*, p. 312.

SPECIAL FEATURE

Engineering and Social Equality: Herbert Hoover's Manifesto

THOMAS A. LONG

"Thus all men become equal before gadgets"
Herbert Hoover (Consulting Engineers, 1963)

In 1922, Herbert Hoover sketched an idealistic vision of American society in his book *American Individualism* [1]. This work was prompted by his strong belief that America had been presented with a momentous opportunity to break from its European past, a past filled with distracting, if not dangerous, "isms" and social institutions. European socio-economic systems constituted a challenge which had to met, and Hoover's little book is his confessional and hortatory attempt to do just that.

EQUALITY OF OPPORTUNITY

In Hoover's eyes, the moral principle which distinguishes American society from the "Old World" is the "ideal of equality of opportunity." And since "the most potent force in society is its ideals" it is of paramount importance that everything be done to guarantee each and every individual both the freedom and opportunity to compete for "that position in the community to which his intelligence, character, ability, and ambition entitle him..." Hoover was so convinced that equality of opportunity was the key to social and individual well-being that he viewed the furtherance of this ideal as *the* test of the rightness of any measure.

Hoover's picture of man is of a being whose most progressive impulse is "the yearning for self-expression, the desire for creation of something." But this most progressive impulse requires "liberty and stimulation" to achievement. Lacking such stimulation, both "the maintenance of productivity and the advancement of the things of the spirit" will come to a halt. In order to prevent such a failure it is necessary to see to it that there is an "ever-renewed supply from the mass of those who can rise to leadership." The *only* way to ensure such a dynamic process of social renewal is by continually expanding equality of opportunity.

Yet expanding equaltiy of opportunity cannot be divorced from engineering efficiency. As an engineer, Hoover had that revulsion for inefficiency which seems such a salient characteristic of the profession. Inefficiency was viewed, at least in part, as a *moral* notion. David Burner, while discussing Hoover's attempts to reorganize a chaotic mining industry, has said, "He wanted a moral economy, which he thought to be the most efficient economy..."[2]

A major driving force behind Hoover's moral economy was the engineer, for the job of dressing the skeleton of pure science with life, comfort, and hope was uniquely the engineer's function. Speaking of his chosen profession as both a social and political force, Hoover expansively wrote, "I hardly need add that it is the engineer who distributed wealth by creation of mass production, and thereby brings the prices of a thousand gadgets to the level of everybody's pocketbook"[3]. It is in *this* sense that "all men become equal before gadgets." By continually finding ways of production and improving on existing products, the engineer enlarges the scope of equality of opportunity. More and more people have access to an increased standard of living, not simply in the sense that more "gadgets" (cars, houses, etc.) become available to them, but also in the sense that, as the *material* conditions of living are improved and extended, individuals can look forward to more recreation and the leisure time to cultivate "the finer flowers of life." If there is one paramount intangible which the engineer substantially helps to contribute to society, it is *hope.* It was precisely this—hope—which Hoover felt was denied the individual by all the European ideologies, whether they were reactionary or radical/revolutionary. In one way or another, all of these ideologies worked against the preeminently desirable process of permanent social flux.

It cannot be emphasized enough that by 1919 Hoover had come to see America as ushering in a new age. This was to be an age in which the public spirit of voluntarism and cooperation would lead to the solution of society's problems through the application of those technical, intellectual, and moral qualitites supposedly embodied in the engineer. For Hoover, the engineer was *not* a mere technician, but someone whose professional training and experience ideally produced both technical expertise *and* the willingness to engage in selfless public service. Occupying a position between capital and labor, the engineer was expected to personify a vocational concern for the public welfare, an idea traceable at least as far back as the mid-Victorian writings of Samuel Smiles.

VOLUNTARY COALITIONS

Nevertheless, though Hoover stressed the centrality of the engineer in American society, his grand vision for America was not a *technocracy.* He was not fascinated by the mechanical or technological for its own sake, nor was he attracted to the socio-political elitism claimed for engineering by influential engineer-reformers during the early 1920s. On the contrary, the "American dream" was to be achieved, at least in part, by the encouragement of voluntary coalitions in which the contributions of the engineer would be vital. But of paramount importance

The author is Associate Professor of Philosophy, University of Cincinnati, Cincinnati, OH 45221.

Reprinted from *IEEE Technology and Society Magazine,* Vol. 3, No. 3, pp. 18–21, September 1984.

in the solution of America's social and economic problems was "a matter of will to find [a] solution... a sense of duty as well as of a sense of right and citizenship." American society should be an on-going experiment, with the attendant trials and errors, and always with a willingness courageously to test every process of national life upon the touchstone of "American Individualism." In Hoover's eyes, what made America great, indeed unique, was not its technology but rather its dynamic, pragmatic approach to problems and its commitment to equality of opportunity.

American Individualism sketches an ideal society with following basic characteristics. The standard of living, which includes education, is continually increasing, with the result that class distinctions, while they may exist, do not serve as sources of social and political domination. The government functions as an "umpire," willing and able to use its regulatory powers to limit economic domination by powerful industrial or business units; and the power of taxation is a means of combatting the contagion of "arbitrary individual ownership" and the "excessive indivdual accumulation of wealth." Finally, some may be surprised to learn that in Hoover's society the right of property is not "inalienable," "inviolable," or "natural," but rather is justified as a "useful and necessary instrument" to promote personal initiative [4]. When the accumulation of private property threatens equality of opportunity, it is the latter principle which is to act as a "militant check" on the former.

But Hoover's ideal is also one of private *voluntarism* as a necessary supplement of governmental action. His society is one in which there is a "rising vision of service" from cooperatives, trade associations, unions, businesses, and professional organizations. Quite simply, Hoover expects Americans' unrivaled penchant for constituting themselves as groups to enhance equality of opportunity by vigilantly seeking efficiency.

When read today, *American Individualism* may seem a naive, simplistic, even blinkered account of reality. Yet this tract, however much it resonates with rhetoric, optimism, and chauvinism, does pose an important question for all professional organizations. The question is simply: What are you doing to enhance equality of opportunity? Hoover is convinced that professions have social responsibilities, and so when they organize themselves each profession must face this most important question about its impact on society.

ENGINEERS AS PROBLEM SOLVERS

Today it often is said of the engineering profession that it exists to solve problems, but that it does not dictate *what* problems are to be solved. To the extent that this claim is true, it serves to qualify seriously Hoover's rather expansive assertion that it is the engineer who distributes wealth. Those persons who decide what problems are to be solved play a critical, and in some sense prior, role in the distribution of wealth. Members of the managerial and bureaucratic classes come to mind as examples. How and where capital investment is to be spent, which social programs are to be given priority—these are issues typically decided by businessmen, lawyers, and politicians, three classes which are hardly exclusive of one another. The decisions taken by these individuals shape both the amount and the pattern of distributed wealth, and so their decisions have a great significance for equality of opportunity.

Thus it is that the engineering profession is seen as the captive of military, industrial, or other forces whose social concerns are all too often narrow or self-serving ones. There is certainly some truth to this view, for while the history of engineering societies shows that some of them did make vigorous attempts to affect reform in large public policy areas (e.g., conservation), these efforts usually could not be sustained. Sometime in the 1920s the engineering profession lost its great enthusiasm for reform and there has been no widespread, concerted effort to recapture this feeling.

What needs to be said is that today the engineer has lost sight of Hoover's ideal of a moral economy. The questions of efficiency which Hoover forcefully raised were very large, long-term ones, national in scope, and sometimes involving entire industries and the natural resources on which they depended. It would be wrong to say that no one is raising such question in the public forum today, but too few of those who do are members of that profession of which Hoover was so proud.

So what can engineers do to promote the equality of opportunity which Hoover viewed as the *ultimate* test of any social policy? Is it sufficient for them simply to continue doing what they are hired to do, all the while having faith that an emphatic insistence on efficiency will produce the moral economy which Hoover saw as a means to America's dream?

No! What the profession badly neeeds is some sense of collectivity which will enable its members to act concertedly on matters of social policy. This means, of course, that the profession must be willing and able to inject itself into the *political* process. It is precisely here that some serious questions arise.

It is sometimes said that engineers have neither the right nor the ability to dictate social policy [5]. Now in one sense it is true that engineers do not have the *right* to dictate such policy, but then neither does any other professional group (lawyers, doctors) possess such a right. This right is given over to elected officials and their appointees. To lack such a right, however, does not mean that engineers should be nothing more than the "faithful servants" of policy-makers. The engineering profession does have the right to speak out either for or against policies which, in its view, inhibit equality of opportunity.

But what about *ability*? Do engineers lack the ability to dictate social policy? Here the inclination of some people is to say that engineering training fails to prepare the professional engineer for the kind of adversarial maneuvering which is characteristic of decision-making in a complex, pluralistic democracy. It is also claimed that engineers are insufficiently trained in the areas of "human relations," and so cannot be expected to grasp all of the complexities involved in successfully setting social policy. An image which has haunted the engineering profession is that of the *technocrat*, skilled in the manipulation of mathematical formulae to control machines and natural processes, but woefully insensitive to human psychology and values.

Two points need to be made here. First, while it may be true that engineers lack the ability and inclination to engage in the kind of adversarial politics which is such a common feature of our society today, there is no evidence that people who do have (and exercise) this ability necessarily contribute to the expansion of opportunity. After all, lawyers, who predominate in our professional political life, are trained either to win cases for clients or to see to it that the client does not lose too badly. When these skills are transferred to the political arena their characteristic use is directed to the satisfaction of special interest groups, with little or no concern for the impact of these interests on equality of opportunity. A training in law will help to prepare a person for adversarial politics, but it is a mistake to think that such training itself will bestow a clearer perception of the public good. No professional training of any kind bestows this wonderful perception.

A second point to be noted is that engineers are typically "realistic," meaning that they eschew perfection in the realization that jobs must get done. But, in a pluralistic society such as ours, is not this approach to work just what is socially desirable? How can a temperament which refuses to demand perfection in the practical realm *disqualify* someone for making public policy decisions?

So what should be concluded about the "ability" argument? The conclusion should be that while engineering training will not prepare someone for the procedural intricacies of contemporary politics—here many lawyers do have the advantage—nevertheless this training need not leave the engineer any less caring about equality of opportunity than any other highly specialized training, nor need it necessarily render this person incapable of pluralistic decision-making.

SENSE OF PURPOSE

It is a mistake to think that the *moral ends* of government are the special preserve of some one profession, where members of other professions graze only at their peril. In fact, these ends do not belong to any group, professional or not. They belong to anyone who happens to be a citizen.

Yet isn't it simply naive to think that engineers could ever achieve the kind of collective sense of purpose which would be necessary for success in the political arena? We have been told often about the diversity of opinion among engineers on political and social issues. This diversity is no doubt a fact, though it is not clear the extent to which this reflects differences about the *ends* society should pursue rather than merely the *means*. After all, "pronuke" and "antinuke" engineers probably want the same social end (e.g., safe, efficient energy), while differing about the consequences of a technology for that end. And is there more diversity of opinion among engineers about "affirmative action" than there is among lawyers and physicians? It seems unlikely.

But even when engineers do differ about ends, that is, about what kind of life is worthwhile, there is one social end which they *should* share. Engineers should be committed professionally to equality of opportunity. Why should this be?

By its very nature, the profession of engineering is committed to technological change, change which typically brings about redistributions in the material resources of society. Furthermore, engineers are in the forefront of redistribution since it is they who must devise the means for it. The mass-produced automobile and inexpensive digital watches ae but two examples of redistributions of material resources which would have been impossible without engineers. Engineers have great power in this sense, and with power comes responsibility.

It was this power Hoover was referring to when he spoke in his *Memoirs* of the engineer clothing the abstractions of science with hope for the future. It is because the engineer occupies a special (unique?) position vis-a-vis the redistribution of material resources that Hoover's image is attractive. But if this image is to be taken seriously, it means that engineers must be especially concerned about their work. Technological change per se is no guarantee of an expansion in equality of opportunity. Such change may serve merely to expand the opportunities of those who already have the most (the Concorde?).

There is no doubt that much engineering does serve equality of opportunity in a socially desirable way (e.g., developments in low-cost housing construction). But if Hoover is right, this is not enough. *All* engineering should serve this purpose. It is too easy to forget that engineering projects are the technological embodiments of social policy decisions, which means that each project is the attempt to further values accepted by those in positions of leadership. How many engineers occupy such positions, and, of those who do, how many share Hoover's concern to promote social flux?

Today the engineering profession needs to rediscover the fact that it is a *political* activity promoting values which may or may not coincide with the type of idealism basic to Hoover's thought. "Equality of opportunity" is, of course, a somewhat vague and controversial expression. Just what conditions must be met before a society can speak of itself as embodying such opportunity is a provocative question. But vagueness can be eliminated at least to some degree through discussion, and controversy needs to be aired. This is just what the engineering societies should be doing—attempting to reach members, students, educators, etc. Only in this way will it be possible for the engineering profession to participate as it should in the emergence of a collective sense of purpose, something so badly needed in today's fractious political arena.

REFERENCES

[1] Herbert Hoover, *American Individualism,* Doubleday, Page and Co., Garden City, New York: 1922, p. 1. Unless otherwise noted, all quotations from Hoover are from this work.
[2] David Burner, *Herbert Hoover*, Alfred A. Knopf: New York, 1979, p. 66.
[3] Herbert Hoover, *Consulting Engineer*, p. 95, Aug. 1963.
[4] This key feature of Hoover's thought aligns him with the nineteenth century British Utilitarians (Bentham, J.S. Mill) rather than with the Locke-Jefferson tradition of the seventeenth and eighteenth.
[5] Samuel C. Florman, "Moral blueprints," *Harper's* p. 32, Oct. 1978.

PART II

Social and Ethical Responsibilities of Engineers

ENGINEERING ethics and professional responsibility have become more relevant to engineering during the last quarter of the twentieth century. As technology and its impacts have become more complex and far-reaching, the importance of responsible engineering decisions to employers and to the public have been underscored. These responsibilities often come into conflict, resulting in ethical problems or dilemmas whose solutions, like other engineering decisions, can benefit from a sound analytical framework.

Ethics may be defined simply as "the rules and ideals for human behavior. They tell us what we ought to do" (1, p. 517). In an engineering context, ethics may be addressed in a number of ways.

Moral thinking and moral theories have a long intellectual tradition. Indeed, "ethics" constitutes an entire branch of philosophy. In recent years, philosophers have evinced a growing interest in applying moral theories to real-world problems, that is, "applied ethics," especially in the professions. In addition to engineering ethics, much attention has been paid to ethics in other professional arenas, for example, business ethics, biomedical ethics, and legal ethics. These fields often overlap with engineering ethics—for example, when engineers are involved in making business decisions or in designing biomedical devices. Another related field of growing importance to engineers and society in general is computer ethics.

As individuals, engineers usually have developed a sense of personal ethics, influenced and molded by their upbringing, socialization, religion, and other factors. As individuals, they are generally no different from other humans in this regard. All too often, however, engineers seem to separate their personal sense of ethics from issues that arise in the practice of engineering. Many seem to feel that moral problems fall outside the scope of engineering or should be left to managers and government officials to solve. It might be said that such engineers too readily "check their ethics" at the door to the office. The field of engineering ethics has emerged in order to focus attention on ethical issues in engineering, as well as to better prepare engineers and engineering students for dealing with such issues.

Consideration of engineering ethics takes place largely in two arenas: research and teaching and engineering practice. As previously mentioned, many philosophers focus their research and teaching activities on engineering ethics and other areas of professional ethics. A common philosophical approach to engineering ethics is to employ moral theories, such as utilitarianism and duty/rights-based ethical theories, to the solution of moral dilemmas in engineering. Utilitarianism is an ethical system that judges an action to be morally correct if its outcome results in the greatest good for the greatest number of people. Duty and rights approaches to ethics, on the other hand, focus on actions themselves, and on whether or not individuals abide by duties to do good and avoid harm, or act out of respect for the moral rights of other individuals. Although these two types of moral theories often result in the same conclusion regarding a particular act, they might produce conflicting conclusions, as when an engineering project built to benefit the public results in the eviction of individuals without their prior consent.

Some engineers disdain ethics as an important topic, especially philosophical approaches which they deem to be too idealistic and distant from engineering practice. Yet a growing number of engineering educators are insisting on research and teaching concerning engineering ethics. Most engineering faculty are from conventional engineering disciplines and are "self-educated" in the matter of philosophical approaches to professional ethics. Some, such as the author, are from nontraditional engineering disciplines that focus on public policy and/or societal issues in engineering. Engineers and philosophers in both the research and teaching areas have begun a collaboration, much of it encouraged by funding from the National Science Foundation (NSF) and private foundations.

Often neglected or played down in the scholarly literature is the significant amount of activity related to engineering ethics among engineers in industry. More often than not, these engineers become involved in such activities through the professional engineering societies. The most visible engineering ethics activity within the professional societies is the promulgation of codes of ethics. In this arena, engineers from industry interact with engineers from academia and, less often, with philosophers engaged in engineering ethics research and teaching. Although there has been an increased trend in recent years to integrate research and teaching in engineering ethics with engineering practice, there is considerable need for further integration. The professional society, which provides a vital link between academia and engineering practice, thus plays a pivotal role.

The code of ethics is the hallmark of the professional engineering society's stance on ethics. Although codes vary from one professional society to another, they typically share common features in prescribing engineers' responsibilities to the public, their employers and clients, and their fellow engineers. In addition, the codes often emphasize such characteristics as competence, trustworthiness, honesty, and fairness. (2) The IEEE Code of Ethics (see Appendix), adopted by the Board of Directors in 1990, is one of the more compact of the current codes.

In addition to maintaining a code of ethics, the professional engineering societies also generally appoint various committees and other bodies charged with dealing with ethical issues. The IEEE, for example, has two such committees at the Board of Directors level: the Member Conduct Committee (MCC) and the Ethics Committee. The MCC's purpose is twofold: to recommend disciplinary action for members accused of violating the code of ethics and to recommend support for members who, in following the code of ethics, have been discriminated against. The Ethics Committee, formed more recently, provides information to members on ethics and advises the board on ethics-related policies and concerns. In some cases, ethics concerns also extend to the technical branches of the professional societies. The IEEE Society on Social Implications of Technology, for example, one of IEEE's 37 technical societies and councils, has engineering ethics and professional responsibility as one of its major focus areas. Professional engineering societies also have other entities concerned with ethical issues in the context of the scope of their activities. For example, committees charged with overseeing the publications of the professional society are often concerned with ethics in publishing, which relates to the responsibilities of editors, reviewers, and authors. Concern for engineering ethics can even extend to student chapters of the professional societies. Some chapters, for example, have cooperated with their home departments in formulating academic codes of ethics modeled, in part, on professional codes of ethics.

Professional societies are particularly important in the case of engineering ethics because engineers are usually employed by large corporations; in contrast, professionals in other fields such as law and medicine have traditionally enjoyed a greater degree of professional autonomy. However, as a result of the corporations' influence over the professional societies, these societies have taken less forceful stances on engineering ethics than some observers would like to see. Nevertheless, the professional society remains a significant organizational force internal to engineering and is potentially capable of promoting and nurturing a sense of ethics and professional responsibility.

In this section, we consider in more detail moral problems in engineering and frameworks for addressing such problems.

References

[1] Wujek, J. W., and Johnson, D. G. 1992. *How to be a good engineer.* Washington, D.C.: Institute of Electrical and Electronics Engineers, United States Activities Board.

[2] Unger, S. 1994. *Controlling technology: ethics and the responsible engineer.* 2nd ed. New York: Wiley.

Chapter 3

Moral Dilemmas in Engineering

A moral dilemma may be defined as a conflict a person experiences between two or more moral obligations in a particular circumstance. For example, an engineer's obligation to protect the public interest might conflict with an obligation to protect the trade secrets of an employer. Moral dilemmas in engineering can take on many forms, including such issues as conflict of interest, bribes and gifts, and failure to credit the work of others.

Engineering Ethics Issues

Many cases in engineering ethics have been highly publicized, usually those involving whistleblowing. Most issues in engineering ethics, though not high profile, can still challenge the typical engineer in the everyday workplace. Engineering ethics issues include the following, adapted from Wujek and Johnson (1):

- *Public safety and welfare*—a key concept in engineering ethics focusing on the engineer's responsibility for the public health, safety, and welfare in the conduct of his or her professional activities. For example, engineering projects and designs often have a direct impact on public safety and the environment.
- *Risk and the principle of informed consent*—assessment of risk in engineering projects and the extent to which public input should be considered in engineering decisions. Technological controversies, often pitting engineers and other technical experts against public interest groups and ordinary citizens, have grown in number and importance over the past two decades.
- *Conflict of interest*—a term that generally applies to situations where an engineer serves more than one client with conflicting interests or has a personal interest in a matter on which he or she is asked to render a professional opinion. Often, even the appearance of such a conflict can undermine the ability of engineers to carry out their assignments in a professional manner.
- *Integrity of data and representation of it*—an issue of great importance since most engineering analyses rely to some extent on collection of reliable data. Falsification or misrepresentation of data has become a major issue in the ethics of scientific research and has played a role in many recent high-profile product liability cases.
- *Whistleblowing*—a term applied to a situation in which an employee "blows the whistle" on unethical or illegal conduct by a manager, employer, or client. Many high-profile engineering ethics cases have involved whistleblowing, which can include actions within and outside of the organization in which the engineer works.
- *Choice of a job*—a choice that can entail a number of ethical decisions, including whether or not the engineer chooses to work on military and defense applications, the environmental record of the potential employer, and the extent to which employers monitor the professional and personal activities of their employees.
- *Accountability to clients and customers*—an important issue concerning such concepts as trustworthiness, honesty, and loyalty, which is often overlooked in light of the attention given to the engineer's primary responsibility to the public.
- *Plagiarism and giving credit where due*—an issue that affects engineering students, their professors, and engineers and managers in the workplace. Failure to give proper credit is not only dishonest but can also affect morale and the integrity of engineering data.
- *Trade secrets and industrial espionage*—topics that underscore the ethical responsibility of engineers to their employers and clients, even when they move on to work for others. Computer software is a hot area in this regard.
- *Gift giving and bribes*—bribes, and their distant cousin, gifts, represent some of the most serious issues in engi-

neering ethics. Virtually every engineer in the course of his or her professional career will have to confront the issue of determining when gifts are acceptable.

• *Fair treatment*—an issue that applies to "civil rights" as well as relations between superiors and subordinates. In addition to being ethically deficient in its own right, failure to treat others on the basis of merit can often have a negative impact on engineering performance.

Some well-documented high-profile cases involving engineers and engineering designs are discussed next.

CHALLENGER

Perhaps the best known engineering ethics case of recent years involves the explosion of the Space Shuttle Challenger in 1986. This case includes a wide range of elements relevant to engineering ethics and professional responsibility, including protection of the public interest, conflicts between engineers and management, integrity of data, and whistleblowing.

The loss of the Challenger resulted from a failure in the design of the vehicle's reusable solid rocket boosters (SRBs). In particular, the O-ring seal that prevented hot combustion gases from escaping through the joints of the SRBs failed as a result of very cold temperatures at the time of launch. Engineers at Morton-Thiokol, Inc., the contractor responsible for the SRBs, had for some time been concerned about the ability of the joints to seal properly but had been unable to get Thiokol management or the National Aeronautics and Space Administration (NASA) to take the problem very seriously. On the eve of the launch of Challenger, faced with unprecedented cold temperatures and the knowledge that the worst previous erosion of an O-ring seal had occurred during the coldest launch to date, the Thiokol engineers attempted to persuade their managers and NASA to postpone the launch until the temperature increased. Initially, the Thiokol managers supported their engineers. However, after NASA management expressed disappointment and serious doubts about the data presented and the judgment of the Thiokol engineers, the Thiokol managers, who were concerned with protecting a lucrative contract, overruled their engineers and recommended launch. At one pivotal point during an off-line caucus, the Thiokol vice-president of engineering was told by one of his superiors to "take off your engineering hat and put on your management hat."

Following the disaster, in which all seven astronauts were killed, President Reagan formed a commission to investigate the accident. During the subsequent hearings, several Thiokol engineers ignored the advice of their managers to stonewall and testified candidly about the events leading up to the disaster. The commission concluded that, in addition to a flawed shuttle design, there was a fatal flaw in NASA's decision-making process. The late Nobel Prize-winning physicist, Richard Feynmann, who served on the presidential commission, went even further in his appendix to the commission's report. In Feynmann's view, NASA's decision-making process amounted to "a kind of Russian roulette" (2).

For their efforts, the "whistleblowing" engineers were reassigned and isolated within the company, a situation that was corrected only after the presidential commission learned of the circumstances. One engineer in particular, Roger Boisjoly, who subsequently took disability leave from Thiokol and was ultimately fired, suffered the typical fate of the whistleblower, including being denounced within the town where Thiokol was located, subjected to death threats, and ostracized within the aerospace industry.

THE BART CASE

The BART case from the early 1970s, though somewhat dated, is of special interest owing to the significant role played by the IEEE. The case involves three engineers working on the design of San Francisco's Bay Area Rapid Transit System (BART) who became concerned about the safety of the system's automated control system for subway cars. Following unsuccessful efforts to have their supervisors rectify the problems, the three took their concerns to a member of the BART Board. Subsequently, the three were fired and ostracized within the industry. A lawsuit by the three was settled out of court but not before the IEEE filed a historic friend of the court brief in support of the engineers. Ironically, the concerns of the three were validated when a train overshot a station shortly after the system became operational, injuring several passengers. The case is useful in illustrating the unfortunate circumstances that all too often envelop whistleblowers. On a more positive level, the case illustrates the important role a professional society such as the IEEE can play in supporting ethical behavior by engineers.

THE DC-10 CASE

In 1979 in Paris, a Turkish Airline DC-10 crashed, costing 346 people their lives; it was one of the worst airliner disasters in history. The accident occurred when an improperly closed cargo door blew open in flight, causing the cabin to decompress and the floor to collapse, thus destroying the hydraulic controls that ran through the floor. An eerie precursor of the Challenger case, the DC-10 case is one in which a design problem was identified early in the production of the aircraft and was recognized as the cause of a near disaster in another failure involving a plane of the same design, but still ignored or dealt with only in terms of a "band-aid" fix. Players in the case include the aircraft manufacturer, McDonnell-Douglas, and the fuselage subcontractor, Convair, both of whom held up design changes in order to protect their economic interests; a

Convair employee who wrote a warning memo that was suppressed by management; the Federal Aviation Administration which was slow to insist on design changes even after the flaw was identified; and Turkish Airlines, which provided inadequate training of the baggage handlers responsible for closing the door.

Hyatt Regency Walkway Collapse

In 1981 two suspended atrium walkways collapsed at the Hyatt Regency Hotel in Kansas City, crushing hundreds of people who were crowding the lobby for a "Tea Dance." One hundred fourteen people died in the accident, and dozens more were seriously injured. An investigation revealed that the design of the supporting structures for the walkways had been altered by the steel fabricator but apparently approved by the design architect-engineers. Moreover, the walkways, as originally designed, did not meet the Kansas City Building Code. The city inspectors were found lax in fulfilling their duties, and the design engineers were criticized for not following through on a commitment to check all of the roof connections following an earlier collapse of part of the roof. The case, which involved substantial litigation, is useful in illustrating the interplay between ethical responsibilities and legal issues. More importantly, the case resulted in a rare delicensing of the two principals of the design firm; following an extensive administrative hearing, they were stripped of their professional engineering licenses by the Missouri Board of Architects, Engineers and Land Surveyors. The case thus suggests the need for stronger coupling between ethical principles and licensing requirements.

Other Engineering Ethics Cases

The ethical dilemmas encountered by most engineers are typically more mundane than the high-profile cases mentioned here. A significant amount of case development has occurred with respect to more commonplace events, including such issues as conflict of interest, trade secrets, and gift giving. For example, the National Society of Professional Engineers' Board of Ethical Review (BER), for educational purposes, publishes fictionalized reviews of actual cases brought to its attention.

A number of philosophers, notably Michael Pritchard (3), have been calling for further development of cases focusing on "good works," cases demonstrating that making sound ethical judgments need not end with a whistleblower being demoted or fired. One such notable incident is the case of William LeMessurier, the noted civil engineer who designed New York's CitiCorp Building. To his horror, LeMessurier discovered after the building was in use that it had not been properly designed to withstand hurricane force winds. Risking his professional reputation and considerable financial liability, LeMessurier went to his partners and to CitiCorp and insisted that immediate action be taken to strengthen the building's structural joints.

Background Reading

In 1982, philosopher *John Ladd* wrote an article entitled "Collective and Individual Moral Responsibility in Engineering: Some Questions;" the article has held up remarkably well. Ladd's critique of the common assumptions of engineering ethics as well as his insightful analysis of the relationship between collective and individual moral responsibility remain significant contributions to the field. Ladd begins by acknowledging that his purpose is to raise questions about the problems of engineering ethics rather than to provide answers regarding proper ethical behavior.

The first topic he turns to—professionalism—raises questions about the often self-serving motive of all those, including engineers, who would claim to be professionals. Ladd also notes two qualities of engineering that distinguish it from other professions: the fact that engineers deal only indirectly with people and the fact that most engineering work takes place in the context of large organizations which typically stifle dissent. The latter point leads Ladd to consider the question of the employee's loyalty to the organization. Ladd rejects loyalty in this context on the grounds that a corporation cannot reciprocate with respect to loyalty because it is a nonperson whose sole end is to increase profit, thereby making the corporation incapable of moral considerations such as loyalty to its employees. Moving to a discussion of whistleblowing, Ladd agrees with those who suggest that we ought not to rely on moral heroes in such situations. More importantly, he is critical of the notion that ethics should be regarded as some form of behavior control, a process that he believes reduces ethics to the same status as law.

Ladd concludes with a discussion of various forms of responsibility, the most important of which he believes is moral responsibility. Moreover, unlike other forms of responsibility, moral responsibility is nonexclusive; that is, it is "everyone's business." This leads Ladd to the observation that the members of the engineering profession, and of society in general, have a collective moral responsibility for safety that is concurrent with the individual moral responsibility of those who have direct involvement in particular safety decisions.

Another philosopher, *Charles E. Harris, Jr.*, refers to the Challenger accident in his 1995 article, "Explaining Disasters: The Case for Preventive Ethics." Harris argues that accidents such as the Challenger might be attributed to lapses in one or more of three realms: engineering, management, and ethics. Indeed, he believes that the three explanations are not mutually exclusive and that a focus on all three ensures that ethical concerns will not be neglected. Such analysis, Harris argues, can lead to the development of a preventive ethics aimed at avoiding future disasters.

Challenger was a failure in all three realms. Poor engineering was manifested in the design flaws in the space shuttle's booster rocket O-ring seals. Poor management was exhibited by Morton Thiokol's failure to heed the legitimate warnings of their own engineers that the O-rings were in need of redesign and were especially vulnerable to the cold weather expected at the Challenger's launch time. Ethical lapses occurred because the astronauts were not provided an opportunity for informed consent to the unusual risks they were facing and because both the NASA and Thiokol managers elected to take unusual risks due to economic and launch schedule pressures.

Harris concludes by arguing for a preventive ethics, parallel to the notions of preventive engineering and preventive management, that would be designed to anticipate and deal with ethical failures before they occur. Such measures as codes of ethics, corporate ethics officers, and procedures for raising ethical awareness would contribute to an environment that seeks to prevent ethical failures before they occur, thus lessening the chances that technological accidents would occur as a result of ethical failures.

Cases for Discussion

In an article that first appeared in 1986, *Gene I. Rochlin* presents a case study of three "'High Reliability' Organizations and Technical Change: Some Ethical Problems and Dilemmas." In order to increase the performance of these systems—air traffic control, electric power grids, and naval air operations—while preserving high levels of safety and reliability, advanced technologies are being implemented. These developments, Rochlin argues, are occurring even though the organizational and societal contexts in which the technologies are to be deployed are not completely understood. This poses ethical and professional dilemmas for the designers, operators, and managers of these systems, whose professional training rarely includes consideration of such organizational and social factors.

These changes are taking place in an environment that relies on technology to perform certain tasks and on human operators to perform other tasks in a particular organizational setting. Operators are generally not eager to abandon the old technology they have become dependent upon. Transitioning to new technology such as "expert systems," despite the intent of relieving operators of routine tasks, nevertheless poses substantial challenges to designers and operators alike. Although the technical problems involved in breaking in new systems are generally well understood, the organizational problems that may arise have received little scrutiny. These problems involve limitations on operator behavior, difficulty in maintaining competence on the old system while learning to use the new system, difficulty in anticipating the technical limitations of the new systems, and overdependence on technology that may hinder recovery from failures in the systems.

Rochlin points out that ethical problems in implementing these systems abound. One such problem is the possibility that their implementation may actually decrease safety and performance by altering the organizational environment. Another problem is the extent to which operators, managers, and designers should be involved in—and held accountable for—the nontechnical parameters of the system design. The unfortunate reality, Rochlin concludes, is that system operators and the public being served are expected to conform without question to the demands of the new technology. Thus, he concludes, those involved in implementing these systems are also confronted with an ethical dilemma regarding how much to apprise the public of this growing dependence on technology.

In "The Bureaucratic Ethos and Dissent," which appeared in 1985, sociologist *Robert Jackall* discusses the difficulties in creating an environment for legitimate critical viewpoints and dissent among managers within large-scale corporate organizations. To illustrate this problem, Jackall discusses three real cases (albeit with fictional names): (1) a health professional, concerned about hearing loss by workers in his company's textile mills, who was unsuccessful in his attempts to get the company to take this problem seriously; (2) a manager for a chemical company who suppressed knowledge of prior worker exposure to chemical hazards; and (3) an engineering manager who was suspended and subsequently fired for continually calling attention to improper management procedures and safety hazards occurring during the cleanup following the Three Mile Island nuclear accident.

The cases illustrate the process by which managers who wish to be successful are coopted into accepting the team view on issues they might find morally objectionable on a personal level. Those who do not buy into this process are unlikely to advance up the corporate ladder. Those who actively fight it by, for example, voicing their dissent from corporate practice are labeled troublemakers and dealt with accordingly. Jackall offers no prescription for change in the corporate environment that would make it unnecessary for dissenters to become moral heroes in this manner, risking their careers for the good of the public (and in some cases the good of the industry itself). The irony is that such dissenters are often made into heroes in our popular culture. Jackall concludes by suggesting that our cultural heroes give us vicarious relief from the unchangeable attitudes of the corporate world.

References

[1] Wujek, J. W. and Johnson, D. G. 1992. *How to be a good engineer.* Washington, D.C.: Institute of Electrical and Electronics Engineers, United States Activities Board.

[2] Feynmann, R. 1988. *What do you care what other people think?* New York: W. W. Norton.

[3] Pritchard, M. 1997. Good works: A positive approach to engineering ethics. Presented at Mini-Conference on Practicing and Teaching Ethics in Engineering and Computing, Sixth Annual Meeting of the Association for Practical and Professional Ethics, Washington, D.C.

FEATURE ARTICLE

Collective and Individual Moral Responsibility in Engineering: Some Questions

JOHN LADD
Department of Philosophy
Brown University

In this essay, I shall examine in a rather general way a number of commonly discussed questions of engineering ethics pertaining to the special ethical obligations and responsibilities of engineers as engineers. However, I shall not attempt to provide specific answers to these questions; instead, I shall raise further questions about the questions themselves. For I believe that it is absurd to try to answer questions about obligations and responsibilities before we are clear about the kinds of questions that we are asking, the context in which they arise and the presuppositions underlying them. It is often taken for granted that philosophers are especially qualified to answer ethical questions, including questions of engineering ethics, and philosophers themselves sometimes welcome the opportunity to play the role of preacher or of ethical guidance counselor. This is a view of the practical value of philosophy that is shared by utilitarians and rights theorists, Rawlsians and Nozickians, and by many others, all of whom are only too ready to hand out answers telling people, in this case engineers, how they ought to act. My own view of the role of philosophy is entirely different; I believe that the most useful contribution that a philosopher can make is to identify, clarify and sort out problems, and in that connection to unmask superstition, bigotry and illegitimate presuppositions. In this last regard, one of my principal objectives in this essay will be to *demythologize* some current notions in engineering ethics![1]

My purpose, then, is *zetetic*, which means questioning, inquiring, doubting.[2] Approaching the problems of engineering ethics zetetically means moving to a second or meta-level and, as I have already indicated, asking questions about the problems themselves. As a propaedeutic to engineering ethics we need to ask questions like the following: What are the ethical problems of engineering? Why are they ethical rather than simply legal, institutional, organizational, economic, or personal problems facing individual engineers? What is added by saying that they are ethical? Into which of these categories do problems connected with whistleblowing fall? How should the distinctively ethical problems be formulated? How do they arise? What do they presuppose? What methodology and what concepts are the best tools for analyzing the ethical issues that are involved in engineering? What is the moral status of a corporation? What is the meaning of "responsibility" and of "collective responsibility?" and so on.

ENGINEERING ETHICS AND PROFESSIONALISM

Many of the ethical problems associated with engineering are connected with the professional status of engineering, and so I shall start off with a few remarks about professionalism in general as a way of providing a background for the problems I shall discuss. We must begin with the fact that it is generally felt that there is something honorable about being a member of a profession and it is often supposed that professionals are idealistic. For this reason, members of a profession are considered to have special duties and responsibilities towards society over and above those of ordinary people. Society has high expectations that professional persons, by virtue of being members of a profession, will be honest, dedicated and responsible, more so than laymen or persons in other occupations. On that account, it is thought quite seemly that professionals be better paid, treated with more respect and accorded a higher status in society than others. All of this most likely applies to engineers considered as professionals.

Many professions have adopted codes of ethics in imitation of the original code of ethics of the Royal Society of Physicians. Indeed, it is often assumed that in order to become a full-fledged profession an occupation must adopt a code of ethics. On the other hand, it has been said about such codes that if a person is really honest and responsible, then he does not need a code of ethics and if he needs one then it will not do him any good. I mention this matter simply to warn against the assumption that a code of ethics can be consulted for answers to questions of engineering ethics.[3]

One frequently overlooked aspect of the ethics of professionalism is that professionals tend to believe that their professional "obligations," e.g., to clients, outweigh their obligations to others, e.g., to the public.

This is a revised version of a paper presented at the Second National Conference on Ethics in Engineering, at the Illinois Institute of Technology, Chicago, on March 5, 1982.

Reprinted from *IEEE Technology and Society Magazine*, Vol. 1, No. 2, pp. 3–10, June 1982.

In other words, their professional obligations are given priority. The most obvious example of this sense of priorities is to be found in legal ethics, where lawyers are expected to treat their obligations to clients as having precedence over their obligations to society, e.g., a lawyer's obligation to a client who is guilty of a heinous crime is thought to outweigh any obligation that he might have to the general public whose interest it is to have criminals convicted.[4] Although this assumption about priorities is not always stated explicitly, it is important to remember that it is often there by implication. Indeed, I sense that engineers, like other professionals, frequently take for granted that their obligations to their clients, i.e., their employers, rightfully take precedence over their obligations to society at large. Whether or not this is or ought to be the case is one of the questions that should be examined in engineering ethics. Another typical feature of professionalism that might be mentioned here is what Parsons calls "affective neutrality," that is, the generally felt requirement that a professional ought to adopt an attitude of neutrality towards the ultimate objectives of his client and ought not to allow his personal judgments about their intrinsic value or disvalue to play any role in determining the services he provides.

In general, it should be observed that the elitism of professionalism and its willingness to sacrifice the public interest to other professional responsibilities attest to the profoundly anti-democratic tendencies of professionalism, which have so often been the target of attack by social critics.[5] In any event, an ethics of professionalism in general, and of engineering in particular, needs to cope with the implications of this charge of elitism, which supposes that we are dealing with the best, brightest and wisest members of society when we discuss professionalism.

I mention all of these points about professionalism here, because it is my impression as an outsider that they play a considerable role in an engineer's perception of himself as a professional and of his professional obligations and responsibilities.

DIFFERENTIAL ATTRIBUTES OF ENGINEERING AS A PROFESSION

The profession of engineering differs from other professions such as medicine and law in two respects that are basic for understanding the particular problems of engineering ethics.

First, unlike medicine and law, whose services are ordinarily directed to the needs of individual persons, the services provided by engineers relate to things, e.g., machines, buildings, equipment, products, etc.[6] Insofar as an engineer has a relationship to persons, it is indirect. For example, he relates to persons as clients who purchase or use his services or as persons who are affected by what he makes (or designs), e.g., workers, consumers or the general public. As a result of not being directly structured around interpersonal relationships, as are medicine and law, the engineer-client relationship is not as central a concept for the ethical problems of engineering as the physician-patient relationship is for medical ethics or the lawyer-client relationship is for legal ethics. Thus, for example, paternalism is not a burning issue in engineering ethics as it is in medical or legal ethics.[7]

A second important difference between engineering and the other two professions mentioned is that almost all of a modern engineer's activity takes place in the context of a formal organization of some kind or other, for example, in bureaucratically organized industrial corporations. Solo practice, which provides the traditional background for medical and legal ethics, is uncommon in engineering. Being part of an industrial organization, the ethical problems confronting the engineer take on another dimension, because many of them are a direct result of this status as an employee in an organization. Thus, the usual supposition that a professional is an independent operator does not hold for most engineers. This facet of engineering as a profession makes it both simpler and more complicated to frame and to deal with the ethical problems associated with the profession.

It is also easy to see that the ethical problems of engineers are more closely bound up with the particularities of the economic system in which they operate than are those, say, of physicians, who, in many respects, face the same sorts of ethical problems in Russia as they do in the USA. These considerations suggest that we really ought to ask: is there (or should there be) a different kind of engineering ethics for engineers working under a capitalist system from an ethics that would be appropriate for engineers working under a communist or socialist system?

One important aspect of having organizations rather than individuals as clients is that, for engineers, their being part of a system or an organization such as a large corporation often encourages a sense of futility and helplessness as far as being ethical is concerned. It is obviously difficult to be ethical when one is powerless. If, and to the extent that being moral requires self-determination and being able to operate independently, engineers, more than physicians or lawyers, may be inclined to think that they can do nothing about a situation which they deplore. They are caught up in a maelstrom and are powerless to influence the outcome.[8] Unlike other professionals, engineers do not live a sheltered existence where one is accountable to no one but oneself. As members of the organization, they are subject to lines of accountability like all the other employees. (Compare in this regard a company physician, who has a degree of "autonomy" in what he does, with a company engineer, who has almost none.) For these reasons, engineering ethics involve wider isues of responsibility of the sort that are encountered in politics and in organizational ethics in general. In this respect, some of the moral dilemmas that trouble ethically sensitive engineers could be compared to situations experienced by persons living under a totalitarian regime, where responsible action involving remonstrance or resistance leads to extermination. I shall return to this problem later.

THE ETHICS OF LOYALTY

Another quite different ethical aspect of the relationship of engineers to formal organizations of which they are members, whether they be public or private, relates to the issue of loyalty. It is often alleged that, as members of a particular organization, engineers have a special duty of loyalty towards that organization and therefore, having that duty, it would be wrong for them to do anything that might harm the organization, i.e., be against its interest. For example, it would be disloyal of an engineer to reveal the secrets of his organization to an outsider, even secrets about unsafe features of products.

In view of the great emphasis on loyalty in discussions of engineering ethics, we need to ask whether or not there is any validity or merit in the concept of loyalty to an organization. We should note right away that the loyalty in question here is not at all like the loyalty that physicians and lawyers are expected to have towards their patients and clients, for the latter kind of loyalty simply amounts to observing the duties of devotion, dedication, zealousness and avoidance of conflicts of interest that are owed to their patients and clients as individuals as a result of their relationship. The loyalty involved in the physician-patient or lawyer-client relationship has often been compared to loyalty between friends.[9] Loyalty to an organization, on the other hand, seems to be quite different from the loyalty that one individual gives to another and that can exist between friends.[10]

In order to see whether and how the concept of loyalty can be applied to an organization, we need to ask a number of other questions: Is loyalty always a virtue? Does loyalty ever permit or require doing something that would otherwise be wrong? What kinds of things can be the objects of loyalty? Do members of organizations, e.g., engineers, have a duty of loyalty to their organizations? If so, could the claims of an organization to loyalty ever justify wrongdoing of some sort?

If we assume that loyalty, as distinguished from blind obedience or servile compliance, is a virtue, then we must inquire into what we mean by "loyalty" in this sense.[11] When regarded as a virtue, that is, as something morally desirable, loyalty is founded on moral relationships of one sort or other between persons, e.g., originally between lord and vassal, but now also between family members, between friends, between colleagues and between comrades. Loyalty to a nation, to a college, or to a family is simply loyalty to the people in them, including perhaps to past and future generations within these communities. As a virtue, it derives its moral value from these interpersonal relationships and consists in thoroughgoing dedication and devotion of a person to what is owed another by virtue of a relationship of this kind. It is odd therefore to speak of loyalty where there is no prior personal relationship, for example, loyalty to a perfect stranger or to someone one admires at a distance but does not have any personal relationship to, such as a movie star. It is even more odd to speak of loyalty to a non-person.[12] Furthermore, if loyalty is a moral virtue, then it cannot be conceived as requiring something that is not due or, even less, something that is not right. There is no virtue in being a "loyal Nazi" or a "loyal member of the Mafia." Such notions, according to my analysis, would in fact be contradictions in terms.[13]

If this analysis of loyalty is correct, and I think that any analysis that takes loyalty to be a virtue would have to be developed along these lines, there are obvious difficulties with the idea of loyalty to a corporation, in the sense of loyalty that implies that it is morally good to be loyal. First, before we could speak meaningfully of loyalty in the context of a corporation, we need to ask: who in the corporation is the object of this loyalty? Is it the managers? the stockholders? one's fellow employees? or all of these? Obviously, loyalty to these different groups requires quite different kinds of conduct, some of which may be inconsistent. If, on the other hand, we choose to say that the object of loyalty is the corporation itself, the corporate entity, we face another paradox. For, quite apart from the issue of corporations being non-persons, something else important for loyalty seems to be missing that might be called "reciprocity." What I mean is that loyalty is thought to be a two-way thing: A is loyal to B and B is loyal to A. Friends are loyal to each other. In this sense, loyalty is a bond tying people to each other reciprocally. Corporate "loyalty" is, in contrast, by its very nature one-way; dedication and devotion can only be in one direction—from the employee to the corporation. A corporation cannot be loyal to employees in the same sense as they are supposed to be loyal to it, not only because it is not a person but also because the actions of a corporation must be conceptually linked to the benefits they bring to the corporation. A corporation can be good to employees only because it is good for business, that is, because it is in its own self-interest. All this is a consequence of the fact that a corporation is logically incapable of having moral attitudes and its conduct can only be understood in relation to the aims of the corporation. As an ethical notion, however, loyalty cannot be founded solely on utility or self-interest.

As I have already indicated a number of times, corporations are not persons in the moral sense. The fact that in law, corporations have the status of persons as far as the Fourteenth Amendment is concerned, does not make them moral persons: law and morality should not be confused. My arguments against the position that corporation are moral persons are based on a particular analysis of the logic of organizational decision-making, according to which organizations are logically incapable of moral decision-making and of moral conduct generally. If, as I contend, rational corporate acts must and can only be logically tied to corporate goals, e.g., profit-making, then such acts cannot be based on moral considerations. Essential to my analysis, it should be observed, is a clear-cut distinction and separation of a corporation as an "abstract" entity from the people within a corporation, who, as individuals, are, of course, persons with the moral responsibilities and rights pertaining to persons.[14]

WHISTELBLOWING

If, as I have argued, the issue of loyalty to a corporation is a red herring, then we need not discuss the ethics of whistlebowing in that connection. However, there are other issues connected with whistleblowing that we need to examine. The cases of whistleblowing that have received wide attention are spectacular. They typically involve preventable disasters, errors and mistakes, and usually some hanky-panky.[15] As has been pointed out, whistleblowing represents a particular kind of solution to a general problem that may have other and better solutions.[16] So we immediately have two issues: the rights and wrongs of whistleblowing and the problem that whistleblowing is supposed to solve. As far as whistleblowing itself is concerned, it is obvious that it is not always good or bad and that it is not always successful or necessary. As De George and others point out, it seems on the face of it undesirable from an ethical point of view to have to solve the kind of a problem that leads to whistleblowing by demanding that individuals be moral heroes. In any event, there are obvious ethical as well as practical objections to solutions of social problems that depend so heavily on individual self-sacrifice.

Turning to the second issue, let us take a closer look at the problems for which whistleblowing is supposed to be the solution, that is, the evils that it is supposed to correct. They include such things as faulty design leading to fatal accidents, as in the Hyatt Regency Hotel and the DC 10 crash. What is the underlying problem? If we assume that in general terms it is how to prevent undesirable states of affairs (evils) of some sort, what particular states of affairs are the ones in question? Is the problem simply: how to prevent preventable disasters? Or, to put it positively, is it simply how to promote safety in engineering? If so, is it the same problem that is addressed in the Engineer's Code of Ethics, when it says that "Engineers shall hold paramount the safety, health and welfare of the public in the performance of their professional duties?" One answer to our question, then, might be that the ethical problem behind whistleblowing is simply how to maximize safety or at least how to reduce unsafety to a decent minimum.

There are two things to say about the problem as I have just defined it. First, it involves a reference to safety. Safety, it should be noted, is the kind of thing that is looked on as a value by everyone.It is like motherhood: no one is against it—in principle. So we start off the discussion of this problem with something that is uncontroversial, for safety is an incontrovertible good, or, if you wish, something whose absence is an incontrovertible evil. It is easy, of course, to see why safety is valued by everyone, for an unsafe bit of machinery, like an airplane or an automobile, might result in one's own death or the death of others who are close. The first point, then, is that our starting place is more solid, so to speak, than the starting place of most ethical problems.

To admit safety as an incontrovertible good is not to say that there are no disagreements about safety. There are disagreements, for example, over the definition of safety, how much safety should be built into a machine, what safety measures are necessary in design, structure, operating procedures, etc.[17] There are also lively disagreements over the costs of safety, over who has responsibility for safety, and over what kind of controls there should be over safety.

The second question about the problem behind whistleblowing is this: granted that the prevention of preventable disasters is a problem, even a pressing or urgent problem, we must determine what kind of problem it is. Is it a moral or ethical problem? Or is it a social problem? a legal, a political, an institutional (organizational) or an economic problem? Or is it simply a practical human problem concerning means and ends, that is, concerning what measures should be undertaken to avoid the evils in question? In any case, why call it an *ethical problem*, unless we are ready to say that any problem whatsoever of means and ends is automatically an ethical problem?

One way that one might answer these questions is to say that safety is an ethical problem simply because it is an important matter that we need to do something about, i.e., we ought to do what we can to prevent preventable accidents. In that case, it becomes the problem of how to prevent accidents, and that surely is an engineering problem. Of course, we might want to broaden the scope of the problem so defined and amend it by including questions about how to get those in power to do something to prevent preventable accidents, that is, to take problems of safety more seriously. In order to solve it, then, we might need to tackle some political, organizational or perhaps even social problems. It is still unclear, however, why we should call the basic problem an *ethical* problem rather than some other kind of problem and why we should call it a "problem" rather than a "task." Are we faced here with something that is in any sense a moral issue and a perplexity? The only thing that we can say for sure is that whistleblowing by isolated individuals is not the answer to the problem of preventing preventable accidents.

On the other hand, we might wish to say that the problem in question is ethical because if everyone in a position to do something about safety were aware of their moral responsibilities in that regard and also were moral (i.e., conscientious), then the problem of preventable accidents would be solved. So construed, the "bottom line" becomes the question of how to get people to do what they ought to do, i.e., to be moral. In the final analysis, however, the way we conceive the problem depends on what our basic concern is: is it with the evil of the disaster or is it with the evil of people, that is, their failure to act to prevent disasters. If it is the latter, then the ethical problem turns into a problem of, say, how to raise moral consciousness about safety. It is solved by overcoming insensitivity or callousness on the part of those who are in a position to do something about safety. (This might, of course, include management.) On this view, being moral (ethical) in these senses is valued as a means of preventing accidents, etc. The problem behind whistleblowing (e.g., concerning faulty engineering) is solved by people becoming more moral, say, through moral education. Accordingly, we have a utilitarian interpretation of the problem itself and a utilitarian answer to it, including a utilitarian view of the value of moral consciousness and moral education. Ethics has been reduced to a means for preventing accidents!

Now, all of this seems to me to be an odd way of thinking about ethics. I want to ask: *whatever became of ethics*? Ethics is treated as if it were a kind of behavior control, an internal behavior control comparable in important respects to external behavior control through law, institutional regulations, social practices, rewards and punishments. As distinct from the latter, ethics is internalized control. But is this what ethics is: internal behavior control? Must we accept the utilitarian answer? Is the problem of how to eliminate preventable accidents simply one of securing moral behavior or one of ethical behavior control?

It might be observed, incidentally, that the assumption that ethics is essentially a kind of behavior control is probably what lies behind the attempts by various professions to codify the rules of their professional ethics. It seems to be taken for granted that if what is ethical is prescribed by a code, then members of the profession will comply and a socially acceptable and desirable outcome will result. Whatever the explicit intent, the underlying purpose of the codes is to create some kind of behavior control analogous to control through law.[18] At this point, I need to make some comments on the use of formal mechanisms of behavior control in the service of ethics.

MECHANISMS OF BEHAVIOR CONTROL: LAW VS. ETHICS

A great deal of confusion in discussions of ethical problems results from the assimilation of ethics to law, institutions, organizational regulations, and other mechanisms of social control. I hardly need mention here the other common fallacy of identifying ethics with the 'value-system' of some group or other, that is, with what John Austin called "positive morality,"—the body of moral beliefs and conventions actually accepted by a person, a group, or a society. Ethics, sometimes called "critical morality," is logically prior to all of these institutions and social mechanisms of control. It is used to criticize, evaluate and weigh the validity and desirability of the norms, rules and principles embodied in such institutions as law and positive morality. Consequently, ethics is prior to or superior to these other systems of norms in that it is used to determine what in them is morally acceptable and unacceptable.

There are a number of other important differences between ethics and formal systems of control such as law and management regulations. By a "formal system of control" I mean a system consisting of formally adopted rules, regulations, procedures and sanctions that are and can be used to control behavior. Usually the rules in question are written down and published for the guidance of those subject to them. For our purposes, perhaps the chief difference between these systems of control and ethics is that in formal systems of control such as law and management regulations some person or body of persons is authorized to create, change and rescind the rules—at will, so to speak. The authority to do this is vested in legislatures, courts, commissions, boards, managers and other officials. The principles of ethics (or morals), in contrast, are not the kind of thing that can be arbitrarily created, changed or rescinded. Ethics cannot be dictated. In old fashioned terminology, the principles of ethics are "discovered" rather than created by fiat. They are established through argument and persuasion, not through imposition by an external social authority.

Another critical difference between the formal systems of control that I have been discussing and ethics is their purpose, for law, corporate regulations, institutional requirements, and other formal systems of control are designed and used to control behavior for various and sundry purposes, which may be good, evil, or indifferent. From an ethical point of view, it is our job to weigh the validity and desirability of these purposes. But just as important from the ethical point of view are the ways and means selected to achieve these purposes. Here we need to ask: which ways and means are legitimate and ethically justifiable and which are not. In particular, we need to ask which sorts of social control are ethically permissible and which are not? In connection with the last question, there is one important means of control that has received insufficient attention in the literature on engineering ethics, namely, the use of secrecy as a means of control. When is secrecy justified and when not? are the current self-serving norms regarding secrecy, e.g., within a corporation and regarding engineering projects, tolerable from the ethical point of view? Is it not possible that the best way to avoid political and organizational corruption in the long run is to make public the plans, projects and purposes behind the operations of our public institutions, government and private corporations?[19]

SOCIAL CONTROL AND THEORIES OF HUMAN NATURE

Ethical theories about proper and improper social control always presuppose, either explicitly or implicitly, a theory of human nature of some kind or other. Theories of human nature in this sense are about what motivates human beings and what ought to motivate them, that is, what would be rational for them to want and to do. Although it is generally assumed that theories of human nature and motivation are descriptive and empirical, in actuality they are always inescapably value-laden, both in regard to how individuals ought to act (if they are rational) and in regard to what are the proper means to get them to act in ways that one wants, i.e., to control them. Different theories of human nature come up with quite different answers to these questions. Thus, according to a Hobbesian self-interest model of human nature, it is reasonable for a person to act for his own self-interest and unreasonable for him to act against it, and for that reason the best and proper way to control others is to manipulate their self-interest, to make it in their self-interest to act in certain ways. This is the typical bourgeois economic view of man and it provides the rationale for many of our institutional arrangements that are designed to control behavior, e.g., of employees.[20] Whatever is in a person's self-interest is, other things being equal, permissible, nay, rational for him to do.

For reasons that I cannot give here, the Hobbesian model does not fit individual human beings very well; it ignores not only the 'irrational' and emotional side of human nature but also its moral aspect. On the other hand, the model does apply very nicely to corporations, at least to the commonly accepted notion of private corporations as propelled by self-regarding drives for profit, power and glory. The whole Hobbesian apparatus can be usefully applied to an ethical analysis of corporations. Like Hobbesian men, corporations are, in theory at least, in constant competition with each other—as in Hobbes's state of nature—and the only effective control over their "rational" voracity is through the manipulation of their self-interest. Accordingly, if we really want to cut down unsafe practices and to reduce the incidence of industrially caused accidents (evils) we have to make it in a corporation's self-interest to take measures to prevent them. It is absurd to appeal to ethics, because corporations are not moral beings. Profits are what count, and so unsafety should be made unprofitable.

I have argued that underlying most thinking about ethics in corporations is a certain mythology, which holds that corporations are persons and that therefore the same theories of human nature, of motivation and of morality apply to them as apply to individual human beings. This mythology leads us either to anthropomorphize corporations and treat them like "nice people" or else to reduce individual human beings to miniature corporations, each in pursuit of his self-interest and in perpetual strife with others for profit, power and glory. Once we rid ourselves of this mythology, we will be able to sort out more clearly and coherently our mutual rights, duties and responsibilities in society in relation to each other.

MORAL RESPONSIBILITY AND OTHER KINDS OF RESPONSIBILITY

Nowhere is the fallacious assimilation of corporations to moral persons more apparent than in prevalent conceptions of responsibility; different senses of responsibility are confused with each other almost as if there were a plot to get rid of moral responsibility altogether. Pursuing the same line that I have already taken in this essay, I shall argue that the concept of moral responsibility, as contrasted with other kinds of responsibility, cannot properly or even meaningfully be attributed to corporations, that is, to formal organizations. But first we have to sort out a number of different senses of "responsibility."

The *locus classicus* for any discussion of responsibility is Hart's essay on the subject.[21] In his essay, Hart lists four different senses of responsibility, which he calls: (a) Role-responsibility, (b) Causal-responsibility, (c) Liability-responsibility, and (d) Capacity-responsibility. The names speak for themselves: role-responsibilities are the responsibilities that go with roles, tasks and jobs; causal responsibility is the responsibility for having caused something to happen; liability responsibility concerns who is to pay for damages, and capacity-responsibility refers to psychological capacities required for legal competence.

It should be noted right away that Hart does not include on his list the most important sense of "responsibility": moral responsibility, that is, responsibility in the virtue sense.[22] He ignores this kind of responsibility because he is interested only in responsibility as it relates to law. Furthermore, it should be observed that all of Hart's four senses of responsibility can be attributed to corporations, for corporations can (a) fill roles; they can (b) cause things to happen; they can be (c) liable, e.g., for damages, and they have (d) the "capacities" that Hart mentions, namely, the capacities of "understanding, reasoning and control of conduct."[23] It is easy to see why Hart's senses of responsibility apply to corporations, for they are essentially legal entities and as such are subject to law, which, as I have pointed out, should not be confused with ethics.

When we come to moral responsibility we are dealing with something quite different.[24] In order to bring out the difference, let me begin by distinguishing between forward-looking and backward-looking senses of responsibility, that is, between responsibility for something that has already taken place and responsibility for what will or might take place in the future. Clearly, liability responsibility is backward looking in this sense and so is causal responsibility: "*Who* and *What* is responsible for the crash of the DC-10?" Role-responsibility may be either past or future: a safety officer's role was, is, or will be to monitor such and such for product safety, etc.[25] Retrospectively, a person can be held responsible (liable) for failure to do what his role required, and prospectively, he is responsible (role) for completing certain tasks and controlling certain kinds of results in the future. It should be noted, however, that role-responsibilities (tasks, jobs) are assigned, e.g. by organizations, and from a moral point of view may be non-moral or immoral, as well as moral. Thus, a public relations official may have the responsibility (job, task) for covering up management's failure to report a hazardous condition.

Now, moral responsibility as I conceive it is forward-looking. It is about what people ought to do to bring about or to prevent future states of affairs. It is based on the duty each one of us has to watch out for what may happen to others or to oneself. As such, it implies concern, care and foresight. To be responsible in this sense is a virtue that cannot be meaningfully predicated of a corporation conceived as a formal organization, that is, as a structure of rules, offices and jobs, etc. Corporations, being nonmoral entities, cannot be virtuous or vicious in the moral sense; only the people in them can be so characterized.

Now, one noteworthy property of responsibility in the moral sense as contrasted with the other senses is that it is *nonexclusive*. In the other four senses, responsibility is exclusive in the sense that to impute responsibility to one thing (X) implies that other things (Y,Z) do not have the responsibility. Thus, if one person has a role-responsibility for something, it follows that other people do not have the responsibility. Similarly, causal and liability responsibility are exclusionary. Moral responsibility, on the other hand, is not exclusionary in this sense; for one person to be responsible does not entail, as it does for the other kinds of responsibility, that other persons are not also responsible. A father's responsibility for his children does not exclude (or negate) the mother's responsibility—or, for that matter, anyone else's responsibility, e.g., the responsibility of the state. In the moral sense, there are some things that everyone is responsible for, and one of these things is safety. Concern with safety is not just one person's job, i.e., his role-responsibility, to the exclusion of others. It is everyone's moral responsibility—varying in degrees only to the extent that one person is better able to do something about it than others. The concept of moral responsibility implies that there are some things that are everyone's business!

COLLECTIVE RESPONSIBILITY

A very significant ethical consequence of the nonexclusiveness of moral responsibility is that, if many people can be morally responsible for the same thing, then there can be such a thing as group moral responsibility, or if you wish, collective responsibility, that is, a responsibility that falls on many people at the same time. In as much as one person's being responsible does not relieve others of responsibility, everyone in a group may have moral responsibility for a certain thing. Thus, the whole family is responsible for seeing that the baby does not get hurt. The whole community is responsible for the health and safety of its citizens. And all the engineers, as well as others, working on a project are responsible for its safety.

Now it should be clear that underlying my analysis of collective responsibility is a distinction between a group of people, a collection or association of individuals, and a formal organization, which is a structure defined by rules, offices and jobs, etc., apart from the people who come and go in the organization. Moral responsibilities, moral virtues and other moral qualities can be ascribed to groups insofar as they pertain to the individuals in them. But since organizations are not persons, they are, as such, beyond the pale of morality. We cannot and should not shift our moral responsibilities onto abstract entities like corporations.

One of the deep problems of our time is that people have followed the lead of philosophers, lawyers and managers and have simply reduced moral responsibility to the other four kinds of responsibility already mentioned. The net effect of this move is to render all responsibility exclusionary and to provide thereby theoretical support for a wholesale abdication of moral responsibility: "That's her job, not mine"; "he did it, not I," etc. We are constantly looking for someone to fix responsibility onto, be it liability or role responsibility. We construe the question of responsibility for engineering errors as a question of fixing responsibility on some engineer, either holding him liable for it (in the past) or assigning him the task of watching out for it in the future. "Divide and conquer" is the motto: if we divide responsibilities like jobs or liabilities we will avoid any trouble and we will know whom to blame.

Our ideology gives us a way to pass the buck as far as moral responsibility is concerned. Against this, it should be pointed out that if everyone in a non-exclusive group with moral responsibility for safety sets out to prevent a disaster, then the world will be much better off than if we simply try to fix a disaster on a single person or if we assign the job (= role responsibility) for preventing disasters to a particular person or outfit and then forget about our own responsibilities. According to the conception of moral responsibility that I have in mind here, there is a sense in which all of us, engineers and nonengineers alike, are responsible for things like the Pinto accidents, say, because we accept a way of life, based on the Hobbesian model, that assumes that what is good for business, anyone's business, is good for us, for society, and correlatively, it is good for business to mind one's own business.

On a broader front, we can see what happens to a society when a sizeable segment of the population abdicates its moral responsibility for the common good, that is, the good of its members, and opts for the principle of minding one's own business (job-responsibility). In Germany, a result of this kind of abdication was Hitler. In the USA, a result of this kind of abdication was Vietnam. And unless we start caring, a future result of our abdication of responsibility will be World War III and a nuclear holocaust. Some would like to blame the engineers for that. But I argue that nuclear weaponry is not simply an engineering problem, that is, a problem for engineering ethics, although it is also that, just as problems of engineering ethics, e.g., concerning safe products and concerning a safe environment, are not simply problems of engineering ethics, but problems for all of us. All of these things are everybody's business.

In conclusion, I want to reiterate what was said at the beginning of this essay: philosophers cannot be expected to provide ready-made solutions to ethical problems in engineering. Instead, following my conception of the zetetic role of a philosopher, I have simply tried to point out a number of questions that need to be asked and some of the pitfalls in ethical thinking about them that ought to be avoided. As far as the latter are concerned, I have tried to show that serious ethical consequences follow from the blithe acceptance of corporations as persons, from the confounding of moral and legal concepts, and from the failure to recognize moral responsibility as a distinctive kind of responsibility that is nonexclusive and that can be predicated of individuals in groups (i.e., collective responsibility) as well as of persons individually. Thus, despite the disclaimers about the practical value of philosophy that I have mentioned, philosophy in the analytic tradition still has an important and perhaps indispensable function in making clear how best to approach the problems that concern us in engineering ethics, even though in the end it does not provide authoritative answers to them.

Notes and References

[1] My position on these questions is set forth in a number of writings. See, for example, "The Poverty of Absolutism," in Timothy Stroup, ed. *Essays in Memory of Edward Westermarck.* Acta Philosophica Fennica. Helsinki, 1982. See also, "The Task of Ethics," in Warren Reich, ed. *Encyclopedia of Bioethics.* New York: The Free Press, 1978, vol. 1, pp. 400-407.

[2] From the Greek *zetein.* It is a word that was used for an ancient school of philosophers known as the Skeptics—otherwise as Zetetics.

[3] For a critical discussion of the notion of a professional code of ethics see "The quest for a code of professional ethics: an intellectual and moral confusion." *AAAS Professional Ethics Project,* eds. Rosemary Chalk, Mark S. Frankel and Sallie B. Chafer. Washington, D.C.: American Association for the Advancement of Science, 1980.

[4] For a forceful statement of this position, see Monroe Freedman, *Lawyer's Ethics in an Adversary System.* Indianapolis: Bobbs-Merrill, 1975.

[5] See, for example, Ivan Illich, Irving Zola, et al. *Disabling Professions.* Boston: Marion Boyars, 1977.

[6] It is often advanced as a criticism of modern medical practice that physicians treat patients as bodies that are like machines needing to be repaired rather than as persons. In this way, medicine becomes a technology and in that respect becomes like engineering.

[7] There is an extensive literature on paternalism in medicine and law.

[8] In this connection, there is a series of rationalizations that are made by members of a bureaucracy to justify their not doing anything about something they think is wrong. For a critical discussion of such attempts to avoid responsibility, see Dennis Thompson, "Moral responsibility of public officials: the problem of many hands." *American Political Science Review,* vol. 74, no. 4 (December 1980).

[9] See Pedro Lain Entralgo, *Doctor and Patient.* Tr. Frances Partridge. New York: McGraw Hill, 1965; and Charles Fried, " The Lawyer as Friend: The Moral Foundations of the Lawyer-Client Relation." *Yale Law Journal,* 85: 1060-89 (1976).

[10.] Hume says in this connecton that virtues like rigid loyalty to persons are "virtues that hold less of reason than of bigotry and superstition." *Treatise,* Book III, Part II, Section X.

[11.] I discuss this concept in "Loyalty," *Encyclopedia of Philosophy,* ed. Paul Edwards. New York: Macmillan and Free Press, 1967, vol. 5, pp. 97-98.

[12] Elsewhere I argue that formal organizations, e.g., corporations, are not persons in the moral sense. See my "Morality and the ideal of rationality in formal organizations." *Monist,* vol. 54, no. 4 (October 1970).

[13] See "Loyalty."

[14] See "Is 'corporate responsibility' a coherent notion?" in *Proceedings of the Second National Conference on Business Ethics.* Ed. Michael Hoffman. Washington, D.C.; University Press of America, 1979, 9, pp. 102-115.

[15] See Alan F. Westin, *Whistle-blowing: Loyalty and Dissent in the Corporation.* New York: McGraw Hill, 1981.

[16] See Richard T. De George, "Ethical responsibilities of engineers in large organizations." *Business and Professional Ethics Journal,* vol. 1, no. 1 (Fall 1981)

[17] See Willie Hammer, *Product Safety Management and Engineering.* Englewood Cliffs, N.J. Prentice-Hall, 1980.

[18] See my "The quest for a code of professional ethics."

[19] I should remind the reader that none of these institutions, according to my analysis, is a person and therefore none has a *moral* right to personal privacy.

[20] See C. B. McPherson, *Democratic Theory.* Oxford: Clarendon Press, 1973, esp. pp. 224-37.

[21] See H. L. A. Hart, *Punishment and Responsibility.* New York: Oxford University Press, 1968, p. 212 et passim.

[22] For an account, see Graham Haydon, "On being responsible." *Philosophical Quarterly,* vol. 28, no. 110 (January, 1978).

[23] Op. cit., p. 227.

[24] Most of the ideas in the following paragraphs are taken from my "The ethics of participation," in *NOMOS XVI: Participation in Politics.* ed. J. Roland Pennock and John W. Chapman. New York: Atherton-Lieber, 1975.

[25] See Hammer, op. cit., chapter 12. □

CHARLES E. HARRIS, JR.

Explaining Disasters: The Case for Preventive Ethics

Crewmembers of the January 8, 1986, Challenger mission, leaving to board the space shuttle. All seven lost their lives following the launch-phase explosion. From front to back: Francis R. Scobee, Judith A. Resnik, Ronald E. McNair, Michael J. Smith, Christa McAuliffe, Ellison Onizuka, and Gregory B. Jarvis.

In 1986, the space shuttle Challenger exploded during launch, taking the lives of six astronauts and one teacher, Christa McAuliffe. The disaster virtually stopped U.S. space exploration for two years. How should this disaster be explained?

Most disasters have multiple explanations. One type of explanation in the Challenger case focuses on the flaws in the design of the field joints in the boosters, or on other engineering failures. Another type locates the problem in improper management practices, either at the National Aeronautics and Space Administration (NASA) or with Morton Thiokol, the manufacturer of the boosters. Another type finds the fault in unethical conduct on the part of NASA or the private contractors. Still other types of explanation might attribute the disaster to an unanticipated convergence of events, or just plain bad luck. I shall confine myself, however, to the first

The author is with the Department of Philosophy, Texas A&M University, 510 Blocker Building, College Station, TX 77843-4237.

Reprinted from *IEEE Technology and Society Magazine,* Vol. 14, No. 2, pp. 22–27, Summer 1995.

three types of explanation: bad engineering, bad management, and bad ethics.

Even though we may know that it is insufficient, there seems to be a natural tendency to focus on a single type of explanation. The result is that these three types of explanation compete with one another in the minds of many people. If an event can be explained in terms of engineering failures, for example, we may think there is no need to look for evidence of improper management or unethical conduct. If there is evidence of incompetent management, why look for engineering problems or ethical improprieties? One reason for this tendency may be that people tend to look for explanations most congruent with their own areas of expertise. Engineers usually look for the explanation of a disaster in bad engineering, most often in faulty design. Managers or management consultants tend to find the explanation in bad management. Ethicists are more likely to look for explanations in terms of ethically improper behavior.

Contrary to this approach, there are good reasons to believe that these three types of explanation are not mutually exclusive. The same disaster can be explained in terms of bad engineering, bad management and bad ethics. One consequence of taking this more pluralistic approach to explaining disasters is that the place of ethical considerations in explaining disasters is not neglected. An appreciation of the importance of ethical failures can, in turn, serve to underscore the importance of avoiding these failures in the future. I shall refer to this effort to isolate the ethical failures involved in engineering disasters and to use this knowledge to prevent such failures in the future as *preventive ethics.*

Before proceeding further, it is important to point out that there is a distinction between an engineering disaster and an ethically improper use of engineering. The Challenger explosion was an engineering disaster: it involved a technical malfunction that had catastrophic consequences. The employment of German engineers to design the gas valves used at Auschwitz was not an engineering disaster. The valves evidently worked all too well. The problem was that the end to which engineering design was directed was unethical. Few people would question the relevance of ethical categories in explaining the tragedy of Auschwitz: the ends toward which engineering design was directed were unethical. For most of us, however, the goals of engineering work in the Challenger project were not unethical. If ethical categories are relevant, they must be relevant in a different way. I am concerned with this second type of situation, not the first.

Three Types of Explanation

What do we mean by an explanation of a disaster? In explaining a disaster, at least two conditions must be met. First, there must be a failure or impropriety of some sort. Since we are limiting ourselves to three types of explanation, we shall be concerned with three types of failure: in engineering, in management, and in ethics. When we say there has been a failure of some sort, we mean that the rules, standards, or canons appropriate to that area have been violated. Thus, to say that a disaster exhibited improper engineering means that engineering standards were violated. To say that a disaster exhibited improper management means that the canons of good management practice were violated. Similarly, to say that a disaster exhibited unethical conduct means that the canons of proper ethical conduct were violated.

A second condition is that the impropriety must have been a contributing cause of the disaster. While there may be times when a single cause is sufficient to explain a disaster, it is more common to find that there are several contributing causes. I shall offer the following as a working account (not a formal definition) of a contributing cause: Event A is a contributing cause of Event B, when Event A is prior to Event B and when the existence of Event A makes Event B more likely to occur. Using these two conditions as tests, we can make a case that the Challenger disaster can be explained in all three ways: it was bad engineering, bad management and bad ethics.

The case for explanation in terms of bad engineering is based on the design flaws in the seal between the sections of the boosters. One of the canons of good engineering design is that static and dynamic situations must be carefully distinguished, and the design must fit the situation. Yet in the case of the Challenger, this canon was violated. The O-ring seal between the sections of the boosters was designed for a static situation, but the flexing to which the seal was subjected in flight meant that it should have been designed for a dynamic situation. This design flaw, along with the unusually cold weather that caused the O-rings to lose some of their resiliency, was perhaps the most obvious engineering explanation of the disaster.[1]

Furthermore, the design flaw was a contributing cause to the disaster. The design flaw made the disaster much more probable. Indeed, apart from this design flaw, the disaster might never

[1] I have used two written sources for my account of the Challenger disaster. One source is the commission chaired by William P. Rogers in 1986 [1]. I shall refer to this as the Rogers Commission Report. The other is Roger Boisjoly's, "The Challenger disaster: Moral responsibility and the working engineer" [2].

have occurred. So both of the conditions of explanation in terms of an engineering failure are fulfilled. There were violations of engineering principles, and these violations made the disaster much more likely to happen.

There is a natural tendency to focus on a single type of explanation.

There is also a case for an explanation of the Challenger disaster in terms of bad management. To say that an event exhibits bad management is to say that it violates the standards of good management.[2] One of the standards of good management is that managers should establish and maintain effective communication with their employees. The reason for this is that good communication not only enhances employee morale, but also furnishes managers with information that is essential in making sound management decisions. In order to enhance communication, managers must do at least two things. First, they must create an atmosphere in which employees can bring up problems without fear of reprisal. Second, they must respond positively to employees when they utilize this freedom to bring up problems. This does not mean that managers must always follow employees' advice, but they must consider and evaluate it carefully.

These requirements of good management practice, which are especially important with regard to professional employees, were evidently violated by Morton Thiokol managers. Roger Boisjoly reports that he alerted Thiokol managers to the problems with the O-ring seal a year or more before the Challenger disaster. He even asked for funding to look for solutions. Not only was his request ignored, but there was evidently an atmosphere of intimidation that inhibited engineers from communicating their concerns freely. On the night before the disastrous launch, Boisjoly and other engineers made their case for a no-launch recommendation to NASA, primarily on the basis of anticipated difficulties with the O-ring seals at the low launch temperatures. After first being accepted by Thiokol managers, this recommendation was later reversed, partially at least as a result of protests from NASA. According to Boisjoly's account, when he objected to the reversal of the original recommendation, his manager (Gerald Mason) looked at him in a way that indicated he was about to be fired.[3]

Improper management was also a contributing cause of the disaster. If Thiokol managers had been more responsive to Boisjoly's early warnings, they might have ordered research aimed at improving the O-ring seal, and an improved seal might well have averted the disaster. If Thiokol and NASA managers had listened to the engineers on the night before the launch, they might have recommended against the launch and the launch might not have taken place. So bad management was also an explanation of the disaster, in that management principles were violated, and the violations made it more likely that a disaster would occur.

Finally, a case can be made that ethically improper conduct was part of the explanation of the disaster. To say that something is unethical is to say that it violates ethical standards. One such standard is the Golden Rule: "Do unto others as you would have them do unto you." One wonders if Thiokol or NASA managers would have been willing to fly in the Challenger themselves, knowing what they did about the problems with the O-rings. I am inclined to say that they would not, and that their action violated the Golden Rule.

Perhaps even more telling is the violation of the standard of free and informed consent. People should be informed about unusual dangers to which they might be subjected and given the chance to consent or not consent to the dangers. According to the Rogers Commission, this canon was not fully honored with respect to the problem created by ice formation on the Challenger, due to the sub-freezing temperatures the night before launch. Although they had been consulted about the ice problem, the crew was not fully apprised of its seriousness [1, p. 118]. There is also no record of the crew's having been adequately informed of the O-ring problem, even though it was known to be potentially life-threatening. These deficiencies can only be considered serious violations of the principle of informed consent. Even though the astronauts knew that they were engaged in a high-risk mission, the principle of informed consent was not thereby rendered irrelevant. The astronauts should have been informed of the unusual problems.

In addition to the violation of widely-accepted ethical precepts, the events preceding the Challenger disaster exhibit other types of ethical deficiencies. The Rogers Commission concluded that Thiokol management reversed its original decision "contrary to the views of its

[2]For further discussion of the engineer/manager relationship, see [3].

[3]Reported in [4].

engineers in order to accommodate a major customer" [1, p. 104]. What explains this reversal? NASA managers expressed extreme displeasure with the original Thiokol decision not to launch, probably due to pressures on them for a quick success.[4] The testimony of Robert Lund, the vice president of engineering at Morton Thiokol, centers around a shift in the burden of proof. Whereas NASA originally adopted a policy that a launch recommendation bore the burden of proof, it had shifted to the position that a no-launch recommendation bore the burden of proof [1, p. 93]. After first agreeing with his engineers, Lund changed his mind, perhaps in response to pressure from Mason. The testimony of Jerry Mason, a senior vice president at Morton Thiokol, centers around the claim that the engineering evidence was inconclusive and that a management decision had to be made [1, p. 773]. There is evidence, then, that both NASA and Thiokol managers may have exhibited ethical deficiencies. One thinks of weakness of will (lack of courage to do what one knows is right), self-deception, and self-interest as likely candidates for these deficiencies.

A good case can also be made that ethical failures were a contributing cause to the disaster. If the managers at NASA and Thiokol had put themselves in the place of the astronauts and had never been affected by weakness of will, self-deception, or self-interest, they would almost certainly have taken the O-ring problems more seriously. The managers would probably have paid attention to Boisjoly's early warnings and ordered further testing, which might have led to the correction of the problem. If managers had held consistently to the canon of free and informed consent, they would have fully informed the astronauts of the O-ring problem. We do not know, of course, whether the astronauts would have decided to fly if they had been informed about the O-ring problem. There is, however, a significant chance that they would have decided not to fly, especially if the information about the O-ring problem had been added to the information about the danger due to ice. Even if they had decided to fly, the managers and engineers might have been more likely to try to correct the O-ring problem, knowing that they must inform the astronauts about it.

One wonders if Thoikol or NASA managers would have been willing to fly in the Challenger themselves, given their knowledge of the O-ring problems.

Thus, if managers and engineers had avoided unethical conduct, they would have been more likely to have made different decisions. This means that ethical failings were present and could be considered contributing causes of the disaster.

Primacy of the Engineering Explanation

If this analysis is correct, there is nothing wrong with saying that bad engineering, bad management and faulty ethics all play a part in explaining the Challenger disaster. Principles of sound engineering, sound management and sound ethics were violated, and these violations could all be considered contributing causes to the disaster. These three types of explanation are not mutually exclusive. One cannot show that one type of explanation is inapplicable merely by showing that another type of explanation is applicable. One cannot show, for example, that an explanation in terms of ethical failings is irrelevant by showing that an explanation in terms of engineering ineptitude is relevant.

But isn't the engineering explanation more fundamental than the others? Even if all three types of explanation are relevant, doesn't the engineering explanation occupy pride of place? The simple answer to this question is, "Yes." The engineering deficiencies seem to be the most crucial, in the sense that the disaster almost certainly would not have occurred if there had been no problem with the O-rings. This cannot be said of the management and ethical failures. Even if management practices and ethical conduct had been exemplary, the disaster might still have occurred. Suppose Thiokol managers had listened to Boisjoly's early warnings about the O-ring deficiencies and ordered further testing and research to resolve the problems. A bad design might still have resulted. Even if NASA and Thiokol managers had listened with an open mind and in a non-intimidating way to the engineers on the night before the launch, they might still have concluded in good faith that the engineer-

[4]According to Roger Boisjoly, George Hardy of NASA said he was appalled by Thiokol's no-launch recommendation. See [2, p. 8].

ing evidence was not compelling. While the managers might have made a mistake, they might not have violated principles of sound management.

The situation is similar with regard to the ethical improprieties. Suppose Thiokol managers had attempted to follow the Golden Rule, so that they ordered further research and development on the O-rings. This research and development could still have issued in a bad design. They might even in good faith have been willing to fly themselves. Furthermore, if the astronauts had been fully informed about the problems with ice and the O-rings, so that the requirement of free and informed consent was met, they might have decided to fly anyway. And if engineers and managers had resisted unethical influences on the night before the launch, they still might have made a decision to launch.

Of course following sound engineering principles does not guarantee that there will be no design mistakes, but it is still true that the disaster probably would not have occurred if the design mistakes had been corrected. By contrast, the disaster might still have occurred, even if the management and ethical failures *had* been corrected. In this sense, then, the engineering failure can be considered the most fundamental or at least the most direct explanation of the Challenger disaster.

Preventive Ethics

It does not follow, however, that management and ethical considerations are irrelevant in explaining the disaster. There is good reason to believe that there were management and ethical failures, and that these were contributing causes to the disaster. We cannot say for certain that eliminating these failures would have kept the Challenger disaster from happening, but it would have made the disaster less likely. This is because eliminating the management and ethical failures would have made the engineering failures themselves less likely to have happened. Of course there are other reasons besides preventing disasters for engaging in sound ethical and management practices, but I am concerned here only with this reason.

Thus, understanding why a disaster happened puts us in a better position to *prevent* similar disasters in the future. This is one of the reasons engineers want to look for the engineering explanation of a disaster. Engineers, like the rest of us, learn from experience. If they can isolate the engineering factors that explain a disaster, they can do something to prevent similar mistakes in the future. Perhaps we could call this *preventive engineering*.[5]

The same thing could be said about management failures. If managers at NASA, Morton Thiokol, and perhaps other private contractors had established a more open and non-intimidating atmosphere for their engineers and had been more adept at listening to the engineers' concerns, remedial measures might have been taken regarding the O-rings and the disaster might not have happened. Perhaps we can call this *preventive management*.

People should be informed about unusual dangers to which they might be subjected, and given the chance to consent or not consent to the dangers.

By similar reasoning we can say that discovering and attempting to eliminate ethical failures can also aid in preventing similar disasters in the future. As we have seen, if managers had been more attentive to ethical considerations such as the Golden Rule and the principle of informed consent and had not succumbed to self-interest or excessive external pressures, they might have taken stronger measures to correct the O-ring problems. Similarly, greater ethical concern and strength of will might have led more engineers to insist that either the O-ring problem be remedied or the Challenger should not fly. Exposing these problems and attempting to eliminate them can make an important contribution to preventing similar disasters in the future. I have already referred to this as part of preventive ethics.

The idea of preventive ethics is not wholly new. Some large health-care organizations employ medical ethicists on the corporate level in order to aid in the formulation of ethically acceptable policies. Management in these organizations apparently believes that operating by ethically acceptable policies may prevent legal and public-image problems and serve as a defense if such problems arise. Promoting codes of ethics and installing ethics officers and procedures for promoting ethical awareness may be a

[5]Steven B. Young and Willem H. Vanderburg develop the concept of "preventive engineering" in [5].

part of this same philosophy.

So far I have focused exclusively on the Challenger case in order to illustrate the ethical dimension of explaining disasters and the concept of preventive ethics. But many other famous cases in engineering ethics also exemplify ethical failures and suggest that the elimination of those failures might have prevented the disasters, or at least made them less probable. Engineers and managers were aware, for example, of the problems with the cargo hatch door of the McDonnell Douglas DC-10, but only one engineer appears to have made any concerted effort to remedy the problem. If managers and engineers had resisted unethical influences or imaginatively placed themselves in the position of passengers in the DC-10, would they have acted differently? If they had, the crash near Paris, France, which killed all 346 passengers, might have been avoided. Ford engineers and managers were aware of the susceptibility of the Ford Pinto to explosion from even low-impact rear-end collisions. Would they have been more inclined to remedy the design defect if they had taken seriously the possibility that they or a family member might have driven the car, or if they had considered informing the public of the danger from rear-end collisions? Similar arguments might be made about the Chevrolet Corvair, the Union Carbide disaster in Bhopal, India, and many other cases not so well known.

A good case can be made that ethical failures were a contributing cause to the disaster.

There is no way of knowing whether greater ethical sensitivity and the absence of impediments to ethically responsible action would have prevented these particular disasters, but it seems almost certain that the presence of these factors can prevent *some* unfortunate and tragic engineering disasters. This is enough to make preventive ethics worthwhile.

References

[1] William P. Rogers, "Report to the President by the Presidential Commission on the Space Shuttle Challenger Accident," Washington, DC, June 6, 1986.

[2] Roger Boisjoly, "The Challenger disaster: Moral responsibility and the working engineer," in *Ethical Issues in Engineering*, Deborah C. Johnson, Ed. Englewood Cliffs, NJ: Prentice Hall, 1991, pp. 6-14.

[3] Charles E. Harris, Jr., Michael S. Pritchard, and Michael J. Rabins, *Engineering Ethics*. Belmont, CA: Wadsworth, 1995, pp. 273-277.

[4] Roger Boisjoly, Massachusetts Institute of Technology, Cambridge, MA, Jan. 7, 1987, videotape record of remarks to audience.

[5] Steven B. Young and Willem H. Vanderburg, "A materials life cycle framework for preventive engineering," *IEEE Technol. & Soc. Mag.*, vol. 11, pp. 26-31, Fall 1992.

T&S

FEATURE ARTICLE

"High-Reliability" Organizations and Technical Change: Some Ethical Problems and Dilemmas

GENE I. ROCHLIN

Abstract—We consider here three exemplary "high-reliability" organizations that perform complex, highly technical, time-urgent tasks upon which many people depend for their safety and well-being. Faced with demands to increase capacity without reducing reliability or increasing operational staff, all three are on the verge of introducing new and advanced technologies to increase system capacity without sacrificing safety or reliability. Yet, few of those involved in engineering the changes understand why these organizations perform as well as they do, or how the structure of the organization will be altered by the incorporation of radically new technologies. This raises a number of ethical problems for system administrators and engineers who are effectively implementing changes in organizational structure through technical change, without understanding which organizational parameters are most critical to maintaining individual and collective performance.

INTRODUCTION

Over the past two years, our research group has been studying three organizations whose responsibilities—enroute air traffic control, utility grid management, and naval air operations—involve the performance of a variety of complex, highly technical, real-time tasks with a very low margin for error. From its inception, our study differs from many, perhaps most, current studies on modern organizations responsible for difficult and sometimes dangerous tasks involving engineering safety, in that the relative success of these three organizations is what first drew our attention [1]. We embarked upon this study in the hope that from it we might learn how to better design and manage the growing class of organizations in both public and private sectors similarly charged with tasks for which the consequences of failure or error can be extremely high.

It is worth noting several other aspects that distinguish the present study:

(1) The safety or reliability of the technologies being managed or controlled is outside its scope. None of the organizations under study is responsible for making aircraft, power equipment, control technologies, etc., safer or more reliable as individual pieces of equipment. That equipment will sometimes fail is simply taken as a fact of life. The risks our organizations deal with are otherwise independent of technical failures; they depend primarily upon social factors such as level, pattern, and density of use, and it is these that operators must manage or control.

(2) Risks are not taken or reduced incrementally, nor can risk be traded for performance on the margin. If the organization performs well, a high level of activity can be sustained at an acceptably low accident or error rate; if not, neither efficiency nor safety can be achieved. Therefore, marginal risk-benefit and static probability analyses are largely irrelevant.

(3) The organizations under study perform well *as* organizations, just as the operators perform well as individuals. Moreover, this is not attributable to externally imposed rules and regulations, which only structure and define authority and tasks. It arises largely from individual and collective acceptance of direct responsibility for safe and efficient operation, which has become internalized in the form of belief structures, values, and goals. In all cases, operators and managers alike are proud of their ability to sustain the highest possible level of operation for extended periods without compromising safety and reliability.

(4) The social and political consequences of failure, particularly in the case of air traffic control, are very high, and the inherent risk of the system without management would be great even at much lower levels of activity. Yet, they receive very little public credit for present performance levels. The present very low error rate has become the norm of social expectation; no measurable increase in accidents or other failures is likely to be accepted, even if operational levels are markedly increased.

All three organizations are now working to the limit of their capacity within the constraints of their present equipment. Yet, each is faced with demands to increase

Gene I. Rochline is with the Institute of Governmental Studies, University of California, Berkeley, CA.

The author wishes to thank the men and women of the Oakland En-Route Air Traffic Control Center, Pacific Gas and Electric Company, and the U.S.S. Carl Vinson (CVN-70), who have been so forthcoming with their information and assistance. Special thanks also to Professors Tod R. La Porte and Karlene H. Roberts of the University of California, Berkeley, for their support, and for the use of their data. A longer version of this paper was presented at the Third National Conference on Engineering Ethics, Los Angeles, CA, May 1985.

Reprinted from *IEEE Technology and Society Magazine,* Vol. 5, No. 3, pp. 3–9, September 1986.

its operational capacity with no major increase in operational resources. In some cases (particularly air traffic control), the pool of potentially qualified operators may also be so limited that no appreciable increase in operator numbers could be achieved under any conditions. In all three cases, the managers and administrators of the system see a transition to "expert systems" or other similar advanced data processing and computational technologies as the only solution capable of buying them enough capacity to provide some breathing space for expansion before demand once again begins to put strains upon them, yet satisfies all three social boundary conditions. Engineers are therefore being asked not just to improve the present system, but to substantively re-create it.

We assume throughout that engineers and engineering firms involved will behave in accordance with the highest standard of traditonal professional ethics, proceeding carefully and thoughtfully in full realization of the importance of the organizations they are working for, and with a high sense of their personal as well as professional responsibility in assuring that the new systems are well designed, thoroughly tested, and reliable [2]. Nevertheless, there is some considerable concern over potential adverse effects as the new technology is introduced. However responsive the engineering may be to instrumental needs, system operators are quite aware that their present high level of performance stems from organizational and social factors that are seldom studied and not well understood.

Any new technology that directly affects operator roles, task definition, the allocation of responsibility, or their relationships to each other, to their managers, and to those whose activities are being regulated, will certainly have indirect and possibly time-delayed effects. The fear is that the nature and importance of those organizational and social parameters central to maintaining high performance levels may not be discovered until performance suffers and errors, failures, and accidents increase.

No responsible engineer would deliberately place the public at risk by introducing a new technology to manage such hazardous and critical tasks as part of an experiment, based on physical or technical principles that were at best only partially understood. Yet, the introduction of radically new technologies into these organizations without understanding the social and organizational principles responsible for their present high performance and reliability constitutes just such an experiment, on a broad and general scale.

This creates a set of ethical and professional dilemmas for the associated engineering community, the operators themselves, and their managers and administrators, that are not well addressed by the traditional canons of their professions. Just what is the obligation of each party to try to comprehend and take into account social and organizational factors that lie well outside the boundaries of their professional training? And how shall they allocate the various levels of responsibility for the processes and outcomes involved?

TECHNOLOGY AND FUNCTION

It is commonplace to note the proper apportionment of capabilities in designing the human-computer interface. What humans do best is to integrate disparate data and construct and recognize general patterns. What they do least well is to keep track of large amounts of individual data, particularly when the data are similar in nature and difficult to differentiate. Computers are unsurpassed at responding quickly and predictably under highly structured and constrained rules. But because humans can draw on a vast body of other experience and seemingly unrelated data (analogous or "common-sense" reasoning), they are irreplaceable for making decisions in unstructured and uncertain environments.

All three of the organizations under study now partition work to exploit the relative advantage of the human operators. Computers and other technical aids are used for data collection and processing, primarily to present the data to operators in a way that allows them to integrate it rapidly. It is difficult, even after long and close observation, to determine just how it is that individual operators can anticipate potential difficulties in a fast moving and complex system. When questioned, all "explain" that they construct a multi-dimensional mental model of the activities around them, using their technological aids as heuristic and information devices. What does not get explained is the degree to which each participant in operations is sensitive to the presence and capability of other operators on duty at the moment, the state and condition of their equipment, the conditions under which they, and those whose activities they regulate, have been operating in the immediate past, experience with other generically similar situations, and so on. The process is not just the sum of individual activities, but an integral, organizational whole [3].

The introduction of new technology into such a milieu is unlikely to follow the traditional path of incremental learning from trial through marginal error, which in principle could minimize general exposure to new hazards. Organizational effects of the new technologies are not likely to manifest themselves immediately. It will take a period of time for the operators and the organization to internalize the new demands and new capacities [4]. This has profound implications for the technical consultants, designers, and installers of the new systems.

At present, all of these organizations depend on what is (at best) mid-1970s computational ability. Air traffic control, for example, still uses monochrome visual displays backed up by limited data processing capabilities. There is no continuous display of altitude nor means for projecting potential flight envelopes (other than the current flight plan). There is, therefore, considerable room for improvement simply by moving to the forefront of computer and display technologies as of 1985. Even so, operators would not consider the transition to be straightforward. A great deal of their ability to form mental images quickly derives from familiarity with their present equipment, and it would take some time to build that experience if they had to deal with new equipment while at the same time maintaining their skills with the old. Each system would use a different approach, different levels of technical sophistication, and far different methods of presenting data and responding to operator inquiries.

Given the room for improvement in present technical capabilities and the difficulty of smoothly implementing any extensive technical change, there is at this time no demand from experienced operators in any of these organizations to move to such radically different or advanced technologies of uncertain specifications and requirements as "experet systems." Pressures for change are external to the operational end of the business, arising from "top management" and its desire to respond to the demands placed upon it to increase the handling capacity of the system (but without decreasing reliability). Even if management were allowed to double or triple the number of operators, they would find it difficult to obtain a large enough pool of candidates. And because management anticipates that any incremental new capacity will almost immediately be absorbed by increasing demands upon the system, it seeks qualitative improvements that will provide enough excess to remove the pressures for some years to come [5]. Unsurprisingly, the engineering and research community sees this as both a challenge and an opportunity to demonstrate the worth of their most exciting cutting-edge technologies.

EXPERT SYSTEMS

Since none of these organizations can increase the number of *operators* rapidly or by much, they seek to greatly increase the number of *operations* per operator. To different degrees, each seeks to leap beyond quantitative improvement in data collection or management to qualitatively more advanced computer technologies, particularly "expert systems" [6].

Whatever that term may be taken to mean in the abstract, it is operationally defined in terms of relieving the operators of some subset of "routine" decision-making. These new technologies will not be "intelligent" in the sense of being able to make new decisions, or "expert" in the sense of being able to re-order their own data collecting or decision rules [7]. The stated objective is to lay the basis for coping with escalating future demands by relieving operators of the stress and workload of demanding, yet routine and repetitive, tasks leaving them free to deal with the most critical and complex decisions.

Taking aside the formal question of what each new technical system is called and how it will be described, the technical objective proposed for each of the organizations under study is to "assist" on-line decision-making within a highly structured and pre-specified domain and range of applicability, with an accuracy and reliability that is promoted as mimicking the best (most expert) of humans performing the same tasks, yet at far greater speed. Whether such systems will be input to the operator, supplementary to operator judgment, or primary with operator over-ride remains uncertain. Indeed, it is possible that each of the actors involved has a quite different belief about the long-term objectives of the new program. *But in no case will the role of the operator's judgment be left untouched,* for the structure and presentation of complex information predetermines the range of probable responses for even the most expert and flexible of operators.

SOME REALITIES OF IMPLEMENTATION

Few new technologies work perfectly in their first major trials, least of all conceptually new ones. Any new system will have flaws. Some will be obvious immediately; others will be discovered only after extensive use in a variety of situations. Perfection is probably beyond reach. "Bugs" and glitches in new technologies are usually worked out in practice through a process of accommodation and adaptation that continues until a level of performance and reliability considered to be acceptable is reached [8].

The process by which the new technology is altered, modified, or "fixed" to overcome conceptual flaws, system shortcomings, and technical imperfections is fairly well understood, though not so systematically studied. What is less generally understood is the way the organization also changes in response to both the advantages and the difficulties of the new technology. For most organizations, this is an important part of the process of technical change. For the organizations under study, this process is more than usually difficult, and potentially quite risky.

Although the bounding rules are externally fixed, the situations and circumstances facing operators are fluid and dynamic, and the new technical systems are supposed to aid the operators in assuring safe and efficient operations within the bounding rules, not to police them. But although new systems are almost always advertised as increasing flexibility as well as capacity, this seems rarely true in practice. The more complex and intricate a technical system, the greater the demand it places on operators and technicians to conform inputs to its internal logic. Thus, the incorporation of a new and more advanced technical system ends up imposing new sets of rules (albeit often subtle and indirect ones) that in themselves can be the source of social and environmental change for the operators.

These are rarely studied and not well understood, so that the outcome for the structure and performance of the organization *as a whole* under the influence of major technological change is at best difficult to predict [9]. Given the high performance criteria and pre-existing high stresses on operators that characterize these organizations, such unspecified and uncertain change in internal structure and values might very well undermine even the most determined technical efforts to ensure that increased operational levels do not entail decreased performance or increased risk.

The impact on the overall system of the inevitable limitations and learning processes of any major technical transition must also be taken into account. New systems can and do break down, or present unanticipated difficulties in operation. In the case of non-critical systems (such as a new accounting package), this is no more than a source of aggravation and delay. In more critical applications, the old system is generally kept on line as a backup until the new can be debugged. But in the case of time-urgent critical systems such as air traffic control or electricity grid management this may not be so easily done. At present, operators are able to practice through simulated breakdowns operating the system with only minimal technical assistance. Given the increased loads projected under new technologies, the backup would have to be the "old" system rather than manual mode. Thus, operators would, for a while, have to maintain their skills at managing traffic under three quite different technical circumstances. Given the mental energy that goes into constructing an operational image to make the present system workable, it is not certain how well that could be done, or how smoothly that transition would take place if breakdowns occurred in peak traffic times.

Any technical system, however advanced, also has certain limitations built in. Expert systems, for example, are programmed with a certain repertoire of responses to a pre-specified range of stimuli. Given an unexpected stimulus, or one outside the range of its programming, they would, in the best case, simply respond to that effect. But if the machine does not recognize that its inputs are out of range, its responses may be difficult to predict, or at odds with what human judgment would prescribe [10]. And for organizations such as these, the prospect of a computer system issuing a set of complex instructions to a fast-moving and interactive system that it cannot properly interpret is, quite frankly, frightening. The system may be "reliable" in terms of its ability to respond upon demand, but how much can the operators "rely" on the appropriateness of that response in other than the most routine situations?

"Intelligent" computer-based technologies such as expert systems are based on an individual and cybernetic paradigm. Their applicability is probably greatest in situations such as medical diagnosis, where decisions must be made by individuals or small groups of people weighing the evidence within specific well-bounded circumstances with structured rules. But the salient decision processes in our cases tend to be collective, judgmental, and heuristic, particularly at times of acute stress. Grid managers, for example, may at any given moment be aware not only of what equipment is on the system and what off, and where extra power might be purchased, but also what that equipment has been doing "lately," who will answer the phone if they call to negotiate additional purchases, and what their current power situation is likely to be.

The explicit purpose of the new technologies is to increase the number and rate of activities, which already strains human capability to manage, to markedly higher levels. The new system will also be designed to fail less frequently than the old. This has two major implications for operators. When the technical system does fail (and most operators are sure that it will from time to time, however carefully designed), both the rate and the complexity of the information presented and the difficulty of managing it in order to reassert control over the system will be considerably greater than at present [11]. Most operators admit that under such circumstances their first strategy will be to shut down as many operations as possible to reduce the overload to tolerable levels. Yet, if the new technology does work as promised to that point, they will have much less practice with coping manually. Thus, technical breakdowns may have far more serious consequences than at present, even if the probability of such breakdowns is greatly reduced.

SOME ETHICAL PROBLEMS

All of these organizations are seeking technical solutions to organizational problems, without any systematic understanding of the character and behavior of the organizations under study either as an organization as a whole or in terms of the collective behavior of the operational subunits. In every case, their primary goal is to increase capacity and reliability, while causing no undue physical or psychic harm either to organizational personnel or the larger public. Yet, in the absence of any satisfactory model or analysis that explains the present low error rate of any of these systems, none of those involved can say with any certainty that present levels of safety can be maintained.

For example, our preliminary inquiry into the Air Traffic Control system has led us to hypothesize that the high personal stress placed on operators may in fact be directly connected to their high performance, particularly in emergencies. The direction of causality (i.e., does the job cause the stress, or do people who like the pressure seek the job) remains an open question, but the linkage seems clear [12]. Thus, any technology designed to relieve that stress, however humane and moral it may seem, must be evaluated carefully in terms of its potential impact on overall system safety, and on the long-term ability of these organizations to continue to attract those persons who are emotionally capable of managing complex activities calmly while under tremendous intermittent stress.

Although all concerned do in some way recognize the importance of such social and organizational parameters, long-term effects on overall safety and reliability are being addressed only in technical terms. Some of the more important ethical considerations are thereby being overlooked by default. Are those who are formally responsible for providing overall system performance and safety adequately represented in the decision process? Are those whose actions will result in organizational changes aware of the possible consequences of their designs and actions? Do administrator s assume they will still be around when unanticipated consequences ensue, or are they primarily interested in visible short-term rewards? To what extent are the hardware and software designers also obligated to try to inform themselves of the importance of nontechnical factors and their potential impacts upon operators and the public? Do these circumstances impose a special burden to ensure that their systems live up to promises and expectations?

Those who promote the new technologies—top management and the designers and engineers of the new hardware and software systems—are not unaware that they are about to profoundly change the operating environment and social charactristics of organizations whose social and psychological parameters are not well understood, but they largely fail to perceive the extent to which current levels of performance derive as much from sociopsychological factors as from technical ones [13]. Given the tendency of engineers to seek optimal and efficient technical solutions, there is a pronounced tendency to view the technical factors as the dominant ones, with the social and organizational "system" in which the technology is put to use being treated as configurable to meet the requirements for putting the technology to use as its designers intend.

The confidence of organizational administrators and the designers and engineers of the new technical systems that these technologies can be implemented so as to increase (or at least maintain) system reliability and safety while increasing capacity is not matched at the operational level. Both operators and their managers have indicated to us their reservations about turning critical decisions that once rested on organizational memory and human judgment over to automated systems. Although they feel that they do understand how their organization works, and how they maintain their low error rate, they also understand that a great deal of that knowledge is experiential. The question of which aspects of what they now do are crucial to reliability and which are largely irrelevant has never been subject to controlled test. Although seldom expressed openly, their uncertainty about the validity of what they think they know manifests itself as a general concern about performance both in the interim—that is, during the transition and training period—and in the long run.

Operators are also more sensitive to those more subtle and less quantifiable aspects of organizational behavior that may appear irrelevant to anyone outside the actual operating system, but may be essential to maintaining operator confidence and morale. For example, the acceptance by the operators of the ultimate responsibility for system reliability (and, in the case of air traffic control, for the lives of the traveling public) may well be dependent upon just those types of "hands-on" operations that put heavy burdens on the operators. If these are indeed linked phenomena, no one component can be altered without shifting or altering one or many other components.

In private discussions, all of the parties involved admit to some unease about the long-term effects of the changes they are putting in place. System designers report that they are turned to for technological tools in areas that cannot be handled safely by computers alone; operators and managers claim that human control will not be able to handle emergencies without shutting the whole system down (if at all); and administrators conclude that the demand stimulated by new capacity will not live up to expectations. All, however, seem reluctant to make a public issue of their concerns. Each has strong norms within his or her professional and organizational environment that make such action suspect as attacking professional or organizational competence.

When the requirements of large, complex technical systems such as are now being planned for these organizations conflict with existing behavioral patterns, it is almost certain that it is the human operators who will be asked to change and adapt to new rules and procedures designed to ensure safe and reliable system operation. Technical specifications tend to be taken as exogenous and immutable; managers and designers almost always place the burden of change upon the presumably more flexible and adaptable human beings [14]. When objections are raised, the very language we have for describing it—"reaction," "resistance," "tradition," *vs* "innovation," and so on, is loaded with a negative and critical effect. This not only obscures and confounds real expressions of concern, it greatly inhibits those professionals who do perceive the more subtle dimensions and implications from speaking out.

Moreover, the pressure to adapt to the demands of the technology extends not only to operators, but to those being served. What is most remarkable in the case of computer technologies to date is the extent to which human beings have been willing to accede to requests for behavioral change. The glimmering perception that value choices are in fact being imposed are all too easily subordinated to the presumed importance of increases in efficiency or performance [15].

In the case of the organizations being studied here, it will be further argued that value change is "necessary" if risk is to be reduced and/or efficiency increased [16]. If the system as modified does not perform up to expectations, it is almost certain that the humans will be asked to change their behavior pattern to make better use of the technology rather than vice-versa. To date, this subordination of values to instrumentalities has almost always been accepted, even when it has been perceived as such [17].

This also raises some rarely analyzed ethical implications for the technical consultants, designers, and installers of the new systems. In effect, both the operators and the public they protect will ultimately be required to adopt and comply with whatever behavior rules turn out to be necessary for safe and efficient system operation. But, until the new technologies are deployed at something like full operational scale, there is no way to anticipate just what those rules might be.

Those who promote these systems usually believe that their main purpose is to add flexibility and increase the availability and capacity of humans to perform critical tasks. But this assumption again ignores the reality of the social context in which technologies are deployed.

Technical systems that are most open and adaptable at the beginning rarely remain so in the long run. Other criteria, such as the drive for greater efficiency, the need for greater standardization, or the inability or unwillingness of managers to handle the uncertainty and ambiguity of changes initiated solely by operators, also come into play [18]. Sooner or later, almost every system is eventually "closed" to a more narrow set of specifications or operating criteria, imposing new and stringent behavioral requirements.

Thus, a second set of difficult ethical and professional questions for those designing and implementing new technical systems arises from the ways in which they ultimately manipulate public behavior and values. To what extent do ethical considerations dictate an obligation on the part of the developer or producer to apprise the public of this particular class of impacts of a new technology? And if so, by what mechanism can such impacts be analyzed and explained?

CONCLUSION

For different reasons and to varying extents, all of those involved in the coming technical transitions share a common conflict between professional ethics, organizational loyalty, and their wider responsibility to inform the public of potential consequences. The problem is not whether it is proper to trade safety for efficiency, or to change values for a perceived advantage. Such social choices are frequent. But they should be made by open and informed political processes that directly involve all those who will be affected [19]. To deprive the public (and their regulatory surrogates) of crucial information is to deprive them of the basis for making a properly moral choice between system efficiency and public safety.

The dilemma is even more acute in those cases where the public has effectively been denying both the level of risk and the nature of the choices they have made. Yet, such is clearly the case for all three of the systems under study here. Public expectation that present levels of performance and safety are the norm, and that any failure or breakdown represents not a statistically inevitable phenomenon but a culpable act, provides no basis for a discussion of possible trade-offs. As a result, those who do see their responsibility as encompassing informing the public are put in the uncomfortable position of raising alarms where none existed before, and therefore of calling their profession, and their colleagues, into question.

We do not think that managers and engineers mean to deprive the public of information, although there is certainly some evidence that scientific and technical elites tend to believe that the public is not well enough informed to judge their work. Rather, they have no means to acquire a better understanding of the broad social context of engineering, the extent to which every new technology that comes into wide use is an on-line social experiment whose consequences are nearly irreversible if not carefully monitored at an early stage of deployment.

It has always been difficult to incorporate indirect and long-term social consequences into technical specifications and systems, particularly in cases where the outcome would entail settling for less than the "best" or most "optimal" technology available. It will not be simple to extend the usual definitions of professional role and responsibility to the broader class of organizational "outcomes" discussed here. The kind of knowledge required cannot be quickly or trivially imparted as a kind of social science whitewash to a finished engineer. Skills would have to be built up through extensive course work or training, and through major interventions in the curricula of scientific and engineering schools.

There are also bound to be those who argue that such intervention would be unwise, that constant worry over the broad, long-term social and political implications of technical change would tend to stifle creativity and intervention. In this view, what is needed is stronger regulation or more intelligent management, so that the unrestrained innovative potential can be tapped and directed without interfering with its creation. The ethical and moral burden would then rest not on the working engineers, but on management, industry, or government. This view seems to us not only naive, but fundamentally misdirected, since it treats engineering as if it were an instrumental rather than a social profession.

Technology is an instrument of social change, and many engineers will, at some point in their careers, arrive at a point where their designs, specifications, or actions directly affect large segments of the public. Traditional ethical discussions have tended to focus on two aspects: short-term effects on individuals and small groups, which traditional ethical and professional canons are meant to address (particularly with regard to direct responsibility for safe operation); and long-term collective societal effects, which are generally regarded as matters for personal rather than professional ethics [20]. But neither of these speaks to the problem of mid-term and collective impacts, which are too precise and direct to be left to individual morality, yet too diffuse and complex to be dealt with within the present framework for determining professional standards and responsibilities.

There are no simple answers to such difficult and troubling questions. Only recently have scientific and engineering activities been perceived to have the kind of broad and collective social implications that require them to be asked. But now that they have been asked, it is most important that they be most generally and broadly debated, within the profession and outside of it. Even if this debate has no "resolution" in the formal sense, the process itself will contribute to the increased awareness of and sensitivity to indirect and subtle social and political effects of new technologies, and to the professional and personal responsibility of engineers and administrators to take them into account.

REFERENCES

[1] One of the few exceptions is the study by Albert Flores on Monsanto and NASA, two organizations that have admirable safety

records. See *Designing for Safety: Engineering Ethics in Organizational Contexts,* ed. A. Flores (Troy, NY, 1982).

[2] This assumption is most strongly supported by our interviews to date. This sharpens the ethical dilemma, since the problem of the moral man faced with the implications of actions taken in good faith are always more complex than those of the immoral one.

[3] One recent study on air traffic control concluded that: "holistic knowing, which comes from an intimate involvement in every detail of the traffic control process, may be necessary to sustain complex control behavior." (Robert Wesson, Kenneth Salomon, Robert Steeb, Perry Thorndyke, and Kenneth Westcourt, *Scenario for the Evolution of Air Traffic Control* (Santa Monica: Rand Corporation, 1981).

[4] J. P. Kotten and L. A. Schlesinger, "Choosing Strategies for Change," *Harvard Business Review,* March-April 1979; R. Beckhard and R. Harris, *Organizational Transitions: Managing Complex Change* (Reading, MA: Addison-Wesley, 1977). For an application relevant to this study, see: Todd R. La Porte, "Nuclear Wastes: Increasing Scale and Sociopolitical Impacts," *Science,* 191 July 7, 1978), pp. 22-29.

[5] In organizational terms, we refer to unused operational capacity as "slack". Slack is a scarce resource that management and operators would like to conserve against future demands, to ease stress, or to provide confidence that there is some internal reserve for emergencies. Administrators and customers, however, tend to regard it as an available resource to be consumed. R. M. Cyert and J. G. March *The Behavioral Theory of the Firm* (Englewood Cliffs, NJ: Prentice-Hall, 1963).

[6] The term "expert" is applied because both the data base and the decision rules are constructed to allow relatively inexpert and untrained operators to replicate the behavior, skills, and reliability of the most respected experts in a given field of research or practice. The best-known examples are DENDRAL, a system for identifying complex chemical structures from a variety of analytic data, and medical systems such as MYCIN and CADUCEUS for diagnosing certain bacterial infections from a structured set of input data on the patient.

[7] However advanced, all computer systems to date treat the world as constructed of pre-programmed elements. However good or quick they are at recognizing complex and changing configurations, they are very ill-equipped to recognize or deal with the introduction of new elements. See, for example, G. R. Martins, "The Overselling of Expert Systems," *Datamation,* Sept. 15, 1984, pp. 44-53.

[8] There is a considerable body of organization literature dealing with the search for "adequacy" on the premise that real optimization is either impossible or far too costly to pursue. See, for example, J. D. Thompson, *Organizations in Action* (New York: Mc-Graw-Hill, 1967); Karl E. Weick, "Organizations as Self-Designing Systems", *Organizational Dynamics,* Autumn 1977, 31-46; Herbert A. Simon, *Administrative Behavior,* 3rd ed., (New York: The Free Press, 1976); J. G. March and H. A. Simon, *Organizations* (New York: Wiley, 1958). Simon's coined term, "satisficing", is probably the most useful and concise description.

[9] Cyert and March, *Behavioral Theory of the Firm,* Thompson, *Organizations in Action,* op. cit. There is an extensive body of theory literature on stimulus and change in organizations, but little of it treats technological change as an independent variable. See also: Abbe Mowshowitz, "The Bias of Computer Technology", in *Ethics and the Management of Computer Technology,* ed. W. Michael Hoffman and Jennifer Mills Moore (Cambridge: Delgeschlager, Gunn & Hain, 1982), pp. 28-40, for a brief but comprehensive discussion of the distributions of costs and benefits across different levels of the organization and in different time frames.

[10] See, for example, Hubert L. Dreyfus, *What Computers Can't Do* (New York: Harper & Row, 1972), especially pp. 184ff. Joseph Weizenbaum *(Computer Power and Human Reason,* San Francisco: W. H. Freeman and Co., 1976, esp. pp. 202-257) has argued further that even if the machines were capable of independent "reasoning", it would be based on an internal logic so foreign to our own experience as to constitute a truly alien intelligence. If so, its judgments are likely to diverge wildly from what humans would do in a similar situation.

[11] C. A. O'Reilly, "Individuals and Information Overload in Organizations: Is More Necessarily Better?," *Academy of Management Journal,* 23 (1980), 684-696.

[12] See, for example, D. Munat & R. S. Lazarus, eds., *Stress and Coping: An Anthology* (New York: Columbia University Press, 1977).

[13] K. H. Roberts, C. L. Hulin, and D. M. Rousseau, *Toward an Interdisciplinary Science of Organizations* (San Francisco: Jossey-Boss, 1978).

[14] Doreen Steg and R. Schulman, "Adapting Behavior," in M. Kranzberg, ed., *Ethics in an Age of Pervasive Technology* (Boulder, CO.: Westview Press, 1980), 183-186.

[15] Langdon Winner, "Do Artifacts Have Politics?" *Daedalus,* 109, No. 1 (Winter 1980), 121-136.

[16] Consider, for example, the profound value change entailed in a free society by the necessity to submit to X-rays, luggage searches, etc., in order to assure safety in air travel.

[17] In onboard observations of the U.S.S. Carl Vinson at sea, for example, we found the emotional depth of personnel sacrificed to structurally and technologically determined characteristics invoked to "get the job done."

[18] A remarkable example from our present study concerns weapons computers on Navy F-14 fighters. These were originally designed to be individually programmable for adaptability and flexibility. But problems of maintenance and standardization became so severe that local programmability was deactivated and the computers centrally standardized.

[19] It has been a principle of ethics since Aristotle that an action is a candidate for moral choice only if it is voluntary and based on all available information. See, for example, *Nichomachean Ethics,* W. D. Ross, transl. (Oxford: Clarendon Press, 1925), Book III, vol. 9.

[20] The most notable example of the latter being the morality of doing military research.

FEATURE ARTICLE

The Bureaucratic Ethos and Dissent

ROBERT JACKALL

I. One of the curious anomalies of American society is our ambivalence toward occupational dissent and dissenters. Congressman James Weaver of Oregon tells a remarkable story about his visit to Three Mile Island right after the 1979 accident. He asked one of the key management spokesmen for the utility company, whose job it was to assure the media and the public that everything was fine, which character the man liked in the film *The China Syndrome*. Without hesitation, the manager replied: "I like Jack Lemmon, the manager who rebelled." The Congressman then asked the manager what he thought of the public relations man for the company in the movie. The manager responded: "I didn't think he was worth a damn." After a pause, he added: "But you know, a week later I was doing the same damn thing." [1] Such inconsistency, while sharply drawn, is not exceptional. Americans are deeply stirred by dramas of resistance and rebellion. Our popular culture, in fact, is filled with images of men and women in different walks of life who buck authority and stand up to the system It is not, of course, just rebellion itself that appeals to us, although as a people with a revolutionary past and residual anarchistic impulses, we have always harbored sympathies for those who swim against the tide, even if this takes the form of just beating the system instead of opposing it. But rebellion for some cause or principle defined as noble particularly compels our admiration. Serpico exposing police department corruption, Woodward and Bernstein fighting the White House to establish the truth, Norma Rae risking her personal safety to help organize a union, and many other figures, both fictional and real, all join, for a time, our pantheon of heroes.

For most men and women, however, such rebellion remains the stuff of distant drama, providing an uplifting but compartmentalized emotional catharsis. Like the management spokesman at Three Mile Island, most people continue to perform their work even when they have misgivings about their assigned roles. The drama of everyday work generally subordinates any urges to emulate dramatized public rebellion. Clearly, a significant reason for this is the as yet incomplete extension of civil liberties into the workplace [2]. The price of public occupational dissent can be, and often is, high. More important, however, in my view, are people's internalized barriers to open dissent that emerge from their dense and intimate knowledge of their everyday occupational milieu, where, among other things, their life chances are cast.

In this essay, I want to examine the issue of dissent in one occupational group in one kind of institutional setting—specifically among managers in large bureaucratically-structured corporations. As it happens, corporate managers, while some privately admire those who do rebel, rarely dissent publicly against their corporations' policies or practices, even when they are troubled by their work roles. The specifically bureaucratic structure of managerial work generates an ethos that, when internalized, makes silence a taken-for-granted occupational norm. Managers' work experiences are, I think, a key to understanding why most people keep silent in the face of occupational abuses, even while they admire rebellion.

Bureaucracy is the dominant institutional form of our mass industrial society, and most people today are increasingly engaged in or affected by bureaucratic work. Virtually every occupation in our society—from physician to assembly-line worker—is specialized, certified, arranged in a hierarchy, and bureaucratized in other ways. Managers' experiences provide a window into the ethos shaped by bureaucratic work because they are the quintessential bureaucratic work group. They not only fashion bureaucratic rules for others, but they are also bound by them; further, since their adminstrative expertise constitutes their livelihood, they are typically not only *in* but *of* the organization. Their work experiences at the nub of bureaucratic structure resonate throughout our occupational and social order.

What is the nature of the bureaucratic ethos, and what kind of code does it produce in managers about dissent? To answer these questions, I shall draw on materials collected in the last few years through fieldwork with managers in large corporations and with occupational dissenters in a variety of contexts (see research note for fieldwork details). I shall concentrate on a few cases in detail both to provide an ethnographic record of the cases and to illustrate the complexity and ambiguity of the problem of dissent in bureaucratic settings.

II. We can grasp key aspects of the nature of managerial work and the ethos it generates by looking at both through the dilemma of a young manager near the bottom of his organization's staff hierarchy. In 1981, Brown (a pseudonym) was a health professional with some supervisory responsibilities in the corporate headquarters of a large textile manufacturing firm. Brown, who had some training in audiology, had principal responsibility for the

Robert Jackall is Associate Professor of Sociology and Chairman, Department of Anthropology and Sociology, Williams College, Williamstown, MA 01267. He is currently completing a book on business, bureaucracy, and morality to be published by Oxford University Press.

Presented in part at the 1984 Annual Meeting of the American Association for the Advancement of Science.

Reprinted from *IEEE Technology and Society Magazine*, Vol. 4, No. 2, pp. 21–30, June 1985.

company's hearing conservation program for production workers.

Noise is inescapable in the textile industry because its basic technology, in particular the loom and spinning equipment, cannot operate without making a certain amount of noise. Old shuttle looms, many of which have been in continuous operation since World War I, are extremely noisy, usually creating a decibel level of 95-105 dB(A) (decibels measured on the "A" scale of a sound level meter, rather than on the older "C" scale [3]) in a weave room of 50 looms. It is important to note that the intensity of sound doubles with every three dB increase. Even with new air-jet shuttleless looms, to which the textile industry is slowly moving because of their greater efficiency and improved cloth uniformity, decibel readings in weave rooms still reach between 93-94 dB(A). Spinning rooms with old equipment register 96 dB(A); with the newer equipment, the readings are 89 dB(A).

In 1981, the Occupational Safety and Health Administration (OSHA) promulgated an amendment to its occupational noise standard specifying that workers whose noise exposure exceeded an 8-hour time weighted average of 85 dB must be included in a hearing conservation program. Along with other industries, the textile manufacturers opposed the amendment; the industry was, in fact, unhappy with the existing standard, which set the maximum 8-hour time weighted average at 90 dB.

Brown disagreed with his industry and strongly supported the OSHA hearing amendment that would lower the critical threshold 5 dB to 85 dB.

Obviously, unimpaired hearing in the voice band pass (300 to 3400 Hz) is of special concern. Without that frequency range, conversational ability becomes severly impaired. Prolonged exposure to high levels of noise, such as occurs in industrial work in textiles, erodes people's hearing at high frequencies first—at between 3500 and 6000 Hz. A person then begins to lose the ability to perceive higher frequency speech sounds such as those associated with the ficative consonants; continued exposure spreads the damage to lower frequency ranges. People come to have problems interpreting sounds and in gauging the intensity and loudness of sounds. Noise-induced hearing loss occurs gradually, and thus often is not consciously perceived until damage is irreversible and becomes evident late in life. With heavy impairment, older people sink into isolation and incommunicability in their retirement.

Within this general context, Brown's work conducting and analyzing the audiograms required by OSHA's existing standard on noise presented him with a serious dilemma. In analyzing incidence data from a representative plant, and by extrapolating to the rest of the firm's mills, Brown discovered that 12 percent of all greige mill (colloquially called "grey mills" in the industry where plain, unfinished, off-white cloth is produced) workers had already suffered hearing loss severe enough to qualify for monetary compensation under state law for as much as a total of $3.5 to $5.7 million. This, however, was only the thunder before a summer storm. The incidence data clearly indicated that another 63 percent of grey mill workers had already suffered substantial, though not yet compensable, damage that could only worsen the longer they stayed in the industry. In brief, the best estimate was that three-quarters of all the grey mill workers in Brown's firm (more than 10,000 employees) had already suffered significant hearing loss. As it happens, the hearing compensation law in the state allows workers to make claims for hearing loss only up to two years after leaving employment. The textile industry in the state was even fighting to have this law reinterpreted to allow claims only while workers were still employed. This point is particularly crucial, since, as I have already mentioned, severe hearing loss that might impair social functioning becomes evident principally in workers' later years.

Brown had known, of course, that hearing loss was endemic to the textile industry, but he was very disturbed by the extent of the damage indicated by his data. He saw only one solution to the problem, namely to make the hearing conservation program already mandated by OSHA a vital force in the grey mills. This meant strictly enforcing workers' use of hearing protection—that is, earplugs—while motivating both supervisors and workers alike with extensive and regular educational programs on the dangers of hearing loss.

I mentioned earlier that there are very few ways to cut down the noise in textile work. Even the newest equipment in weaving and spinning keeps noise at 90 dB or above, the level at which prolonged exposure will lead to permanent damage. The use of physical space in what is essentially a machine-tending industry, where each worker in, for instance, weaving or spinning tends a large number of machines, obviates any practical and reasonably affordable engineering remedies. Even if the large spaces presently used could be broken up into smaller units, a step that would completely alter the social organization of work, this would baffle noise only in the far field of sound, not in the near field where workers tend individual machines. Moreover, in the weave rooms where the noise problem is greatest, the extensive and difficult training of weavers (in Brown's firm, this lasts thirteen weeks with only a 32 percent success rate) makes job rotation impracticable. Brown wrote a report in which he detailed his analysis of the incidence data and proposed the extensive and regular educational programs he felt were necessary. For him, this was a clear moral issue, and he felt that he had to act as the conscience of the company.

Brown's report and his suggestions did not fare well. He was enmeshed in a hierarchical and political situation typical of all the large corporations I have studied. Brown reported directly to the senior vice-president for human resources, who reported to an executive vice-president, who reported to the chief executive officer (CEO) of the company. Reporting relationships are crucial in the corporate world. The person to whom one reports—that is, one's boss—in great measure controls one's fate; despite its formal impersonality, authority in corporations is actually quite personally administered and experienced.

Bosses depend on subordinates not only to get work done but to protect them from egregious blunders. In turn, a boss is expected to extend a measure of protection to his subordinates, guarding them against other higher authority figures where possible and elevating a favored few when and if he himself is promoted. Over a period of time, patterns of fealty—of reciprocal though asymmetrical obligations—develop between bosses and subordinates. And because of the interlocking reporting system, these patterns of fealty link managers up and down the organizational ladder. Patronage relationships overlay and intertwine with boss-subordinate fealty patterns. Patrons are usually powerful managers in the higher echelons of a bureaucracy. When one has a patron higher up in an organization, located either vertically or diagonally from one's own position, one takes on additional though more ambiguous fealty obligations [4].

Brown stood at the margin of the intricate social structure of his organization. He was, first of all, unlucky in his boss. The medical director was an older man who had completed a long career in the military and then had begun a second life with the textile company, a relatively common pattern for company doctors in southern textile firms. The director saw himself as a guardian of the textile industry's interests against increasingly vociferous and hostile critics. For him, the hearing issue, like the cotton dust problem, was simple. If workers did not smoke cigarettes, he argued, they would not have pulmonary disorders such as the alleged disease byssinosis; if they wore earplugs, they would not go deaf. The director took Brown's report and rewrote it extensively to emphasize the financial liability of the company, while only mentioning the desirability of more educational programs. Brown was caught in a fealty relationship with all of its obligations—especially those of protocol—but with little likelihood of support for his work or of future rewards. Nor had Brown been adopted by a powerful patron higher in the organization, a Godfather who could fight for him in forums to which Brown himself had no access. In Brown's case, the logical patron would have been his boss's boss, the senior vice-president for Human Resources, who was known to be favored by the new CEO and destined for higher things. However, it was almost certainly the senior vice-president who buried even the recast report on the hearing issue, and with it any opportunity that Brown might have had for a patron-client relationship with him.

Brown never heard about the report from his superiors again, and it took him a while to understand that he was opposed by powerful interests rooted in the occupational group structure of the company. The group of top executives, first of all, had two concerns. The company faced potentially enormous liability costs if those workers who were already severely damaged were fully alerted to their hearing problems. There was, in addition, the even greater potential liability threat of the large reservoir of workers whose hearing was steadily worsening to the compensable level. Second, workers' health benefits could not, on this issue, be linked with any productivity gains. Corporate executives always desire such linkage. This had been achieved, though with a great deal of ideological subterfuge, in addressing the cotton dust issue [5]. Here, however, even the latest technology could not solve the hearing problem, and implementing vigorous educational programs, in addition to raising the liability specter, could only reduce production time. Plant managers, who would have to implement any educational programs, echoed the latter concern, and most that Brown contacted wanted nothing to do with anything but the most perfunctory required annual notification to workers. Group meetings do not produce cloth, the principal criterion by which plant managers rise or fall. Staff in related health fields, sensing top management's reluctance to act in the absence of forceful external compulsion from OSHA, also refused to help; they had their own programs to protect.

In addition, Brown had never gained access to the numerous managerial cliques, networks, or alliances that crisscrossed the formal occupational structure of his organization. Such alliances grow directly out of, though are not bound by, one's present and former patterns of fealty and patronage. In the textile company, as elsewhere, alliances were deep ties of loyalty and understanding forged and tempered by common work, struggles against common enemies, and favors traded over time. Such alliances are not only crucial in gaining protection during times of crisis — an organizational upheaval, for example—but in advancing group and individual interests. These managerial cliques informally make their own criteria for admission which may vary considerably depending on their locaiton in the organization. The universal criteria, however, are one's ability to further a group's interests and one's adeptness at establishing personal rapport and comfort with clique leaders. Comfort means meshing personal styles and achieving the familiarity, predictability, and ease that come from sharing taken-for-granted views about how the world works. Without belonging to one or more of what we may call these managerial circles, one cannot go very far nor be effective in the corporate world.

Brown was not included in such a circle. This is not to say that other managers were unfriendly to him; one of the hallmarks of the textile company was a smiling, courtly geniality extended to everyone. But Brown's weak fealty links with his boss, and the refusal of his boss's boss, the senior vice-president, to extend patronage to him, even though the senior vice-president had a sizable barony of clients in this network, made him suspect. The fact is that, although the company publicly pointed with pride to its employment of a health professional with training in audiology, Brown's moral squint on the hearing issue made others uncomfortable. The only publicly acceptable way to discuss such an issue is in rational/technical, emotionally neutral terms like "liability consequences," the "trade-off between noise reduction and efficiency," or the "linkage of compliance with regulation to productivity improvement." Brown was in a double bind. He could not gain access to the managerial circles that might make his opinions count in large measure because he defined the hearing issue as a moral concern; if, however, he were to frame the issue as a rational/technical problem, the powerful interests rooted in the occupational group structure of the hierarchy that I have already noted seemed likely to oppose him on every count.

Brown's case illustrates the intricate social structure of managerial work. The puzzle that Brown encountered—one he never solved—emerged out of the intertwined and overlapping group structures of the hierarchy. At one level, Brown came up against a series of vested interest groups produced by the bureaucratic segmentation of work—executives focused on avoiding immediate liability, plant managers on making cloth and other professionals on preserving bailwicks. At another level, his lack of strong fealty and patronage ties isolated him in a social context where personal ties with powerful others are crucial; without the protection and borrowed influence such ties afford, one is subjected to the impersonal norms of bureaucratic structure. At still another level, Brown had no allies to champion his ideas.

The code that emerged from this intricate structure was, in addition, exactly counterposed to Brown's insistence on seeing the hearing issue in moral terms. Brown's moral stance threatened others by making claims on them that could impede their ability to read the drift of social situations. The managerial code subordinates any independent moral judgement to the social intricacies of the bureaucratic workplace; it thus brackets and makes irrelevant—indeed, dangerous—any notions of morality that one might hold outside the workplace. What matters instead are crucial relationships to significant others, both individuals and groups, within the workplace. Since these relationships are always in flux, morality is always situational, always completely relative. With no agreed-upon fixed moral points, right and wrong get defined by those individuals or cliques powerful enough to make their judgments stick.

Brown's frustration, isolation, and moral unease finally led him to apply for completely different work in the corporation, an office manager's job. He was, however, passed over for this position and subsequently left the company. The OSHA Hearing Conservation Amendment did become a final rule [6]. The only significant change in the textile company's practice is that now the firm has to give workers, instead of requiring them to buy, replacement earplugs. These cost 25 cents a pair.

III. The structure and code of managerial circles require further explanation than Brown's outsider status and consequent puzzlement can provide. Let us consider the case of a manager deeply imbued with the bureaucratic ethos who is suddenly confronted with a dilemma. Smith, (a pseudonym), an upper-middle level business area manager for a large chemical company, is widely respected for his business skills and, in certain key circles of his company, for his trustworthiness. One day, he was in conference with two of his peers on the managerial ladder but in a different product area, when a newly-hired technical aide, who reported to one of these other managers, dropped by with an inquiry. The aide had been doing some extra reading and noticed that a particular substance used in making one of his division's products was a very potent carcinogen when inhaled in dust form. The aide did not know how the substance was processed at the company and asked casually if any dust were emitted in production. As it happens, Smith's managerial colleagues had been polishing the substance in question in open drums preparatory to other use; the polishing gave off dust that workers inhaled. Moreover, they had employed this process for years, never suspecting that anything was wrong. Smith and the other three men were immediately aware of their guilty knowledge and knew that they had to act swiftly. They ceased operations in the plant and sent the workers home. In the next few days, they had the drums encased so that the polishing of the substance emitted no dust. They also decided not to tell the workers the facts of the case. They simply said that a decision had been reached to alter the production process.

Smith's reasons for not informing the workers are important, and offer some sharply drawn insights into the code of managerial circles. Smith argues:

1) The workers have already been exposed to the chemical. No known medication will remedy whatever damage the exposure has caused, and it will do the workers no good to know about the dangers they may face. In Smith's view, the analogy here is the case of the heavy cigaretter smoker who quit the habit upon hearing the Surgeon General's first report on the health hazards of smoking. Nothing the smoker does subsequently can alter the damage done while smoking.

2) If, however, some of the workers involved do develop cancer and, say a decade later, file a compensation suit, Smith feels he will have to wrestle with himself to know what to do. He points out that he was basically a bystander in this affair, and not really responsible for—that is, immediately in charge of—what happened. He sees himself as more of a witness to events than anything else. He might, he says, call one of the other managers involved and ask if he were willing to come forward. He thinks it unlikely that any would.

Smith, too, would be unwilling to risk his career in business—the inevitable result of coming foward, he feels—just "to give a widow compensation." Other managers and other managerial cliques are always on the lookout for others' mistakes, and will pounce on anyone foolish enough to admit them. Even if others restrain themselves from an immediate attack, their knowledge of another's mistakes is ammunition for the future. One of Smith's own maxims, in fact, in being on the lookout for others' errors is to "lay in the weeds, with rocks, and wait." And, of course, one who exposes a colleague's errors and makes him vulnerable to others evinces only a fundamental untrustworthiness.

3) It is unlikely, however, that any workers affected could ever piece things together. First, there is nothing in writing. Second, Smith feels sure that everyone involved will simply deny knowledge and claim that the process was altered solely for production reasons.

4) Finally, he says:

"The basic rule is that you hope that these kinds of things never occur. Nobody wants to hurt people. Nobody would ever consciously plan to do something that would endanger people. But when things happen, well, you cover for yourself and your company."

He goes on to say:

"The thing that makes. . . the corporation work at all is the support we give to each other no matter what happens. . . . We have to support each other, and we have to support the hierarchy. Otherwise you have no management system."

I should note that Smith's and his colleagues' actions were totally at variance with formal policies later adopted by their corporation. In the aftermath of an environmental catastrophe in the 1970s, the chemical company set a number of guidelines, including a commitment to inform workers of hazards they might confront. The company has, in fact, done so in several cases. However, the crucial variables in such situations seem to be: 1) how public the knowledge about a hazard is; 2) whether the people affected by a hazard—here, workers—have independent access to that knowledge; and 3) what a company's recent public history on such issues has been and how that history has shaped internal corporate structure and politics. Of particular importance is the relationship between professional staff working explicity in these areas and the business areas of a company properly speaking. Business areas normally establish the tone, tempo, and ethos of a corporation, monitored or checked by professional staff. As it happens, Smith's case preceded the chemical company's adoption of elaborate formal policies on environmental, safety, and health issues. But those same policies, and the professional staff that have implemented those commitments, have come under assault at the chemical company in the last few years under the pressures of economic hard times, conservative regulatory triumphs, and a backlash by line managers. This suggests that Smith's ethic, rooted in the intricate social structure of managerial work, might prove more enduring than formal policies, the implementation of which ebbs and flows with external pressures.

Where Brown was marginal to the managerial circles of the textile firm, Smith is enmeshed in one at the chemical company. Managerial circles bind managers to a code of loyalty—a code that extends, however, only to oneself, to one's fellows, and to the idea of organization itself. Covering up one's own mistakes and those of one's allies is central to that code, as it is in many other occupations and professions.

In Smith's case, his taken-for-granted assumption that covering-up is right and proper emerges directly from: 1) the factional structure of bureaucratic work, where conflict and competition between rival managerial interest groups, cliques, and alliances makes open admission of mistakes simply foolhardy; 2) the typical demands of clique leaders for comfortable relationships with their colleagues, here translated as trust and reliability; 3) the knowledge that, by diffusing responsibility, fragmenting knowledge, and relying on private verbal agreements rather than documentation, bureaucracy separates people from the consequences of their actions and makes real accountability unlikely, and 4) the ready availability of accounts accepted as plausible in Smith's world to excuse and justify action, if only to himself. Among many other accounts, Smith appeals to inevitability, arguing that since the workers' fates are sealed he cannot reasonably be expected to jeopardize himself as well. Moreover, in a calculation that brings all these assumptions and rationales together, Smith gauges that the penalty for violating the code of managerial cliques is expulsion from the circle of trust at the core of such groups. And he chooses to act expediently to avoid that consequence. Bureaucracy rewards behavior that conforms to its own internal logic, thus linking institutional needs and personal advantage.

IV. This kind of alertness to expediency—seeing the links between external exigencies, institutional logic, and personal advantage—is at the heart of the bureaucratic ethos; it is also at the nub of the problem of dissent in bureaucratized occupational contexts. To pose the problem sharply, dissent typically points to principles outside of the immediate occupational milieu and demands adherence, or at least attention, to them. Expediency, however, demands and rewards solutions to problems on their own terms and in accordance with the social rules of one's occupational milieu. The tension between principle and expediency, and some of the ambiguities of that tension, are cast into bolder relief in the case of Joe Wilson (pseudonym), a manager who insisted on adherence to external principles and ended up taking a public stance against his bosses.

Wilson was trained in marine engineering in the Merchant Marines, where, among other work experience, he had key responsibilities on a nuclear vessel. After leaving the Marines, Wilson worked in a variety of engineering fields, including the space program, eventually returning to the nuclear field. After work with nuclear submarines, he became the Senior Systems Engineer for a large nuclear plant, and later Plant Superintendent at another nuclear station.

In June 1980, Wilson began working for General Public Utilities, Nuclear (GPUN) at Three Mile Island, Unit-2 (TMI-2). He quickly rose through three levels of management and eventually became Site Operations Director supervising between 260 and 340 employees and, at various points, between 47 and 100 contractors.

The structure of authority on Three Mile Island was multilayered and complex. In March 1979, of course,

TMI-2 had suffered a serious accident that had worldwide negative repercussions for the nuclear power industry. Since TMI-1 was also inoperative due to its own problems, almost all the effort on the Island was directed toward the cleanup of TMI-2, a task seen as urgently necessary in light of the vast national and international media coverage and FBI and Congressional investigations that followed the 1979 accident.

There were essentially three organizations operating on the Island after the accident: GPUN; Bechtel Corporation, the principal contractor to GPUN for the cleanup; and the Nuclear Regulatory Commission (NRC), which had to approve each step of the cleanup. In the background was General Public Utilities (GPU), the parent corporation of GPUN. GPUN had a chief executive officer who was also a board member of GPU, a president, and an executive vice-president. The last was director for both TMI-1 and TMI-2. Reporting to this executive vice-president were two deputy directors for TMI-2, one from the GPUN organization and the other from Bechtel Corporation. Wilson, as Site Operations Director, reported to both of these men and, in turn, had a number of people reporting to him, including directors and managers of key operations. Parallel to Wilson were other directors also reporting to the two deputy directors; some of these directors were Bechtel people, others were GPUN employees. Overseeing the whole operation was the NRC, which had a Program Office on site. The basic ambiguity of the authority structure, one should note, stemmed from the incorporation of Bechtel empolyees into GPUN's regular line of authority. This resulted from a reorganization on the Island in September 1982, which saw leadership in the cleanup pass from GPUN to Bechtel.

The ambiguity of authority relationships after September 1982, was to become one of Wilson's principal concerns. But even from the earliest days of his employment at TMI, Wilson was worried about other management practices. In fact, he says that his first instinct upon coming to TMI and seeing all of GPUN's organizational and operational problems was to get out. He did not leave, however, not least because he had spent the better part of his working life in the nuclear power field and was deeply committed to the development of nuclear energy. He felt that cleaning up TMI properly was crucial to the industry's future. During his time on the Island, he worked an average of 60 to 70 hours a week and instituted a number of important managerial control systems. These included, to take but two examples, a tracking system to reduce backlogged work requests from Site Operations. In a year and a half, backlogged requests were reduced from 3700 to 350. Wilson also introduced a Unit Work Instruction (UWI) system that involved sign-offs up and down the line to fix responsibility for work done. This program was part of a commitment to the NRC to address serious safety concerns in the aftermath of a radiation contamination incident caused by crucial drains out of a containment building being merely taped rather than plugged shut.

From Wilson's standpoint, his inclination to thoroughness became harder to implement after Bechtel's integration into the management hierarchy and its ascendancy to leadership in the cleanup. Wilson became increasingly concerned about a number of issues. First, Bechtel's management both above and parallel to him in other sectors of the organization began pushing very hard to meet schedules to satisfy Department of Energy (DOE) demands that the cleanup be accomplished promptly. There had indeed been long delays in the cleanup, and both the DOE and NRC were nervous about the possible effect of these delays on other nuclear programs. During one period, in fact, the DOE allocated monies to GPUN only on a task basis as each project was completed.

Wilson felt that one of the key reasons for the delays was the lack of any integrated schedule for all departments working on particular projects; he made this point many times both verbally and in writing. One of his superiors, the GPUN deputy director, responded to one such protest with a note asking what Wilson was trying to prove and saying that Wilson was setting a bad example. Integrated schedules never were developed. Instead, as Wilson sees it, Bechtel addressed the scheduling delays by instituting a series of shortcuts around established procedures. These shortcuts were so numerous and obvious that Wilson feels the NRC must have either approved or tacitly acquiesced in them [7]. To take but one example, Bechtel substituted its own "work package system" for Wilson's UWI system. Bechtel's work packages required either no signatures at all for work done or only Bechtel personnel signatures. This system did speed work, but it also clouded responsibility for the accomplishment of important procedures.

Wilson also had broader concerns in this area of responsibility and exact delineation of authority after Bechtel's ascendancy in 1982. Wilson felt that he and others legally responsible for decisions were overruled or bypassed by other managers and departments. Bechtel's ascendancy to prominence in the cleanup had prompted, it seems, a loosening of authority and fealty relationships throughout GPUN's structure, a typical development in hierarchical organizations as people try to establish new alliances to protect themselves. On issue after issue—from a dilution in the authority of the designated emergency director (in the event of a catastrophe) to increased unclarity about the lines of responsibility for significant repairs—Wilson felt that his authority had been diminished but that his responsibility—and, as he defines it in retrospect, his "blamability"—remained unchanged. He pushed hard-to-clarify lines of authority, and was partially successful in doing so only toward the end of his tenure at TMI.

Wilson also saw the kinds of issues he was raising come to a head in a dispute, later widely publicized, about the safety of a piece of equipment called a polar crane, which is wholly contained and operated inside the nuclear reactor building. The polar crane is normally used during refueling to lift off the 163-ton nuclear reactor vessel head, and then to remove the rods that make up the fuel for the nuclear reaction. In cleaning up TMI-2, the crane would have to have been used repeatedly—first, to remove the four missile shields, ranging in weight from 32 to 440 tons, that protect the reactor's steel top; then to remove the steel top itself, and finally to clean the rubble and debris from the reactor core, partially destroyed as a result of the 1979 accident. The crane had not been used since the 1979 accident, and Bechtel was placed in charge of refurbishing it and load-testing it for safety.

Even after Bechtel had refurbished the crane at a cost of several million dollars, key engineers reporting to Wilson, one of whom was a Bechtel employee, felt that the crane might have suffered structural damage due to radiation and to rust caused by water on the floor of the containment building. These engineers felt that Bechtel's proposed tests for the crane's safety and reliability were technically inadequate and posed a possible hazard to public health and safety.

In basic terms, the issue was this: Bechtel wanted to perform the test required by the NRC code on the crane while doing, in stages, the actual lifts necessary to get on with the cleanup. The tests would begin with lifts of the four missile shields. After preliminary lifts, the shields would then be set to one side in order to construct, along with materials in the building, enough weight to load-test the crane fully before proceeding to lift the 163-ton reactor head. Although the crane was originally designed to do 500-ton lifts, it would be tested for only about 200 tons. Site Operations argued that, if management were wrong and the crane failed and dropped one of the missile shields, there could be a repetition of the 1979 accident. Site Operations wanted a lift made of other materials, either already in the containment building or brought in by workers, to make sure that, to begin with, the crane could sustain a 32 to 40 ton weight. If so, it could then proceed with removing the missile shields and building the test weight for the reactor head.

Wilson and his Site Operations staff were particularly concerned because, when the polar crane was finally used, it would be Site Operations who used it. They were fully aware, of course, that given the lack of any tracking system in the corporate world, the rule is that whoever is in charge of something at the moment is responsible, and potentially blamable, for whatever might go wrong. If the polar crane failed, it would be seen as the fault of Site Operations. Site Operations felt that Bechtel's deputy director at TMI, in particular, had exerted pressure to prepare procedures without adequate data and without giving Site Operations enough time to review the program fully. Wilson asked repeatedly for the documentation of crane modifications and for an analysis to verify Bechtel's plan was safe [8]. He never received such an analysis, and felt that his requests were brushed aside.

The NRC complicated the dispute in some ways. It seems to have become customary for top GPUN and Bechtel officials to submit preliminary materials to NRC officials on site to see if there were any problems. NRC often gave informal approval to work requests before internal reviews within GPUN were completed. When Site Operations disputed technical issues with top management, it was told that the NRC had no problem with particular technical specifications, and that it was curious that Site Operations should object. On the polar crane issue, the top official of the NRC later characterized Wilson's and his engineers' concerns as stemming from a philosophy that emphasized procedural matters rather than a focus on final goals. This characterization was echoed by GPUN management, which stressed that what was at issue in the polar crane dispute was not procedures but results. At a certain point, they said, decisions had to be made to resolve technical disputes, and work had to proceed toward what everyone acknowledged to be a worthwhile goal—the cleanup of TMI-2.

As these kinds of disputes intensified, Wilson documented his own and his staff's objections on a variety of issues with ever greater thoroughness, regularly sending memoranda on disputed issues to his two immediate superiors. Wilson knew that his bosses were unhappy with such written objections, but he felt that he had little choice except to register his concerns in this way. He had come to see his protests and his insistence on proper procedures as a moral issue. In his view, not only did public health and safety actually depend on upholding procedural safeguards, but just as important, the appearance of upholding them was crucial to the long-term success of the nuclear industry.

Once, when Wilson wrote a memo to the GPUN deputy director about radioactively contaminated sewage being trucked out of the plant and disposed of illegally, his boss replied that he did not need such a memo from Wilson. It was, his boss said, not constructive and wasted his own and Wilson's time. Finally, on February 7, 1983, Wilson requested a meeting with his bosses' boss, the GPUN executive vice-president, to discuss among other things his safety concerns; this meeting was scheduled for February 25, 1983. On February 17, Wilson and one of his engineers expressed their fundamental disagreement with Bechtel's handling of the polar crane. Then on February 24, Wilson was suddenly suspended from his post, on the grounds of conflict of interest. As it happens, Wilson was the part-owner of a consulting firm that had one client, a nuclear plant in another state; during his employment at GPUN, Wilson had received no income from his consulting work. Wilson's secretary was told to report to the site's Stress Control center, and she was later fired. One of the engineers under Wilson, who had been particularly vociferous on the polar crane issue, was also asked to report to Stress Control for a neuropsychological examination; he was later transferred to a non-nuclear GPUN plant. Another engineer, the Bechtel employee, was later suspended and then transferred across the country.

During his month-long suspension, Wilson continued to raise his safety concerns. He met with the GPUN executive vice-president, with special investigators who report to the GPU board of directors, with a member of the Safety Advisory Board for TMI, and with NRC officials on site. There was no investigation during this period into Wilson's involvement with his consulting firm. After the month suspension, Wilson was fired.

Both Wilson and his engineers went public with their concerns, with one of the engineers being singled out for the greatest media attention. Top GPUN officials maintained to both the press and to a Congressional committee that Wilson was fired because of a conflict of interest; one official told the Congressional committee that had Wilson not been associated with the consulting firm, he would still be working at TMI. However, *The New York Times* cited unnamed corporate officials' characterization of Wilson as someone who "was not a team player." [9]

As things turned out, the NRC's Office of Investigations launched a special inquiry into the whole TMI-2 management situation after Wilson and his engineers went public. The office released a report on September 1, 1983, that found that not only were the dissenters' allegations substantiated, but that they were "illustrative rather than exhaustive." [10] In particular, the report critiques Bechtel's shortcuts around proper procedures and GPUN management's failure to "responsibly monitor Bechtel's work and hold Bechtel accountable." [11] Some weeks later, the chairman of the NRC, in a letter to the chairman of GPU, put the matter in sharp focus:

> "In the past, the Commission (NRC) has clearly stated its position advocating a safe and expeditious cleanup. Your organization has stated its commitment to the same goals. However, it appears that in the interests of expediency, proper management controls may have been compromised." [12]

The immediate meaning of expediency here is the swift—that is, expeditious—accomplishment of what has to be done. High levels of organizations always exert pressure, of course, to do what has to be done. Managers in other corporations that I interviewed about Wilson's situation stress this meaning of expediency. They too see the cleanup at TMI-2 as the overriding issue in this situation. In their view, GPUN's and Bechtel's general management practices, as well as their plan to load-test the polar crane, were wholly "reasonable," the code word in management circles for practical. To insist as Wilson did on meticulously following proper procedures, whether mandated by regulation or not, could only "lay logs in the path" of getting the job done at all. Besides, they point out, the local Program Office of the NRC went along with the shortcuts, and they are supposed to be the procedural guardians in such cases.

As things turned out, the NRC did eventually approve a modified version of Bechtel's plan for the polar crane [13], which, these managers feel, clearly indicates that the NRC opposed Wilson all along. The investigation and public reprimand of GPUN was all for public show. Wilson's emphasis on proper procedure is clear evidence, they say, of a "military mind," of "nitpicking," of "straining at a gnat," and of "being out in left field." To make an omelet, one must be willing to break eggs; results are what count; the proof is in the pudding. An insistence on procedure betrays a zero-risk mentality, one of the crippling banes of our society.

Of course, as suggested earlier, expediency has other levels of meaning. Particularly in our society, which has enshrined the virtue of self-interest, expedient action—doing what has to be done—is almost always tied to personal advantage of some sort. Tasks defined as necessary are yoked to personal fates. At TMI, as I noted earlier, monies from the DOE, and with them the fate of many employees and executives, depended on rapid accomplishment of necessary tasks. However, to reap the personal advantage available in necessary tasks, one must also act expediently within the rules of one's social milieu. The managers that I interviewed about Wilson reserve their sharpest criticism of him for this area—that is, for what they see as his violation of the rules of managerial circles. In particular, they stress:

1) One does not make one's view of a technical issue into a matter of priniciple. It is the prerogative of authority to resolve technical disputes. Whether Wilson liked it or not, Bechtel had won the power struggle and they had the right—that is, the power—to call the shots on the cleanup. What is required of managers is a flexibility that enables one to bend with the prevailing wind. One has to recognize that one can be beaten even when one is right, so whether Wilson was right or not is irrelevant. What mattered was that people in authority, who always have the "extra vote," decided, for their own reasons, that Wilson and his people were wrong. As these managers see it, Wilson should have accepted his defeat gracefully, told the engineers who reported to him to drop the matter, and declared his willingness to do whatever he could to help expedite things.

2) Instead, Wilson committed what are cardinal sins in managerial circles. He tried to fix responsibility for what was happening, a tactic certain to shatter the trust that what we might call "cooperative nonaccountability" requires. Worse yet, he tried to fix responsibility on the "up side," when of course, when things go wrong, responsibility and blame are normally fixed on the "down side."

3) Finally, Wilson put things into writing in a world that fosters ambiguity by its insistence on talk as the basic mode of negotiation and command. Talk lends itself more readily than the written word to backtracking, evasion, and subterfuge, all important virtues if one is to do what has to be done. These managers are not at all surprised that Wilson's superiors reached for whatever pretext they could find to fire him; they too would have gotten rid of him at the first convenient opportunity. Sunday school ethics, they say, do not help managers to cut the sometimes unpleasant deals necessary to make the world work.

V. In such ways, the bureaucratic ethos creates internal barriers to dissent among managers. The wise and ambitious manager comes to know that the only way for him survive and succeed in the precarious and intricate social world of the corporation is to be admitted into key managerial networks. Without the personal affiliation and social approbation such acceptance signifies to other managers, he has no chance to establish the social standing necessary to gain any credibility, and therefore any influence, in his organization. He learns to read adroitly the criteria for admission to key cliques and, precisely because he is ambitious and talented, he subjects himself willingly to the asceticism the bureaucratic ethos demands. He learns to wear the public faces his corporation requires and, more important, he internalizes the code of everyday moral rules-in-use that form the norms of his world. This means, on one hand, adopting a certain flexibility of moral perspective, in which the principles and verities of other spheres of his life are bracketed altogether or subjected to a subtle process of redefinition.

In such a process, private notions of morality give way to hard pragmatic judgments, shaped principally by the dynamics of the bureaucratic occupational milieu. The internal rules and social context of one's managerial circle become the principle moral gauges for action. In Smith's case, this meant a willingness to cover up for himself and others when things went wrong, because he knew, as do other managers, that without the consent to the bureaucratic ethos that such silence betokens, one's organizational effectiveness in the thickets of one's bureaucracy is at an end. This explains to some extent the suspicion in the corportate world towards dissent in general, even on issues devoid of any ostensible moral import. Managers' emphasis on the importance of personal comfort with others is simply the public vocabulary to explain their pervasive anxiety about whether others can be trusted. Men and women who regulary raise problems become thought of, by those in power, as problems themselves.

The reason for this is not hard to discern. The bureaucratic world places a premium on expedient action—doing what has to be done while keeping one's own eye on the main chance. Those who act without complaint, even if their actions may be judged by another standard to be irrational, are thought of as doers; those who raise objections with any regularity, even if their dissent may be judged by another standard to be rational, are thought of as "naysayers" or "crepehangers." In the back rooms of the bureaucratic milieu, managers are paid, and paid well, to bring sometimes obdurate technology and always difficult people together to make money. This uncompromising task demands the willingness to make continual compromises with conventional verities. Principles, and those who raise them, do not fare well in back rooms.

We accept the images of occupational dissenters that the cinema screen or the journalistic profile extol before us because these men and women seem to cut through the thickets of bureaucratic work life and act with clear moral resolution and character. Their actions provide vicarious revolt against the seaminess of everyday organizational contexts. Most of the dissenters that I have studied, however, see themselves in a more ambiguous light—as men or women caught in circumstances neither of their making nor of their choosing that for some reason they cannot accept. Sometimes they adhere to an old-fashioned professional ethic; sometimes they are haunted by their role, whether intended or not, in potentially causing harm to others; sometimes they insist on an epistemelogy that honors the search for truth; sometimes they are simply stubborn. They choose, most often with great reluctance, principle over expediency. In doing so, they pit themselves against an ethos that transforms ethics into etiquette, turns dissent into a kind of heresy, and makes silence a cardinal virtue.

REFERENCES

[1] See United States House of Representatives, Committee on Insular Affairs, Subcommittee on Energy and the Environment, *Oversight Hearing: Current Status of the Three Mile Island Nuclear Generation Station, Units 1 and 2.* 98th Congress, 1st Session (April 26). Washington, D.C.: Government Printing Office, 1983.

[2] See Alan F. Westin, "What Can and Should Be Done to Protect Whistleblowers in Industry," in *Whistleblowing! Loyalty and Dissent in tthe Corporation,* ed. Alan F. Westin et al. (New York: McGraw Hill, 1981), pp. 131-167.

[3] See Robert Alex Baron, *The Tyranny of Noise* (New York: Harper Colophon, 1971), pp. 40-41. The "C" scale measures sound pressures on a flat scale. The "A" scale was adopted when it was discovered that humans are not as sensitive to lower frequencies as to the higher. The "A" scale is the measure used in the textile industry.

[4] For an ellaboration of the social structure of managerial work, see Robert Jackall, "Moral Mazes: Bureaucracy and managerial work," *Harvard Business Review,* Vol. 61 (September-October, 1983), pp. 118-130; and "The Moral Ethos of Bureaucracy," *State, Culture and Society,* vol. 1, no. 1 (Fall, 1984), pp. 176-200.

[5] There is a brief review of some of the ideological contortions surrounding the cotton dust issue in Robert Jackall, "Moral Mazes: Bureaucracy and Managerial Work." For an analysis of how the industry was able to link compliance with the cotton dust standard with increased productivity, see Dr. Ruth Ruttenberg, *Compliance With the OSHA Cotton Dust Rule: The Role of Productivity Increasing Technology.* For the Office of Technology Assessment. (Washington, D.C.: Ruttenberg, Friedman, Kilgallen, and Associates Inc. March, 1983).

[6] Department of Labor, *Occupational Safety and Health Administration, Occupational Noise Exposure; Hearing Conservation Admendment; Final Rule. Federal Register,* Vol. 48, No. 46, Tuesday, March 8, 1983, Part II.

[7] Wilson's suspicion in this regard was confirmed as a result of later investigation conducted by the NRC's Office of Investigation, which found that: "The NRC has contributed to TMI-2 organizational problems by not acting in a conventional regulatory mode for the facility." See Nuclear Regulatory Commission, Office of Investigations, *Three Mile Island Nuclear Generating Station, Unit 2 Allegations Regarding Safety Related Modifications, Quality Assurance Procedures and Use of Polar Crane. Docket No. 050 320. Summary.* September 1, 1983.

[8] On January 20, 1983, Wilson wrote a memo to one of his two superiors, the GPUN Deputy Director, trying to ensure that documentation for Bechtel's refurbishment of the crane would be provided to Site Operations. The memo concludes: "If the

crane is turned back over to Site Operations, it is expected that adequate documentation will be provided to ensure its requalification." See Nuclear Regulatory Commission, *Op. Cit.* (Docket No. 050 320), *GPU Nuclear Inter-Office Memorandum.*
[9] Richard D. Lyons, "Crews at Reactor Criticize Cleanup," *New York Times,* March 28, 1983, pp. 1f.
[10] Ben B. Hayes, Director, Office of Investigations, Nuclear Regulatory Commission, Memorandum for Chairman Palladino, *Subject: Three Mile Island NGS, Unit 2 Allegations Regarding Safety Related Modifications and QA Procedures* (H-83-002), September 1, 1983, p. 1.
[11] *Ibid.*, p. 2.
[12] Nunzio J. Palladino, Letter to Mr. William Kuhns, Chairman of the Board, General Public Utilities Service Corporation, October 7, 1983, p. 2.
[13] See *Safety Evaluation By the Office of Nuclear Reactor Regulation, Three Mile Island Program Office, Facility Operating License No. DRP-73 GPU Nuclear Station Unit-2 (TMI-2). Docket No. 50-320. Refurbishment of the Reactor Building Polar Crane, Load Test and Recertification for Use.* November 18, 1983.

RESEARCH NOTE

From 1980-1983, I did field work in several corporate settings. This included preliminary work in a small chemical firm and in a large defense contractor, and intensive work in the chemical company of a large conglomerate, and in a large textile firm. During this period, I also did field work in a large public relations firm; this work is still continuing. My basic methodology has been both nonparticipant and participant observation, and especially intensive semi-structured interviews with managers and executives at every level of management. I have interviewed more than 125 managers. I am indebted to the National Endowment for the Humanities, which provided me with a Fellowship for Independent Research in 1980-1981, and to Williams College for several small research grants.

The field work with organizational dissenters began in the summer of 1982 and is still continuing. To date, I have completed 13 case studies, interviewing 18 dissenters in the process. I have also done a number of other case studies that have proved to be too ambiguous to provide much understanding of the nature and motivation of organizational dissent. I have presented the materials on organizational dissenters to several key long-term informants in the corporate world, whom I have known and been interviewing for more three years. I am indebted to the Wenner-Gren Foundation for Anthropoligical Research for their finanacial support of my studies of dissenters. The Educational Fund for Individual Rights in New York, NY, and the Governmental Accountability Project in Washington, D.C., have given me considerable help in this project.

ACKNOWLEDGMENTS

I want to thank Joseph Bensman and Janice Hirota for their thoughtful and helpful comments.

POSTSCRIPT

Since this paper was delivered at the American Association for the Advancement of Science Meeting in May 1984, there have been a number of developments. In July 1984, GPUN successfully used the polar crane to lift the head from the TMI-2 reactor. In August 1984, GPUN inspectors spotted a problem with a manual brake release on one of the crane's two redundant brake units; the brake release had been installed in November 1982 during the refurbishment of the polar crane. According to a GPUN press release, this was done "... under the supervision of Bechtel personnel. The mechanisms were fabricated and installed without normal engineering oversight because of administrative deficiencies that existed while an integration of GPU Nuclear and Bechtel management was being completed at TMI-2. Those deficiencies have been corrected." (GPU Nuclear, News Release, January 10, 1985, # 5-85N) The crane was taken out of operation and the hand release mechanisms removed. The crane was later put back into service and GPUN claims to have instituted an expanded program of preventive maintenance and preoperational testing for the crane.

On February 10, 11, and 12, 1985, *The Philadelphia Inquirer* ran a three part series on TMI that reports in a journalistic way many of the events described in this paper. GPUN responded to the series both with press releases and with a two-page advertisement that appeared in several central-Pennsylvania newspapers disputing the *Inquirer's* facts and interpretations. These materials are available from GPU Nuclear, Public Information Service, Three Mile Island Nuclear Station, P.O. Box 480, Middletown, PA 17057.

In its series, the *Inquirer* reports that the manager that I call Wilson (the pseudonym is at his request) received an undisclosed monetary settlement from GPUN and was last reported working as an inspector at a nuclear facility in the southern United States. GPUN continues to maintain that this manager was dimissed only because of a conflict of interest.

Chapter 4

Frameworks for Engineering Ethics

Moral Thinking and Moral Theories

MORAL theories form the basis of traditional approaches to the philosophical study of ethics. For a theory to be useful, ethicists argue that it should be verifiable, consistent, and present a reasonable accounting of what is good (1).

Underlying moral theories is the concept that in order to make moral judgments, a person must be an autonomous moral agent, capable of making rational decisions about what should be the proper course of action in confronting a moral dilemma.

Philosophers often begin discussions of moral thinking (adapted from 2) by dismissing three sorts of "theories" that individuals often employ but that ethicists generally agree are not useful moral theories. The three rejected theories are divine command ethics, ethical egoism, and ethical conventionalism.

Divine command ethicists hold that a thing is good if God commands it to be so. Philosophical arguments against this theory can be quite complex. Suffice it to say here that divine command theory must ultimately rest on faith and cannot be verified by purely rational means. In rejecting divine command theory, ethicists are in effect drawing a distinction between religion and ethics. This is not to say that religion is irrelevant but only that ethics as a rational system of moral decision making can be conceptualized apart from any considerations of religion.

Ethical egoism, which holds that a thing is right if it produces the most good for oneself, is easily rejected as a workable moral theory on the basis that it is not generalizable. In other words, if everyone operated solely out of their own self-interest, there would be no basis at all for morality. This argument is not always easy for engineers to grasp, especially since our economic system is based on a similar theory—that is, individual pursuit of profit will benefit everyone in the long run. Here again, however, the point is that ethical systems and economic systems are not the same thing and clearly will not always produce the same conclusions about whether or not an action is good.

Ethical conventionalism, also known as cultural relativism and situational ethics, holds that a thing is good if it conforms to local convention or law. This theory fails to provide a reasonable accounting of what is good—numerous examples can be cited of actions that, though acceptable within the framework of the actors, are morally unacceptable to most individuals. To argue for ethical conventionalism is, in effect, to argue that ethics has no objective meaning whatsoever. This theory is quite popular nonetheless and often surfaces in discussions of engineering ethics in an international context.

What then are the useful moral theories? The two most prevalent theories are utilitarianism, developed by John Stuart Mill and Richard Brandt, and duty-based theories which derive from Immanuel Kant and John Rawls. Some ethicists also favor rights-based theories advocated by John Locke and A. I. Melden, which are closely aligned with duty-based theories and virtue theories (Aristotle and Alasdair MacIntyre) (adapted from 2).

Utilitarianism is an ethical system whereby an action is considered good if it results in the maximization of utility, which is defined as the greatest good for the greatest number of people. Act utilitarianism evaluates the consequences of individual actions, whereas rule utilitarianism, favored by most philosophers, considers generalizable rules, which if consistently followed, would result in the greatest good for the greatest number of people. Utilitarianism is a popular moral theory among engineers and engineering students. Indeed, it has its analog in engineering decision making in the form of cost-benefit analysis, wherein a project is deemed to be acceptable if its total benefits outweigh its total costs.

It is also consistent with simplistic notions of democracy as being characterized merely by "majority rule." One problem with cost-benefit analysis, however, as well as the utilitarian thinking on which it is based, is that the distribution of costs and benefits is not considered. A new highway or bridge, for example, may produce the greatest good for the greatest number; however, those bearing the costs of relocation may not share equally in the benefits of the project.

Utilitarianism's major competitors, duty- and rights-based ethical theories, take the distribution question head-on by focusing not on an act's consequences but rather on the act itself. Individuals are thought to have duties to behave in morally correct ways. Similarly, people are moral agents whose basic rights should not be infringed. In this manner, rights-based theories can be seen as the flip side of the more prominent duty-based theories. In each case, however, the focus is on the act itself rather than on the consequences as in utilitarianism. One problem with duty-based theories is how to handle situations in which there are conflicting duties. Such situations frequently arise in engineering ethics, wherein engineers may have duties to themselves, their families, their employers or clients, and the public in general.

A final moral theory is virtue ethics, which focuses on qualities thought to be found in virtuous persons, such as loyalty, dependability, and honesty. Such theories often appeal to those with strong religious convictions since the virtues are similar to those expounded on in religious doctrine. Virtues also frequently appear in the language of engineering codes of ethics.

One great challenge of engineering ethics is to learn how to distinguish the various types of moral reasoning and to know when to apply the different theories. For example, in most questions involving engineering projects, utilitarianism might be an adequate theory. However, if the projects or designs represent substantial risks to individuals who are unlikely to benefit much from them, then duty/rights-based theories may be more appropriate.

Many contemporary philosophers believe that formal discussion of abstract moral theories is not necessary in doing applied ethics; indeed, it might be counterproductive by turning engineers away from consideration of ethics. Utilitarian and duty/rights concepts, it is argued, can and should be presented in the layperson's terms. Such concepts are often implied in engineering codes of ethics.

Engineering Codes of Ethics

Codes of ethics can serve various functions, including education, encouragement of ethical behavior, and elevation of the profession's public image. Codes can also provide the basis for disciplinary action regarding unethical conduct (3). Many critics of codes of ethics charge that their primary purpose is to create a positive public image for the profession and that the codes are largely self-serving (4).

Although engineering codes of ethics have existed for about a century, only in the last several decades has responsibility for the public health, safety, and welfare gained prominence in the codes. Most modern codes, however, now state that this is the primary responsibility of engineers, thus bringing the major thrust of the codes into closer compliance with philosophical notions of ethics in both the utilitarian and duty/rights traditions. Nevertheless, provisions still remain in some codes that might be interpreted as merely self-serving.

Not uncommonly, for example, codes contain provisions barring public criticism of other members of the profession. (See, for example, article 9 of the IEEE code contained in the Appendix.) Unger (3), among others, believes that such provisions stifle dissent within the professional society. A famous 1932 case involving the American Society of Civil Engineers concerned the expulsion of two members who publicly accused another member of participating in corrupt activities. Though vindicated by the outcome of a criminal trial, the engineers' memberships in the society were never restored.

Not all engineering codes of ethics are as succinct as the IEEE code. Perhaps the most extensive is the Code of Ethics for Engineers of the National Society of Professional Engineers (NSPE) (see Appendix), a multidisciplinary organization that represents registered professional engineers. The NSPE code includes four elements: a preamble and three sections entitled "Fundamental Canons," "Rules of Practice," and "Professional Obligations." The section on Professional Obligations contains considerable elaboration of many points in the canons and Rules of Practice. The code also contains brief commentary on a prior prohibition of competitive bidding that the NSPE was ordered to remove by the U.S. Supreme Court in connection with antitrust litigation. Although the NSPE regards competitive bidding as a violation of professional standards, others, including the courts, have interpreted it as a self-serving measure designed to limit competition for engineering services. Rarely, however, have the courts become involved in settling such disputes over the codes, which are largely constructed and maintained by the professional societies themselves.

The IEEE and NSPE codes are representative of the two extreme formats in which codes can be developed. Unger (3) cautions against codes that are either too short or too long. The dangers of the former are the possibility of important omissions and the lack of specific guidance to engineers; the dangers of the latter include overprescription, thus leading to a code that is cumbersome to read, and the possibility of loopholes if important issues are inadvertently left out.

One potential weakness of engineering codes of ethics is their multiplicity—many professional societies have their own unique code. For some critics this suggests a lack of a consistent sense of ethics among engineers and a source of dilemmas for individuals who belong to two or more societies with conflicting codes. However, efforts to create a unified engineering code of ethics, through such organizations as ABET or the American Association of Engineering Societies, have heretofore failed.

Another important issue relating to codes of ethics is the extent of their applicability in different cultures. This issue is growing in importance as most of the major U.S.-based engineering societies are global in organization or are becoming more so as time passes. A typical argument, for example, is that in some cultures bribery is an accepted, even expected, form of doing business. Such arguments are persuasive to many on practical grounds and are also an impetus for adopting the posture of ethical conventionalism. Others argue for the universality of codes of ethics. These are difficult, though not necessarily insurmountable, questions. One way to gain greater understanding of these problems, which the IEEE has adopted, is to ensure that the organization's ethics committee has adequate representation from regions other than the United States.

Support for Ethical Engineers

In many high-profile ethics cases such as those discussed in Chapter 3, engineers and others who have blown the whistle on unethical behavior have often had to pay a high price for their ethical stance, including demotions, firings, ostracism within the industry, and even threats to life and limb. Many believe that it is unreasonable to expect engineers to be "moral heroes" in this manner (5). Consequently, a great deal of attention has been focused on providing support for ethical engineers, the notion being that members of society have a collective responsibility for nurturing ethical behavior (see Ladd's background article in Chapter 3).

In recent years, the trend has been toward establishing management practices that encourage internal dissent within corporations. For example, many corporations now have ethics officers or ombudpersons who provide a confidential channel for airing ethical concerns within the company. It may be unrealistic, however, to rely too heavily on businesses to encourage such behavior. As a number of philosophers have noted, businesses are not moral agents, but rather are motivated by economic profit (see Ladd's background article in Chapter 3). Encouraging businesses to "do the right thing" usually means seeing to it that it is in their economic self-interest to do so.

One means of doing so would be to enforce strong regulatory penalties for unethical behavior among corporations. Unfortunately, since the early 1980s there has been a strong antiregulatory climate in the United States. And even when regulations are in place, their enforcement often involves the cooperation of the industries being regulated. Another avenue would be stronger product liability legislation, but here again the trend is in the opposite direction, with implications for engineering ethics that will be discussed at greater length in Chapter 6.

Another governmental approach for supporting ethical behavior by engineers and others would be to establish stronger legal protection for whistleblowers. Although some existing state and federal laws provide support for whistleblowers in certain circumstances, a National Employee Protection Act, such as that proposed by the Government Accountability Project (6), would help ensure that all employees who become legitimate whistleblowers would be shielded from employer reprisals.

The engineering community itself is perhaps in the best position to provide greater support of ethical conduct by members of the profession. Appropriate responses by the professional engineering societies include taking seriously the promulgation of engineering codes of ethics, providing legal and financial support for whistleblowers, and giving awards for noteworthy ethical conduct. Ultimately, however, to be effective (as Unger notes in the background article in this chapter), the professional societies may need to seek means of sanctioning employers who punish their engineering employees for acting in the public interest.

The IEEE took a major step toward providing such support in 1995 when it established an Ethics Committee that reported directly to the Board of Directors. Before then, ethics support was left to the Member Conduct Committee, which has a dual function of member discipline and ethics support and which, until recently, was largely inactive. Since its inception, the IEEE Ethics Committee has established an Ethics Hotline, promulgated guidelines for ethical dissent, and begun to draft more detailed guidelines for interpreting the IEEE Code of Ethics.

One problem with relying too heavily on the professional societies for providing support for ethical engineers is the level of influence, mentioned earlier, which business wields over the professional societies. As E. T. Layton points out (7), many leaders of the societies are senior members who have moved from technical engineering duties into business management roles within their companies. In addition, many companies fund and support the participation of their employees in the professional societies.

After a year of operation, the IEEE Ethics Hotline was suspended by the Executive Committee of the Board of Directors, and efforts by the IEEE Ethics Committee to establish an ethics support fund to aid engineers exhibiting ethical behavior have run into numerous roadblocks. Although opponents usually argue on the basis of liability concerns, impacts on the tax-exempt status of IEEE, and other reasons, some also feel that an ethics hotline and ethics support fund would put IEEE in the position of mediating disputes between members and their employers who are often corporate sponsors of IEEE activities. On the other hand, such resistance to ethics support can often be worn down by the persistent activism of members of the professional societies. This has been witnessed, for example, by the IEEE's establishment of a board-level ethics committee and its early role in filing a friend of the court brief in support of whistleblowing engineers in the BART case.

In closing this chapter, it should be noted that calls for greater support of ethical engineers are not intended to suggest that engineers should not be expected to exercise their own moral judgment. As I have noted elsewhere (8):

> In arguing for greater corporate responsibility, better legal protection for whistle blowers and more active support for ethical conduct on the part of engineering societies, I do not wish to discount the importance of individual moral responsibility. I suspect that in moral risk taking, as in engineering design, nothing can be risk free, and that responsibility ultimately must rest with individuals who have both the knowledge and opportunity to protect the public health, safety and welfare. As a society, however, I believe we have a duty to such individuals to insure that the risks of whistle blowing are reduced to an acceptable level. (p. 619)

Background Reading

In his 1996 essay on "The Public Health, Safety, and Welfare: An Analysis of the Social Responsibilities of Engineers," *Michael C. McFarland* draws on social ethics to examine the social responsibilities of engineers, using examples from the nuclear power industry. McFarland begins with the premise that engineers' responsibilities must be seen in a broader social context than the actions of individual practitioners (see Part III). In citing the work of others, McFarland agrees that individuals have an obligation to come to the aid of another when there is a critical need, they are in proximity to the problem, they have the ability to help without undue damage to themselves, and there is no other source of help. To this list, however, McFarland adds the importance of organized social structures and a history of common action.

Turning to nuclear power, McFarland argues that engineers meet three of the conditions previously cited for requiring coming to the aid of others—in this case, members of the public potentially exposed to the hazards of nuclear power. With regard to acting effectively, however (i.e., the third condition), engineers lack the social structure to exercise their ethical responsibilities without doing damage to their own livelihoods. McFarland then critiques and finds ultimately inadequate (albeit for different reasons) various responses of engineers to the issue of nuclear power. He addresses issues such as the view that engineers and other technical experts are best situated to make judgments about the risk of nuclear power, whistleblowing, organized opposition through such groups as the Union of Concerned Scientists, and action through professional engineering societies. In light of these inadequacies, McFarland calls for new institutional structures that would provide for independent technical input regarding difficult questions such as nuclear power, as well as consensus decision making involving various technical specialists and other stakeholders, including members of the general public.

In his 1987 article, *Stephen H. Unger* asked, "Would Helping Ethical Professionals Get Professional Societies into Trouble?" Unger, an engineering professor who served as chair of the IEEE Ethics Committee during 1997–1998, answers his own question with a resounding "No!" His argument is based on the premise that engineers with ethical concerns are often disregarded or disciplined by their managers, and there are few laws in place to adequately protect such individuals. Unger believes that the professional engineering societies have an important role to play in providing support for engineers endeavoring to adhere to high standards of professional conduct.

Basing his remarks on a comparison to such organizations as the American Association of University Professors (AAUP), Unger dismisses the major claims regarding why it might be dangerous for the professional societies to play such a role. He suggests that professional societies consider several procedures: investigating particular cases and publishing reports regarding improper employer conduct; censuring offending employers and advertising this action; publishing ethical ratings of employers; establishing legal defense funds for ethical dissenters in the engineering profession; and presenting awards for engineers who place their career in jeopardy by upholding ethical standards.

In response to the claim that such actions would pit managers who are members of the engineering societies against nonmanagers, Unger notes that such procedures would provide protection for ethical engineers and managers alike. With respect to the legal liabilities involved in such procedures, based on the experiences of the AAUP and the Consumers Union, Unger argues that lawsuits against the professional societies would be very unlikely and that they would be winnable in any case, provided that careful procedures are followed in administering the programs. Lastly, Unger attacks the argument that professional society support for ethical engineers would undercut employer support for the societies. He notes that the current level of support is based on the employers' economic self-interest—that is, retaliating in a vengeful manner would be like cutting off their nose to spite their face.

Cases for Discussion

"Technological Hazards and the Engineer" are discussed in a 1986 article by engineer *Roland Schinzinger*. Defining "hazard" as the potential for harm (i.e., a hazardous product is unsafe), Schinzinger discusses a series of cases designed to illustrate hazardous products and processes and ways to mitigate hazards either through prevention or designing effective response mechanisms. As the cases illustrate, hazards can have many origins, including poor design, poor workmanship, inadequate maintenance, and poor operating procedures.

Because engineers are involved in many of these situations, and because regulations and liability laws have only a limited effect on preventing possible harm, Schinzinger notes that engineers need to be actively involved in hazard mitigation. Standard engineering practices such as designing in safety factors and risk-benefit analysis are useful tools but are quite complex in practice and are often subject to great uncertainties. Ultimately, questions of hazards and safety reduce to decisions regarding the level of risk. Such discussions, though uncomfortable to many engineers, require their technical input.

In order to deal with these problems, as well as with the personal and institutional influences that sometimes make engineers reluctant to become involved in safety concerns, Schinzinger suggests that engineering be viewed as an experimental process. Like other experiments, such a process would involve informed consent from the parties exposed to hazards. Engineers should also be involved in promoting more effective safety standards. Though difficult to accomplish in practice, Schinzinger believes this form of participatory technology, along with an approach that favors prevention over response, are keys to minimizing technological hazards.

In 1985 attorney *Alfred G. Feliu* contributed an article, "The Role of the Law in Protecting Scientific and Technical Dissent." Feliu reviews a number of legal cases that have arisen with respect to whistleblower protection in federal health, safety, and environmental law. Feliu points out that over the years the traditional "dismissal-at-will" legal doctrine that permitted employers to dismiss employees without cause was modified with various exceptions. The two most relevant are public policy exceptions in some states and the enactment of federal whistleblower protection laws. Feliu also notes that some states have enacted free speech laws for private employees that parallel laws protecting free speech of public employees.

Feliu elaborates on two cases, one involving a medical physicist employed at a state hospital and the other involving a chemist, both of whom suffered reprisals for reporting violations of federal regulations. In the first case, the physicist won in court and was reinstated with back pay, whereas the scientist in the second case lost his court case. Feliu concludes with his opinion that the law is inadequate in fostering a climate of scientific and technical dissent in the corporate world, a climate he believes is needed to ensure the vitality of the scientific and technical enterprise. The legal mechanisms that exist are cumbersome, too issue-specific, and too narrowly focused on administrative matters rather than the substance of the dissenters' claims. On the other hand, the law does serve an educational purpose, sets standards, and punishes extreme conduct among employers. Through such means, Feliu believes, public and management policymakers may eventually perceive the value of scientific and technical dissent in the workplace.

References

[1] Panichas, G. 1990. Personal communication.

[2] Martin, M. W., and Schinzinger, R. 1989. *Ethics in engineering*. 2nd ed. New York: McGraw-Hill.

[3] Unger, S. 1994. *Controlling technology: ethics and the responsible engineer*. 2nd ed. New York: Wiley.

[4] Ladd, J. 1980. The quest for a code of professional ethics: An intellectual and moral confusion. In *AAAS professional ethics project: professional ethics activities in the scientific and engineering societies*, ed. R. Chalk, M. S. Frankel, and S. B. Chafer, 154–159. Washington D.C.: American Association for the Advancement of Science.

[5] Wujek, J. H. 1996. Must engineers behave *heroically*? Presented at Frontiers in Education Conference, Institute of Electrical and Electronics Engineers and American Society for Engineering Education, Salt Lake City, Utah.

[6] Chalk, R. 1988. Making the world safe for whistle-blowers. *Technology Review* (January): 48–57.

[7] Layton, E. T. 1986. *The revolt of the engineers*. Baltimore, MD: Johns Hopkins University Press.

[8] Herkert, J. R. 1991. Management's hat trick: Misuse of "engineering judgment" in the Challenger incident. *Journal of Business Ethics* 10: 617–620.

FEATURE ARTICLE

The Public Health, Safety and Welfare: An Analysis of the Social Responsibilities of Engineers

MICHAEL C. McFARLAND, s. j.

Abstract—What obligations does an engineer have to protect the public interest in the creation and use of new technologies? How can the engineer best act so as to fulfill his or her responsibilities to the public? This paper considers these questions from the point of view of social ethics, by means of case studies of engineers in the nuclear power industry. An ethical framework is presented that allows us to define the social responsibilities of engineers. The modes of action generally available to engineers for fulfilling those responsibilities are then analyzed. All of these are judged to be inadequate, leading to the conclusion that unless the decision-making structures for the use of technology are changed, engineers will continue to be frustrated in their ability to ensure the responsible use of technology.

1. INTRODUCTION

If you were to observe someone being mugged on the street outside your house, you would certainly have some obligation to try to prevent it, at least to the point of calling the police. Now suppose that you were aware of a group of people being "mugged" by the misuse of a certain type of technology—for example, by having a toxic waste dump placed in their town, or by being put in danger by shoddy design or construction of their automobile. Would there be an obligation to try to prevent that? What if the people were not necessarily being injured, but were being put in danger of suffering a catastrophe—for example, by having a poorly-designed nuclear power plant placed a few miles from them? Is there an obligation in that case to try to make the situation safer for them? Suppose you were partly responsible for the development and use of the technology in question. How might that affect your obligation? Do technologists have any special obligation to protect people from the technologies they create?

The purpose of this paper is to give an account of the social responsibilities of engineers and, more importantly, to explore ways in which they might best fulfill those responsibilities. Most of the writing on engineering ethics in the past has focused on the responsibilities of the individual engineer to employers, clients, and users—responsibilities that could be met adequately by loyalty, competence, and integrity on the job [1]. My thesis, however, is that engineers must be seen in a wider social context. Unless their work is seen in its relation to society, an adequate account of their ethical responsibilities cannot be given, as some of the most recent literature on engineering ethics has recognized [2]. Furthermore, unless individual engineers learn to act in collaboration with others, both inside and outside their profession, they will not be able to meet those responsibilities.

We will investigate the social responsibilities of engineers by looking at their role in the development of a specific technology. The technology I have chosen is nuclear power generation. This is especially well suited to our purposes for a number of reasons. First, nuclear power technology has very serious social implications. On the one hand, it offers the possibility of a long-term, relatively clean and inexpensive power source. On the other hand, it carries great risks with it, especially those connected with the release of radiation into the environment [3]. Second, many in our society have become deeply concerned about the implications of nuclear power, and almost every aspect of the problem has been the subject of intense debate. Third, technical issues have played an important part in the debate, issues that require a great deal of technical training and knowledge to understand. Fourth, nuclear power is a good paradigm for large technological projects, illustrating all the financial, governmental, scientific and social forces that influence such projects. Finally, nuclear power has been an issue for long enough that there is a very large body of literature on all aspects of the problem.

We will analyze the role of engineers in the development, use and criticism of nuclear technology. In particular, we will look at some of the ways in which engineers have tried to fulfill their responsibilities, and analyze the strengths and weaknesses of each response. This will lead to some conclusions about ways in which the society in general and the engineering professions in particular must move so that engineers might better fulfill their social responsibilities.

2. ETHICAL CONSIDERATIONS

Most engineers would, in theory, recognize the existence of broader obligations, beyond those of doing their jobs well. The difficulty is in discerning those obligations in particular circumstances and understanding how they apply to individual engineers. A study done by John

The author is with the Department of Computer Science, Boston College, Chestnut Hill, MA.

Reprinted from *IEEE Technology and Society Magazine,* Vol. 5, No. 4, pp. 18–26, December 1986.

G. Simon, Charles W. Powers and Jon P. Gunneman on the social responsibilities of universities regarding the investment of their funds provides some useful guidelines [4]. The authors begin by analyzing the case of Kitty Genovese, the woman who was stabbed to death outside her Queens apartment while at least thirty-eight of her neighbors looked on. It is generally agreed that the refusal of all the neighbors to help represented a serious failing. But what created the obligation to act when none of the neighbors was responsible for causing the situation? When is there an obligation beyond that of not harming another? From consideration of this case, the authors identify four factors that must be present in order for there to be an obligation to come to the aid of another. They are:

(1) Critical need: some fundamental right or good must be threatened.
(2) Promixity: this is "largely a function of notice," but it also involves role relationships. "We do expect certain persons and perhaps institutions to look harder for information about critical need. In this sense proximity has to do with the network of social relations that follow from notions of civic duty, duties to one's family, and so on." ([4], p. 24)
(3) Ability to help, without damage to self and without interference with important duties owed to others.
(4) Absence of other sources of help.

Further consideration of the Genovese case can lead us, I believe, to two other important points. First of all, part of the reason that the neighbors did not help the woman out was their sense of fear, isolation and helplessness. This does not excuse them from at least calling the police, of course. But their options were severely limited by their inability to take common action. If the thirty eight of them had rushed the assailant at once, they certainly could have overcome him, or at least scared him off, without great risk to any one of them. This shows the power, indeed the necessity, of organized social action in meeting social obligations. Secondly, the reason the neighbors did not act together was that they had no history of common action and no institutional basis for it. If there had been a neighborhood association or, better yet, a "crime watch" organization, it is much more likely that the neighbors would have come together and done something [5]. This shows the importance of having adequate social structures to enable and to support common action.

The importance of social structures and their place in our democratic system has been the subject of much recent literature, both secular and religious [6]–[8]. Consideration of this social dimension of the human person widens our sense of responsibility to include those obligations that we share by reason of the groups to which we belong; but it also presents us with new possibilities for responding to those obligations and thus increases our freedom.

3. THE SOCIAL RESPONSIBILITIES OF ENGINEERS

It is not difficult to see how conditions (1), (2) and (4) in the analysis of Simon, Powers, and Gunneman, apply to engineers in the nuclear industry.

With regard to (1), there is certainly a *critical need* either to make nuclear power safer or to find alternative energy sources. Nuclear power does offer benefits that make it extremely attractive if it can be produced safely. Nuclear fuel is relatively inexpensive and abundant in the United States, and its use in preference to coal could actually save lives becuase of reduced air pollution [9]. Yet there are many serious dangers connected with nuclear power. These include health hazards due to radiation encountered during the mining and processing of nuclear fuel [10], the possibility of a catastrophic accident at a nuclear power plant [11], and the long-term effects of nuclear wastes [12].

Regarding point (2), engineers have a special *proximity* to the problem. Their potential contribution to the responsible use of nuclear power is unique and irreplaceable for a number of reasons. One is that engineers, because of their special training and expertise and their involvement in the design and maintenance of nuclear power plants, are best qualified to clarify the technical issues in the nuclear power debate. They can define the risks involved and give the best available estimates of the costs of reducing or eliminating those risks.

Another reason why the contribution of engineers is so important is that they are the first to see and to evaluate the dangers and risks of the technology. They work with and evaluate the systems while they are still in the design stage, before an irreversible commitment has been made to them. Thus the engineers, and they alone, are in a position to recognize potential dangers while something can still be done about them.

The final contribution engineers can make is to propose and explore alternatives to the current technology that might avoid some of its problems. It is a fundamental ethical principle that the best way to handle ethical dilemmas, where a certain amount of evil must be accepted in order to do good, is to find imaginative alternatives that minimize the evil effects or avoid the conflict entirely. If engineers could find alternatives to current nuclear technology that do not have many of the safety problems that it has, they would have made a considerable contribution to the nuclear power debate. In fact, David Freeman, the director of the TVA, a large user of nuclear power, has predicted that the future of nuclear power in the U.S. depends on "our choosing a more 'forgiving' nuclear reactor design and then standardizing it." ([13], p. 273)

The fourth condition of Simon, Powers, and Gunneman is the absence of other sources of help. Indeed, engineers do often excuse themselves from responsibility on the assumption that existing institutions, such as government regulation or the discipline of the marketplace, will take care of any problems. In the case of nuclear power, however, these institutions have not performed in a way that inspires great confidence.

From the very beginning, the Atomic Energy Commission, which was given the responsibility of regulating nuclear power, was also deeply involved in its promotion. As a result, substantial questions about safety were not raised, or, if they were, were swept under the rug. While a licensing procedure for nuclear power plants was set up, no specific standards for safety were set, and the main interest seemed to be in expediting the approval process in order to get plants on line as quickly as possible [14]. A study done by Steven Ebbin and Raphael Kasper on the nuclear regulatory process in 1974 found that "the licensing process is one which is geared to the promotion of nuclear power plants. The hearing process was originally designed to provide a mechanism to inform the local citizenry of the benefits of a nuclear power generating plant." ([15], p. 5)

It was only after the Three Mile Island accident that the Nuclear Regulatory Commission, the successor to the AEC, began demanding a number of design changes in nuclear power plants, both those already in operation and those being designed and built. These changes have had an effect, making the plants perhaps three to six times safer than before [16]. Yet they have been costly because they have come so late. Certainly they do not make up for the years of neglect of safety issues. Nor have they served to win back public confidence in nuclear power or its regulators.

None of this is to be taken to mean that government regulation is unnecessary, or that it will always perform as badly as it has in the case of nuclear power. But it does show that for commercial power, as for any large-scale technology, it cannot be taken for granted that the government will be able to protect the public adequately from the risks of the technology.

Nor is there any basis for supposing that without government regulation the nuclear industry would have sought solutions to the problems of nuclear safety or that consumers would have demanded that the cost of electricity be raised in order to make nuclear power safer. The history of private industry in the U.S. has shown that both producers and consumers have to be forced to accept responsibility for the long term effects of economic activity. ([8], p. 39) The nuclear industry in particular has showed itself quite content to enjoy government protection from such "external" costs as the risks to public health and safety from possible radiation releases, the cost of nuclear fuel processing, and the problems of disposing of nuclear wastes. In fact, such protection was a necessary condition for the involvement of private industry in the development of nuclear power [17].

4. POSSIBLE RESPONSES

The remaining condition of Simon, Powers, and Gunneman is the third one, the ability to act effectively. This presents a much more difficult problem. It often seems that existing social structures limit seriously the ability of engineers to act on behalf of the public safety without undue risk to themselves. To see how this is so, we will now look at four ways in which engineers have tried to respond to the ethical problems raised by nuclear power. By studying these responses, we hope to learn something about the possibilities for engineers' involvement in social problems and the limitations on it.

4.1 Control by the Technical Elite

The first response is what might be called technical paternalism. It springs from the attitude that the issues involved are purely technical and that the public should simply step aside and let the engineers and scientists decide them. In the early years of nuclear power, this took the form of the "technical fix." The assumption was that a well designed nuclear power plant would not pose any danger to the public. If a safety problem arose, it would simply be fixed. "While many of the hazards of radiation were clearly apparent in 1946, no one viewed the potential threat of an evolving nuclear technology as intolerable or something good engineering could not acceptably control." ([14], p. 22). In the commercial nuclear power industry, this was based on a reliance on multiple backup safety systems, designed to handle all imaginable contingencies. "'Defense in depth' bespeaks nuclear engineers' confidence that scientific method, carefully applied, can permanently triumph over human fallibility. The key assumption behind this approach is that nuclear experts can anticipate each and every sequence of events that might lead to serious accidents. Armed with that foreknowledge, they can construct systems that no combination of human error or national catastrophe can compromise." ([13], p. 176]). The problem was that it never quite worked. Critics kept finding holes in the safety systems, and accidents kept occurring with alarming frequency. Each emergency, of course, was met with more technology. "However, each attempted solution seemed to bring with it further problems, while continued scientific research and experience kept bringing to the surface problems that were either unsuspected or had been considered remote possibilities." ([18]. p. 45)

Finally, with the greater public awareness of the problems of nuclear safety, the industry had to admit that there were risks. To justify these risks, they attempted to "prove," by a supposedly objective calculus, that the benefits outweighed the risks and therefore made them worth taking. This approach is standard practice for an engineer. As a top manager in the nuclear industry put it, "An engineer is a practical man. He is resigned to using science for the betterment of mankind. And he's used to making tradeoffs. Cost-benefits are put into him from day one of engineering school—if you do this, you'll get that, is it worth it?" ([13]. p. 180)

A prime example of this type of analysis is a study on reactor safety commissioned by the AEC and coducted by a committe chaired by Professor Norman Rasmussen of MIT. The study used current reliability analysis techniques to estimate the probability of failure of a certain type of reactor and the consequences of such a failure. It had unquestionable value as an engineering tool, pointing out weaknesses in the reactor safety systems and providing a basis for comparing alternative reactor designs [19]. However, the AEC also used its calculated probabilities and consequences as the basis for estimating the absolute risk of human life imposed by the present generation of nuclear reactors [20]. Since this was found to be less than 0.1 percent of the total risk due to all man-made and natural causes, it was assumed to be acceptable.

This paternalistic approach does have certain advantages, of course. For one thing, some problems actually can be solved by applying proper engineering techniques. For example, in the first commercial power plants, the release of radiation during routine operation was a major concern. New standards and the redesign of the existing plants took care of that. "The limits on the allowable release of radioactivity from nuclear power plants have now been reduced to the point where few critics think that the issue is worth their time anymore." ([21], p. 38)

Nevertheless technological paternalism brings dangers of its own. First of all, it tends to overlook the uncertainties in technological determinations. It assumes that all the factors are well-defined, well-understood, and quantifiable. "[Nuclear engineers] tend to ignore uncertainty and concentrate only on that part of the knowledge base that is subject to quantifiable manipulation. As a result, a point of view emerges whereby technical analyses are expected to resolve the problem. Because of the wide bands of uncertainty on safety improvements, we need different kinds of decisions that aren't technical." [22], p. 60)

In the case of the Rasmussen Report, for example, later analyses have shown that for various reasons the actual "best estimate" of the risk to human life due to reactor accidents could be higher than the one calculated in the report by several orders of magnitude [11]. This would make the risk more than what the public has found acceptable in similar situations, so that there is no basis for pronouncing the risk acceptable on the basis of Rasmussen's study [23], as the AEC claimed when the report first came out.

The second problem with a cost-benefit analysis is that it is not as "objective" or "value-free" as its proponents claim. Like any utilitarian calculus, it contains hidden assumptions about the value of competing goods and how they are to be compared [24]. For example, there must be some assumption about the economic value of a human life, so that threats to human life can be compared with other "goods." Furthermore, even if cost-benefit analysis were an acceptable way to determine overall whether a certain risk was acceptable, it is not sensitive to inequities in the distribution of risks. It simply finds the course that brings the largest net benefit. A cost-benefit analysis might find it desirable to expose a small and powerless minority to grave danger, so that the rest of the society could benefit. That would be a gross injustice, of course, and totally unacceptable for that reason, but the cost-benefit analysis by itself would not reflect that fact.

The insensitivity of cost-benefit analysis to justice and other nonquantifiable values is one reason why the nuclear debate has become so intractable. Many of those who challenge nuclear power do not take issue with the facts and figures of the engineering analyses but with the underlying assumptions, especially the assumption that human life can be quantified or have a price tag put on it. In other words, the real issues are ethical ones, and they must be decided on the level of ethics, of value judgments, not as a result of technical calculations. "Scientists... cannot determine when something is safe or safe enough, because that is a matter of preference or judgment. Does the group want to live with the risks described by the scientist as accompanying the product, pay for reducing the risk, or, instead, forego the product?" ([14], p. 3)

The third objection to paternalism is that it violates the right to self-determination of those who are affected by the use of technology. Lurking behind the paternalistic ideal is the assumption that technically trained people should make all the decisions on the use of technology, since they are the ones who best understand it. This has certainly been the attitude of the nuclear industry and its regulators. According to a former legal counsel for Babcock and Wilcox, "The public should rely on the NRC to carry out the task it is charged with: protecting the public. The public must have confidence that these people are doing their jobs, and if they're not, to get new people to do it. But to have housewives coming into these highly sophisticated technical decisions is ridiculous." ([13], p. 152). Yet no group of technical experts has the right to impose significant risks on the public without their consent. The risk bearers have the right to decide if the risk is worth bearing.

These and other objections to risk assessment procedures such as those used in the Rasmussen Report are detailed in a recent study by K. S. Shrader-Frechette [25].

Engineers have an obligation to be involved with issues of public safety and welfare; but that does not entitle them to control the decisions to be made. Rather they must help the public decide what they want and whether it is worth assuming the burden required to get it. As the National Academy of Sciences' Panel on Technology Assessment has said, "Decisions affecting the course of technology, and hence the course of history, require the broadest possible public participation and should not, even if they could, be delegated to narrow elites." ([3], p. 50)

4.2 Going Public

What can engineers do when they are aware of a serious safety problem, or some other misuse of a technology they are involved with, and they are blocked in their attempts to seek remedies through normal channels? One possibility is to speak out publicly in the hope of bringing enough pressure to bear to force action on the problem. This is commonly known as "whistle blowing." Sometimes this is accompanied by resignation, on the grounds that the protesting engineer can no longer in good conscience be associated with the project. Even if the engineer is willing to stay on, he or she is often fired.

There have been a number of whistle-blowing incidents in the nuclear industry. One involved Peter Faulkner, a systems application engineer with Nuclear Services Corp. In March, 1974, Faulkner presented a paper that raised questions about the adequacy of the design of one of General Electric's reactor containment structures to a Senate subcommittee hearing on energy research and development. The paper also expressed concern about mismanagement and inadequate attention to safety in the nuclear industry. Faulkner did this, he wrote later, because "through my studies of safety, management and reliability problems, I became aware of many engineering deficiencies in nuclear power systems that were already on the market. [26]" Three weeks later, Faulkner was fired. He continued to speak out, however, and in 1975 GE began its own investigation of the adequacy of the containment design. The next year, the manager of the study, Dale Bridenbaugh, and two other engineers, Greg Minor and Dick Hubbard, resigned and began speaking out on nuclear risks. They testified before the Joint Committee on Atomic Energy that "the cumulative effect of all design effects and deficiencies in the design, construction and operation of nuclear power plants makes a nuclear power plant accident, in our opinion, a certain event." ([13], p. 92)

Engineers like Faulkner and the "GE Three" are caught in a tragic conflict between their loyalties to their jobs and their employers, on the one hand, and their "wider loyalties" to the public good, on the other. That they have chosen to honor the wider loyalties, even at great personal expense, is commendable. In doing this, they are being faithful to their responsibilities as engineers. "A determination that a project entails an unjustifiable risk to the public is a matter of engineering judgment and general values. Having made such a judgment, a decision to withdraw from the project clearly falls in the realm of engineering ethics." ([27], p. 56)

Sometimes whistle blowing can be effective. Faulkner, the "GE Three," and other engineers and scientists who have "gone public," have been able to bring their message to the public through Congressional hearings, the electronic and print media, and Nuclear Regulatory Commission licensing hearings. Moreover, they have been able to lend scientific substance and credibility to the protests of activists. Certainly these engineers cannot be dismissed by their opponents as either ignorant of the technical issues or as anti-technology.

Whistle blowing may be the best available response for an individual under certain tragic circumstances. Nevertheless, it cannot in general be accepted as an adequate way for the engineering profession to meet its social responsibilities. For one thing, it is unjust to the engineer who speaks out and then must accept the consequences. Faulkner, for example, not only was fired, but was never able to get another job in the nuclear industry, even though he applied to 67 corporations for employment. Whether he was blackballed or not, at the very least companies regard with suspicion anyone who might make embarrassing allegations about them, no matter how justified. Harassment and hostility are the standard response to whistle blowing. As a result, very few are willing to take the chance and speak out. Faulkner's colleagues agreed with him on the substantive issues of his allegations but had "no desire to rock the boat."

Another problem is the polarization that is almost inevitable in a whistle-blowing situation. The organization against which the charges are being made regards the action as a threat and an act of disloyalty and reacts accordingly. It is usually intent on both intimidating and discrediting the whistle blower. The individual, meanwhile, has taken a big risk in taking a stand and now is enduring unjust harrassment because of it. As a result, positions harden and personal animosities enter into the debate. It becomes almost impossible for either side to be objective and fair. This often obscures the original point of the whistle blower's challenge, so that the important concerns that inspired the challenge become lost amidst the charges and counter-charges.

A number of authors have proposed procedures for whistle blowers that are designed to keep the discussion as fair and rational as possible, prevent overreaction, keep alive the possibility of resolving the dispute, and provide some measure of protection for the whistle blower [27]. These are not likely, however, to overcome the fundamental limitations of whistle blowing as a way of responding to social needs. The fact that the whistle blower acts alone, taking on a large organization without supporting structures and usually without being part of an organized opposition, both puts him or her at great risk and limits the impact of the protest.

4.3 Organized Opposition

When Robert Pollard, a project manager with the Nuclear Regulatory Commission, resigned from the NRC because of its "poor attention to safety concerns." He joined the staff of a group called the Union of Concerned Scientists (UCS). Through the auspices of the UCS, he gained much greater access to the press than he might otherwise have had, and he was able to contribute his considerable expertise to a UCS study of the NRC's suppression of information on reactor safety problems. Thus by joining forces with other similarly concerned technical people, Pollard was able both to soften the effects of his loss of a job and to increase the effectiveness of his protest. This illustrates the value of engineers coming together and organizing their efforts to correct the misuse of technology.

The Union of Concerned Scientists is an excellent example of such an organization. It was formed by a group of professors and graduate students at MIT in 1969 "to encourage more humane use of scientific and technical knowledge." The safety of the nuclear power industry has been a major concern of the organization, along with alternative energy sources, and nuclear arms limitation.

The UCS has contributed to the nuclear power debate in three ways: through analysis of some of the key technical issues; through public education; and through participation in legislative and administrative hearings and other phases of the regulatory process.

The UCS has done a number of expert studies on the safety of nuclear power, including a study of the many dangers involved in the preparation of nuclear fuel and the disposal of nuclear wastes [10]; an analysis of the fire at the Brown's Ferry nuclear plant, which exposed the laxity of the NRC in enforcing safety standards [28]; and a very careful and detailed critique of the Rasmussen Report on reactor safety [11]. The last, along with the responses of a number of other technical groups, eventually caused the NRC in 1979 to repudiate many of the findings of the Rasmussen Report.

The UCS has participated aggressively in the regulatory process, both on its own behalf and in concert with groups of concerned citizens. It has been an intervenor in reactor licensing hearings and has petitioned the NRC to close some plants because of safety problems. These maneuvers have brought some improvements in reactor safety features. The UCS has also intervened in general NRC hearings on the reliability of the Emergency Core Cooling System (ECCS), a key safety system in nuclear reactors, asking for more through testing and an independent evaluation of the system. As a result, some progress has been made toward making the ECCS safer and more reliable [29].

The participation of the UCS in such hearings has been very important because citizens' groups who have tried to challenge the nuclear power industry often have not been able to marshal enough technical expertise to make a credible showing. "The entire proceeding [of an NRC hearing] is reminiscent of David versus Goliath. The intervenor's counsel sitting alone, usually without adequate technical assistance, faces two or three AEC attorneys, two or three attorneys for the applicant, and large teams of experts who support the AEC and applicants' attorneys." ([17], p. 134). Therefore a group of scientists and engineers like the UCS, who are committed to working for the responsible use of technology and the protection of the public safety, and who are willing to use the political process to do so, fills a critical need. Not only do they ensure that the important technical issues regarding public safety are raised and adequately debated, but they also help give a voice to the people who are being asked to bear the risk and who therefore have a right to be heard in the debate. "Technical expertise is a crucial political resource in conflicts over science and technology. For access to knowledge and the resulting ability to question the data used to legitimize decisions is an essential basis of power and influence." ([30], p. 15)

By organizing themselves and by allying themselves with the public, therefore, engineers can overcome some of the problems inherent in paternalism and whistle blowing. But other problems remain. One of the strengths of a group like the UCS, its independence, can also be a limitation. The fact that the organization is seen as an "intervenor," a group that is outside the normal decision-making process, gives it less credibility and influence than it might otherwise have. The UCS has met many obstructions, and has lost on a number of issues because of unresponsive administrators or judges. This marginalization also means that the outside group must scramble for resources and information, whereas the establishment forces are well-funded and have access to all the data and analyses collected by the nuclear industry and the NRC. The fact that the UCS and similar groups have accomplished as much as they have is a tribute to their determination, persistence and political sophistication. It certainly cannot be attributed to the fairness of the regulatory system.

Questions could also be raised about the wisdom and propriety of trying to settle technical issues in a highly charged political atmosphere. It is true that decisions on the use of technology are necessarily political; yet they should also be well-informed, and sometimes technical clarity is lost in the heat of battle. Scientific objectivity is a fragile thing. "The boundaries of the problem to be studied, the alternatives weighed, and the issues regarded as appropriate—all tend to determine which data are selected as important, which facts emerge." ([30], p. 16). This has led one major study of the regulatory process for nuclear power to question the effectiveness of the current adversarial system used to make decisions on reactor design and siting, waste disposal and so on.

> It is the difference in the goals of scientific method and the adversarial system that lead us to the conclusion that for the purpose of determining technical efficacy and identifying the impacts of nuclear power plants on safety and on the environment, adversary science and resolution of scientific issues in a trial type of proceeding must fail to provide an adequate resolution of scientific and technical issues. ([15], p. 222)

As we have seen, the engineer has a specific contribution to make to problems on the use of technology precisely in his or her role as engineer. But it is not possible to fulfill that role adequately without room for objectivity, cooperation and the free flow of information. If the system does not allow that, it cannot be said to be protecting the rights either of the public or of the engineer, and changes are called for.

4.4 Professional Societies

The one institution through which engineers act corporately as members of their profession, and which has a broad base of support, is the professional society. Most of the engineering specialties have such societies. They are generally very involved in technical activities such as the publication of research, and sponsoring of technical conferences, continuing education, and the development of technical standards. They also are involved in lobbying for government action on issues that affect the welfare of their members, such as the improvement of engineering education, increased funding for engineering research, the free flow of scientific information, and more liberal standards for Individual Retirement Accounts. The focus of the societies, then, is more often on the profession itself than on the profession's responsibilities to the public. As Judith Swazey said, in a speech to the AAAS Workshop on Professional Ethics, "Professions tend to operate as fairly narrow self-interest groups when push comes to shove, that in the end are more apt to protect the guild than to protect the public interest when there is a conflict." ([31], p. 63)

Most engineering societies have codes of ethics. These tend to concentrate on the responsibilities of individual engineers to serve employers and clients with competence and integrity.[31] Nevertheless, there has been a growing awareness of the profession's wider responsibilities to the public. For instance, while the 1912 code of the American Institute for Electrical Engineers said that "the engineer should consider the protection of a client's or employer's interests his first professional obligation, and therefore should avoid every act contrary to this duty," the current code of the Institute of Electrical and Electronics Engineers (IEEE) says that the responsibility to employers or clients is limited by the responsibility to "protect the safety, health and welfare of the public." The code of the Engineers' Council for Professional Development is even stronger: "Engineers shall hold paramount the safety, health and welfare of the public." ([32], p. 8). These acknowledgments of the broader social responsibilities of engineers are important; but to date little has been done to put them into practice. A study by the American Association for the Advancement of Science on the ethical activities of professional societies found that "the professional scientific and engineering societies have not developed in-depth programs addressing the ethical implications of their members' work." ([31], p. 8)

There are two areas where professional societies could take practical measures to "protect the safety, health and welfare of the public." First of all they could lend their technical expertise to those who are questioning the use of certain technologies; or they could act as independent evaluators of controversial technical issues such as those involved in reactor safety or the disposal of radioactive wastes. If the technical expertise were to come from the professional societies, with their broadly based membership, their prestige, and their independence, it might avoid some of the problems of partisanship that have arisen with organizations whose sole purpose is to challenge certain technologies.

The IEEE took one small step in this direction by agreeing to do an independent accreditation of laboratories that test and certify safety equipment used in nuclear power plants. After inviting the IEEE to do this, however, the NRC refused to go ahead with the plan [33]. This indicates that existing institutions and attitudes still must change before any very meaningful involvement is possible.

The other action the professional societies can take on behalf of social responsibility is to protect whistle blowers from reprisals from their employers. A number of proposals have been made in that regard. For example, Stephen Unger has urged that engineering societies work for an environment where the engineer can "function as a professional rather than a cog in a machine. . . . Where his professional judgment is overruled in serious matters involving the public welfare, and where protests to his management are to no avail, the engineer should be free to take his cases to professional societies, government, or any other appropriate forum. [34]." The society could investigate the case and publish its conclusions, identifying the employer if it is found at fault. The society could also perhaps give legal aid.

The IEEE, for one, does have a section in its bylaws in which it pledges support for "any member involved in a matter of ethical principle which stems in whole or in part from such member's adherence to the Code of Ethics, and which can jeopardize the member's livelihood, compromise the discharge of such member's professional responsibilities, or which can be detrimental to the interests of the IEEE or of the engineering profession." This provision has been invoked once in support of a member. The employer was found at fault and was publicly identified. ([31], p. 93f). To date, however, the professional societies have hesitated to take strong united action in support of their members, so that they have limited influence in such cases. "Employers may have little incentive to settle disputes with the professional society, which often lacks the motivation, evidence or resources to pursue more serious options such as publicizing the violation and naming the offending employer, or taking legal action." ([31], p. 57)

The engineering societies do have the potential for providing an institutional base for engineers to act on social issues. But they are still a long way from realizing that potential. There are various legal and institutional obstacles, to be sure. But the major problem seems to be internal. Becuase of the pluralism of the engineering profession, engineers' preference to concentrate on technical issues and stay away from "people problems," and the tradition of putting clients' and employers' interests before the public's, engineers have to date been slow to take united action on issues of social responsibility.

5. NEW DIRECTIONS

As the above study has shown, engineers should not simply act as individuals in meeting their social responsibilities, but should look for ways to foster collaboration, both within the profesion and with those outside the profession. Collaboration within the profession is important especially when engineers must challenge the established forces of government and industry, not only to make the protest more effective, but also to provide protection and support for those who might have to endure reprisals because of the challenge. Collaboration with the general public is important because decisions about the use of technology involve value judgments as well as technical knowledge. Therefore the obligation of engineers to protect the public does not mean that they, or any elite group, should decide what risks are worth taking or how they are to be distributed. Engineers should rather help the public make responsible, well-informed decisions on these issues.

One reason that such collaboration has not occurred as much as it should is that existing social structures have not supported it. Certainly the government and the private sector are too committed to the technology to tolerate a really critical view of it. Organizations such as the UCS have accomplished a great deal. But, through no fault of their own, they have been kept out of the decision-making process as much as possible, so that an adversarial situation is almost inevitable. Professional societies have not yet shown themselves capable of sustained, effective action on behalf of the public welfare.

There is a need, then, for new structures, for new ways of making decisions about the development and use of technology. These structures should first of all allow independent technical evaluations as part of the decision-making process. These evaluations should be free both from pressures to bias the results to serve some particular interests and from the distortions of the adversarial process. Without this freedom it is very difficult for engineers to maintain their integrity and to make their unique contribution to the responsible use of technology. Ebbin and Kasper, for example, after studying the present regulatory system for nuclear power, concluded that there was a need for independent "technology assessment teams" of experts who could evaluate the consequences of a technology so that the public debate could at least be well-informed [15].

Second, the structures should be collaborative. They should not only allow interdisciplinary teams of scientists and enginers, but should also bring these technical people together with other interested parties, including representatives of government, industry, labor and, of course, the public. This is the only way to ensure that all legitimate interests are represented and that the different sides begin to appreciate one another's point of view. This is crucial for any kind of fair and reasonable resolution of the problems. If the technical people do not understand what the real vaues at stake are, for example, they are not likely to define the technical issues in a way that is helpful for the discussion. And if the public and other nontechnical interests do not appreciate both the possibilities and the limitations of technical expertise, they will not be able to evaluate the alternatives intelligently.

One way to foster greater collaboration between engineers and other interested parties is to use a consensus model of decision making. In this model, all the relevant interests in effect sit around a table and hammer out a solution that everyone can live with. This is different from an adversarial process, where opposing sides make as strong a case as possible for their chosen alternatives and a supposedly neutral third party decides between them. In a consensus model, many conflicting interests are represented, but all those involved ultimately have the same goal: to find a solution that satisfies all interets. Thus there is a great deal of give and take, with the forming of alliances and the shifting of positions, in an attempt to build a consensus around a particular solution. All interests get a fair hearing, at least, and there is a better chance either to make the necessary compromises and tradeoffs or to find creative alternatives that better satisfy the competing interests. In this type of many-sided discussion, too, engineers and scientists can take part as independent parties not aligned with any one side yet very much in dialogue with everyone.

One example of such a broadly based consensus process was the Kemeny Commission that was convened to investigate the accident at Three Mile Island. The commission included a biochemist, a nuclear engineer, a physicist, a state governor, an ex-presidential aide, an industrialist, a trade unionist, a sociologist, a journalism professor, an environmentalist and a woman not employed outside the home, commonly called a housewife. According to Kemeny, "I think it would be wrong for the Commission to consist of a bunch of engineers. We are sort of a national jury on this issue and would you want the jury to consist of experts? I think it would be a mistake; you want the average American to believe what this Commission says." ([35], p. 137)

The Commission did in fact produce a credible report. It was endorsed by groups as diverse as the Union of Concerned Scientists [29] and *Nuclear News*, a publication of the American Nuclear Society, which has always been most supportive of the nuclear industry [36]. Investigating an accident is somewhat different from deciding whether and how to use a certain technology, of course, but the idea of a "national jury," representing a broad range of interests, to assess the technical facts, weigh the value judgments, and recommend a course of action on the use of a particular technology is a promising one.

These are only rough suggestions. Much more investigation, discussion and development will be needed to build the necessary structures. But until the mechanisms and institutions are in place that give engineers the freedom and opportunities for collaboration that they need in order to respond to social needs, the public will be without an important source of protection from the risks of technology, and engineers themselves will not be able to fulfill some of the most important duties of their profession.

REFERENCES

[1] Alger, P., N. A. Christensen, and S. Olmsted, *Ethical Problems in Engineering*, New York: John Wiley and Sons, 1965.

[2] Fruchtbaum, H., "Engineers and the commonwealth: notes toward a reformation," *Ethical Problems in Engineering*, A. Flores, Ed. Troy, NY: Center for the Study of the Human Dimensions of Science and Technology, 1980, p. 253-261.

[3] Keating, W., "Politics, energy and the environment: the role of technology asessment," *American Behavioral Scientists*, vol. 19, no. 1, Sept./Oct. 1975, pp. 39-42.

[4] Simon, J. G., C. W. Powers, and J. P. Gunneman, *The Ethical Investor: Universities and Corporate Responsibility*, New Haven: Yale University Press, 1972.

[5] McMillan, G., "Neighborhood raises a cry and a rape is prevented," *The Boston Globe*, vol. 227, no. 14, Apr. 24, 1985, p. 21.

[6] Adams, J. L., *On Being Human Religiously: Selected Essays in Religion and Society, Max Stackhouse, Ed., Boston: Beacon Press, 1976, pp. 57–88.*
[7] Berger, P. L. and R. J. Neuhaus, *To Empower People: the Role of Mediating Structures in Public Policy.* Washington, DC: American Enterprise Institute, 1977.
[8] Lodge, G. C., *The New American Ideology.* New York: Alfred A. Knopf, 1975.
[9] Jordan, W., "Nuclear energy benefits versus risks," *Physics Today,* May 1970, p. 33.
[10] Union of Concerned Scientists, *The Nuclear Fuel Cycle: A Survey of the Public Health Environmental, and National Security Effects of Nuclear Power.* Cambridge, MA: MIT Press, 1975.
[11] Kendall, H. W., *et al., The Risks of Nuclear Power Reactors: A Review of the NRC Reactor Safety Study WASH-1400. Cambridge, MA: Union of Concerned Scientists, 1977.*
[12] Walker, C. A., L. C. Gould and E. J. Woodhouse, Eds, Too Hot to Handle: Social and Policy Issues in the Management of Radioactive Wastes. New Haven: Yale University Press, 1983.
[13] Hertsgaard, M., *Nuclear, Inc.: The Men and Money Behind Nuclear Energy.* New York: Pantheon Books, 1983.
[14] Rolph, E., *Nuclear Power and the Public Safety.* Lexington, MA: D. C. Health, 1979.
[15] Ebbin, S. and R. Kasper, *Citizen Groups and the Nuclear Power Controversy: Uses of Scientific and Technological Information.* Cambridge, MA: MIT Press, 1974.
[16] Phung, D. L., "Technical note: LWR safety after TMI," *Nuclear Safety,* vol. 25, no. 3, May/June, 1984, pp. 317–323.
[17] Del Sesto, S. L., *Science, Politics and Controversy: Civilian Nuclear Power in the United States, 1946–1974,* Boulder, CO: Westview Press, 1979.
[18] Falk, J., *Global Fission: The Battle over Nuclear Power:* London: Oxford University Press, 1982.
[19] Weatherwax, R. K., "Virtues and limitations of reactor safety," *Bulletin of the Atomic Scientists,* vol. 31, Sept. 1975, pp. 29–32.
[20] Rasmussen, N., "The safety study and its feedback," *Bulletin of the Atomic Scientists,* vol. 31, Sept. 1975, pp. 25–28.
[21] Von Hippel, F., "A perspective on the debate," *Bulletin of the Atomic Scientists,* vol. 31, Sept. 1975, pp. 37–41.
[22] Sugarman, R., "Nuclear power and the public risk," *Spectrum,* vol. 16, no. 11, Nov. 1979, p. 60.
[23] Rowe, W., "Risk analyst sees need of vote of confidence on nuclear power," *Spectrum,* vol. 16, no. 11, Nov. 1979, p. 60.
[24] MacIntyre, A., "Utilitarianism and cost/benefit analysis," in Donald Scherer and Thomas Attig, Eds., *Ethics and the Environment.* Englewood Cliffs, NJ: Prentice-Hall, 1983, pp. 139–152.
[25] Shrader-Frechette, K. S., "The conceptual risks of risk assessment," *IEEE Technology and Society Magazine,* vol. 5, no. 2, June, 1986, pp. 4–11.
[26] Faulkner, P., "Exposing risks of nuclear disaster," in *Whistle Blowing: Loyalty and Dissent in the Corporation,* Alan F. Westin, Ed., New York: McGraw Hill, 1981, pp. 39–54.
[27] Unger, S., "How to be ethical and survive," *Spectrum,* vol. 16, no. 12, Dec. 1979, pp. 56f.
[28] Ford, D. F., H. W. Kendal, and L. Stye, *Brown's Ferry: The Regulatory Failure.* Cambridge, MA: Union of Concerned Scientists, 1976.
[29] "15 years of UCS," *Nucleus (a quarterly report of the UCS),* vol. 6, no. 1, Spring, 1984.
[30] Nelkin, D., Ed, *Controversy: Politics of Technical Decision.* Beverly Hills, CA: Sage Publications, 1979.
[31] Chalk, R., M. Frankel and S. Chafer, *Professional Ethics Activities in the Scientific and Engineering Societies.* American Association for the Advancement of Science, Washington, D.C., 1980.
[32] Baum, R., *Ethics and Engineering Curricula.* Hastings-on-Hudson, NY: The Hastings Center, 1980.
[33] "Nuclear laboratory accreditation: IEEE presses NRC for action," in *The Institute,* vol. 7, no. 1, Jan. 1983, p. 1.
[34] Unger, S. "Engineering societies and the responsible engineer," in *The Social Responsibilities of Engineers Annals of the New York Academy of Sciences,* vol. 196, no. 10, Harold Fruchtbaum, Ed., 1973, p. 435.
[35] Nelkin, D., "Some social and political dimensions of nuclear power: examples from Three Mile Island," *American Political Science Review,* vol. 75, no. 1, March 1981, p. 132.
[36] Payne, J., "Kemeny commission: a bright amber light," *Nuclear News,* vol. 22, no. 15, Dec. 1979, p. 25.

FEATURE ARTICLE

Would Helping Ethical Professionals Get Professional Societies Into Trouble?

STEPHEN H. UNGER

Abstract—Various possible troubles that may befall engineering societies defending the ethical practice of engineering are outlined. These include lawsuits by irate employers and the loss of support for the societies. It is pointed out that properly conducted activities in this realm are not only very well justified but are highly unlikely to result in damage to the societies involved. Attacks made in the spirit of vengeance are far more likely to damage the attacker than the intended victim. Reference is made to the experience of other types of organizations that have remained unscathed, even though, at first glance, they appear to be highly vulnerable to legal harassment.

INTRODUCTION

We have seen in recent years numerous cases of harm being caused to individuals and to society as a whole through badly implemented technology. The Chernobyl, Challenger, and Bhopal disasters come to mind immediately, but there have also been many lesser instances involving such matters as releases of toxic chemicals into the environment and defects in the design of automobiles.

A common factor in such cases is that the problems were known in advance to some individuals, usually to engineers in the organizations involved. It is also often the case that these people called attention to the problems and their possible consequences, but that their warnings were ignored. Engineers who persist in efforts to get such problems corrected despite negative reactions from their managers are often punished. Hence, as in the case of the Challenger episode, the engineers frequently stop short of going over the heads of management when their professional judgments are arbitrarily overridden.

In certain types of cases, there are laws to protect those who act to correct hazardous situations, but these are, at present, rather limited in effectiveness, as well as in scope. (The Morton Thiokol engineers, for example, would not have been covered.) It would be very helpful to broaden and strengthen these laws, and efforts along those lines are now being made.

It has been suggested that it would be very appropriate for engineering societies to play an active role in these matters by backing up engineers who, when seeking to adhere to high standards of professional conduct, find their careers jeopardized by managerial actions. The model pointed to here is that of the American Association of University Professors (AAUP), which has for more than half a century been an effective force in safeguarding academic freedom for university professors.

Among the ideas proposed for implementation by the societies are the following:

1. Carefully investigating specific cases and publishing reports, where justified, that show how an employer has acted improperly toward an engineer in cases of the type referred to above. (Central to the AAUP approach.)
2. Explicitly censuring the employer in such cases, and calling the matter to the attention of the profession in general. (Also part of the AAUP procedure.)
3. Publishing ratings of employers indicating the extent to which they encourage ethical behavior on the part of their engineering employees.
4. Establishing legal defense funds to help ethical engineers who find it necessary to resort to the courts.
5. Giving awards to engineers who uphold ethical principles in situations where this places their careers in jeopardy.

The IEEE has already established machinery for implementing items 1 and 5. However, the mechanism related to item 1 has been fully exercised in only one case. There are several explanations for the fact that the IEEE (and perhaps other engineering societies) has not taken a strong position on the matter of supporting ethical engineers.

Since a portion of IEEE's membership consists of upper-level managers and owners of businesses, there is some feeling that it would be divisive to have the IEEE appear to take sides in disputes between members. Rather than analyze this position in detail, I will simply point out here that, as is understood by many managers who are IEEE members, measures to encourage ethical behavior by engineers and to discourage improper behavior by managers serve the interests of good managers as well as the interests of ethical engineers and of society as a whole. This is because the existence of such measures has the effect of alleviating pressure on honorable managers to cut corners to keep up with less scrupulous competitors. It also helps them resist efforts by *their* managers to get them to do things that they feel are improper.

Other explanations of the lack of vigorous action by engineering societies are concerned with costs that may be incurred in the process. These include direct expenses, such as the need for added staff and legal services; the possibility of being faced with extraordinary legal costs

The author is with the Computer Science Department, Columbia University, New York, NY.

This article was presented at the Electro '87 Conference on April 7, 1987. It has been revised for publication.

Reprinted from *IEEE Technology and Society Magazine*, Vol. 6, No. 3, pp. 17–21, September 1987.

in defending against lawsuits brought by companies and individuals in conflict with the engineers being defended; the possibility of losing such suits and, thereby, being required to pay large sums in compensatory or even punitive damages; and the possibility of losing the support of employers of engineers. In the following sections, the reality of these dangers is assessed, and means for minimizing them are proposed.

LAWSUITS

There are two basic parts to this problem. First, to what extent is it likely that a society involved in defending the engineer's rights to practice ethically might be found guilty of unjustifiably causing significant financial losses to individuals or corporations? If it can be shown that the answer to this question is that such an event is, or can be made to be, extremely unlikely, then at least the threat of having to pay huge amounts in damages could be dismissed. But the matter does not end there. The second part of the problem is the extent to which suits might be filed against the society that, even if unsuccessful, would still be costly to defend against. An organization could conceivably be bankrupted by lawyers' fees and other costs entailed in fighting off attacks by vengeful employers. Both of these issues are addressed below.

HOW TO AVOID LOSING

The processes for defending ethical engineers can gain respect and become effective only if they are, and are perceived to be, executed in a careful manner, with strict attention to accuracy and fair play. Their objectives must also be, and be perceived as, the raising of the moral level of the practice of engineering. They cannot be effectively used to further the selfish interests of the members of any group.

If, for example, reports are published based on investigations that misrepresent facts, or that do not fully explore pertinent arguments in support of the management position, then engineers reading these reports will soon learn that they cannot be trusted, and, hence, that it would not be sensible to take them into account when considering job offers. If the support machinery is used to pursue selfish economic interests of society members, then it will be considered as only another weapon in economic warfare, to be attacked or defended according to one's own economic interests. (I do not mean to imply here that there is no proper place for organized efforts by engineers to increase their incomes. The rights and wrongs of such efforts should be considered in the same way as corresponding efforts of other groups, such as physicians, nurses, or airline pilots. But any mechanisms for promoting economic interests should be kept separate from those intended to promote engineering ethics.)

It happens that the same characteristics that make support efforts effective, principally truth and fairness, also serve to protect these efforts against successful litigation. There are two components to this protection. First, in American courts, truth is an absolute defense against charges of libel or slander, or causing actionable damage in the marketplace. (This is *not* the case in many other countries, including some democracies such as Great Britain.) Hence, if the reports published by the support committees are free of factual errors, then there is no basis for successful damage suits. Nor is a report summarizing, in a factual manner, the responses of engineers to a questionnaire pertaining to the way their employers deal with dissenting professional opinions a very good target for a lawsuit.

But, human enterprises being what they are, is it really possible, even with the best will, to guarantee that, over a long period of time, with many reports being issued, that absolutely *no* factual errors will creep in? The question answers itself. Perfection, while a worthy goal, is not a wholly attainable one. Mistakes *will* be made. Not surprisingly, this fact is taken into account by our legal system. As a second line of defense, if it can be shown that an error, even a damaging one, occurred despite sincere, *diligent* efforts to be accurate, and if the thrust of the enterprise in which this error was made was one operating on behalf of a good cause, then the likelihood of being found at fault is very small.

What is needed then is a carefully designed set of procedures to guard against error, incorporating a variety of mechanisms for cross-checking the essential ingredients of the report. (While great care should be taken to make the reports completely accurate, obviously some errors are of little consequence, while others might be of a critical nature.) In this respect, we can draw upon the experience of the AAUP for an extremely powerful tool that they have used to ensure both accuracy and fairness.

After doing the best they can to ascertain all of the relevant facts and arguments, the AAUP investigators submit drafts of their reports to all of the parties concerned for their review and comments. Thus, if there are significant errors or omissions, these can be corrected by the parties concerned. Where there may be disagreement as to the validity of certain statements in the report, it is revised to include the opposing views, perhaps with some arguments supporting one or the other, so that the reader can decide for himself. If, at some later date, it is argued (say, in a courtroom) that some crucial misrepresentation of the situation is embodied in the published report, a powerful defense is available; namely, that the parties concerned all had the opportunity to call attention to the error(s) prior to publication. A clear case can be made that the publishers were *not* negligent in their effort to get at the truth.

One might still argue, that, even with all these efforts, one cannot rely upon people to carry out such procedures perfectly, and that, as a result of such imperfections, including the imperfect implementation of the law in real courtrooms, an engineering society would eventually suffer a crushing defeat. Fortunately, we can again draw upon the experience of the AAUP, this time not for a specific technique, but rather for experimental evidence as to the validity of the above-outlined procedure for avoiding courtroom losses. The bottom line here is that, after over half a century of publications that pull no punches in assigning fault for improper treatment of faculty members, the association has never been successfully sued for damages.

Is this an artifact due to the milieu in which the AAUP operates? Is it because colleges and universities are administered by softhearted, gentle people who would not wish to become involved in such sordid activities as libel suits? I can almost hear the laughter of fellow university employees. But let us turn to a very different realm for further evidence leading to the same result.

Consumers Union (CU) has been evaluating consumer products and publishing detailed critical and comparative reports ranking specific commercial products for about 50 years. Each month, perhaps a hundred different brand names are mentioned. Clearly, being ranked low in the list of competitive brands or, worse yet, being listed as "not acceptable" can have significant effects on sales of an item. This, of course, would mean real, possibly very large, losses in income for the firms involved. Now we are in what everyone would have to admit is the "real" world. The basis for CU's evaluations are laboratory tests and judgments by panels of experts or, sometimes, by panels of just ordinary users of the product being evaluated.

CU makes every effort to be accurate and fair, but, given the scale on which they operate, it would be impossible to avoid all kinds of errors, ranging from mistakes in the reading of instruments to editorial errors in their publication. Often subjective judgments are being made by the panels. Certainly, if there were ever a natural battlefield for lawyers to wage warfare in, this is it. Obviously CU must maintain a vast legal establishment to keep them from being destroyed by enraged manufacturers of such items as "not acceptable" toilet tissues. Even when defended by the best legal talent, one would expect that CU would periodically suffer major defeats and have to pay huge amounts in damages.

Not so. While lawsuits against CU are occasionally filed, they seldom cause much trouble. Only *once* did they lose, and in this instance the judgment against them was for less than $16,000—and even *this* decision was overturned on appeal. Most of the time spent by CU in court is in the role of *plaintiff* against companies who try to use its ratings in their advertising.

What is the explanation for this record? Here we encounter another aspect of our legal system that is also very pertinent to the topic of defending ethical engineers. As does the AAUP, CU makes every reasonable effort to be fair and accurate. They make a practice of correcting errors when they find them or when others call attention to them. (They do *not*, however, submit reports to manufacturers for comment prior to publication—this would be too unwieldy a practice in their case.) But the courts recognize that CU is not furthering the selfish ends of any group—unless we define "consumers" (which, of course, all of us are) as such a group. They are considered as promoting the welfare of the public. Hence, when they make an honest mistake, they are not penalized in the same way that other types of organizations would be. For example, a sensational commercial magazine that publishes a colorful story written to attract attention and promote sales had better be extremely careful not to malign anyone with incorrect statements. Should a damaging story be found to contain any significant errors of fact, even if they were not grossly negligent in checking out the statements, they would be vulnerable to major libel suits.

But even if defeat in the courtroom is unlikely, what about the danger of winning a series of victories so costly as to lead to financial ruin? Indeed, a legal maxim is that "anyone can sue anybody over anything." Thus, an angry individual or corporation might file a suit even if there is little chance of victory, simply to punish the organization that exposed some instance of wrongdoing. The fact is that such vengeance suits have, in fact, *not* been filed against either AAUP (it has been sued only once—unsuccessfully) or CU. Some explanations for this, which apply very directly to the case of engineering societies, follow.

REVENGE CAN BE BITTER

Let us now imagine that an IEEE committee, operating in a careful and fair manner, has published a report supporting an engineer and showing the employer of that engineer in a bad light. Suppose, to be more specific, that the issue involved a protest against inadequate testing of some safety-related product. Outraged by the fact that their shoddy practices have been publicly exposed, the company president decides that no amount of stockholders' money is too great to spend in an attempt to entangle the IEEE in a costly lawsuit, even though the company's lawyers have indicated that the chances of victory are nil (for reasons along lines discussed in the previous section). Ignoring their advice, he files suit against the IEEE for $20 million in damages. What happens next?

Before the case even comes to trial, the IEEE attorneys initiate pretrial discovery proceedings. They subpoena all of the company's files having to do with the testing of not only the product directly involved in the case, but of all other related products. They call as witnesses in the proceedings (to testify under oath) all of the managers, engineers, and company officials involved in these processes. In the course of this operation, they turn up not only a wealth of evidence corroborating further the original claims made by the engineer (previously they were able to establish only that his technical position was "plausible"), but also discover hard evidence of a number of other parallel cases of slipshod practices, in some cases, violating state safety codes.

By the time the trial begins, the press and TV news departments, previously indifferent to the fate of a "mere" engineer, are closely following developments as horror stories concerning the company's products begin to take shape. At this point, the chairman of the board begins to see the handwriting on the wall and the president of the firm decides that perhaps he is not spending enough time with his family—he tenders his resignation. The new president orders the suit dropped—but now the matter cannot be ended that easily, as IEEE attorneys file a motion against the company, demanding compensation and punitive damages for their having misused the court system by bringing to it a case obviously without merit. Charges are also filed against the company's attorneys for having participated in this activity.

Thus, as illustrated in this hypothetical case, an organization has a great deal to lose by resorting to the courts in a situation where they are basically in the wrong. The damage that they might have suffered as a result of the publication of a critical report by a professional society, or by an unfavorable rating by such a society with respect to the disregard of their engineers' professional rights, may be greatly amplified by the uncovering of additional evidence concerning previously known cases, the existence of other, previously *unknown* instances of similar, or even quite different abuses of either engineering employees or of the public welfare. There will almost certainly be a great deal of publicity about all of these matters. In addition, they may wind up paying court costs and legal fees for *both* sides, as well as punitive damages. I suggest that these are the sort of considerations that account for the fact that CU and the AAUP have not been impeded by harassing litigation. If properly organized, there seems to be no reason why ethics support operations carried out by engineering societies would not enjoy similar success in avoiding legal entanglements.

LOSING EMPLOYER SUPPORT

At present, engineering societies receive significant backing from those who employ engineers. Some employers pay society dues for their engineers, the great bulk of advertising revenue from society publications comes from the same organizations, and employers often encourage engineers to attend professional society meetings on company time, with all expenses paid. In addition, many engineers carry out society business on company time, often using office services while doing so. Thus, if employers of engineers were suddenly to terminate all forms of support for engineering societies, the latter would certainly feel the pinch. How likely is it that, by instituting strong ethics support measures, a society might provoke such a response?

As in the case of kamikaze lawsuits, it is not hard to show that any organization adopting such a tactic would be damaging itself far more than it could possibly damage a society. To begin with, one should ask why organizations, particularly those whose purpose it is to make money for the owners, are *now* supporting engineering societies to the extent suggested above. The short answer is that they do so because they themselves profit from this.

In the most general sense, organizations that employ engineers clearly benefit from progress in technology, and that, in essence, is the reason for the existence of professional engineering societies. Through such activities as the publishing of technical journals, the organizing of technical meetings and exhibits, and participation in standards committees, engineering societies play a major role in the development and promulgation of new and old technology. Hence, it is little wonder that users of technology are anxious to support them.

On a more detailed level, organizations benefit directly from each type of support mentioned here. They advertise their products in the publications read by those who make the decisions about purchasing them. If they are interested in having their engineers apply the latest technology to their products, it is clear that encouraging them to be active in engineering society technical activities would be clearly beneficial.

On the other hand, if any organization decided to stop sending its engineers to technical meetings and to terminate other forms of support for society membership, they would be handicapping themselves in the engineering job market. An organization that acquired a reputation for being hostile to engineering societies on the grounds of the latter's support for ethical engineers would thereby incur resentment on the part of many in the profession. This ill will could easily be translated into a loss of sales when engineers must make choices among products that do not differ much in price or performance.

The support supplied by employers stems from self-interest, and it is not likely to be withdrawn as a result of temper tantrums that would have serious consequences to the all-important bottom line.

CONCLUSIONS

Providing strong support for the ethical practice of engineering is a very worthwhile and appropriate endeavor for engineering societies. As in the case of most significant activities, there are indeed risks involved. It is quite possible to bungle the job and get the society involved in various kinds of trouble. But the degree of competence necessary to avoid trouble is by no means out of the range of those who run engineering societies.

If you are interested in doing something that may have legal implications, do *not* ask a lawyer if it is legally safe. The answer is bound to be that this activity is fraught with hazards. This would apply if what you wished to do was as simple as buying a house. What you must do is to ask the lawyer *how* to do what you wish in a manner that minimizes legal risks. Some general principles for minimizing the risks of ethics support have already been pointed out. An additional point is made below, and, doubtless, a good lawyer would be of great value in making sure that the detailed procedures were both efficacious and minimally dangerous from a legal standpoint.

In order to emphasize that they are not merely acting in the interest of their own members, the AAUP defends the academic freedom of *all* professors, whether or not they are members of the association. Engineering societies should do likewise. (This *is* done in the case of the IEEE SSIT Award for Outstanding Service in the Public Interest, but it is *not* now done by the IEEE Member Conduct Committee.)

Ethics support should be carried out prudently, but vigorously, by engineering societies. In particular, when threatened with lawsuits, they should make it known in advance that bluffs will be called and that the full resources of the law will be utilized.

The time is ripe for action in this area. It is up to engineers, standing together in their professional societies, to see to it that they are no longer subjected to agonizing choices between sacrificing either conscience or career.

FOR FURTHER BACKGROUND

S. H. Unger, *Controlling Technology: Ethics and the Responsible Engineer* (New York: Holt, Rinehart & Winston), 1982.

M. W. Martin and R. Schinzinger, *Ethics in Engineering* (New York: McGraw-Hill), 1983.

FEATURE ARTICLE

Technological Hazards and the Engineer

ROLAND SCHINZINGER

Abstract—The nature and variety of technology hazards; how engineers and society assess the hazards and react to avoid harm; the need for a new style of engineering.

INTRODUCTION

Along with its many benefits, technology also produces numerous and often unforeseen hazards. This calls for a conscious effort to defensively design and operate our technological products.

A selection of case studies will be used to establish a taxonomy of hazards based on their origins, potential for harm, types of victims, and means of avoidance. The incentives and disincentives for hazard mitigation will be examined, leading to a discussion of how society reacts and how engineers can respond.

CASES ILLUSTRATING A VARIETY OF HAZARDS

The term "hazard" is used here to denote a peril, danger, or risk, and not an accident which has actually occurred. In this discussion, then, something hazardous is potentially harmful, but does not imply the actual presence of harm. Appropriate antonyms would be safe and reliable; they are attributes of a product, not of an external force.

If a product is hazardous—such as a car with faulty brakes or a boat with a leaky hull—we may think that a harmful outcome is only a matter of chance. But if we extend the concept of a product to encompass a more complex technological creation, then we should think in terms of a system which ought to have safeguards built into it. A car should have redundant brakes; even a small boat should have life vests and, if it is taken out very far, a raft and radio contact. In the case of the automobile, the safety feature can actually avert an accident; in the case of the boat, the safety feature will not prevent the vessel's sinking, but it will provide a means of safe escape for those in the boat. Hereafter, the term "product" will refer to a system large enough to accommodate its own safety features if such exist.

When safety measures are absent or inadequate, lives may be lost. In many cases, product and system failures are also accompanied by heavy financial loss. Harmful events can be prevented frequently through mitigation measures, even though the hazards, as such, remain. In other situations there are no safety measures which can be applied, and the hazard itself must be removed.

A selection of illustrative examples follows. Engineers may be familiar with many of these examples and wish to skip them. It would nevertheless be beneficial if they asked themselves what really posed the hazard and how harm could be averted, either by design or by providing safe exit. The cases are 1) aluminum house wiring which starts a fire, 2) an oil rig which capsizes, 3) an illness of unknown nature and origin which is traced to pipes leaking PCB, 4) careless use of computer-based police files, 5) the shipwrecks of the *Titanic* and the *Andrea Doria,* and 6) toxic gas leaks resulting in illness at Seveso and death at Bhopal.

Aluminum wiring [1, 2]

The house was quite new when a fire destroyed it. While many fires are unjustly blamed on electric wiring, there was no doubt in this instance. As had been the case in several other fires in development tracts in this California city, aluminum conductors were at fault. In the early 1960s, and especially during the Vietnam War, the price of copper had risen and aluminum had been used increasingly as a conductor for electric wiring. It comprised 15 percent of the branch-circuit wiring produced in 1972. Aluminum is not as good a conductor as copper, but its deficiency can be compensated for by using a larger diameter wire, which is still less expensive than copper. The hazard presented by using aluminum arises from its tendency to flow under pressure and to form a poorly conducting oxide layer on its surface. Poor contacts result where wires are fastened to outlets, usually in concealed locations. The subsequent arcing during current flow had started this and many other fires. The home owner was unaware of the type of conductor material used and of the potential danger. The city had failed to inform its residents that pigtailing copper wires to the end of the aluminum conductors or using improved outlets and the like could reduce the danger of fire. It was to take several more years before the city decreed that homes could not be wired with aluminum because even with improved connection devices and techniques, the high quality of workmanship, inspection, and maintenance available in industrial settings would be lacking in the residential sector.

The hazard of aluminum wiring in homes can be summarized as follows: defective design (faulty specification of material with respect to installation and maintenance practices); hazard hidden from view; damage in the form of loss of property; loss sustained by end user of product; some remedy available in form of retrofit (Cu

---

Roland Schinzinger is with the Department of Electrical Engineering at the University of California, Irvine.

Reprinted from *IEEE Technology and Society Magazine,* Vol. 5, No. 2, pp. 12–16, June 1986.

pigtails at end of Al wires) or, better yet, redesign (not specifying Al for homes); both enforceable through regulations, no simple safeguards once arcing occurs.

The Oil Rig Alexander L. Kielland [3]

Built by the French firm CFEM as an oil drilling platform and delivered to Phillips Petroleum Company of Norway in 1976, the 10,000-ton Kielland had been used only as a 'flotel' (floating hotel) by the time it capsized in 1980 on location in the North Sea. It was capable of accommodating 343, but on the fateful night of March 27th, it held only 212 men of whom 123 perished. What had caused the rig to list suddenly and then capsize? It was a poor weld on a brace which had not been considered sufficiently critical to the rig's structural integrity to warrant thorough and frequent inspection. There were so many deaths because what lifesaving equipment there was had been poorly placed or could not be used once the platform was listing heavily. Here we have an example of an occupational hazard brought on by poor workmanship (the weld) which was not mitigated by defensive design (redundancy, safety provisions). There were also inadequate means of escaping this platform (as most other platforms at that time).

Kanemi's Rice-Oil [4]

In 1968, about 10,000 Japanese were stricken by a mysterious disease which caused its victims general distress, loss of hair, disfiguring skin acne, and discoloration. An investigation of 121 patients and 121 healthy individuals, paired by age and sex, revealed that the only difference between the two groups was in the amount of fried foods consumed. The problem was traced to the Kanemi Rice Oil Company; but not until autopsies on death victims were performed could the poisonous ingredient of the rice oil be determined. It turned out to be PCB (highly toxic polychlorinated byphenyls) which had leaked out of corroded heat-exchanger tubes into the rice oil while it was being heated to remove objectionable odors. (Incidentally, Kanemi was replenishing about 27 kilograms of PCB a month because of the leak.)

This case can be summarized as having its origin in faulty maintenance, being difficult to trace, causing damage in the form of illness and death, affecting end users, and being avoidable by regulations prohibiting the use of PCB anywhere in food processing plants.

Computer-Based Police Files [5]

The three young men who bought gas at a filling station outside of Paris, France did not look trustworthy to Mr. Nicholas, the station owner. Something about their clothes, the mended license plate, and the hurriedly scrawled signature on the check spurred to him to call the police after the trio had left. Sure enough, police files queried by a computer revealed that the car had been stolen. A special police team was dispatched, and the stolen car was intercepted at a red light. Two plainclothes policemen approached the suspects, the only one in uniform remaining in the patrol car. One of the policemen had a readied machine gun, the other a .357 Magnum. One of the young men, Marcel Seltier, reported, "We did not understand anything. We saw the one with the gun aim at Claude. A moment later a shot rang out. The bullet went through the windshield and hit Claude's face just under the nose. We thought they were gangsters."

This is the case of trigger-happy policemen who in 1979 dispensed justice on the spot under the erroneous impression that the suspects were dangerous. Undoubtedly the lawmen discharged their weapons in response to some inadvertently suspicious movements on the part of the three young men still in the car. The car had indeed been stolen several years earlier. In the meantime, it had been recovered and resold by the insurance company. Here the news of a potential danger to society—thieves on the loose—was reported with dispatch, but it was accepted at face value without further investigation. The result: a more grievous harm. The earlier cases had to do with faults in design, material, manufacturing, or maintenance. Here we have a case of faulty use or operation. The automated data retrieval system is capable of rapid searches through vast files, fast communication, and fast decisions; but no checks were built into it to verify accuracy and to protect the rights of those named in the files. There was no way the innocent bystander could escape the effect of an inadvertent operator error or oversight.

Fast Ships At Sea [6, 7]

The *Titanic* sailed on during the night at high speed through seas where icebergs had been sighted. It was on its maiden voyage, the owners were eager to have the ship break the speed record, and the builders had promised an unsinkable hull. When the ship hit an iceberg at an awkward angle, more than the expected number of hull compartments were damaged. When the "Titanic" began to sink, only half of those aboard could be saved because there were not enough life boars. (There was lifeboat space for only 1200 persons—30 percent of the ship's full capacity and 52 percent of the persons on board during the maiden voyage.) The number of lifeboats had been dictated by outdated regulations which did not foresee vessels the size of the *Titanic.*

The accident was caused by overconfidence, and the loss of life was due to lack of foresight. Another example of overconfidence is the collision of the *Andrea Doria* and the *Stockholm* half a century later. The captains relied on their radar to warn them of approaching ships in fog, but they did not consider the time it would take for a big ship at high speed to change its course. Loss of life was much less, but again there was trouble with hull design: the *Andrea Doria* was supposed to float evenly after taking on water. Instead, it listed and capsized, and thus could not be salvaged. These episodes bring to mind a saying attributed to John A. Shedd: "a ship in harbor is safe, but that is not what ships are made for!"

Lack of Containment

Visit an oil tank farm and you will notice that surrounding each tank is a wide moat big enough to hold the tank's contents should it rupture. This was a lesson which had to be learned the hard way. For instance, 2.5 million gallons of molasses poured through the streets of Boston in 1919 when a large wooden storage tank suddenly burst. Numerous people drowned in the sticky fluid, which reached up to the second floor of many buildings [8]. Today we hear of high technology firms in Silicon Valley storing their toxic wastes in corroding tanks which allow leakage to the water table! [9]

Toxic liquids can be confined from leaking fairly easily. Poisonous gases are much more difficult to contain. It can be done, as was demonstrated by the containment structure at Three Mile Island, which was not breached in a major way during the reactor accident of 1979; however, the cost is immense. Thus, it should not surprise us that, in 1976, dioxin escaped from a chemical plant in Seveso, Italy, operated by a subsidiary of the Swiss firm Hoffman LaRoche, and that many people fell severely ill [10]. It came as a shock that no one was prepared for such an eventuality. Hence, countermeasures could not be effected quickly, and medical diagnoses were severely hampered.

Eight years later, in December 1984, several tons of methyl isocyanide escaped from a tank at the Union Carbide plant in Bhopal, India. Turning into a gas, it spread over the unsuspecting town affecting its inhabitants in a dreadful way very much as would nerve gas. The lesson of Seveso had not been learned: workers, townspeople, and doctors alike had not been taught how to escape the effects of the gas and to treat the injured [11]. While Bhopal's death and injury cases are being prepared for court hearings, the Seveso case is nearing its end. In the aftermath of the accident, a large batch of contaminated chemicals had to be safely disposed of. Packed in barrels, they were transported by an Italian subsidiary of the German firm Mannesman across the border into France. There they disappeared, only to be found again in leaking barrels hidden away in a jumble of similar refuse after inquiring reporters had trailed them through a maze of equally alarming dump sites in several countries.

The above cases were selected to indicate the many possible origins of hazards. House fires caused by aluminum wiring point to poor design; the sinking of the oil rig was due to poor workmanship; the PCB got into the rice oil because of poor design and inadequate maintenance; Claude was shot when a policeman failed to question a report generated and transmitted by a computer; poor practices in operation also caused the sinking of the *Titanic* and the *Andrea Doria*. Some accidents are avoidable, others perhaps not. But in most cases a means of safe escape or treatment of injury can be provided. Such measures were lacking in the oil rig case, the police file case, the *Titanic* case, and in the Seveso and Bhopal cases. The next question is why such oversights can occur and how society and the engineer react.

SOCIETY'S RESPONSE

It has been shown how failures of technological products can pose a danger to life, physical well-being, civil and human rights, property, and the environment. The causes can be traced to ignorance or errors in designing, manufacturing, testing, operating, and maintaining the product, or in safeguarding its environment. All of these activities usually involve engineers in one way or another. The parties affected may be individuals or larger groups of people; they may be workers who handle the product during manufacture, end users or owners, or innocent bystanders. Avoidance of hazards can include retrofit measures, redesign, and regulation of design and use of the product. Furthermore, failures are not necessarily harmful when safeguards are in place to deflect their impact.

The variety of possible harmful outcomes and the fact that engineers do not intend these to happen should alert us to the uncertain nature of the engineering process. Yet such uncertainties do not excuse the engineer and others involved in the technological enterprise, such as managers, technicians, or salespeople, for allowing the public to use the products with mistaken and uninformed trust.

Hammurabi settled the problem of poor structural design or workmanship by decreeing that a builder whose house walls collasped and killed the owner (or his son) should pay with his own (or his son's) life. Considering the shifty nature of the sands along the Tigris and Euphrates, this was a tough code for the early builders. Because such draconian measures, and even lesser ones, would have slowed down modern industrial growth, only gross cases of actual neglect were prosecuted until the turn of the century. In a modified way, Hammurabi's legacy lives on in today's tort law. The "off-with-his-head" approach has given way to protracted litigation and financial compensation for the victims. Today the concept of strict liability even does away with the need to prove neglect on the part of the manufacturer, unless punitive damages are sought as well. But justice is not necessarily well served. For instance, it is not until after death, injury, or property loss has occurred that damages can be claimed in court. This may provide a lesson for the future, but it does not prevent the current harm. And if a case is settled in favor of the plaintiff, a large company such as Johns Mansville is allowed to declare bankruptcy to escape further claims for damages by workers suffering from asbestosis. Other manufacturers find ways to export their hazardous products or move their toxic manufacturing processes abroad.

Tort law can protect individuals or classes of consumers from repetition of harmful activity. A regulatory approach that protects the public from hazards before widespread damage occurs is necessary. An elaborate regulatory structure has greatly improved safety, but it does not necessarily shield us from the unknown hazards of new technologies. Rules cannot be written fast or comprehensively enough to cover all future eventualities. Attempts to do so result in documents which are difficult to follow in intent but not so difficult to circumvent. Consider, for example, the hazardous collection of dust in grain elevators, where sparks from electrostatic discharges produced by blowing the grain through pipes can cause explosions. It is known that "grain dust is more explosive than coal dust or gun powder, but for most of the century the grain handling industry has treated it as though it were not much worse than cobwebs. According to a 1983 study, workers in this industry can expect each year to endure over 2000 fires, 26 explosions, 950 injuries, and 24 deaths [12]. It may take several more years before agreement can be reached on what fraction of an inch constitutes an acceptable layer of grain dust.

Publicity may alert the public to the threat posed by selected products, but people exhibit a surprising amount of trust in most products. After all, does not industry's incessant complaint about overregulation indicate that hordes of picky examiners are on the job? This overconfidence makes it difficult for the engineer to raise signals of caution without sounding like an alarmist.

THE ENGINEER'S REACTION

Good engineers are aware of the many uncertainties which abound in design, manufacture, testing, sales, and operation. The laws of nature and human behavior are not fully known, yet the designer must proceed with the task at hand and complete it under constraints of money and time. In manufacturing, the precise nature of the materials supplied, the dimensions actually produced, and the quality of joints cannot be precisely assessed. In sales and operations, the varieties of use, loading, and maintenance to which the product will be subjected can, at best, be guessed. Thus it is that the engineer will build safety factors into the product. If an oil drilling platform is rated for a maximum load of, say, 1000 tons, then it is considered advisable to design the platform to hold as much as, say, 2000 tons. If the operational load limit (OL) = 1000, and the limit for which the structure is designed, its design limit (DL) = 2000, then there exists in this case a safety factor of $SF = DL/OL = 2000/1000 = 2$.

Safety factors often range from 1.1 to 3.0, but the numbers can be highly misleading. If the uncertain nature of the load during operations of the platform is considered, one must introduce a probability distribution which takes into account the variation in maximum loads that any particular platform may experience. The static and dynamic nature of the load, its position, wind, and other factors are not easily foreseen. Accordingly, we must replace the single value OL by a density function as shown in Fig. 1a.

Similarly, a product's DL does not fall on a single value. Variations in the materials used in manufacture, differences in workmanship, wear during use, and unbalanced loads (as in the oil rig case)—all these conspire to give the product-as-built a load capacity distributed about the DL as shown in Fig. 1a. We can no longer speak of a safety factor here. What should concern us now is

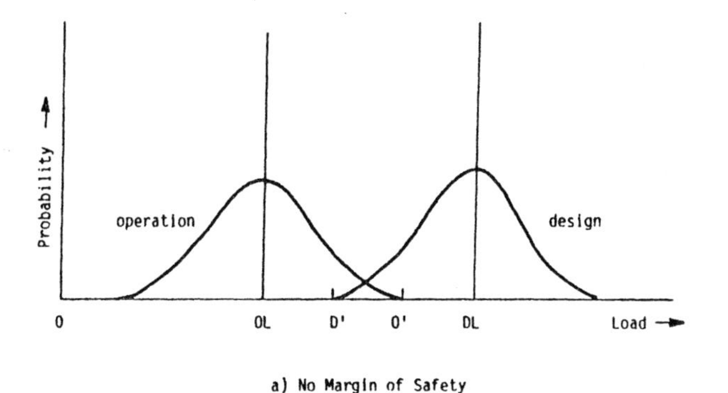

a) No Margin of Safety

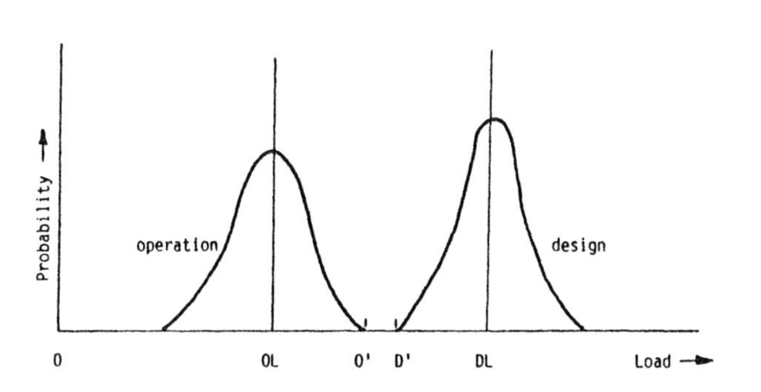

b) Margin of Safety = D' - O'
(with design/manufacture and operation more tightly controlled)

Fig. 1 Probability distribution of loads.

the margin of safety given by D'-O', provided that D' > O'. Case I in Fig. 1a allows no margin of safety. Case II (Fig. 1b) represents a safer product, although safety was most likely purchased at a higher price.

It should be appreciated that an analysis carried out as described still leaves much to be desired. As soon as probabilistic considerations are introduced, there arise questions regarding the proper density functions and (in)dependence of events. Most importantly, there remains in many cases the fact that even though serious harm may be very improbable, it is nevertheless possible. This mandates designs which provide a fail-safe cessation of operation and even escape procedures in the worst case. At some point all improvements and safety measures available to the designer will have been incorporated as necessary, and the design is complete. But will the product be safe enough? This raises the question of how safe is safe enough.

Many engineers will be faced with the requirement of preparing or commissioning an environmental impact report (EIR) as part of an engineering project. Such reports are demanded by public agencies so that they can assess the possible effects of a project on the community and the environment. The decision maker will then have to weigh the potential risks of the project against its potential benefits. Such risk-benefit analyses are often provided by the project's proposer, either as a requirement or as a means to avert anticipated objections to the project. It has been proposed as a standard activity in all engineering jobs.

Risk-benefit analysis is an important tool, but it is not

a cure-all, and has potential pitfalls [13]. One which should be mentioned here may be called the "two-party" aspect of risk and benefit: all too often the persons placed at risk are not the same as those who benefit from the use of a product. Clearly there are many parties with divergent interests involved in any evaluations of this sort.

Safety has been defined as a level of acceptable risk [14]. What is acceptable will remain largely a matter of subjective perception, or at best of consensus among the many parties at interest. Engineers generally feel uncomfortable or incompetent in the give-and-take of adversarial discussions, but when the limits of technical analysis have been reached, that is the way the discourse continues. Nevertheless, as well informed experts on a project, they should be involved, because the real uncertainties begin to emerge as a product is put to use.

IMPROVING SAFETY

Large engineering projects are usually examined for risk on a case-by-case basis by appropriate public agencies. Repetitive projects can be handled on the basis of a standard, approved design. (Indeed, nuclear power plant safety could have benefited considerably from such an approach.) Not covered by routine investigation are novel or low-volume products which in manufacture or use may have side effects which are not widely known, may be difficult to recognize, and whose origins are hard to trace. This category includes numerous cases of products which either incorporate toxic chemicals or use them in their manufacture. Engineers aware of the hazardous nature of such products and processes have the responsibility either to remove hazards of which they are aware, or to alert those who could be subjected to any remaining hazards, and then to set in motion procedures for the safe disposal of toxic waste.

It is easy to suggest that engineers shoulder such responsibility. It is more difficult to make it happen. Most engineers are employed and conditioned to work in large organizations where division of labor also results in division of responsibility. An engineer might say, "I assume that safety, risk, and all such matters are someone else's responsibility. And if not in the company, then in government! Let the government hire safety engineers. We vote, and that takes care of our responsibility!" If assigned to act on safety, engineers will do so. If not, they might consider it a waste of their organization's time to busy themselves with matters outside their immediate sphere of operation.

To induce engineers to take a broader view of their obligations to society, it has been suggested that engineering be viewed as an experimental process [15, 16]. The uncertainties associated with most engineering activities—from design to operation—make it one. That the subjects of the experiment are human beings makes it a serious undertaking. Indeed, it is the responsibility of the experimenter to secure the informed consent of such human subjects. That it may take some effort to inform all experimental subjects and to secure their consent does not absolve the experimenter of the duty to try, as physicians and drug manufacturers are required to do [17].

A recognition of the experimental nature of engineering can make engineers appreciate more fully the public's worries, render the interaction between engineers and their critics less painful and more fruitful, and give the engineer the impetus to work for improved safety. Timely recommendations for changes in a design before it is frozen, formation of a task force which can quickly intervene when a hazard threatens to turn into a calamity, and organizing safety measures on a public scale are examples of responsible reactions to risk. Engineers may have to form groups, either on the job or through their engineering societies, to accomplish such goals. If necessary, they even ought to suggest new regulations—which they should help draft, instead of leaving the writing solely to persons less skilled in technology. There is no longer an excuse for complaining about bad regulations and joining the chorus "there-ought-to-be-a-law." The key is what may be called participatory technology, which involves all interested parties [18]. In terms of design and operation, preventive or defensive technology should be the key [19]. As in medicine, preventively removing a potential hazard is more effective—even if less glamorous—than performing a heroic rescue after the accident.

REFERENCES

[1] Howard Falk, "Aluminum Wire: The Heat is On," *IEEE Spectrum,* p. 44, Dec. 1974.
[2] *IEEE* "Whatever Happened to Aluminum House Wiring?" *IEEE Spectrum,* p. 17, May 1984.
[3] Victor Bignell and Joyce Fortune, *Understanding System Failure,* Manchester University Press and The Open University, 1984.
[4] Jun Ui, *Polluted Japan,* Tokyo: Jishu-Koza, 1972.
[5] Jacques Vallee, *The Network Revolution,* Berkeley: and/or Press, 1982.
[6] Walter Lord, *A Night to Remember,* Holt, Rinehart and Winston, 1955; Bantam Books, 1956.
[7] Alvin Moscow, *Collision Course: The* Andrea Doria *and the* Stockholm, G. P. Putnam's Sonc, 1959.
[8] Beryl Frank, *Great Disasters of the World,* New York: Galahad Books, 1981.
[9] Paul Jacobs, "Birth Defect Study Shows Link to Toxic Spill," *Los Angeles Times,* pp. 1 and 28, Jan. 17, 1985.
[10] John G. Fuller, *The Poison That Fell From the Sky,* Random House, 1977.
[11] Mark Whitaker, *et al.,* "It Was Like Breathing Fire," *Newsweek, Special Report,* Dec. 17, 1984.
[12] Eliot Marshall, "Deadlock Over Explosive Dust," *Science,* vol. 222, pp. 485-487, discussion p. 1183, 1983.
[13] Kristin Shrader-Frechette, "Conceptual Risks of Risk Assessment," this issue.
[14] William W. Lowrance, *Of Acceptable Risk,* Los Altos, William Kaufmann, 1976.
[15] Roland Schinzinger and Mike W. Martin, "Informed Consent in Engineering and Medicine," *Business and Professional Ethics Jl.,* vol. 3 No. 1, pp, 67-77, 1983.
[16] Mike W. Martin and Roland Schinzinger: *Ethics in Engineering,* McGraw-Hill, 1983.
[17] Roland Schinzinger and Mike W. Martin, "Engineering as Social Experimentation," *Proc. of ASEE Annual Conference,* vol. 2, Am. Soc. for Engineering Education, pp. 394-398, 1980.
[18] James D. Carroll, "Participatory Technology," *Science,* vol. 171, pp. 647-673, 1974.
[19] Ruth M. Davis, "Preventative Technology: A Cure for Specific Ills," *Science,* vol. 188, p. 213, 1975.

FEATURE ARTICLE

The Role of the Law in Protecting Scientific and Technical Dissent

ALFRED G. FELIU

Dissent in private employment, it is assumed by those who want to keep their corporate noses clean, is to be avoided at all costs. For a nation free thinkers with a taste for the underdog, we are notably intolerant of outspokenness in the workplace. The First Amendment of the Constitution, the cornerstone of American civil liberties, limits *government* interference with free expression, but does not reach the activities of large, "government-like" private corporations. Management discretion, rather than government restrictions, define the bounds of dissent in corporate America.

Scientists and engineers are particularly burdened by this state of affairs. Trained to rely on their professional skills and senses, and required by statutes and ethical codes to act in the public interest, scientists and engineers make problematic employees. Management's need for flexibility is clear; the public's interest in independent-minded professionals is compelling. The resulting tension seems inevitable and hopelessly incurable.

The law has traditionally sided with management when dissent spilled over into (perceived) insubordination. In recent years, however, the law has increasingly intervened in defense of the responsible employee dissenter in situations in which the employee is acting to further an acknowledged public interest. I will review these legal developments, focusing particularly on one area of legal protection of special significance to scientists and engineers—namely, whistleblower protection in federal health, safety, and environmental protection legislation enacted since 1970. A review of the leading cases in the area, keeping in mind the scope of the protection offered and the strengths and weaknesses of this type of anti-reprisal legislation, leads to the conclusion that, despite recent developments, the law, by its nature and by the nature of the problem, is an inadequate tool for protecting scientific and technical dissent in the corporation and for fostering a workplace in which the expression of unorthodox or minority points of view are not only tolerated, but encouraged.

Alfred G. Feliu is a labor and employment law attorney associated with the law firm of Finley, Kumble, Wagner, Heine, Underberg, Manley and Casey in New York City.

Presented in part at The American Association for the Advancement of Science Annual Meeting, May 26, 1984.

BASIC ASSUMPTIONS

Three basic assumptions underlie this analysis.

First, that dissent in the workplace is not, like fair treatment or due process, an innately positive value to be applauded in all instances. One employee's dissent may very well be a co-worker's or manager's contentiousness. To be worthy of legal protection, a dissenting view should be: sincerely held and offered in good faith, based on credible data or information, and related to the safe and adequate performance of the job. Chronic naysayers tend to make speeches rather than exchange views or discuss problems. Neither the law nor management need strain to protect the merely contentious worker.

Second, in order to properly perform their jobs and by the very nature of their professions, scientists and engineers are on occasion required to express a differing professional view. Expressing their professional opinion on a subject within their expertise is often the essence of what they are paid to do, and the structural engineer, industrial chemist, or marine biologist who fails to bring to management's attention a differing view affecting the success of the project may properly be subject not only to dismissal, but to professional disciplining as well.

Finally, the expression of a differing professional opinion is not whistleblowing, although an ignored or rejected dissenter may at some point resort to blowing the whistle. Dissent is the expression of a differing view; whistleblowing is the disclosure to one in authority of a violation of law or threat to public health and safety in the hope of correcting the ill. Whistleblowing, by definition, is an extraordinary act, usually involving going out of established channels with a concern related to the public interests. An employee who takes a differing view to a company official empowered to address such concerns is less the whistleblower and more the customer at the company dispute-resolution store.

With these assumptions in mind, a review of the legal setting in which scientific and technical dissent occur is appropriate.

LEGAL BACKGROUND

Since the late 19th century, employers have been able to dismiss workers at will without incurring liability [1]; no reason need be offered the employee. Even a "bad" reason would not confer a right of redress on the employee. Underlying the employment-at-will doctrine is the freedom of contract notion that both parties benefited by being able to freely sever the relationship. Employees can move to a better position if one becomes available; employers, in turn, gain the flexibility to respond to shifting business needs. The preceived equality of this bargain faded with the pre-industrial workplace. Workers' ties to a job today, such as seniority rights and pension and medical benefits, further render illusory the assumed free mobility between jobs. Nonetheless, the employment-at-will doctrine remains the centerpiece of American employment law.

Reprinted from *IEEE Technology and Society Magazine*, Vol. 4, No. 2, pp. 3–10, June 1985.

Employer discretion has, however, been somewhat restricted in ensuing decades. Congress, in the 1930s, 1940s, and 1950s, limited employers' ability to retaliate against workers who collectively seek to protect their rights through unionization [2]. Further job protection was provided to workers who asserted their rights under minimum wages/maximum hours laws [3]. In the 1960s, both the federal and state levels emphasized the protection of group rights, and employers were precluded from discriminating on the basis of race, color, sex, nationality, religion, age, and, to a lesser extent, handicap [4].

In the 1970s, the focus shifted to the individual rights of employees. For example, privacy legislation was enacted in some states guaranteeing workers access to their personnel files and restricting the use of polygraphs by employers [5]. Two developments particularly inform this discussion, namely, the development of a public policy exception to the employment-at-will doctrine, and, second, the enactment of statutory whistleblower protection.

1. Public policy exception

Courts embraced the dismissal-at-will doctrine over a century ago. In the 1970s, courts began to create limited exceptions to the doctrine where a particular discharge violated a significant and identifiable public policy. For example, damages were awarded to employees who were dismissed for performing jury duty [6], for filing workers' compensation claims [7], and for rejecting the sexual advances of a supervisor [8]. Whistleblowers were also protected in a number of notable cases. The following whistleblowers who were discharged were allowed to sue their ex-employers: a quality control inspector who urged his employer to comply with Connecticut's food and drug act [9]; a West Virginia bank employee who furnished information to bank auditors regarding overcharged accounts [10]; an Illinois employee who reported co-worker thefts to the police, and then assisted in the subsequent investigation [11]; and a sales representative who refused to participate in an illegal scheme to fix retail gas prices in California [12]. The following trilogy of cases from New Jersey involving professional or health care employees is instructive.

In 1978, Frances O'Sullivan, an X-ray technician at Washington Memorial Hospital, was discharged for refusing to perform catheterizations on patients, a procedure that she was neither trained nor permitted—under New Jersey's Medical Practices Act—to perform. The New Jersey Board of Nursing issued a cease and desist order preventing further such actions by the hospital, but lacked the authority to reinstate O'Sullivan. A New Jersey appellate court ruled that an employee who is asked to perform an illegal act may not be terminated for refusing to do so [13]. The court found this to be particularly true where patient care was implicated, an area already extensively regulated by the government.

The most significant New Jersey case in this area arose out of the constructive discharge of the Director of Medical Research at Ortho Pharmaceutical Corporation, Dr. A. Grace Pierce. The New Jersey Supreme Court ruled that an employee whose dismissal violates a significant public policy may sue for wrongful discharge [14]. Unfortunately for Dr. Pierce, the facts of her case were found not to fit within the newly created exception to the dismissal-at-will rule.

Dr. Pierce was the only physician on a research team developing a new liquid treatment for chronic diarrhea in children and the aged. She objected to the use of large amounts of the suspected carcinogen saccharin in the drug. When she refused to give in to management's demand that she continue work on the drug, she was demoted and told she had no future with the company. She resigned and sued.

In court, Dr. Pierce argued that the standards of her profession demanded that she not continue work on the drug. In particular, she pointed to the Hippocratic Oath. The court, while recognizing an employee's right to sue for wrongful discharge, concluded that the Hipprocratic Oath was an insufficient basis upon which to rely. Instead, the court required that the employee's action be based on specific legislation, regulations, court decisions, or, in certain instances, professional codes of ethics. Professional codes that were self-serving or administrative in character would not be sufficient. Actions based on personal morals rather than a professional code were also outside of the new protection. The court commented in this regard that employees responding to "a call of conscience should recognize that other employees and their employer might heed a different call." [15] In Dr. Pierce's case, the court determined that she was acting out of the mandates of her conscience and not in response to the ethical mandates of her profession. The court noted that there was no immediate threat to the public in that the drug had not yet been marketed, nor even approved by the FDA. In ruling against Dr. Pierce, the court expressed its fear that "[c]haos would result if a single doctor engaged in research were allowed to determine, according to his or her individual conscience, whether a project should continue." [16]

The third case is an interesting application of the *Pierce* decision. A New Jersey appellate court allowed a pharmacist who claimed he was discharged for forcing his employer to comply with the state Board of Pharmacy regulations to sue for wrongful discharge [17].

Sidney Kalman was the pharmacist-in-charge in a Grand Union store in Paramus, NJ. The store wanted to keep the pharmacy closed on July 4th while the remainder of the store was open. Kalman checked with the state Board and determined that the pharmacy was required to be open when the store was open. The pharmacy remained open on July 4th, and Kalman was fired when he appeared for work the next day.

Kalman was able to persuade the court that the relevant regulation implicated a significant public policy. He argued that an unsecured, unsupervised drug counter in a large store created the risk that potentially hazardous drugs would be sold by unqualified persons. The court noted that both the state regulations and the Code of Ethics of the American Pharmaceutical Association required Kalman to report the store's attempt to allow the pharmacy to remain unattended. The court concluded that this was "an instance where a code of ethics coincides with public policy." [18]

New Jersey is just one of a growing number of states, now approximately 20, that have recognized a public policy exception to the "dismissal-at-will" rule. New York is perhaps the most notable example of a state that has expressly refused to create such a limitation on employer's discharge rights. Just last year, the highest court in New York State rejected a suit by an assistant treasurer alleging a fraud of over $50 million resulting from illegal account manipulation by high management [19]. The court determined that such a radical departure from a century's worth of case law should await action by the state legislature. [Note that the employment-at-will doctrine was created and has been dutifully applied by the courts for over a century, and that half the other states modified the doctrine judicially.]

In sum, the public policy exception is just that—an exception to the general rule, and a limited one at that. Scientists or engineers whose expressions of differing views get them into personnel trouble will be unlikely to make use of this legal remedy, unless they meet the following three conditions: (1) they get themselves fired; being transferred, harrasssed, demeaned, or ignored is generally not enough; (2) they raise an issue clearly implicating a recognized and substantial public policy; and (3) they are willing to pursue litigation that will most likely take one to three years to complete. The public policy exception is more appropriately suited for martyrs than dissenters.

2. Whistleblower protection laws

The legislatures of three states—Michigan, Connecticut, and Maine—have enacted whistleblower protection legislation that covers workers in the private sector [20]. This type of legislation protects employees against discrimination on the job for having reported a violation of law or regulation to a government body, or for having assisted in a government investigation. These laws parallel the protection offered by the court-created remedy just discussed, with one exception—the employee need not be fired, but merely discriminated against, to support a claim. These laws are so new that little can be said about the type of cases that have been filed. More such legislation can be expected in the coming years.

EMPLOYEE FREE SPEECH

As mentioned earlier, the First Amendment's free speech guarantee is not enforceable in the workplace. That is not to say that you lose your First Amendment rights upon entering the workplace; just that, in exercising your rights, you may lose your job.

What would a workplace with an enforceable First Amendment look like? Somewhat like a government workplace (putting aside for the moment other variables). Free speech rights are guaranteed to public sector employees. A differing opinion on an issue of public interest may be protected against retaliatory acts by superiors. For example, a federal appeals court in New York recently ruled that the director of nursing at Harlem Hospital, a municipal facility, could not be discriminated against because she disclosed illegal or wasteful practices at the hospital [21].

Two very recent developments indicate that the ether of free speech may be entering the corporate atmosphere. First, a federal appeals court in Pennsylvania ruled in October 1983 move that an employee may not be fired for refusing to join his employer's lobbying effort [22]. In that case, an insurance agent was urged by his employer to lobby the state legislature for enactment of certain no-fault insurance law revisions. The employee was discharged, and was allowed to sue under the public policy exception described above. Those scientists and engineers active, for example, in the anti-nuclear movement or other political/social causes may find solace in such developments in the law.

Second, Connecticut has become the first state to enact statutory free speech protection for private sector employees [23]. Discipline or discharge for the exercise by employees of their constitutional free speech rights is prohibited insofar as the speech does not substantially or materially interfere with job performance or the working relationship between management and the employee.

The Connecticut statute tries to balance management's broad discretion in personnel matters with workers' right of expression. This is a difficult task. Employees who report violations of law are protected; the persistent complainer with a personal gripe that interferes with job performance presumably would not be.

The day-to-day experience of professional employees seems in special need of such protection. The toxicity of certain chemical substances may or may not be anticipated; the design problems presented by wind sheer for jumbo jets may not be easily resolved. Unless there is only one acceptable answer to the question posed, differing views within the professional staff must not only be expected, but encouraged.

To complete this overview of recent legal developments, attention to the legal protection for scientific and technical dissent in private sector employment—namely whistleblower protection, found in all major federal environmental protection and health and safety legislation enacted since 1970—is appropriate. The substantial experience gained under these legislative provisions supports the broader thesis on what we can and should expect of the law in relation to scientific and technical dissent.

WHISTLEBLOWER PROTECTION IN FEDERAL LEGISLATION

The Occupational Safety and Health Act was enacted in 1970 with the laudable goal of ensuring a safe and healthful workplace for all Americans. Section 11(c) of the Act prohibits an employer from discharging or discriminating against an employee who exercises "any right afforded by" the Act [24]. Workers who complain to OSHA about what they believe to be hazardous conditions on the worksite, or who refuse to perform an assigned task because of a reasonable apprehension of serious injury, are protected under this provision. Similar protection exists in the Federal Mine Safety and Health Act Amendment of 1977 [25]. Workers making claims under both these whistleblower provisions tend to be blue collar and on the construction site, working in factories or, in the case of the Mine Act, in the coal mines.

The whistleblower protection that is most appropriate for use by scientists and engineers is found in the environmental protection and nuclear safety legislation enacted through the 1970s, such as the Clean Air Act [26], the Federal Water Pollution Control act [27], the Safe Drinking Water Act [28], the Toxic Substances Control Act [29], and the Energy Reorganization Act [30]. The cases under these provisions are instructive.

What type of employee has made effective use of these laws so far? Those in highly regulated work settings working in safety sensitive positions—for example, a quality control inspector or welder at a nuclear plant [31]; a chemist at the water pollution treatment plant [32]; a coordinator of environmental safety reports for a major coal company [33]; and a radiation safety officer at a hospital [34].

In order to be protected, the employee must raise an issue or perform an act in furtherance of the purposes of the federal environmental protection or nuclear safety legislation listed above [35]. The employees need not directly contact a government official or agency, but must do something more affirmative than merely discussing the topic over lunch with a buddy [36]. For example, a protected act may be the filing of a nonconformance report in a nuclear plant [37], discussion of safety issues with an on-site inspector [38], or establishing a policy for staff of apprising state regulatory officials of improper waste discharges at a treatment plant [39]. In expressing the view or taking the action, the employee must be acting to further the purposes of the legislation, although he or she need not specifically rely on the statute. Just as one can buy an item on sale without knowing it is on sale and still be entitled to the discount, protection against whistleblowing in certain settings can benefit the unknowing.

The employee need not be right on the merits, nor need the claim be meritorious. The claim need only be objectively reasonable [40]. Dissent would be unduly inhibited if the dissenting view were required to be correct, particularly when issues do not lend themselves to precise calculation, as is typically the case with scientific and dissent.

The scope of employer action triggering the protection of the legislation is broad. Basically, any discriminatory personnel action will suffice, such as an unjustified bad performance review [41], a baseless denial of a leave application [42], the retaliatory transfer of the employee [43], the withdrawal of a previously approved raise [44], or, most dramatically, the firing of the dissenting employees [45].

Are there ways in which protection to which an employee may be entitled may nonetheless be lost? Yes, several ways.

As with all rights, the protection offered by this legislation is most often lost out of ignorance of its existence. Unknown rights cannot be asserted. With no outreach effort by the Department of Labor, which enforces these laws, widespread ignorance of this form of protection may be assumed. Tardiness may also result in loss of this protection, as employees are allowed only 30 days to file a claim [46].

Some employees have lost their rights by coupling their protected acts with abuses of their own. For example, a quality control inspector whose reports of serious noncompliances with specifications were ignored finally reported the problem to the Nuclear Regulatory Commission [47]. When his employer learned of the report, the inspector was reassigned. Two days later, the employee was found trying to remove the personnel records of 15 dismissed quality control inspectors. He was immediately suspended for this act of self-help, then terminated. The judge reviewing the case concluded that the inspector's act of removing the files constituted sufficient reason for his dismissal apart from his protected activity of reporting hazards to the NRC.

Finally, workers hired to monitor their employers' compliance with government regulations are prone to losing the protection to which they would otherwise be entitled by performing their job too well—by crusading rather than cooperating. A company-hired safety inspector is still the employer's employee. An insubordinate employee can be fired whether he or she is a clerk/typist or the director of environmental health. A safety inpector who nitpicks [48], an electrician who complains of a hazard not within his bailiwick [49], or a supervisor who antagonizes staff or seems to derive undue pleasure in discovering flaws in underlings' work product [50] can expect unsympathetic hearings from a judge following allegedly discriminatory treatment. As an example, a drinking-water inspector, although competent, was reassigned following numerous complaints about his attitude [51]. He was characterized as overbearing, volatile, as one who came on too strong, and who tended to require unnecessary procedures and remedies in his work. As described by the judge, the engineer tended to "extrapolate" beyond the point that his employer could support him [52]. The judge concluded that the transfer was designed to further enforcement of water system regulations which had been hindered by the engineer's attitude. Thus, a troublesome employee who injects personality into his scientific and technical dissent runs the risk of confusing the issue upon legal review, allowing form to prevail over substance, personality over performance. The legal issue must not be confused with the substantive technical issue upon which the dissenting view is based.

CASE STUDIES: RICHTER AND THOMAS

Two of the more notable cases arising under this legislation brought by professionals involve a medical physicist for a state cancer hospital and a DuPont chemist. A review of these cases and the factors prompting different legal results may be instructive.

1. Dr. Clifford Richter [53]

Ellis Fischel State Cancer Hospital in Columbia, MO, was licensed by the NRC to use radioactive materials for medical therapy purposes. Highly radioactive irridium seeds were implanted in a cancer patient in December 1977 to be removed prior to her discharge from the hospital. Over three months later, Dr. Clifford Richter, the Chief Medical Physicist and the Radiation Safety Officer at the hospital, discovered that four seeds had been mistakenly left in her body. As Radiation Safety Officer, Dr. Richter was required to report this incident to the NRC, which he did. It was later discovered that a suture needle had also been left in the patient. The presence of irridium seeds in the patient's body was hazardous not only to her, but to those with whom she came in contact after her discharge. After a review of the incident, the NRC concluded that the failure to remove the irridium from the patient's body violated a condition of the hospital's license to use radioactive materials, and sanctioned the hospital.

Dr. Richter, in his four years with the hospital, had consistently received outstanding ratings in reviews conducted by his peers. The hospital administration had requested a $10,000 raise for him just prior to his report to the NRC. (Dr. Richter never received that raise.) Six days after his report to the NRC, the hospital executive committee met to discuss the incident. Dr. Richter, though a member, was not invited. The committee directed Dr. Richter to clear further reports to the NRC through the hospital administration. At the same time, the Department of Medical Physics which he headed was abolished as an independent entity and made a division of the Radiation Therapy Department. Dr. Richter's next three job performance evaluations became progressively less favorable. Seven months later, the position of Chief Medical Physicist was eliminated; Dr. Richter was terminated two months after that.

Dr. Richter filed a claim with the Department of Labor alleging a violation of the employee protection provision of the Energy Reorganization Act. An administration law judge and later a federal appeals court found that the hospital had discriminated against Dr. Richter because of his actions in furtherance of the aims of that law—in this case, the safe medical use of radioactive materials. Dr. Richter was reinstated to his position as Chief Medical Physicist, awarded back pay and attorneys' fees, and had all adverse references in his personnel file removed. This case serves as a paradigm of how the law may serve to protect scientific and technical dissent.

2. Lovick Thomas [54]

Lovick Thomas, a chemist and DuPont employee for 35 years, prepared material safety data sheets at a DuPont chemicals, dye, and pigments laboratory. In September 1978, Thomas first expressed the view that DuPont's zircon mineral sands products presented a radioactive hazard. He later asked whether the alleged hazards should be included on the material safety data sheet on the substance. After its own review, DuPont determined that mention of the hazard was not necessary, and because Thomas disagreed, preparation of the material safety data sheet would be reassigned. Thomas received a satisfactory rating on his next performance review. His reviewer wrote, "Tends to be overly defensive vs. criticism and argues opposing views excessively—should concentrate on major priorities of job. Should take a more active role in moving his program." [55] The reviewer also noted that Thomas should consider an assignment with another division, the pigments group.

Thomas persisted. He wrote letters and memoranda to supervisors regarding new OSHA regulations in the area, and his continued belief in the hazards of zircon sands. Finally, Thomas indicated that he felt compelled to go to OSHA with his concern. Thomas was soon after reassigned to the pigments division, in part because he was dissatisfied with his recent performance ratings. Eight other members of the professional staff were reassigned at the same time. Thomas filed a claim alleging discrimination under the Toxic Substances Control Act.

As stated by the reviewing judge, the issue was not "whether Mr. Thomas is right or wrong, but rather whether he was discriminated against because he threatened to, or did, report a hazardous material made by DuPont." [56] The judge found no basis for the discrimination claim. Discrimination, the judge noted, means treating similarly-situated individuals differently. Thomas' performance ratings, while not what he thought he deserved, were not shown to be discriminatory. Further, his reassignment, contemporaneous with that of others, was not shown to be the result of his threat to go to OSHA with his complaint. The judge also noted that his salary had not been decreased. Finally, Thomas had earlier acknowledged that he would accept reassignment to the pigments division. The judge offered no opinion on the strength of Thomas' views regarding zircon sands, a question over which he had no jurisdiction.

LAW AS AN INADEQUATE TOOL

This review compels the conclusion that the law is an inadequate tool to foster and support scientific and technical dissent in the private sector. The law does have a role to play in protecting those employees who are retaliated against for expressing a differing view. The more significant factor, however, in assuring that differing views will be encouraged and listened to in the American workplace is a new management ethic that recognizes employees' views and experiences for what they are—a valuable resource upon which to draw. This does not require that a view contrary to office orthodoxy be adopted—only that it be listened to and rejected on the merits, if appropriate. The law can encourage this development primarily by focusing public attention on the issue—for example, through large damage awards to deserving employees—but cannot mandate it.

Why is the law inadequate? I suggest three reasons. First, the law is a cumbersome tool. Litigation is expensive and time-consuming, and drains the resources of the parties and the courts. Barring settlement, a case can be expected on the average to take two years to complete. Answers to personnel decisions cannot realistically wait that long. Also, legal action creates adversaries where once colleagues stood. Further, an effective employment relationship can rarely be reestablished after litigation, so that the seemingly inviting remedy of reinstatement is illusory. Finally, with the law as unsympathetic to employees' rights claims as it is at the current time, despite recent developments, it will be difficult to find a lawyer to take a case with so little law in its favor and so limited a prospective financial reward.

Secondly, legal remedies tend to be (for lack of a better term) issue-specific—that is, concerned only with the immediate question of whether the claim presented sets forth a violation of law as defined by statute or court decision. The policy question of whether dissent in the corporation should be encouraged or protected is not before a court on review. Only the subject of the speech, and not the speech itself, is protected. In contrast, the First Amendment protects generic "speech" and not particular speeches. A speech by an American Nazi party official and one by a Girl Scout troop leader are both entitled to the same protection. Only the time, place, and manner of the speech may be regulated. With employee free speech, it is the substance of the speech that determines its entitlement to protection. One of the ironies of the current state of the law is that to be protected, an employee is usually required to do the extraordinary, such as blow the whistle in a very public way, which in turn hastens the breakdown of the employment relationship. Under the current law, this public display may be required to entitle an employee to protection. If the worker stays within channels, works diligently within the system, and fails to leave a paper trail, he or she may be without a remedy.

Conversely, employers who respond dramatically, for example, by promptly and vehemently retaliating against an employee who performs a protected act, are more likely to be sued and lose than the subtle retaliator who waits a period of time and then makes life so miserable for the employee that he or she resigns in frustration. Of all the recent legal developments, only the Connecticut free speech law holds out the potential of focusing the issue on protecting dissent, and not merely selected dissenting views. Whether it and similar legislation will succeed will not be apparent for years.

Finally, the law is an inadequate tool because the legal issue does not parallel the substantive professional dispute prompting the dissenting view. An amazing metamorphosis takes place: a simple technical dispute involving, for example, the discharge of an effluent into a stream, becomes, for a judge applying existing law, a fact question involving such matters as motives, work histories, and the intent of Congress. Just as a rape trial seems to victimize the victim by requiring her to account for her sexual history and proclivities, so a whistleblower or wrongful discharge suit turns on matters unrelated to the substance of the original dispute. As the judge in the *Thomas* case said, the issue is not whether the employee was right or wrong, but rather whether there was discrimination under the law [57]. Thomas lost, despite very sympathetic facts—an experienced chemist speaking out forcefully about a perceived hazard. There is a lesson in that for scientists and engineers who expect to win a lawsuit simply because they feel that they are right on the merits of the case, believing the scientific/technical issue encompasses the legal issue. Dr. Richter, in contrast, won his case because 1) his actions furthered the purposes of a particular federal law, 2) the hospital's actions were egregiously discriminatory, and 3) the hospital's actions were in direct retaliation for his protected activity. The fact that Dr. Richter's claims proved meritorious was of little consequence in the lawsuit. Legal rights and scientific truth need not cross paths. This fact became particularly clear to me in speaking to the Chief Administrative Law Judge for the Department of Labor [58]. When asked if the complex technical issues often present in these cases turned the matter into a battle of experts, he assured that it did not. The sole issue for the judge is the question of discrimination, and side trips into the intricacies of a scientific or technical disagreement only get in the judge's way. In legal terms, evidence offered on the substance of the technical dispute, depending on the circumstances, may not be allowed into evidence as simply not being relevant to the discrimination question, which is the focus of the lawsuit. The law and science make, at times, awkward bedfellows.

PROPER ROLE OF THE LAW

The proper role for the law, then, is that of teacher, standard-setter, and, when necessary and solely for its therapeutic value, punisher of flagrant violators.

The law unquestionably can prohibit discrimination on the basis of race, sex, or age, factors for which no legitimate grounds exist. The law can prohibit the use of lie detectors in the workplace, since management has no overpowering need to use them to run its business effectively. The law can even protect against hazards in the workplace so that management cannot improperly threaten worker safety. To mandate the recognition of the value of dissenting views in the workplace, however, is a more delicate task requiring a less blunt tool than the law. The law has long recognized that management must be given broad prerogatives in running its business as it sees fit. A company's acceptance and recognition of differing views in the workplace has traditionally been viewed solely as a question of management style. The law cannot effectively mandate an open workplace marked by the free and willing exchange of views. To require management to hear is not to assure that it will listen.

Elsewhere in this issue, Robert Jackall has persuasively demonstrated how the bureaucratic ethic, which extolls fealty to one's immediate superiors and the perception of being a team player, works to dilute personal and moral responsibility for corporate decisions [59]. The idealogue, the stickler for detail, the noble dissenter—the resister in one form or another—is discredited, shunned, and inevitably (in personnel manager's jargon) "separated" in this setting.

The law, in contrast, has a tradition of the noble dissenter—Oliver Wendell Holmes, Louis Brandeis, William O. Douglas—whose views in dissent have come to be adopted by future generations. Litigation, adversarial in nature, lends itself naturally to prevailing and dissenting views. The American workplace typically does not. Production schedules, to those who live by them, determine continued employment, not retroactive acceptance of their views. Making widgets and making law are not comparable, at least in this respect. The physicist, the marine biologist, and the chemical engineer, working in the corporate setting, are more often in the role of widget-makers rather than lawmakers. Dissenting views like Dr. Pierce's and Mr. Thomas' are likely to be resented rather than respected.

FREE SPEECH AS GOOD POLICY

The laws reviewed above are, to a limited extent, able to protect the dissenting employee. One form of protection, the statutory whistleblower protection found in major federal environmental and health and safety legislation, is particularly suited to the needs of the dissenting scientist and engineer working in certain highly regulated industries. This protection has the further advantage of being relatively efficient and inexpensive. Nonetheless, less than half of the claims filed that have not already been withdrawn or dismissed result in a finding of discrimination. In the three years since enactment of the Michigan whistleblowing protection law, only a handful of lawsuits have been filed, and none as yet have resulted in a reported court decision. This is not to say that such legislation is not desirable or worthy of support. Rather, it is to emphasize the limited nature of the protection it promises.

These laws do, however, establish public policy, setting standards against which behavior should be measured. Laws in the employment environment, as elsewhere, have a way, after an initial period of assimilation, of becoming policy. What major respected chemical concern today would not make a serious effort to assure that it produced a safe product while providing a safe workplace for its workers? Fourteen years after enactment of OSHA, occupational safety and health have become part of the cost of doing business and an integral part of corporate planning, and not merely a liability concern of counsel. So, too, with Title VII of the Civil Rights Act of 1964 and employment discrimination. Equal employment opportunity has become part of the personnel management function of any good, profitable firm, not simply because it is the law, but because it is good policy.

Free expression in the workplace must follow the same course. We as a society can enact, and should enact, further legislation to protect legitimate employee dissent. Such legislation has a way of publicizing the need for protection and of punishing in a very public way egregious (or unlucky) offenders of the law. Publicity is good. Public awareness is good. Legislation in this area, as in the occupational safety and health and employment discrimination areas, will eventually lead, I believe, to good management policies and practices. Fair treatment of employees is also being encouraged by competition from abroad. Progressive companies are instituting a variety of employee participation models, such as quality control circles. Some have gone further and established internal dispute-resolution systems in which scientific and technical disagreements may get a full airing. Further, publicity surrounding the enactment of whistleblower protection or free speech legislation may encourage both the general public and management policy makers to value more highly free expression in the workplace.

But what of the individual scientist or engineer in the interim? Differing views are too infrequently welcome in the workplace. With or without legal protection, the question of whether to speak out is an intensely personal one depending on the nature and importance of the issue and the strength of the convictions and charater of the prospective dissenter. Noble causes do not require noble characters, only individuals with strong senses of personal responsibility. We may, with Professor Stephen Unger, hope for "an enhanced sense of responsibility on the part of the individual." [60] We should also work towards making the expression of a dissenting view easy rather than extraordinary or bold. Dissenters should not have to be martyrs, sacrificed for the cause of full disclosure.

Until that time, the expression of dissenting views will be left to the courageous and the foolhardy. And until that time, the law should be broadened, as it has been in the state of Connecticut, to come to the aid of the brave and foolhardy who do what it would seem should be expected of every employee—namely, to express his or her opinion on a work-related issue without fear of reprisal.

REFERENCES

[1] *See, e.g., Payne* v. *Western & A.R.R.*, 81 Tenn. 507, 519-20 (1884), *overruled on other grounds, Hutton* v. *Watters, 132 Tenn. 527, 179 S.W. 134 (1915).*

[2] National Labor Relations Act, 29 U.S.C. §151, *et seg.*

[3] Fair Labor Standards Act, 29 U.S.C. §215

[4] Civil Rights Act of 1964, 42 U.S.C. §2000e; Rehabilitation Act of 1973, 29 U.S.C. §794

[5] Personnel file statutes: Cal. Lab. Code §1198.5; Conn. Gen. Stat. Ann. §31-1238 (b); Pa. Stat. Ann. tit. 43 §1321. Polygraph statutes: Mass. Gen. Laws Ann. ch. 149, §19B; N.J.S.A. §2A:170-90.1.

[6] *See, e.g., Reuther* v. *Fowler & Williams, Inc.*, 255 Pa. Super. Ct. 28, 386 A. 2d 119 (1978).

[7] *Frampton* v. *Central Indiana Gas Co.*, 260 Ind. 249, 297 N.E.2d 425 (1973).

[8] *Monge* v. *Beebe Rubber*, 114 N.H. 130, 316 A.2d 549 (1974).
[9] *Sheets* v. *Teddy's Frosted Foods, Inc.*, 179 Conn. 471, 427 A.2d 385 (1980).
[10] *Harless* v. *First National Bank in Fairmont*, 246 S.E.2d 270 (W. Va. 1978).
[11] *Palmateer* v. *International Harvester Co.*, 85 Ill.2d 124, 421 N.E.2d 876 (1981).
[12] *Tameny* v. *Atlantic Richfield Co.*, 27 Cal.3d 167, 164 Cal. Rptr. 839, 610 P.2d 1330 (1980).
[13] *O'Sullivan* v. *Mallon*, 160 N.J. Super. 416, 390 A.2d 149 (1978).
[14] *Pierce* v. *Ortho Pharmaceutical Corp.*, 84 N.J. 58, 417 A.2d 505 (1980). *See* Feliu, "Discharge of Professional Employees: Protecting Against Dismissal for Acts Within a Professional Code of Ethics," 11 *Colum. Hum. Rts. L. Rev.* 149 (1979/1980).
[15] *Id.* at 75, 417 A.2d at 514.
[16] *Id.*
[17] *Kalman* v. *Grand Union Company*, 183 N.J. Super. 153, 443 A.2d 728 (1982).
[18] 443 A.2d at 730.
[19] *Murphy* v. *American Home Products*, 58 N.Y.2d 293, 461 N.Y.S.2d 232 (1983).
[20] Mich. Stat. Ann. §§17.428.1-9; 1982 Conn. Pub. Acts §2-289; Me. Stat. ch. 452 §821 *et seq.*
[21] *Rookard* v. *Health and Hospitals Corporation*, 710 F.2d 41 (2d Cir. 1983).
[22] *Novosel* v. *Nationwide Insurance Co.*, 721 F.2d 894 (3d Cir. 1983).
[23] Conn. Pub. Act 83-578 (Oct. 1, 1983).
[24] 29 U.S.C. §660.
[25] 30 U.S.C. §§815, 820(b)
[26] 42 U.S.C. §§7401, 7622.
[27] 33 U.S.C. §1367.
[28] 42 U.S.C. §300j-9.
[29] 15 U.S.C. §26622.
[30] 42 U.S.C. §5851.
[31] *See, e.g., Atchison* v. *Brown & Root, Inc.*, 82 ERA 9; *Crider* v. *Pullman Power Products Corp.*, 82 ERA 7.
[32] *Ray* v. *Harrington*, 79 SDWA 2; *Ray* v. *Metropolitan Government*, 80 SDWA 1.
[33] *Murphy* v. *Consolidated Coal*, 83 ERA 4.
[34] *Richter* v. *Ellis Fischel State Cancer Hospital*, 79 ERA 1, *aff'd*, 629 F.2d 563 (8th Cir. 1980), *cert. denied*, 450 U.S. 1040 (1981) (hereinafter "*Richter*").
[35] *See* 29 C.F.R. 24.2
[36] *See Murphy* v. *Consolidated Coal*, 83 ERA 4.
[37] *See Atchison* v. *Brown & Root, Inc.*, 82 ERA 9.
[38] *DeFord* v. *Tennessee Valley Authority*, 81 ERA 1; *Hedden* v. *Conam Inspection*, 82 ERA 3.
[39] *Fischer* v. *Town of Steilacoom*, 83 WPCA 2.
[40] *Id.*
[41] *Richter, supra*, note 34.
[42] *Hanna* v. *School District*, 79 TSCA 1, *rev'd*, 657 F.2d 16 (3d Cir. 1981).
[43] *DeFord* v. *Tennessee Valley Authority*, 81 ERA 1.
[44] *Richter, supra*, note 34.
[45] *Cotter* v. *Con Edison*, 81 ERA 6.
[46] *Greenwald* v. *City of North Miami Beach*, 587 F.2d 779 (5th Cir.), *cert. denied*, 444 U.S. 826 (1979).
[47] *Dartey* v. *Zack Company*, 82 ERA 2.
[48] *Murphy* v. *Consolidated Coal*, 83 ERA 4.
[49] *In the Matter of William Wood*, 79 ERA 3.
[50] *Mackowiak* v. *University Nuclear Systems*, 82 ERA 8; *Murphy* v. *Consolidated Coal*, 83 ERA 4.
[51] *Bauch* v. *Landers*, 79 SDWA 1.
[52] *Id.* at p. 5
[53] *Richter, supra*, note 34. *See Science*, March 1980.
[54] *Thomas* v. *DuPont*, 81 TSCA 1.
[55] *Id.* at p. 3.
[56] *Id.* at p. 5.
[57] *Id.*
[58] Meeting with Chief Administrative Law Judge Nahum Litt, April 19, 1982.
[59] Jackall, Robert, "Moral Mazes: bureaucracy and managerial work," *Harvard Business Review*, Sept./Oct. 1983, p. 118.
[60] Unger, Stephen H., *Controlling Technology: Ethics and the Responsible Engineer*, p. 137 (Holt, Rinehart & Winston, New York), 1982.

PART III

Engineering Ethics and Public Policy

THE ABET Engineering Criteria 2000 provide that engineering students have "an understanding of professional and ethical responsibility" and "the broad education necessary to understand the impact of engineering solutions in a global and societal context." The first provision has traditionally fallen under the rubric of "engineering ethics," while the second, to the extent it has been addressed by engineering, has often fallen under the banner of "engineering and public policy."

Engineering ethics and engineering and public policy are both relatively new fields that have grown in stature and impact over the past two decades. Although each has a unique focus, in both cases engineers are challenged to move beyond traditional disciplinary boundaries and combine knowledge and methods from other disciplines with technical knowledge and methods. As would be expected, engineers have been involved in developing both fields through research, education, and the activities of professional engineering societies. In addition, applied ethicists, working on their own or in conjunction with engineers, have made significant contributions to the theory and practice of engineering ethics, as have social scientists in the area of engineering and public policy. Although a number of individual engineers, as well as the major professional societies, have been active in both realms, few formal efforts have been made to achieve greater integration and cross-fertilization of the two fields.

Langdon Winner, well known for his critical analyses of technological development (see background article in Chapter 1), is equally critical of traditional approaches to engineering ethics (1). Such approaches, Winner argues, focus almost entirely on specific case studies of ethical dilemmas to the exclusion of larger issues relating to the development of technology and to the career choice of engineers:

> Ethical responsibility . . . involves more than leading a decent, honest, truthful life, as important as such lives certainly remain. And it involves something much more than making wise choices when such choices suddenly, unexpectedly present themselves. Our moral obligations must . . . include a willingness to engage others in the difficult work of defining the crucial choices that confront technological society and how to confront them intelligently. (p. 384)

Similar critiques of engineering ethics have been raised by others, including engineers such as Vanderburg (2) who draws a distinction between "microlevel" analysis of "individual technologies or practitioners" and "macrolevel" analysis of "technology as a whole."

A number of ethicists have attempted to categorize levels of ethical analysis relating to engineering (see Table 1). De George, for example, as reported by Roddis (3), distinguishes between ethics in engineering and ethics of engineering, ethics in engineering referring to the actions of individuals and ethics of engineering to professional relationships and the engineering profession's responsibilities to society. Ladd argues (4) that professional ethics can be delineated as micro-ethics or macro-ethics depending on whether the focus is on relationships between individual engineers and their clients, colleagues and employers, or the collective social responsibility of the profession. Mclean (5), an engineer, utilizes three categories: technical ethics, dealing with engineering design decisions; professional ethics, dealing with business and professional issues; and social ethics, dealing with sociopolitical decisions concerning technology.

As indicated by the column headings in Table 1, combining these various schemes suggests that an individual engineer can confront ethical dilemmas in at least three distinct, albeit overlapping, roles that I call the personal, professional, and public roles. The personal role means the role of an engineer as an individual actor engaged in making technical decisions; the professional role refers to ethical rights and responsibilities that arise in the relationships entered into with others of the same profession, clients (including the public) or employers; and the public role entails the involvement and responsibilities of engineers, individually or collectively, in relation to questions of broad technology policy.

While some work in engineering ethics has focused on the public role of engineers, Winner's critique is essentially correct. The primary focus of engineering ethics, whether by ethicists, engineers, or professional engineering societies, has been on the personal and professional roles of engineers. Examples of this focus are case studies such as the DC-10 case, the Challenger disaster, and the Hyatt Regency skywalk collapse (see Chapter 3), where individual decisions by engineers and managers are called into question. Similarly, engineering codes of ethics, though stressing moral responsibilities to the public, tend to be aimed at the duties and rights of individual engineers in carrying out such responsibilities. In the traditional approach to engineering ethics, public policy is often considered only tangentially, as in the case of legislation to protect whistleblowers or the ethical responsibilities posed by environment, health, and safety regulations.

The public role of engineers, which encompasses the

TABLE 1 ETHICAL RESPONSIBILITIES IN THE ENGINEER'S PERSONAL, PROFESSIONAL, AND PUBLIC ROLES

	The Engineer's Roles		
Source	Personal	Professional	Public
De George, as reported in Roddis (3)	*Ethics in engineering*: actions of individual engineers	*Ethics of engineering*: the role of engineers in industry and other organizations, professional engineering societies, and responsibilities of the profession	
Ladd (4)		*Micro-ethics*: professional relationships between individual professionals and other individuals who are their clients, colleagues and employers	*Macro-ethics*: problems confronting members of a profession as a group in their relation to society
Mclean (5)	*Technical ethics*: technical decisions and judgments made by engineers	*Professional ethics*: interactions between engineers and other groups	*Social ethics*: technology policy decisions at the societal level

responsibilities of engineers and engineering with respect to technology policy at the local, national, or international level, has been a central focus of efforts falling under the banner of Engineering and Public Policy (EPP). EPP approaches, however, tend to be long on policy and economic analysis and short on ethical analysis—at least of the sort that would connect back to the engineer's personal and professional roles. As will become apparent from the activities of professional engineering societies and the beliefs of individual engineers (which will be examined in some of the following chapters), these groups and individuals often make the connections between ethics and policy in a shallow manner or not at all.

This state of affairs is problematic for a number of reasons. From a societal viewpoint, we need policies that are ethical and ethical viewpoints that are sensitive to social problems and issues. For example, as will be discussed in Chapter 6, we should question a product liability policy that might make it more difficult for engineers to perform their jobs in an ethical manner, thus compromising public safety. On the other hand, an ethical stance that all technology should be risk free on the grounds that engineers have a duty to avoid harm would clearly run counter to societal needs and economic realities.

From the individual's viewpoint, engineers need ways of dealing in a consistent and holistic manner with ethical issues that arise in their various roles. In the absence of integration of ethical considerations from their personal and professional roles with issues that may arise in their public roles, engineers might become confused or complacent regarding the importance of ethics and its connection with public policy.

The need for this integrated approach evolves from the recognition that the social implications of technology permeate the three spheres in which the engineer operates: personal, professional, and public. Limiting the focus of engineering ethics to the personal and professional spheres and public policy to the public sphere (the conventional approaches) can leave the engineer (and the engineering student) with a disjointed sense of his or her role in addressing important issues of ethics and public policy. The result can be public policy positions by engineers and professional engineering societies that are lacking in ethical foundations.

In order to illustrate these problems, this section of the book, following a general discussion of technology policy and ethical concerns, considers five major public policy issues that have caught the attention of engineers and professional engineering societies. Each of these cases—risk and product liability, sustainable development, globalization, health care, and telecommunications—will highlight ethical concerns related to the policy issues.

References

[1] Winner, L. 1990. Engineering ethics and political imagination. In *Broad and narrow interpretations of philosophy of technology: philosophy and technology* **7**, ed. P. Durbin, reprinted in D. Johnson 1991. *Ethical issues in engineering,* Englewood Cliffs, NJ: Prentice Hall, 376–385.

[2] Vanderburg, W. H. 1995. Preventive engineering: strategy for dealing with negative social and environmental implications of technology. *Journal of Professional Issues in Engineering Education and Practice* 121: 155–160.

[3] Roddis, W.M.K. 1993. Structural failures and engineering ethics. *Journal of Structural Engineering* 119: 1539–1555.

[4] Ladd, J. 1980. The quest for a code of professional ethics: An intellectual and moral confusion. In *AAAS professional ethics project: professional ethics activities in the scientific and engineering societies*, ed. R. Chalk, M. S. Frankel, and S. B. Chafer, 154–159, Washington, D.C.: American Association for the Advancement of Science.

[5] Mclean, G. F. 1993. Integrating ethics and design. *IEEE Technology and Society Magazine* 12 (3): 19–30.

Chapter 5

Technology Policy and Ethical Issues

Technological Catastrophes

A multidisciplinary exploration of technological catastrophes can be used to illustrate the interrelation between engineering ethics and public policy. Although the study of technological catastrophes begins with a technological failure (1), it inevitably leads to an examination of public and private organizations, individual and social values, and a critical assessment of one of our most important contemporary issues—technological risk.

Technological catastrophes clearly lend themselves to multidisciplinary analysis. A familiarity with the technology itself is essential to an understanding of why such accidents occur and what their effects are. Similarly, the behavior of complex organizations is a relevant topic. The economics of risk and risk management also plays an important role. Finally, social and behavioral aspects of how risks are perceived have a great bearing on the success or failure of risk assessment and risk communication.

A number of technological catastrophes have occurred over the past 15 years, including the Bhopal chemical leak in 1984, the 1986 explosions of the space shuttle Challenger (see Chapter 3) and the Chernobyl nuclear reactor, and the Exxon Valdez oil spill in 1989.

As will be discussed in Chapter 6, risk and methods of risk assessment are topics that involve fundamental value judgments. The impacts of technological catastrophes also have value dimensions, as, for example, in the cases of Bhopal, where Third World exploitation is an issue, and Chernobyl, where the impacts are international, cross-cultural, and intergenerational in scope. The social and ethical responsibilities of individuals and of corporations are also brought to light by technological catastrophes, for example, through the whistleblowing engineers in the Challenger case and in Exxon's response to the Valdez spill.

An array of human, organizational, technical, and external environmental factors underlie the causes and impacts of technological catastrophes (see Table 2). Human factors, often mistakenly regarded as the sole cause of technological accidents, generally entail some form of operator error or lack of proper training for the operators, as in the case of Chernobyl when the operators shut down virtually all of the plant's safety systems during an on-line test, or when the intoxicated captain of the Exxon Valdez left an inexperienced third mate at the helm of the ship. Organizational factors include such items as production pressures and cost-cutting, which resulted in most of the plant safety systems in Bhopal being either inadequately designed or inoperable at the time of the accident, and overly rigid organizational hierarchy, which is exemplified by NASA's failure to focus on serious safety concerns raised by the engineers closest to the design of the space shuttle. Technological factors are typified by poor designs such as the booster rocket O-rings on the space shuttle and the unstable Chernobyl-style reactor. External environmental factors include regulatory laxness, typified by the Coast Guard's deteriorating vigilance with regard to oil tanker traffic in the port of Valdez, and lack of emergency preparedness, a failure of both Union Carbide and the government of India with respect to Bhopal.

The nature of "catastrophe" is subject to various interpretations, for example, in terms of human safety (as in the case of Bhopal and Chernobyl), environmental impact (Chernobyl and Exxon Valdez), and the role of advanced technology in society (all of the cases but especially the Challenger). Perrow's notion of "catastrophic potential," which focuses on potential harm to large numbers of people (3), particularly innocent bystanders and future generations, can be contrasted to Shrivastava's concept of "industrial crisis," which looks beyond the initial accident to consider its broader social and economic impacts (4).

The lessons learned from technological catastrophes (5)

TABLE 2 CONTRIBUTING FACTORS IN TECHNOLOGICAL CATASTROPHES (Adapted from 2)

Factors	Examples
Human	Misperception, misjudgment, negligence, lack of skills
Organizational	Systems, strategies, structures, culture, resources, skills
Technological	Design, equipment, procedures, operators, supplies
Environment	Regulations, infrastructure, preparedness, risk management policies

entail both policy and ethical elements, including options for living with complex technological systems (or modifying or abandoning them in some instances), and the relative merits of preventive versus response strategies. In Perrow's study of "normal accidents" (3), for example, he concludes that some technologies are not worth the risk (nuclear power and nuclear weapons), whereas others should be tolerated but subject to careful regulation and modification where possible. In contrast to Perrow's argument that highly complex and tightly coupled systems have internal contradictions that render fail-safe management of these systems impossible, Pool (6) notes that there are many examples of such systems, including aircraft carrier operations and air traffic control, that are operated safely (see background article by Rochlin in Chapter 3). Clarke (7) also favors prevention, arguing in the case of the Exxon Valdez that meaningful response to large oil spills is impossible. Consequently, much more attention to preventive strategies is warranted, such as double-hulled tankers, better navigational and communication systems, and improved operating procedures.

Other areas of actual or potential technological catastrophes include airline safety, dam failures, computer network crashes, breach of biological containment, and high-speed police pursuits. Although each entails matters of public policy, they also raise unique sets of ethical issues for designers, operators, and users of these systems.

Technology Assessment

The tools and methods used in the conduct of policy analysis also highlight the relationship between public policy and ethics. Technology assessment, for example, is a form of policy analysis used to anticipate the impacts of emerging technologies or technological problems (including, but not limited to, indirect, delayed, or unanticipated impacts), in order to identify potential policy options in response to the impacts/problems. Technology assessments are typically concerned with an important public policy problem with significant technical components and have the goal of developing policy options for solving the problem. Projects typically require assessment of economic, social, environmental, and ethical aspects of the technology under consideration.

According to Coates, one of the early theorists of technology assessment, the process has six fundamental assumptions (8):

1. Technology assessment is a policy tool.
2. Technology assessment is likely to be iterative and part of an interlocking set of studies.
3. New technological knowledge creates new ignorance.
4. A major policy need is the organization of certainty and uncertainty to define effective strategies and tactics for managing any particular technology.
5. More information and analysis, rather than less, promotes better decisions.
6. In the long run, indirect and unanticipated effects of a technology are often more significant than the immediate or planned consequences.

One critique of technology assessment, however, is that it is excessively technical in character, to the neglect of social and moral values. For example, Skolimowski, a philosopher, suggested the following "laws" of technology assessment (9):

1. No system can adequately assess itself.
2. The more satisfactory the assessment is from the quantitative point of view, the less valuable it is from the social point of view.
3. All genuine assessment must terminate in value terms.
4. The "real expertise" in technology assessment is social and moral, not technical.

Technology assessment, however, need not fall into the trap predicted by Skolimowski. For example, the study "Forecasting the Applications and Environment of Technology to the Year 2010 and Beyond" (FAME 2010+) focused on the reciprocal relationship between the development of multimedia technology and society (10). Rejecting a technological determinism approach often found in technology forecasting, the FAME 2010+ study incorporates social construction of technology concepts into development of a three-layered model of technological innovation in the multimedia arena which focuses on components, systems, and applications. In contrast to a top-down policy approach, FAME 2010+ focuses on specific technologies and their interrelationships across the three layers. Based on an examination of the application of multimedia in three sectors—retailing and shopping, education, and work—the study cautions that inflexible technological "path dependencies" are likely to develop unless closer attention is paid to user needs and wants as revealed through a process of social learning. Thus, while rejecting technological determinism, the study acknowledges that in many cases the technical culture has set the tone for innovation.

Other topics suitable for technology assessment include low-level radioactive waste disposal, First Amendment issues and the Internet, alternative fuels vehicles, digital libraries, artificial organs, and intelligent transportation systems. In each case, the policy issues raised by the technology or technological problem are accompanied by an array of ethical questions relating to such issues as risk distribution, freedom of speech, right to privacy, and access to health care.

In 1996, the U.S. Congress eliminated its own Office of Technology Assessment (OTA) and consolidated its functions into other congressional support offices. Though originally entangled in partisan disputes, OTA had emerged over the years into one of the most respected and authoritative sources of research in the science and technology policy arena. A joint venture of both houses, it ably provided Congress with an independent voice on impending technological developments and has been used as a model for establishing similar agencies in other industrialized countries. To many observers, elimination of OTA suggests the Congress's lack of concern for the need to continue to develop its in-house capability to respond to the challenges posed by technological innovation. Such actions run counter to the more proactive role for government in setting the tone for science and technology policy advocated by Kline and Kash in the case study section of this chapter.

Integrating Ethics and Public Policy

The issues discussed above and in the remaining chapters of this section highlight the importance of integrated approaches to engineering ethics and public policy, including the need for innovative pedagogies in engineering ethics, the professional societies' increased attention to consistency in their treatments of ethical and public policy issues, and policy analysts' and policymakers' increased understanding of the fundamental importance of professional ethics.

Examples of integration of engineering ethics and public policy are the preventive approaches to engineering and ethics advocated by Vanderburg (11), Harris (see background article in Chapter 3), and others. By focusing on proactive measures to avoid narrowly conceived engineering designs and circumstances that precipitate ethical dilemmas, preventive approaches circumvent the problem of reducing engineering ethics to a series of individual cases. At the same time, they make engineering students and practitioners aware of the broader environmental and societal contexts in which engineers operate.

Educational innovation aimed at stronger integration of engineering ethics and public policy can occur at the levels of both curricular reform and classroom methodology. At the curricular level, Devon (12), for example, advocates an integrated sequence of design courses that stress the interdisciplinary nature of engineering "in terms of the technological, ecological and social systems within which it operates" (p. 9). In a similar vein, Mclean (13), noting that "engineering is *all about* ethical questions" (p. 30), argues for a thorough integration of ethics and the engineering design curriculum.

Professional engineering societies need to engage in efforts to consolidate their public policy and ethics activities. One approach to this problem would be to revise their codes of ethics to incorporate a broader appreciation of the public role of engineers, as a complement to the public responsibility of engineers in their personal and professional roles, a concept that is already present in most of the codes. For example, the IEEE Code (see Appendix), in addition to the standard admonition for engineers to uphold the public safety, health, and welfare, has a provision that, by calling on engineers to "improve the understanding of technology, its appropriate application, and potential consequences," requires moral responsibility in the engineer's public role. Taking this notion even further, Vesilind (14) has proposed that the ASCE code be replaced with a Code of Professional Responsibility that would require engineers to consider public viewpoints on such issues as risk and to include the public in decisions concerning their projects.

The professional societies should also ensure closer linkage between their bodies that make policy recommendations and their bodies that oversee ethics concerns. IEEE, for example, has been considering mechanisms by which policy statements could undergo rigorous technical review by the appropriate experts within the organization. Why not also subject policy recommendations to ethical review by the organization?

The still growing field of engineering and public policy would be strengthened by more serious attentiveness to ethical issues, particularly micro-ethical concerns that often escape the attention of policy analysts and policymakers. It is possible and desirable, for example, to include considerations of public viewpoints and perceptions in the conduct of risk assessments (see background article by Shrader-Frechette in Chapter 6) and to include social indicators among the sustainability metrics utilized in decision models evaluating technology choice for sustainable development (15).

Although this chapter has only scratched the surface of the complexities of technology policy, the point has been to demonstrate that, in both substance and method, policy analysis is inextricably linked to social and ethical issues. Consequently, if engineers are to have an impact on public policy decisions, they must become conversant in the accompanying social and ethical issues.

Background Reading

The notion that ethics and public policy are closely linked is apparent in the 1994 article by *Stephen Cohen* and *Damian Grace*, "Engineers and Social Responsibility: An Obligation to Do Good." Grounding their argument in ethics, Cohen and Grace suggest that the social responsibility of engineers extends beyond the avoidance of doing harm to an obligation to do good. Moreover, they argue that social responsibility, as suggested in modern engineering codes of ethics, should be an integral part of engineering practice.

Rejecting conventional approaches to engineering ethics that rely merely on prevention of harm (e.g., minimization of risk), Cohen and Grace argue for a social rationale for the engineering profession that requires promoting the public good. In seeking professional autonomy, the engineering profession cannot hide behind the suggestion that their sole role is to follow agendas set by others. In order to fulfill their obligation to promote good, individual engineers and the engineering profession collectively must be aware of the social context in which their work is performed and give voice to their views on social aspects of technological development. For example, the authors are highly critical of professional engineering societies (such as the Australian Institute of Engineers) which have disciplined members for calling to the public's attention the actions of other members that have compromised the public good or which downplay social responsibility on such matters as environmental quality because of a conflicting obligation to employers or clients.

The duty to do good, they argue, requires engineers and the engineering profession not only to minimize and prevent harm in the conduct of their work but also to understand and participate in "shaping the social agenda." This duty is an intrinsic characteristic of the engineering profession and should be stressed throughout engineering education and professional practice.

E. J. Woodhouse and *Dean Nieusma* consider the role of expertise in technological decisions in their 1977 article, "When Expert Advice Works, and When It Does Not." They begin with the assertion that "both experts and users of expertise fail to grasp the role of expertise in decision making." This problem arises, they suggest, because of two faulty models commonly used to conceptualize the role of expertise. The first view, which they call "the simple theory of expertise," holds that experts objectively undertake technical analyses and leave value and political judgments to others. A second cynical view, which arises in response to obvious flaws in the first, is that expertise only serves wealthy and powerful elites.

The authors suggest that an improved model that would benefit both experts and users of expertise should include considering sociopolitical complexities that affect using expertise, acknowledging disagreement among the stakeholders in the problem, and recognizing and dealing with uncertainty. The last point, they argue, ties together the first two, and so they spend the remainder of the essay focusing on ways experts can help other stakeholders cope with uncertainty. Expert roles might include designing response strategies for technological accidents, providing more flexible options for accomplishing tasks, and expediting feedback mechanisms. One significant aspect of this approach is that it can actually benefit from disagreements among experts and between experts and nonexperts, for even conflicting views can be useful in coming up with flexible responses for dealing with uncertainty.

The authors conclude by noting that corporate interests often do hold the upper hand in activities requiring expert assistance. Consequently, they argue for a leveling of the playing field by the experts themselves, through such mechanisms as taking on powerful interests or providing pro bono services for community groups. At the very least, the imbalance in marshalling expert resources must be openly acknowledged if a more workable model of expertise is to emerge.

Cases for Discussion

S. J. Kline and *D. E. Kash* provide a general case study of U.S. technology policy in their 1992 article "Do We Need a Technology Policy?" They take as their starting point the "growing disjuncture between the nation's position in research and its (weaker) position in commercial technology." They make clear from the outset that the government's role should be in assisting and stimulating technological development, not in doing it—a task they believe should be left to the private sector. With the exception of defense, medicine, and agriculture, the United States has had no deliberate technology policy. Such a policy is hard to maintain, the authors argue, because of the current bias toward laissez-faire government and the inadequacy of predictive theory—including contemporary economic theory—to guide technology policy.

The authors therefore turn to a historical examination and make six key points regarding technology policy: (1) technology and policy are interdependent—the only question is whether or not a nation's technology policy will be deliberate; (2) "high technology" is inseparable from other forms of technology; (3) technology develops in a process of continuous innovation; (4) both process and product are important; (5) innovation differs among commercial and industrial sectors; and (6) innovation often takes place in sociotechnical systems that extend beyond single companies and nations. Based on these observations, they outline a series of four vital interrelated functions for government to fulfill with respect to technology policy: (1) setting a favorable climate for technological innovation in such areas as labor and capital, organization, and markets; (2) surveying global developments in technology policy and technological innovation; (3) coordinating complex technological activities; and (4) filling critical gaps in the innovation process that are unlikely to be tended to by commercial means. Only by making these four functions national priorities will the United States be able to mount technology policy in other sectors that match the successes in defense, medicine, and agriculture.

John E. Ullmann, a professor of management, provides a case study of the U.S. rail system in his 1993 article "Task for Engineers—Resuscitating the U.S. Rail System." In calling for an integrated high-speed electric rail system for both passenger and freight traffic, Ullmann cites the following benefits: job creation; revitalization of equipment industries and conversion from defense production; reduction of traffic congestion and pollution; and decreased reliance on imported oil. Unlike many European and Asian regions, the United States has dragged its feet in confronting these problems.

In a brief review of the history of U.S. rail transportation, Ullmann laments the way a once substantial resource has been disassembled due to poor management, greed, and neglect. Nevertheless, he argues, a few good routes remain, which means it will not be necessary to completely start over should we choose to resuscitate the system. Critical of current piecemeal plans to improve the nation's rail system, Ullmann argues for integrated development and improvement of routes and operations that are suited to markets, technology that builds on the currently available base rather than exotic new technologies such as maglev and a revival of the U.S. rail equipment industry. Ullmann concludes with a discussion of some of the obstacles to his plan, including the considerable financial costs, the current trend of U.S. railroads toward purchase of diesel locomotives, and the difficulty in siting new facilities due to community resistance.

References

[1] Schlager, N., ed. 1994. *When technology fails: significant technological disasters, accidents and failures of the twentieth century*. Detroit: Gale Research.

[2] Shrivastava, P. 1994. Technological and organizational roots of industrial crises: lessons from Exxon Valdez and Bhopal. *Technological Forecasting and Social Change* 45: 237–253.

[3] Perrow, C. 1984. *Normal accidents*. New York: Basic Books.

[4] Shrivastava, P. 1987. *Bhopal: anatomy of a crisis*. Cambridge, Mass.: Ballinger.

[5] Roush, W. 1993. Learning from technological disasters. *Technology Review* (August/September): 50.

[6] Pool, R. 1997. When failure is not an option. *Technology Review* (July): 38–45.

[7] Clarke, L. 1990. Oil-spill fantasies. *Atlantic Monthly* (November): 65–77.

[8] Coates, J. 1974. Some methods and techniques for comprehensive impact assessment. *Technological Forecasting and Social Change* 6: 341–357.

[9] Skolimowski, H. 1976. Technology assessment in a sharp social focus. *Technological Forecasting and Social Change* 8: 421–425.

[10] Stewart, J., and Williams R. 1998. The co-evolution of society and multimedia technology: issues in predicting the future innovation and use of a ubiquitous technology. *Social Science Computer Review* 16 (3): 268–282.

[11] Vanderburg, W. H. 1995. Preventive engineering: Strategy for dealing with negative social and environmental implications of technology. *Journal of Professional Issues in Engineering Education and Practice* 121: 155–160.

[12] Devon, R. F. 1991. Sustainable technology and the social system. *IEEE Technology and Society* 10 (4): 9–13.

[13] Mclean, G. F. 1993. Integrating ethics and design. *IEEE Technology and Society Magazine* 12 (3): 19–30.

[14] Vesilind, P. A. 1993. Why do engineers wear black hats? *Journal of Professional Issues in Engineering Education and Practice* 119: 1–7.

[15] Herkert, J. R., Farrell, A., and Winebrake, J. J. 1996. Technology choice for sustainable development. *IEEE Technology and Society Magazine* 15 (2): 12–20.

STEPHEN COHEN AND DAMIAN GRACE

Engineers and Social Responsibility: An Obligation to Do Good

Is social responsibility a normal part of good practice? Should it be? In the face of some claims to the effect that engineers not only need not, but should not, pay special regard to social responsibility, we investigate what is involved in a claim that engineers do have a duty of social responsibility; and we urge that there is an important sense in which social responsibility should be seen as integral to the performance of engineers as individuals and engineering as a profession. This sense involves thinking positively of social responsibility, that is, not only as avoiding harm but as an obligation to do good. Social responsibility for engineers should not be merely reactive but, in contemporary jargon, "pro-active."

Stephen Cohen is with the School of Philosophy, and Damian Grace with the School of Social Work, at The University of New South Wales, Sydney, 2052 Australia.

Special Duty to Society

It is not self-evident that engineers or any other professionals have a social responsibility. As individuals, we all have moral obligations, but beyond these, what special requirements are placed upon us by virtue of joining a certain profession? Of course engineers, doctors, psychologists, teachers, and others are required to perform their professional duties competently. That is an ethical requirement; and it is also a legal requirement. But it is not what people usually mean when they talk about social responsibility. They mean something like a *special* duty which professionals have to society simply by virtue of being members of a profession, or a special duty which a profession as a whole has to society.[1] *Pro bono* work, for example, is considered by the legal profession to be socially responsible in meeting the needs of the disad-

[1]See, for example, the usages of [1].

Reprinted from *IEEE Technology and Society Magazine,* Vol. 13, No. 3, pp. 12–19, Fall 1994.

vantaged. There are two aspects to this characterization of social responsibility: the first suggests that social responsibility is an imposition upon a profession and its individual members; and the second suggests a remedial or prosthetic function. We urge that this characterization is inadequate on both counts.

The first tenet of the Australian Institution of Engineers (IEAust) *Code of Ethics* obliges member engineers to place the public interest and safety ahead of all other considerations and obligations.[2] It requires members to adopt this concern not as something over and above the profession's requirements, and not as remedial (analogous to the legal profession's commitment to *pro bono* work), but as integral to the central requirements of the profession *per se* — integral to what the profession itself is about. This provision is found also in similar codes, sometimes in a stronger form, such as that of the IEEE Code of Ethics (Art. IV.1), which not only enjoins members to "protect the safety, health and welfare of the public," but to "speak out against abuses...in the public interest."[3] Are such requirements, which oblige members to keep an eye clearly focused on the public interest, morally justifiable? Are they morally mandated? Are they ones to which the profession should be seriously committed? These questions are of the utmost importance, because if the answer to any of them is "no," then such a requirement will amount to hardly anything at all.

It is not self-evident that engineers or any other professionals have a social responsibility.

Philosophical Roots

The movement for social responsibility in science grew after the Second World War, when many of the scientists associated with the Manhattan Project worried about the global implications of making the bomb. It was certainly a major issue during the Vietnam War and, with increasing concern about damage to the environment, "social responsibility" has become an integral part of scientific and engineering endeavor. For all that, it is not clear what is implied by the phrase, either for the individual or the professions.

John von Neumann, the famous mathematician, apparently believed that a scientist could be socially irresponsible with honor.[4] Individual scientists should not have to shoulder the burden of social guilt for the results of their discoveries. In an analogous context, however, the view which emerged from Nuremburg is that individuals can indeed bear responsibility for the evil that comes from their acts, even if they are only part of a system which promotes that evil.

In both types of situation, the attribution of individual responsibility is, in many cases, questionable. First, the problem of "many hands" occurs [5]. Scientists and engineers work in teams, and the attribution of responsibility for the overall results to each of the participants is difficult. Notoriously, groups reach decisions that their members might not take individually. Secondly, "groupthink" typically is responsible for perceiving and articulating whatever justification there is for the project. A critical perspective is difficult for an individual to take, let alone sustain, in the course of "normal" engineering work under an accepted paradigm.[5]

In business, limited liability is an accepted way of inducing investors to put up capital. Further, a veil of anonymity applies to company directors to enable them to make decisions without fear of personal liability. These hitherto accepted practices in business are increasingly coming under attack, and individuals are now sometimes held legally accountable for actions done in the corporate name.[6] Nevertheless, individual instances emphasize the norm which, at present, is to make investors and directors responsible in a deliberately limited way for adverse consequences of their decisions.

Milton Friedman has argued that the notion of social responsibility in business is objectionable. Managers and directors owe a fiduciary duty to shareholders, not to society or putative stakeholders. We elect legislators to make policy in democracies. For nonelected officials to do so is a violation of the democratic mandate, and allows the injection of private decisions, values,

[2]The eight tenets of the Code are reproduced in the Appendix [2].

[3]For this and other codes see [3].

[4]See the remarks of Gleick in [4].

[5]John Ladd gives a critical analysis of these issues in [6]. See also, [7].

[6]For example, continuing to trade where directors know that a corporation's assets are insufficient to cover its liabilities.

and priorities into public life. A legislator has to consider the reactions of many parts of society, and seldom has the luxury of indulging personal whims, preferences, or values. This is a desirable constraint. In contrast, people of conscience (those who would include social responsibility as part of their job descriptions) have no constituency to answer to: they are defending their personal integrity, which, ironically, is responsible not to society but only to themselves as individuals. This may be individually satisfying but it is not, according to Friedman, socially justifiable. It is not mandated, and it is not democratic [8]. There are two things here then: the first is the questionable fairness or desirability of placing the burden of social responsibility on individuals. The second is the wisdom of placing it upon groups or organizations whose continuing benefits are important to society.

There is a serious problem in some discussions of social responsibility: they address the notion of "social responsibility" as though it were some kind of add-on to one's standard moral perspective. The status of social responsibility as often described is what ethicists call "supererogation." Heroism is supererogation. So is the work of Mother Theresa, and those like her who remain unnamed. To a greater or lesser extent, many people perform such works. They (the works) are not unusual, but are above and beyond the call of duty. As such they are of high moral worth, but cannot (should not) be required. Supererogation is like this; but is social responsibility? This is not how the ethical responsibilities of engineers are presented in the literature.

Social Experimentation

An important characterization in arguing the special ethical duties of engineers is that devised by Martin and Schinzinger. They characterize engineering as a form of social experimentation — literally, as well as figuratively. No project is totally free from risk, and so is in at least this sense, an experiment.[7] And, inasmuch as people will be affected by the experiment, they should be considered in the undertaking of engineering projects; and, Martin and Schinzinger argue, their "informed consent" should be obtained, as in any other experiment.[8]

Although Martin and Schinzinger's insight is important, it will not do as a justification of social responsibility. The characterization of engineering as experimentation illustrates the ordinary ethical requirements of the profession. It does not capture any special requirement of "social responsibility" as a duty of the profession. Put simply, society would not accord engineering the status of a profession — society would not value the enterprise of engineering — if engineering did not consider in the ordinary run of things the stakeholders affected by its processes, projects, and products. The social experimentation model does not mandate the taking on board of special responsibilities by a profession. It does not require social responsibility in any significant sense.

There is an important respect in which this social-experimentation model is correct, and an equally important respect in which it is not correct at all. It is correct in its characterization of a social-responsibility requirement as not an extra duty, an add-on of the profession. It is correct in viewing social responsibility as an integral part of what the profession is about: a foundational duty, a first principle which should operate in the practice of engineering. However, the model stops too short. In considering the practice as one of social experimentation, the model looks only at a requirement that harm be avoided, risks minimized, etc. This is a "negative" aspect of the profession. Martin and Schinzinger are correct to look globally, or generically, at the overall practice of engineering, for the purpose of establishing social responsibility. What they have not done, however, is adopt a broad enough perspective from which to look at the overall practice. Their model is too limiting. Instead of focusing only on the possible harmful results of an experiment and a concern to avoid or minimize harm, they should have focused as well on the point of the experiment in the first place.

Looked at in this way, we can appreciate the artificial limits that their perspective imposes. The point of an experiment is not to minimize the harmful results of the experiment itself. The point of the experiment is to do something positive. Martin and Schinzinger have asked the question, "As experimentation, is there any ethical concern which, generically, the practice of engineering should take into account?" And, they have answered, "yes; minimisation of risk, (etc.)." Suppose we widen the scope of the ques-

[7]See [9]. This is, of course, an extension of the concept of experiment. While a project might have an experimental character, its safety should be beyond reasonable question. The notion that projects are experiments should then be construed partly as a metaphor. Experiments are not normally done for the good of their subjects, whereas engineering is done for precisely this objective. On the other hand, experiments have stipulated protocols, and engineering projects can be made to satisfy specific criteria, e.g. those proposed for environmental protection by [10].

[8]The role of informed consent in taking account of the interests of stakeholders in the enterprise of any organization is an important topic, to which we can only allude in this article. See, for example, [11].

tion, and ask, "As experimentation, is there any morally significant purpose which, generically, the practice of engineering should take into account?" Very importantly, the answer here is also "yes." And this is very considerably different from Martin and Schinzinger's "yes." This question asks that we consider the "social rationale" for the profession, and it asks whether or not we can identify some generic morally significant features of the profession by reference to that rationale.

Social Rationale

Let's consider the question very simply. One might ask what the social rationale is for the profession of engineering. The answer is, let us suppose, to apply scientific knowledge to the public good. Importantly, the answer is not simply, "to build bridges" or "to deliver power," to provide blinkered technical solutions to narrowly conceived technical problems. If there is a general moral to this, then it is something like "promotion of the public good" (where this can be articulated in enforceable terms) is an integral part of the profession.[9] And that, as we shall see in the examples below, is a broader, more demanding requirement than that which is suggested by the Martin and Schinzinger model of social experimentation.

A requirement to promote good is not simply another way of stating a requirement to avoid or minimize risk. To speak only about avoidance or minimization of risk is to view the profession as purely and solely instrumental toward the achievement of whatever aims are set for it by whoever employs the profession or any of its members. It is to say, "Perform the task which is set for you; and in performing it, take care to avoid or minimize risk." As such, there is an instruction to the profession to operate according to other agendas as they are presented to it. Minimization of risk functions as a constraint on what this profession can do in applying its expertise to the problems with which it is presented. This is a very different requirement from a putative requirement to promote good. A requirement to promote good is an agenda of its own. It is a statement of a positive duty on the part of the profession; not simply a constraint on its dealing with whatever (else) is presented to it.

Perhaps we can present this as three possible conceptions of the relation between engineering as a profession and society as a whole:

1) *None.* Engineering's proper regard is as purely instrumental, with no constraints at all. Its task is to provide purely technical solutions to problems.

2) *Protection.* Engineering's proper regard is as in 1), except that it must be concerned, as a profession, to minimize risk to the public. The profession is to operate on projects as presented to it, as an instrument; but the profession is to operate in accordance with important constraints, which are integral to its performing as a profession.

3) *Promotion.* As a profession, engineering has a positive social responsibility to try to promote the public good; not merely to perform the tasks which are set for it, and not merely to perform those tasks such that risk is minimized or avoided in their performance. Rather, engineering has a positive social responsibility, a responsibility to assume some initiative toward promoting the public good.

With increasing concern about the environment, "social responsibility" has become an integral part of scientific and engineering endeavor.

If we consider the social rationale for the profession, for the purpose of gaining some insight into how the profession should see its role, then in narrow terms, engineering simply cannot be conceived of as 1). For example, it cannot be the case that the social rationale of civil engineers is to build bridges *simpliciter*. Bridges are no good to society unless they can be traversed safely. The social rationale is, rather, "to build *safe* bridges." In this respect, even when considered as an instrument only, the profession must have an eye to the public welfare. But there can be a very important question about how far this concern must extend, and whether it is appropriate for it to be a concern only as in 2), or whether it should extend to 3).

On some occasions where engineers have been criticized for apparently disregarding the public good, a response has been something like this:

[9] A parallel point has been made in [12].

"Look, we're not the ones who are failing to look out for your interests. It's not that engineers have their own agenda which they (we) are running, despite the public interest. Rather, we simply don't have a vote in determining the public agenda. In this respect, it is utterly appropriate that we limit our view of the situation — that we operate to a great extent with tunnel vision concerning our activities."

At this point, we do not want to go all the way back to a discussion of Friedman. However, we do want to allow that, to some extent, it is correct to say that engineers (like other professions) should indeed operate according to someone else's agenda, rather than their own. But, we equally want to suggest that there is a limit to the validity of this claim (particularly when it is used as a "defense" or to pass the buck). That limit is that the profession carries with it a social responsibility not merely to minimize risk but to positively provide benefit.[10]

There is another point related to this one. Professions — and engineering is certainly no exception — demand (and must be allowed) a reasonably high level of autonomy in the carrying out of their activities. By its nature, the profession is best placed to police, judge, and, to an important extent, direct its own activities.[11] This being so, it must also be admitted that it can be only partly true to say that the profession does not have its own agenda and that it should be allowed to operate with tunnel vision. The profession, as a profession, is not merely the blind instrument of others. Neither is the profession, as a profession, concerned only with technical prowess. The profession itself is very well placed to offer a well-considered, well-informed view concerning the appraisal of its various possible activities — not merely in terms of practical and technical feasibility, and not merely in terms of the risks those activities run, but in terms of the desirability of the activities themselves.

If it is not the case that therefore engineers should determine the social agenda (Friedman's point), then what significance can there be to our claim that the profession should take some serious role in identifying the desirability or undesirability of some (any) proposed engineering activity? It is this: in the name of social responsibility, there is a requirement on the profession and its professionals precisely to be aware of and to consider the overall social situation. It is a duty for the profession and its professionals to have a position about the merits of whatever enterprise it is called on to work in. This itself requires adoption of a perspective considerably wider than either technical feasibility or awareness and appraisal of risks. And this itself places responsibility on the profession and its members to become informed of the relevant social concerns present in any proposed engineering activity.

Further, it may be a duty of the profession and its professionals to *take* a position, as well: that is, a duty not merely to *have* an informed opinion from a social perspective, but to *voice* it — perhaps publicize it — in one's professional capacity. This is not to say that engineers should set the social agenda; rather it is to say that they must offer a considered opinion about the social agenda. This requirement rests on the profession and on its members individually. The latter cannot dodge a personal onus by putting responsibility onto the profession as a whole. The other side is that the profession as a whole cannot remove the responsibility from the individual professionals. In neither direction is the relation between the profession as a whole and the professionals who operate in it one of vicarious responsibility: individuals are accountable as professionals; the profession is accountable as the profession. Neither can correctly claim that social responsibility resides exclusively in one of these locations over the other. These are requirements of social responsibility. They rest with the profession as a whole, *and* with the individual professionals within the profession.

Two Examples

In Australia, for example, recognizing and assenting to this responsibility — resting on the shoulders of the profession and on its members as individuals — could go some distance toward avoiding situations concerning how far and to what effect individual engineers may, in their own right, look to Tenet 1 of the IEAust *Code of Ethics*, and then proceed to offer advice or criticism accordingly, in the name of the "welfare, health, and safety of the community". Sharon Beder instances two cases of members of the IEAust who spoke out in the public interest, and in the process criticized other engineers [14]. In the first, John Tozer, a consulting engineer criticized the engineering staff of the council of Coffs Harbour, a resort city about 400 km north of Sydney. The basis of the criticism was the recommendation of the council's engineers that there was no viable alternative to the classic Australian solution to the problem of sewage disposal — pumping it out to sea. The recommended site of the outfall was a prominent and

[10] Cf. the remarks of Brown in [13].

[11] Over 2000 years ago, Plato called our attention to the fact that the best safe-designer is also surely to be the best safe-cracker. They know best what goes to make for a good safe; and they know best how it is that safes can go wrong when they do.

beautiful headland called Look-At-Me-Now. The council engineers complained to the Association of Consulting Engineers Australia — which has adopted the IEAust Code — alleging that the criticism had breached the Institution's Code, specifically that Tozer had harmed the reputation of the council engineers intemperately with their employer. The Association censured Tozer, finding that "he did intemperately and publicly criticize the complainants," and it refused to renew his membership [14, p. 37]. The IEAust published his name and offense in its journal, *Engineers Australia*. Neither of the professional bodies investigated the substance of Tozer's complaints, only the manner in which he made them.

Individually and collectively, engineers have a duty not only to minimize harm, but to do good.

In the second case, Richard Herraman, a South Australian Government civil engineer, wrote to the *Adelaide Advertiser* criticizing the Electricity Trust policy of tree lopping to reduce the risk of bushfires from overhead power lines. In doing so, he suggested that Trust engineers might be breaching the Code of Ethics by giving inflated cost estimates for placing the lines underground. Some of the Trust's engineers complained to the IEAust that Herraman had "set himself up as arbiter of the code of ethics and implicitly cast aspersion on other members of [the IEAust]." The Institution published Herraman's name and offense in *Engineers Australia* [14, p. 38]. Again, there was no investigation of Herraman's claim that underground cables might be provided more cheaply than the Trust had suggested.

These cases suggest two interesting comparisons with the IEEE Code. The first is that the IEAust Code does not contain an express provision for engineers to speak out in the public interest, even though it specifies an overarching responsibility to the community. The second is that the IEAust does not investigate cases as the IEEE does, and therefore is limited in the kinds of conduct which it can designate as unethical, the breadth of its provisions notwithstanding [15]. Ironically, the public interest provision comes first in the IEAust Code, yet its effect is blunted by the difficulty of interpreting it in a way which does not bring censure upon individual engineers who seek to behave ethically. It is ironic too that the IEAust has taken the view that ethics is cognate with responsibility in the negative sense — avoiding harm — rather than in the positive sense of the individual seeking to do good. Until it assesses claims of a breach of the Code against claims to be protecting — let alone fostering — the public good, the IEAust will not have understood the full meaning of social responsibility.

Professional Etiquette

There seems to be a common tendency for professions to insist on the proprieties of professional etiquette at the expense of individuals to speak out on controversies [16]. If we are right and social responsibilities cannot be confined to the profession as a whole, then professional associations, like the IEAust, should not only be circumspect in disciplining members who cite their code of ethics in taking a public interest position on a matter, but should look at ways of supporting and promoting such independence. Stephen Unger considered the most likely objections to professional societies supporting individuals in an article in this *Magazine* some years ago and concluded that they were not well founded [15]. Professional societies should encourage professionalism in all its aspects, even if that means taking a public stand on controversial issues. The problem of reconciling tenet 1 and tenet 5 in the IEAust code is a case in point: the latter provision softens the obligation to protect the public welfare by requiring members to "act as faithful agents or advisers" of employers or clients. Tenet 5 is reasonable enough in principle, but in practice it has been used against individuals who have taken a public interest position.

Tenet 5 sits particularly uneasily with tenet 1 in cases where, for example, the engineer believes that an employer has not paid sufficient attention to, or commissioned a properly prepared, environmental impact study. The implementation of the IEAust's recently released *Environmental Principles For Engineers*, would be compromised if the Institution were perceived to be responsive to complaints about those who try to defend the environment, but to be passive on public interest concerns about the environmental impact of engineering projects.[12] In other words, there is a danger that professional associations can appear to be means of controlling individuals, of deterring or preempting socially responsible behavior (socially responsible decision-making) on the part of individual members.

[12]Beder provides an example of this in the case of the Sydney Harbour Tunnel [14].

Basic Rules

Several years ago, the philosopher William Frankena suggested some very basic rules of ethics as a) do no harm, b) prevent harm where possible, c) remove evil, and d) do good [16]. Typically, social responsibility theorists, including Martin and Schinzinger, concentrate on the first of these (do no harm), and occasionally on the second and third (preventing and removing harm). We are suggesting that the second and third should be a mainstream concern of engineering, and that the profession should indeed have an eye on the fourth (do good), as well.

Engineering's purpose is to bring about effects which are considered to be good in some way or other. Social responsibility is not an extra requirement on engineers, something additional to the professional practice of engineering. It is, rather, a central requirement, integral to the practice. We admit that there might be some problem in articulating exactly what this requirement amounts to in any particular situation; but that there is this requirement — involving at least *having* a position and *taking* a position — seems to us to be important and unproblematic.

A claim that engineering integrally involves social responsibility might seem to speak against the desirability of specifically designated education in social responsibility for engineers. The claim would seem to imply that it is sufficient to teach engineers the necessary competencies of their profession and the laws and policies governing it, because, after all, social responsibility is one of the necessary competencies. Drawing such an inference, however, would be to make the mistake of those who forsake professional ethics because they believe that it implies that some remedy is necessary for the normal courses they teach and for the normal practice of the profession. When questioned about the absence of a business ethics subject in his management school, one director replied, "What do you think we teach them — unethical business?" This is an inappropriate response, based on an incorrect perception. Medicine teaches ethical medicine but also teaches medical ethics. The presence of a separate subject, say, "Ethics in Engineering," as part of an engineering course, does not imply that professional practice education is anything less than ethical. Ethics relates the intrinsic requirements of professional competence, excellence, and respect for client and stakeholder interests to the extrinsic moral demands of society. In an educational context, social responsibility becomes a heuristic device, best considered in a separate subject so that its scope within practice may be better appreciated. It need not and should not on this account, appear as a separate and dispensable annexure to engineering practice. The requirement of social responsibility does not make supererogatory demands upon practitioners, but sets out the ordinary demands of professionalism. Social responsibility requires that a profession examine its social "license," its public role. And this requires an awareness of and sensitivity to the intrinsic demands of professionalism as well as social and moral concerns which are more broadly based.

Duty to Do Good

We have urged against Friedman-type objections that engineers have no business being socially responsible. Our position is that engineers, both individually and collectively, have not only a duty to minimize harm, but, according to the very rationale of their profession, a duty to do good. If doing good is intrinsic to engineering, then the realization of that good inescapably involves a commitment to understanding and having an important voice in shaping the social agenda; and this social responsibility belongs irreducibly both to practitioners and to the profession. Awareness that social responsibility resides in *both* these locations should be part of the educational process for engineering students, and should play an important role in the everyday activities of practicing engineers as well as in the activities of professional engineering societies.

Appendix

Code of Ethics
The Institution of Engineers, Australia

Tenet 1: The responsibility of members for the welfare, health and safety of the community shall at all times come before their responsibility to sectional or private interests, or to other members.

Tenet 2: Members shall act so as to uphold and enhance the honour, integrity and dignity of the membership and the profession.

Tenet 3: Members shall perform work only in their areas of competence.

Tenet 4: Members shall build their reputation on merit and shall not compete unfairly.

Tenet 5: Members shall apply their skill and knowledge in the interest of their employer or client for whom they shall act as faithful agents or advisors.

Tenet 6: Members shall give evidence, express opinions or make statements in an objective and truthful manner and on the basis of adequate knowledge.

Tenet 7: Members shall continue the development of their knowledge, skill and expertise throughout their careers and shall actively assist and encourage those under their direction to do likewise.

Tenet 8: Members shall not assist, induce or be involved in a breach of these Tenets by another member.

References

[1] Richard De George, *Business Ethics*, 3rd. ed. New York: 1990, p. 177.

[2] The Institution of Engineers, Australia, *Code of Ethics*, Barton, ACT, 1992.

[3] Stephen H. Unger, *Controlling Technology*. New York: Holt, Rinehart, and Winston, 1982, App. IV.

[4] James Gleick, *Genius*. London, U.K.: Abacus, 1993, p. 182.

[5] Dennis F. Thompson, "Moral responsibility of public officials: The problem of many hands," *Amer. Pol. Sci. Rev.*, vol. 74, pp 905-916, 1980.

[6] J. Ladd, "Collective and individual moral responsibility in engineering: Some questions," *IEEE Technol. & Soc. Mag.*, vol. 1, pp. 3-10, Mar. 1982.

[7] J. Ladd, "Morality and the ideal of rationality in formal organisations," *Monist*, vol. 54, pp. 488-516, 1970.

[8] Milton Friedman, "The social responsibility of business is to increase its profits," *New York Times Mag.*, Sept. 13, 1970, reprinted in T. Donaldson and P. Werhane, Eds., *Ethical Issues in Business: A Philosophical Approach*, 2nd. ed. Englewood Cliffs, NJ: Prentice Hall, 1983, pp. 239-244.

[9] M.W. Martin and R. Schinzinger, *Ethics in Engineering*, 2nd. ed. New York: McGraw-Hill, 1989, ch. 3.

[10] H. Rolston III, "Just environmental business," in *Just Business*, T. Regan, Ed. New York, NY: Random House, 1984, pp. 324-359.

[11] Stephen Cohen, "Stakeholders and consent," *Bus. & Prof. Ethics J.*, to be published.

[12] D.G. Johnson, "The social and professional responsibility of engineers," in *Ethical Issues in Engineering*, D.G. Johnson, Ed. Englewood Cliffs, NJ: Prentice-Hall, 1991, pp. 210-218.

[13] G.E. Brown, "Can scientists 'make change their friend'?," *Sci. Amer.*, p. 116, June 1993.

[14] S. Beder, "Engineers, ethics and etiquette," *New Scientist*, Sept. 25, 1993, pp. 36-41.

[15] S.H. Unger, "Would helping ethical professionals get professional societies into trouble?," *IEEE Technol. & Soc. Mag.*, vol. 6, p. 17, Sept. 1987.

[16] A.J.D. Bellett, "Social issues in the ethics of expert advice," *Austl. Quart.*, Autumn 1993.

[17] W. Frankena, *Ethics*, 2nd. ed. Englewood Cliffs, NJ: Prentice-Hall, 1973, p 47.

T&S

When Expert Advice Works, and When it Does Not

Both experts and users of expertise fail to grasp the role of expertise in decision making. Of course, expert advice is essential in complex human activities, but those who follow that advice also sometimes regret it. Under what conditions does expert advice and action tend to produce satisfactory outcomes? Recent studies of technological controversies and governmental regulatory procedures reveal some of the conditions under which expert advice may work well, and conditions under which it does not [2]-[8].

Before considering these situations, it is useful to review two mental "models" that obstruct clear thinking about what users of expertise should expect from professional knowledge — and what professionals should seek to contribute to technological controversies and other situations requiring expertise.

Two Models of Expertise

If everyday thinking about expertise were distilled, it would run something like this: Satisfying majority will is the proper object of democratic governments, but most citizens have only vague ideas about what they want, and too little information, skill, time, or motivation to understand the details of automobile airbags, NASA budgets, or tradeable pollution permits. Citizens therefore delegate decisions about these matters to elected officials, who set general policy based on negotiations with each other and with relevant interest groups and businesses. Administrative agencies handle the details. Throughout this process, technical experts employed by business, government, and interest groups provide advice on factual matters and techniques, leaving value judgments largely to others.

Likewise, in this mode of thought, consumers are acknowledged to know very little about the technical issues involved in designing, producing, transporting, and disposing of goods; but

[1] Although this work is devoted primarily to science, engineering, and other certified forms of expertise, we make no effort to bound the subject or define the term "expertise," because attempts to do so uniformly prove either wrongheaded or vacuous. As one recent review of the literature puts it, "There appears to be no single 'expert way' to perform all tasks.... An expert is someone capable of doing the right thing at the right time" [1].

E.J. Woodhouse is Associate Professor of Political Science, Department of Science and Technology Studies, Rensselaer Polytechnic Institute, Troy, NY 12180-3590; email:<woodhouse@rpi.edu>. Dean Nieusma is a mechanical engineer now completing the Ph.D. degree in Science and Technology Studies at Rensselaer Polytechnic Institute, Troy, NY 12180-3590.

Reprinted from *IEEE Technology and Society Magazine*, Vol. 16, No. 1, pp. 23–29, Spring 1997.

they can "vote" with dollars or Deutschmarks for certain items among the mix available in the market. Business executives and consumers, government officials and citizens decide what is to be done, while technical professionals are hired to do the R&D, design production machinery, and otherwise provide necessary supporting techniques.

Existing in disorganized fashion in many minds, this stylized schema might be called the simple theory of expertise, simple because it operates unproblematically: labor is divided neatly between experts, government officials, business executives, and others; tasks are accomplished straightforwardly without uncertainties or inaccuracies; each participant can play his or her role effectively. The simple theory was never an adequate description of politics, economics, or expertise, of course, but it had more plausibility prior to the 1960s than in recent decades. Three Mile Island, Bhopal, NIMBY siting controversies, safety recalls of automobiles, and the general disappointments and stresses of life in the late-twentieth century have made it a good deal more difficult to endorse a naively optimistic view of expertise [9]-[10].

For starters, neither business executives nor government officials are much trusted by citizens, so questions are bound to arise about the work of technical experts who are part of business or government. Simple views of professional knowledge also have been undermined by highly publicized cases of conflicting courtroom testimony, by splits between pronuclear and antinuclear scientists and engineers, and by the more general perception that technical experts willingly participate in corporate and governmental malfeasance such as the Department of Energy's now-acknowledged stonewalling regarding contamination at its nuclear weapons production facilities [11]-[13].

As defects in the simple theory of expertise become obvious, many people overreact and shift toward a second, more cynical view of expertise that fits with their cynical views of big business, bureaucracy, and elected officials. One element of this cynicism is a despair of ever getting a straight answer from professional experts; as a former U.S. senator put it, "I wish just once some scientist would come in here and tell us whether or not nuclear power is or isn't safe" [14]. Noting that many experts disagree with each other and that they tend to align with their client's or organization's stance, some people jump to the inference that statistics can be made to say anything and that most experts are venal. The result, according to the cynical theory: Expertise serves only the affluent and powerful.

Some social scientists come close to accepting the cynical theory, albeit in a more subtle form. They point to the fact that many scientific claims can be deconstructed if the stakes are high enough and sufficient resources are devoted to finding ambiguities in research methods or interpretations. The regularity with which this happens in high-stakes political controversies can be interpreted as casting doubt that science ever really can provide guidance on complex policy questions [15]. Others suggest that experts are helpful only under relatively unusual circumstances: when the scope of a problem is narrow, when data collection and analysis is complex, and when the task involves merely the application of settled policy rather than basic policy formulation [16].

Unlike those who think cynically about expertise, seasoned observers of expertise in action are not so easily misled. They are more able to acknowledge the human elements in the operation of expertise, recognizing that some nontrivial errors, misjudgments, and even outright corruption occur in every arena of human life — all the more so in systems as complex as those of modern government and business. Practicing professionals and other close observers of expertise in action can acknowledge as well great inequalities in existing political economies, without embracing the most extreme forms of the cynical theory of expertise.

Many people, however, clearly are unable to maintain a balanced view; indeed, we suspect that a majority of government officials, ordinary citizens, and even some technical professionals now carry around a poorly integrated mix of *both* the simple and the cynical mental models. If so, might this be partly because a better, more sophisticated model is not readily available? Clear thinking is always assisted by some framework, and neither everyday life nor the professional literature on expertise provides enough help to make sense of the complexities surrounding contemporary expertise.

Rethinking Expertise

Experts make enormous contributions to problem solving, but they could make even better ones if thinking on the subject could be improved. Part of this would depend on users of expertise learning how to play their roles better, including becoming more adept in the questions they direct toward experts. And part would depend on experts developing a more sophisticated understanding of how they can best fit into complex problem solving that draws on, but goes beyond, their professional knowledge. One important step in moving toward such greater sophistication is to develop a checklist of the main issues that need to be addressed by an improved mental model.

A starting point is to remind ourselves that complex endeavors ordinarily call for value judgments and for tradeoffs among conflicting considerations (such as speed versus safety, efficiency versus effectiveness). At such tasks, there are no true experts; so it ordinarily makes sense insofar as feasible for those who are going to be affected by the decision to make the choices [17].

When a homeowner hires a mason to craft a new brick walkway, for example, and asks that it curve gradually from porch to driveway, the expert might legitimately advise that a curvy walkway would cost twice as much as one built with straight edges. It would be inappropriate, however, for the mason to say that such a walkway could not be built (thinking to himself that it would be too expensive), for that would be taking on a value judgment appropriately left to the homeowner. In fact, craftspersons routinely do step over the line and usurp such judgments: centering replacement windows or sinks instead of asking where the client would find the item most useful or appealing, hanging a door so it works efficiently but looks funny (or the reverse), quoting a lower price using shoddier materials instead of a range of prices with materials of different quality.

These mundane examples are replicated in complex technical settings, but it becomes harder to trace who exactly is making the judgments. Nuclear engineers searching for economies of scale in the mid 1960s quickly scaled up reactors to 1000 MW, for example, instead of allowing time for extended public debate and perhaps experimentation with smaller reactors with less potential for core breaches (and, therefore, less problematic safety systems) [18]. But how can one blame the nuclear engineers, who of course were working for senior management at General Electric and Westinghouse? Utility executives eagerly sought the larger reactors; moreover, the Atomic Energy Commission raised not a peep about the new reactor designs [19]. It is apparent, then, that a more sophisticated understanding of expertise must somehow take into account the fact that professional knowledge normally is deployed via complex political-economic arrangements.

Consider another case, where engineers and other professionals currently are preparing to construct a new plastics plant on the banks of the upper Hudson River. The firm's CEO says that the work will go ahead "regardless of public opposition," and citizens will have no opportunity to decide whether to pay more for plastics produced farther from a fragile ecosystem and an urban population center. The chemical engineers and others designing the plant's inner workings obviously have no authority to relocate the new facility; and if the architect-engineering firm were to protest, another bidder surely would be sought. Yet, from the viewpoint of local citizens, the plant could not be built without the active collaboration of the various technical professionals. How might one take such social conflicts and other complexities into better account?

According to cynical theory, expertise serves only the affluent and powerful.

Acknowledging Disagreement

One ingredient is to acknowledge that there typically will be social conflicts and disagreements in esoteric issues involving multiple experts and multiple users. A long-held hope in western civilization has been that losers might accept an outcome (and the expertise leading to it) if their loss could be justified by an overriding criterion everyone finds acceptable. No such criterion has emerged in several thousand years of philosophical and practical (and, recently, economic) analysis: majority will breaks down in the face of minority rights, the "public interest" turns out to have multiple interpretations, and other criteria likewise fail to win anything approaching universal assent. Whereas almost everyone can agree on grand principles such as "sustainable development," agreement quickly breaks down when specifics must be addressed: is it worth a one percent increase in the unemployment rate to diminish pollution by five percent? Cost-benefit calculations can provide insights on such tradeoffs, but analysis inevitably fails to add up to *one clear-cut answer* compelling to all persons and groups with differing stakes and values [20].

The simple theory of expertise fails to recognize this problem, mistakenly assuming that calculation can achieve agreed solutions. The cynical theory does "solve" the problem of disagreement — by assuming the worst: the affluent and powerful will win, usually with the aid of their paid technical cronies. A more sophisticated understanding of expertise, while not denying the obvious fact of inequality, must explain why disagreements sometimes are dealt with relatively fairly and effectively. And it should explain how this could be brought about more frequently.

Coping with Uncertainty

Experts and users of expertise tend to concentrate their hopes and resources on turning up facts that can make uncertainty disappear. Even when the range of legitimate disagreement thereby can be narrowed, however, it often is not narrowed soon enough to be of much practical use. Thus, after more than a decade of intensive research, climate models are much improved. But in another sense, a majority of atmospheric scientists remain about where they started: believing there will be climate warming from combustion of fossil fuels and release of other greenhouse gases, but not knowing (and disagreeing with each other regarding) how much warming, how soon, and what should be done about it.

Far from being unique, this sort of outcome is common in technological-social arenas where neither definitive experiment nor accumulated experience can dispel uncertainty.[2] The remainder of our analysis focuses primarily on uncertainty, and on the roles experts can play in helping to cope with it. The issues surrounding disagreement and the deployment of expertise in a complex political economy turn out to be intimately intertwined with the approach one takes to uncertainty.

Toward a More Sophisticated Model

Whereas the simple mental model of expertise denies uncertainty, disagreement, and the problem of value judgments, and the cynical schema supposes that these difficulties are doomed to violate the public interest, how should a more sophisticated approach deal with these matters? Recent thinking about technological decision making explores the possibility that scientific and other experts should aim at helping political participants *cope* with uncertainty and disagreement [18], [20], [21].

Professional knowledge can prove crucial as one or more persons begin to consider taking action in the face of high uncertainty. First, if the unknowns include potentially severe negative consequences, engineers or other relevant professionals often can help develop ways to reduce, mitigate, or contain the worst possible consequences. Thus, the National Institutes of Health required that early biotechnology research be conducted in special laboratories with decontamination chambers, special door seals, and other means of containing microorganisms [4]. New supertankers are constructed so that if one hull is penetrated another can keep petroleum from leaking into the ocean. Such initial precautions will not prevent errors (dropping a petri dish containing a re-engineered bacterium, running aground an oil tanker), but can dramatically reduce the *consequences* of such errors. Experts almost always are in a better position to envision such precautions than are ordinary decision makers.

A second role for experts is to help problem solvers evolve more flexible options. Because of high uncertainty in many complex new endeavors, it often is necessary to learn by doing — as did General Electric Medical Imaging via numerous false starts in formulating its new line of business in nuclear magnetic resonance imaging (MRI) [22]. In learning by doing, there is a risk that by the time negative feedback becomes available, a course of action may have become quite resistant to change — deeply enmeshed in careers, in organizational routines, and in the expectations of customers or clients. By the time the dangers of vinyl chloride became apparent, for example, it was involved in nearly $10 billion of economic activity annually. In framing policy moves, therefore, the odds of eventual success go up if early trials retain sufficient flexibility to be altered readily through learning from experience.

One way experts can promote flexibility is by crafting ways of proceeding that allow costs to be borne gradually; such "payment on performance" allows expenditures to be redirected as learning develops [23]. If payment has to be made through large, up-front capital investments, misguided investments typically will be irrecoverable, and future options are likely to be unduly limited. Thus, expendable rockets would have been much easier to revamp than NASA's billion-dollar shuttles [24]. The $200 billion in sunk costs of civilian nuclear power is another poignant illustration.

Experimenting in a limited geographical area, simultaneous trials of two or more alternative approaches, and phasing in a new endeavor during a learning period are among other pathways that promote flexibility and thereby help cope with uncertainty. Experts also can prod political participants and other problem solvers to try more creatively to envision scenarios under which flexibility might prove valuable. Automotive engineers, for example, were in a good position to have cautioned about the desirability of flexibility in government regulations and industry designs for automobile airbags. As developed, there was no provision for easy

[2]Certainty can be approached in narrowly bounded endeavors based entirely on time-tested methods, such as installing a typical household fuse box. This is not, however, the sort of case where expertise is at all problematic or in need of social science analysis.

retrofitting with lower-potency gas cylinders if it turned out that too many children were harmed by the 200 mi/h explosive force of the original design.

A third way experts can help ameliorate potential problems caused by uncertainty is to figure out how to speed up feedback. Adroit adjustments cannot be made until feedback from policy trials reaches those with authority to make a change. In fact, feedback often takes too long, allowing accumulation of unfortunate results, as when harmful effects of chlorofluorocarbons (CFCs) were not persuasively documented for nearly half a century after their initial use. Rather than accepting slow feedback as an immutable fact of life, appropriately targeted expertise could be invaluable in assessing the comparative time lag required for learning about different options under consideration.

Some regulatory endeavors do make concerted efforts to speed up learning. In the United States, after numerous bad experiences from chemicals such as PCBs and DDT, the Toxic Substances Control Act mandated that all new commercial chemicals be screened by the Environmental Protection Agency (EPA) prior to marketing. Chemical toxicologists went a step further, advising EPA senior administrators of the possibility of setting priorities among the 1200 Premanufacture Notices the overworked agency was receiving annually. The key was to use structure-activity relationships to identify proposed chemicals with molecular similarity to existing chemicals known to be "bad actors" [4]. The U.S. Food and Drug Administration likewise long has required elaborate premarket testing and approval of new pharmaceuticals, and medical devices now are subject to screening. Such testing is simply a way of speeding up negative feedback instead of waiting for it to emerge naturally over a longer period and with greater damage.

The above problems and possibilities apply wherever expert advice is sought, especially in political life. First, since we do not want to step over a cliff while "learning from experience," it makes sense to protect against unbearable risks whenever feasible, e.g., by building containment shields around nuclear reactors [19]. Second, because people and organizations often resist change if it is costly and inconvenient, it makes sense to arrange new endeavors so they can be changed fairly readily when negative feedback warrants it — instead of hoping that no change will be necessary or that those in authority automatically will be receptive to change. Third, because negative consequences accumulate and sometimes multiply when learning takes a long time, it makes sense to call on experts for advice about how to speed up the learning process via testing, strategic monitoring, and other techniques.

Expert Disagreement and Political Conflict

If factual uncertainties and value conflicts often cannot be eliminated or even markedly reduced, then conflict and disagreement are bound to arise and sometimes will dominate social problem solving. Whether the disagreements are among experts or among users of expertise, or both, this need not prevent expertise from being useful.

As part of governmental debate on regulating energy industries in Switzerland, for example, the Swiss Federal Council put together an expert committee to provide advice. Members drawn from industry, environmental groups, universities, and civil service worked for several years but could not agree on a recommended course of action. They did, however, come up with four scenarios on how to address the problem, and their public report helped government officials, interest groups, and ordinary citizens to think more deeply about the issue. In fact, setting out several possible trajectories for energy policy arguably provoked more thought than would have been achieved by a consensus position [25].

In 1976-1977, while the evidence linking CFCs and ozone depletion still was controversial, the U.S. Congress and EPA debated what, if anything, to do about the problem. Aspects of the Molina-Rowland theory of chlorine-catalyzed ozone destruction had been validated, mostly in the lab, but much remained unknown. "At this point, political judgment had to be exercised. Granted that CFCs would have some effect on atmospheric ozone, was the effect significant enough to regulate? Various interest groups, scientists, and governments had substantially different answers to this policy question.... Some $400 to $450 million per year in chlorofluorocarbon product sales were at stake in the United States alone" [4, p. 81].

Few other national governments contemplated action, and the Chemical Manufacturers Association opposed any move, claiming it would be "the first regulation to be based entirely on an unverified scientific prediction" [26]. Nevertheless, members of Congress decided to take action, because they found an approach that was relatively cheap and easy: ban usage of most CFCs used in aerosol sprays (because there was a readily available substitute valve), but leave CFC refrigerants and other ozone-depleting chemicals untouched pending further evidence.

The ozone case was unusual in the scope and magnitude of the risk, as well as in the extent to which citizens and government officials had to decide whether to accept esoteric knowledge claims of atmospheric scientists without any tangible basis from which to work. But the basic predicament is a common one: some participants want to be more cautious than others in coping with uncertainty. Agreement among the relevant experts, or even agreement about who counts as a relevant expert, is not necessary in such cases for expertise to be helpful. All that is necessary is for a subset of political participants to be persuaded to consider action against a risk identified by one or more experts, and for a combination of expert analysis and ordinary reasoning to identify a course of action that can win majority assent. Pragmatic moves of that sort emerge from well-structured democratic negotiations; improving the political use of expertise depends in no small part on technical professionals and users of expertise reorienting their expectations toward the search for policy moves of that sort, moves circumventing rather than attempting to eliminate uncertainty.[3]

In effect, the CFC researchers answered the question, "If some partisans wanted to be more cautious in protecting against uncertainty in a particular problem area, what policy options would be available to them at an expense that might be low enough to win majority assent?" That same question always can be addressed, even when there is considerable disagreement among expert communities and/or among the political partisans using (or rejecting) expertise.

Expertise for Whom?

We now are in a position to tackle perhaps the toughest problem of expertise in action: How to deal with the fact that most professional experts must earn a living by working for business executives or government officials allied with business, who often wish to use expertise for tasks not designed to promote the goals of workers, consumers, taxpayers, or those who seek to preserve the environment? The simple theory of expertise ignores this problem, and the cynical theory is overwhelmed by it. Can we do better?

Large corporations enjoy a two-fold privileged position in market-oriented societies: 1) They exercise the initiative in economic activity, with government playing catchup as problems become apparent; 2) For political activity, businesses have disproportionate funding, access, and expertise to help sway elections, regulation, and litigation [27]. The combination of freedom to innovate in the economic sphere coupled with resources for disproportionate influence in the political sphere reduces the chance for nonbusiness interests to dispute effectively. If relevant concerns, ideas, and expertise are not brought to bear, then significant angles on a problem may be neglected or underemphasized, thereby reducing the overall intelligence of political negotiation and economic action.

One of the most important contributions experts can make to social problem solving, therefore, is to develop and deploy knowledge that helps level the playing field among social interests. Many environmental scientists, for example, have implicitly or explicitly done this in challenging agribusiness interests regarding pesticides. Some epidemiologists likewise have done *pro bono* work for communities affected by cancer clusters in the northeastern United States and in Louisiana's cancer alley [28]-[29].

On first inspection, concerns arise that such a stance could reduce the technical merit of the resulting work. On reflection, however, practicing professionals will recognize that working for someone actually is their usual situation, and that it need not lead to invalid results. More professionals work for the corporate sector, federal weapons labs, and mission agencies like NASA, moreover, than are ever likely to turn up doing research and advocacy on behalf of ordinary people. At worst, then, any biases introduced by partisan analysis on behalf of have-nots could serve only to reduce the imbalance in how expertise currently is allocated among potential clientele groups.

Put another way, professionalized expertise in the twentieth century arguably has become captured to an indefensible extent by large organizations and by the elites who run them and benefit disproportionately from them. No one knows very well how to change this, but those who want to think straight about expertise must at least choose not to duck the issue.

Beyond Oversimplification and Cynicism

Evidence and ideas accumulating from practical experience and from social science research provide the raw materials for moving beyond both the simple theory of expertise and its converse, the cynical theory. Although much further work remains to be done, we have begun here to sketch a more sophisticated way of thinking about the conditions under which expertise is useful.

We start with the widely accepted idea, from the simple theory of expertise, that value judgments should be left to users of expertise rather than being made by experts themselves. This

[3]For a complementary analysis of negotiations among regulatory agency administrators and scientific advisory committees, see [3].

turns out to be much more difficult to put into practice than ordinarily recognized, because technical professionals are enmeshed in business and government settings where they act within parameters set largely by senior management. Second, contrary to the simple theory, disagreement is inevitable; and, contrary to the cynical theory, disagreement by no means ruins the utility of expertise. In fact, disagreement can invigorate political debate, drawing in additional participants, raising new considerations, broadening the range of solutions examined, and otherwise strengthening the intelligence of democratic learning. Third, the tradition of targeting professional knowledge largely toward the reduction of uncertainties turns out to be questionable when applied to complex endeavors, because uncertainties often cannot be narrowed sufficiently or in a timely way.

A more sophisticated understanding of expertise faces up to these realities as design constraints, instead of wishing them away (as does the simple theory) or becoming fatalistic (as does the cynical theory). Experts can and should promote better outcomes by helping nonexperts cope better with uncertainty. That includes helping to devise initial precautions against unbearable risks, envisioning means of retaining flexibility to allow time for learning from experience, and speeding up feedback to promote learning. Because most of social life involves trial-and-error learning by doing, we believe that more expert attention paid to coping with uncertainty is one of the main ways of improving not only the usefulness of expertise but the success of human endeavors more generally.

Professional knowledge can hardly be used optimally if some people have ready access to it while others do not. Because of the unusual status of the business sector, the weakness of elections as a method of holding government officials accountable, and marked economic and political inequality among the world's citizens and nations, there is little realistic prospect any time soon of arranging a genuinely fair allocation of expertise among all humans who need assistance. But acknowledging that professional knowledge generally is deployed in ways sanctioned by those with political and economic influence opens the way for inquiry and experiment aimed at retargeting expertise wherever possible toward the service of ordinary citizens. If a sophisticated understanding of expertise gives a central place to the goal of retargeting professional efforts toward ordinary citizens, it is conceivable over an extended period that we might revitalize the Enlightenment project of drawing knowledge systematically into the service of human progress [30].

References

[1] K.J. Holyoak, "Symbolic connectionism: Toward third-generation theories of expertise," in *Toward a General Theory of Expertise: Prospects and Limits,* K.A. Ericsson and J. Smith, Eds. New York, NY: Cambridge Univ. Press, 1991, pp. 301-335.

[2] D. Nelkin, "Science controversies: The dynamics of public disputes in the United States," in *Handbook of Science and Technology Studies,* S. Jasanoff *et al.*, Eds. Thousand Oaks, CA: Sage, 1995, pp. 444-456.

[3] S. Jasanoff, *The Fifth Branch: Science Advisers as Policymakers.* Cambridge, MA: Harvard Univ. Press, 1990.

[4] J.G. Morone and E.J. Woodhouse, *Averting Catastrophe: Strategies for Regulating Risky Technologies.* Berkeley, CA: Univ. of California Press, 1986.

[5] S. Hill, *Democratic Values and Technological Choices.* Stanford, CA: Stanford Univ. Press, 1992.

[6] B.G. Peters and A. Barker, Eds., *Advising West European Governments: Inquiries, Expertise and Public Policy.* Pittsburgh, PA: Univ. of Pittsburgh Press, 1993.

[7] F. Fischer, "Citizen participation and the democratization of policy expertise: From theoretical inquiry to practical cases." *Policy Sci.*, vol. 26, pp. 165-187, 1993.

[8] B. Bimber, *The Politics of Expertise in Congress: The Rise and Fall of the Office of Technology Assessment.* Albany, NY: SUNY Press, 1996.

[9] B.A. Williams and A.R. Matheny, *Democracy, Dialogue, and Environmental Disputes.* New Haven, CT: Yale Univ. Press, 1995.

[10] D. Koehn, "Expertise and the delegitimation of professional authority." *Amer. Behavioral Sci.*, vol. 38, pp. 990-1002, 1995.

[11] S. Jasanoff, *Science at the Bar.* Cambridge, MA: Harvard Univ. Press, 1996.

[12] S. Solomon and E.J. Hackett, "Science and law: The case of *Daubert v. Merrell,*" *Science, Technol., & Human Values,* vol. 21, 1996.

[13] A. Weiss, "Shaping sensible systems: Politics, organizations, and the construction of technology," Ph.D. diss., Rensselaer Polytechnic Inst., Troy, NY, 1995.

[14] Senator John O. Pastore, in U.S. Congress, Joint Committee on Atomic Energy, hearings on "The Status of Nuclear Reactor Safety," Washington, DC, 1973.

[15] D. Collingridge and C. Reeve, *Science Speaks to Power: The Role of Experts in Policy Making.* London, U.K.: Frances Pinter, 1986.

[16] M.H. Bernstein, *Regulating Business by Independent Commission.* Princeton, NJ: Princeton Univ. Press, 1955.

[17] R.A. Dahl, *Democracy and its Critics.* New Haven, CT: Yale Univ. Press, 1989.

[18] J.G. Morone and E.J. Woodhouse, *The Demise of Nuclear Energy?: Lessons for Democratic Control of Technology.* New Haven: Yale Univ. Press, 1989.

[19] D. Okrent, *Nuclear Reactor Safety: On the History of the Regulatory Process.* Madison, WI: Univ. of Wisconsin Press, 1981.

[20] C.E. Lindblom and E.J. Woodhouse, *The Policy-Making Process,* 3rd ed. Englewood Cliffs, NJ: Prentice Hall, 1993.

[21] D. Collingridge, *The Social Control of Technology.* London, U.K.: Frances Pinter, 1980.

[22] J.G. Morone, *Winning in High-Tech Markets: The Role of General Management.* Cambridge, MA: Harvard Bus. Press, 1992.

[23] J. Pressman and A. Wildavsky, *Implementation.* Berkeley, CA: Univ. of California Press, 1974.

[24] D. Collingridge, *The Management of Scale: Big Organizations, Big Decisions, Big Mistakes.* London, U.K.: Routledge, 1992.

[25] C. Mironesco, "'Expert' and 'political' elements in official scientific advice on Swiss nuclear power," in *The Politics of Expert Advice: Creating, Using and Manipulating Scientific Knowledge for Public Policy,* A. Barker and B.G. Peters, Eds. Pittsburgh, PA: Univ. of Pittsburgh Press, 1993.

[26] J.A. Tannenbaum, "Fluorocarbon battle expected to heat up as regulators move beyond aerosols," *Wall Street J.*, p. 38, Jan. 19, 1978.

[27] C.E. Lindblom, *Politics and Markets.* New York, NY: Basic, 1977.

[28] P. Brown, "When the public knows better: Popular epidemiology challenges the system," *Environment,* vol. 35, pp. 16-20, 32-41, Oct. 1993.

[29] R.D. Bullard, *Dumping in Dixie: Race, Class, and Environmental Quality.* Boulder, CO: Westview, 1990.

[30] L. Marx, "Does improved technology mean progress?," *Technol. Rev.*, vol. 90, pp. 33-41, 71, 1987.

T&S

U.S. DOD

Technology and public policy are inherently interdependent in industrial societies. In the United States technology policy in the defense area has led to many spin-offs into commercial products.

Do We Need a Technology Policy?

The current debate on United States technology policy is driven by the same factors that launched the debate on U.S. industrial policy in the 1970s: the nation's trade balance remains negative; productivity growth is slower than in many competitor nations; and U.S. firms continue to lose market share in many product sectors. These trends are occurring even though the United States remains the world leader in fundamental research. This growing disjuncture between the nation's position in research and its position in commercial technology suggests that a reexamination of the government's role in the commercial area is overdue.

Article originally titled: " The Functions of a Government Technology Policy: What Should it Do?" S.J. Kline is Woodward Professor of Mechanical Engineering, and Values, Technology, Science & Society, Department of Mechanical Engineering, Room 500k, Building 500, Stanford University, Stanford, CA 94305-3030. D. E. Kash is Hazel Chair of Public Policy, The Institute of Public Policy, 4400 University Drive, Pohick Module, George Mason University, Fairfax, VA 22030-4444.

This reexamination requires clarity concerning the meaning of "technology" and "technology policy." Technology as we use it denotes systems created by humans to carry out tasks they could not otherwise accomplish. Examples include the creation and use of airplanes, computers, telephones, and pharmaceuticals by complex sociotechnical systems, that is organizational systems composed of tightly coupled hardware and social parts. Technology policy comprises actions by government aimed at assisting or stimulating delivery of competitive goods or services by sociotechnical systems.

We seek to contribute to the reexamination of policy by identifying and defining the essential functions of a technology policy. For this purpose it is important to emphasize that an effective government role involves assistance and stimulation but not doing. We have been unable to identify any case in which government run (as opposed to owned) industries have been able to carry out technology development over an extended period of time effectively enough to meet commercial competition. At the same time government assistance, or resistance, often makes the difference between success and failure in the increasingly globalized marketplace.

Reprinted from *IEEE Technology and Society Magazine*, Vol. 11, No. 2, pp. 18–25 & 31, Summer 1992.

U.S. Policy

The U.S. has long had self-conscious technology policies in defense, medicine, and agriculture. The present debate is occurring because many observers see now a need to develop similar policies for commercial technologies [1], [2]. Advocates of a self-conscious U.S. commercial technology policy believe that: 1) internationally competitive technologies are essential if the nation is to maintain its standard of living; and 2) government support and stimulation is important to maintaining competitive industries.

U.S. development of policy which supports commercial technologies has been seriously inhibited by two conceptual problems. The first flows from an American ideology which celebrates the virtues of free markets, and thus suggests any government involvement in the market is inherently imprudent. The core theme of this ideology is that capitalist societies must be *laissez-faire* in structure and operation [3]. This powerful value position places the burden of proof on those advocating government support of commercial technology.

The second conceptual problem results from lack of any predictive theory that can provide guidance on which government actions would be effective in supporting innovations and competitive technologies [4]. Theory in this context includes numerical models run in computers. The lack of adequate predictive theory is exemplified by the extraordinary variety of approaches taken to technology policy by the OECD (industrially developed) nations and various states in the U.S. [5], [6].

Both conceptual problems are exacerbated by continuing efforts to use inadequate economic theory. Most of this theory assumes the neoclassical model of a free market with buyers and sellers who are rational economic actors making decisions on the basis of three static variables: labor, capital, and resources. As Porter documents, these static variables are grossly inadequate for understanding competitiveness [7]. In the U.S. policy-making context, this inadequate theory has caused confusion in two ways. First, it gave support to the value based *laissez-faire* ideology. Second, it supplied a purported explanatory and predictive model of real world economic behavior which leads to simplistic conclusions. Never mind that the limits of this neoclassical economic theory have been the focus of sufficient volumes to fill whole sections of libraries; its impact on the technology policy debate remains powerful [3], [7], [8].

We will not add to the debate concerning the adequacy or inadequacy of this economic model. Rather, we assert that quantitative estimates of the complexity of the sociotechnical systems with which technology policy must deal place those systems well beyond the reach of any economic or other type of current, or foreseeable, predictive theory [4]. In the absence of predictive theory where does one go for guidance on the key factors technology policy must address? We see no alternative to careful empiricism, that is, looking at what has happened in the past over a wide range of experiences and summarizing the lessons those experiences teach us. This summary is the next task.

Lessons from Experience

First, whether by direction or indirection, advanced nations have had technology policies since at least the middle of the last century because technology and public policy are inherently interdependent in industrial societies. In the United States, these links have been manifest over history via many kinds of government actions, for example: patent policy; the land grant acts for railroads; the creation of the land grant universities via the Morrill Act; the protection of the steel industry via tariffs; and in more recent times many spin-offs from defense technologies into commercial products, and a broad support program for medical technologies. Indeed, there are so many actions of this type that it is hard to count all the agencies and all the types of public policy decisions which have affected technological enterprises in the past. The question facing the United States is not whether to have a technology policy, but rather whether the policy will be formulated openly and directly or in an *ad hoc* and indirect way via Congressional and agency decisions whose primary purpose is other than fostering competitive technologies.

Second, there is no meaningful way to differentiate among high technology, other technology, and related services, which is useful in regard to fostering competitiveness. However defined, high technology is continuously changing; what is high technology today may not be high tech tomorrow. What was called high technology in the semiconductor industry has totally changed several times in the past two decades. The revolutionary "lean production" methods pioneered at Toyota Motors between 1950 and 1970 have become the centerpiece of competitivieness during the 1980s for manufactured articles with a high piece count, thus outmoding what for decades had been the most advanced methods of manufacture. Lean production involves "inskilling" rather than "deskilling" workers, and thus leads to far more rapid innovations of both processes and products. The formation of well integrated networks of small companies, as in the Italian shoe and fabric industries, leads to the same speed-up of changes in competitive practices [7], [9]-[11]. Both systems allow easier and more rapid integration of new materials components and subsystems into nominally mature technologies. In the auto industry, for example, we may one day find competitors producing bodies of composite materials, and engines made of ceramics which reduce costs and improve reliability. Since the automotive industry (in a broad sense) accounts for roughly 1/6th of the U.S. economy, technology policy cannot be restricted to a set of "critical technologies," nor can it be successfully operated by a set of general guidelines or decision rules [1], [12].

Third, the critical characteristic of contemporary technology as it impacts competitiveness is continuous innovation. Some find it useful to divide innovations into three time scales: 0-2 years, 2-10 years, and more than 10 years, in order to emphasize that continuous efforts need to include not only incremental and radical innovations but also long-range fundamental research [13]. We live in a world in which continuous improvements in quality, performance, and production efficiencies are the norm; firms which stand still will, in the not very distant future, fall behind. To put this differently, more and more sectors of the international marketplace are becoming structured around innovation-to-obsolescence cycles in products, processes, and services. In such sectors, any specific technology policy will become obsolete, sometimes in no more than a few years. Technology policy must therefore be concerned with structuring a flexible policy-making system which is quick to sense and respond to changes.

Fourth, commercial success is normally enhanced if both process and product are considered as matters for innovations. This lesson is particularly important for the U.S. because many of the recent successes of foreign competitors appear to have resulted from a heavier emphasis on process innovations.

Fifth, the sources of and the routes to innovations differ among industrial sectors. In many sectors, small cumulative innovations are critical to commercial success; this is particularly true in relatively mature industries such as consumer electronics and automobiles. In these sectors, innovations often depend more on inputs from design, development, production, and marketing than from research, especially fundamental research. In other sectors, such as pharmaceuticals and chemicals, success rests much more directly on radical innovations and long-range research. In these sectors, a tight link between R&D, production, and marketing often becomes very important.

Sixth, the sociotechnical systems which create goods and services in the late 20th century frequently take the form of networks or webs which transcend single companies and national boundaries [14]. These networks are fluid in action and minimally hierarchical; this makes it possible to synthesize the expertise and skills necessary to accomplish a wide variety of innovations. The networks serve a number of needs; the most critical is effective flow of information and knowledge. Thus technology policy must be self-conscious about the need to foster network building in order to maintain competitiveness in many situations. In other words, technology policy must recognize the need for cooperation and coordination among activities in government, industry, universities, and nonprofit institutions and aid in removing barriers to their interaction.

With this empirically based appreciation of the general circumstances and processes needed for effective innovation and competitiveness, we now turn to an identification and discussion of the functions which technology policy needs to carry out. Our focus on the basic functions of technology policy reflects two perceptions: 1) moving to specific proposals before there is good understanding of the necessary functions can be dysfunctional; and 2) the basic functions have not been adequately clarified in past U.S. debates concerning technology policy.

Four Functions of Technology Policy

If, as suggested above, successful technologies require interdependent efforts by government, industry, university, and nonprofit institutions, then what are the key functions which government needs to carry out? We suggest four: 1) climate setting; 2) surveying; 3) coordinating; and 4) gap filling.

▾ *Climate Setting*

Technology is both embedded in and impacts most aspects of industrial societies. As a result, many areas of public policy impact technology. If the general body of public policy does not create a hospitable climate for technological enterprises and innovations, then the larger and

more successful enterprises are likely to seek other homes, and there is little that a specific technology policy can do to maintain competitiveness. A large diversity of elements make up a hospitable (or less than hospitable) climate. A primary element of the climate is clear identification and wide acceptance of the idea that commercially competitive technologies are necessary to a high standard of living and, therefore, technological innovation constitutes a national need of high priority. To gain this wide acceptance, national leaders need to communicate this priority repeatedly by word and deed.

A reexamination of the government's role in commercial technology is overdue.

The diversity of what climate setting entails can be understood by considering the role of policy in the creation of a hospitable climate for each of the three components of a hardware producing-using system: inputs such as labor and capital; organizational systems; and the international marketplace.

An important input is "patient" (long-term investment) capital. Thus the extent to which governments make such capital available is an important climate-setting factor. Many comparisons of the current situation of the U.S. with that of Japan show not only that the Japanese enjoy great advantages in the area of capital but also that this advantage is in large measure the result of conscious government policy. The range of government actions which affect the capital climate include: taxes; depreciation schedules; regulations concerning securities and corporate behaviors; government procurement processes; budget deficits through their effect on interest rates and availability of credit; and many others.

A second critical input is people with many kinds of skills ranging from research to production and marketing. Public education provides the substructure on which the creation of many of the necessary skills can be built. Beyond public education, the government plays a role through forecasts of personnel shortages in key areas, in the regulations which support or deter the retraining necessary because of innovations in products and processes, and in social support of the workforce through such programs as medical insurance and tax laws regarding retirement set asides.

Americans, during most of the period since World War II, have subscribed to the view that most of the ideas supporting innovations came from scientific research. Support for long-range research was seen, therefore, as the primary role of government in supporting technological competitiveness. It is now widely recognized that in many sectors innovation is heavily originated from ideas arising in marketing, production and design processes [15]-[21]. The belief that research was the source of innovation, led to federal support of fellowships for graduate students in the sciences and in engineering research areas which deemphasized the critical manufacturing sector. This support gradually shifted the focus of engineering faculties away from design and production toward research. By the 1980s, this created a significant imbalance within the engineering education system [22].

The point is, federal policies have and do affect everything from where both public and private R&D monies are spent to the nature of engineering education, to financial incentives for (or against) patient capital, to job security (or insecurity), to the nature of the social support system, to the potential for capturing profits from patented devices, and to other matters. None of these government actions alone constitutes a controlling element of climate; taken together, they determine a major part of the climate for innovations crucial to competitiveness.

A hospitable climate for innovative organizational systems is one which provides for both stability and flexibility. For organizations or networks to compete in markets structured by innovation to obsolescence cycles, the organizational systems must focus on the long as well as the short term. Inasmuch as innovations frequently require new mixes of skills, retraining of personnel, and the capacity to acquire new skills must be facilitated. Fear of job loss, mergers and takeovers, and demands for short-term returns on investments create environments which are not conducive to innovations. All these matters are affected by government actions which deal with fiscal policy, labor unions, health care, limits to liability, tax regulations regarding retraining of workers, plus many others. Here again, many actions by government contribute to the climate for technological innovation.

The last component of climate setting concerns the market. In many sectors, the market is increasingly international. Over time, an increasing number of nations made a trade surplus a high priority national goal. Many instruments can be used in pursuit of this goal, such as surveying of opportunities by diplomats, financing of purchases of foreign companies to acquire advanced technology, and trade barriers of many types. In the international market, the U.S. is now only one among a number of economically important nations. Insofar as few other nations share the U.S. definition of free markets, it is important that U.S. trade policies reflect the realities of the international market and not a theoretical notion of how the market ought to function [3].

In sum, creation of a hospitable climate for technological innovation involves a very wide range of governmental actions. Most of the decisions on these actions take place in committees of the Congress and in agencies with missions little concerned with creating an hospitable climate for commercial technological innovation. This situation can be addressed only by continuous leadership in both the public and private sectors. A new national consensus must be evolved concerning the importance of technology if the necessary climate is to exist.

▼ *Surveying*

The need for continuous innovations, coupled with the lack of adequate theory, requires technology policy include continuous surveying of the global situation. Under these conditions, surveying is a necessary foundation not only for climate setting but also for coordinating and gap filling. In the absence of predictive theory, the only way to improve policy is to observe continuously and gain experience on how government actions affect such essential activities as: development, research, production, and the use of organizational systems and networks. Surveying of this sort requires much more than mere data collection. Effective surveying requires long-term individual and organizational learning. Such learning requires a stable organizational setting. Only long-term assessment of the status and trends in technologies worldwide can provide the knowledge needed to inform the making of technology policy. The wisdom and insight needed to inform policy of this complexity cannot be gained over the short term.

Chalmers Johnson's classic study of Japan's Ministry of International Trade and Industry (MITI) supports this view [23]. MITI, in its early days, made a number of serious mistakes, but over the long term it has learned to be far more effective. More recently, Fumio Kodama identified the ability of MITI to sense and respond quickly to changing conditions as the reason for its continuing importance [24]. MITI's efforts have been guided by experience gained through continuous surveying and not by theory. Currently, The White House Office of Science and Technology is playing a role in surveying commercial technologies as demonstrated in the recent Critical Technologies Report. However, over the longer term, the executive branch of the federal government has had only an on-again, off-again system for surveying needs in the technology policy area.

This intermittent surveying seems to reflect three influences. First is the continuing power of an ideology hostile to any government role in commercial technologies; second, the dominant position of the U.S. in technological industries from World War II until about 1970 seemed to make explicit actions unnecessary; third, the now largely discredited belief that support of long-range science was by itself enough to maintain the U.S. predominance in technological enterprises.

The U.S. has long had technology policies in defense, medicine, and agriculture.

The preceding discussion suggests the need for a far more permanent and stable surveying function for commercial technologies. This view is supported by the record of agriculture, medicine, and defense in the United States. In these areas, a specific agency of the federal government carried on continuous surveying at least since World War II, and used the results to create a policy assisting both incremental and radical innovations. As noted above, these three sectors deliver the majority of products which generate the nation's largest trade surpluses [25, pp. 73, 75].

Surveying commercial technologies is now an essential function because our strongest competitors do it and use the information to guide the next stages of innovation. Without a U.S. surveying function, our competitors enjoy superior intelligence. As Richard Nelson, one of the nation's shrewdest observers of technologically-based economic competition said, our eroding position is not because the U.S. has become less effective in innovation, but rather because our competitors have become far more effective [26].

Other nations became more effective for many reasons, and one of them is sophisticated surveying capability. Foreign competitors know that leadership in technological enterprises needs the kind of surveying the U.S. has long maintained in medicine, agriculture, and

defense. A stable, ongoing, sophisticated, surveying function is, and will remain, central to an effective U.S. technology policy.

▾ *Coordinating*

The need for coordination flows from two characteristics of much of contemporary technology: current technology is often complex, and it is frequently the product of complex synthesis. Many of the most recent advances in technology arose from fusion of two or more preexisting technologies, or from movement of know-how into areas to which it had not been previously applied. Examples include the integration of computers, machine tools, new software, and changed social arrangements as bases for flexible manufacturing. Fiber optics was achieved by the fusion of know-how in plastics with knowledge about the effects of frequency on transmission within the telecommunications industry. Improved cameras arose from the movement of bar code technology from retail sales into camera designs. A recent example of the creation of a wholly new industry through fusion of older technologies is the creation and rapid diffusion of FAX devices.

This use of ever more diverse expertise and skills in single innovative projects requires the creation of new organizational arrangements which facilitate rapid and flexible coordination. When needed expertise is dispersed among a variety of organizations, assistance from governments may be needed to coordinate these activities, as in many activities in the space program, and in the Manhattan project which created the atom bomb. Governments can, for example, foster such networks via cooperative projects, incentives, and appropriate antitrust policy. A good example of such a network is the system whereby the U.S. commercial aircraft industry used government funded R&D to develop advances in military aircraft. A negative example is the failure of the U.S. machine tool industry to integrate computers with machine tools into saleable products as rapidly and effectively as the Japanese networks did, with a consequent large decline in market share. Part of this failure seems to have come from the splintered nature of the U.S. machine tool industry and part from the lack of computer expertise in that industry. In the absence of any agency or network with a capability and interest in creating the necessary coordination of knowledge and organizations, failure was the result. The needed coordination could have very probably been accomplished without great difficulty by the U.S. government had it had appropriate surveying and an appropriate agency to facilitate coordination.

The coordination function appears common to all nations which are successful in commercial innovation. Contrary to conventional wisdom, governmental coordination often played a central role in the history of U.S. technologies. Among the obvious examples are the creation of the railroads, the highway system, and the air transportation system. In the post World War II period, coordination was essential to success in the aerospace/defense, agricultural, and medical sectors [25].

What specifically is meant by the government coordinating function? Usually it implies government efforts to assure that various organizations which can contribute to a technology know of each other and can interact without serious barriers. In some cases, the government can take actions to forge important missing links. With few exceptions, government coordination does not mean direction of the networks by government. To put this differently, top-down control is usually the opposite of what is needed. Once goals have been set, effective work of coordination requires fostering strong lateral communications among the active elements in the network. This implies a process for repeated formation of consensus among the groups with the relevant forms of expertise.

An effective government role involves assistance and stimulation but not doing.

▾ *Gap Filling*

Gap filling is the clearest and the most controversial of the technology policy functions. Gap filling implies the supply of functions or of organizational arrangements that do not exist and are not likely to be filled by commercial organizations. Gap filling has been common in the U.S. in defense, agriculture, and medicine, but has been rare and *ad hoc* in the commercial sector. So long as the U.S. led the world in a very large number of commercial technologies, there was little if any need for gap filling, but that is no longer the case. Gap filling, remains nevertheless highly controversial because it raises the specter of "picking winners." Picking winners connotes at least two negative attributes: 1) supporting commercially unsuccessful products or processes; and 2) giving some companies commercial advantage over others. Gap filling need not, however, involve picking winners in this sense for reasons discussed below. In fact, there are several gap-filling activities which can be very important to the success of commercial technological enterprises that are relatively free of the negative attributes.

Historically, gap filling by governments has taken three forms: direct, contract, and cooperative. The direct mode commonly involves the creation of new organizations by the government to accomplish specific innovations. Such organizations include, for example, agricultural research stations; the National Bureau of Standards; the National Cancer Institute; and the wind tunnel test facilities of the former National Advisory Committee on Aeronautics (NACA). There is general agreement that many of these historic gap-filling functions were quite successful. For example, the wind tunnels of NACA were critical in developing the U.S. aircraft industry, and could not have been afforded by the fledgling aircraft companies in the early stages of the industry. The standards and the tables of properties of materials produced by the National Bureau of Standards played important roles in many industries.

The contract mode involves government funding of other organizations to do specific tasks. It is the norm in military procurement. One such project involved Air Force funding of the first numerically controlled machine tool to improve aircraft production. Orphan drugs are another example. The original support for the development of computers and for production of semiconductors came from the military. Significantly, when IBM first looked at computers, it was not convinced of their commercial utility. Thus IBM played no role in the very early stages, but began developing computers after several military sponsored projects demonstrated their potential, and reduced the risks.

A classic situation which warrants contract support by government is when a missing bit of knowledge or a missing process prevents innovations in larger systems. A large part of the government supported work which can be called "technology-induced-science" carried out in university research groups and non-profit institutions falls into this category. A few examples out of many are the work on high temperature superconductors, on improved processes for clean-up of emissions in combustion systems, and on miniaturization of semiconductor circuits.

In the 1960s and 1970s numerous commercial advances were spun off from contract supported defense innovations. These ranged from the creation of whole industries such as computers, semiconductors, jet aircraft, and microwave devices, to a large number of more detailed advances. These examples illustrate the "trickle-down" effect. In earlier decades, when there was little foreign competition, such trickle down effects were sufficient to create leading positions for U.S. industry in many technologies. However, the trickle-down effect is not adequate in the face of serious foreign commercial competitors. Given strong competition, trickle down is at best inefficient and at worst insufficient as the source of commercial innovations.

An organization which is noted for its successes in filling gaps employing the contract mode is the Defense Advanced Research Projects Agency (DARPA). DARPA has a charter to do gap filling in the military sector. DARPA surveys technologies, and makes projections about technological trajectories. It then funds high risk projects which have high potential for improving military systems. Since it is a part of the U.S. Department of Defense (DOD), DARPA exists within a network which does continuous surveying of technologies and which has connections to many U.S. institutions. The rules by which DARPA chose projects and created many successes are in no way secret; they were recently summarized by Craig Fields, a former director [27]. DARPA stands as a powerful counterpoint to the notion that government inevitably makes wrong choices in supporting technological trajectories of future importance.

The cooperative mode in the U.S. is of more recent origin, and its initiation was heavily influenced by patterns developed in Japan (e.g., the VLSI project) and in Europe (e.g., the EUREKA project) where there are long traditions of cooperative work. In the cooperative mode, government acts to support companies in creating a consortium in order to work on a problem seen as important to the nation. A recent U.S. experiment is SEMATECH, a consortium focused on the development of semiconductor production processes.

Gap filling via consortia created to work on generic technologies most closely approaches picking winners. But this form of gap filling is often attractive for two reasons. First, support for generic technologies is warranted in areas of important promise in which the cost and risk of the R&D necessary for the next round of innovations are too great for single firms to assume. The function of government is then to aid in reducing the risk and the cost to a point at which individual companies can take over. Second, it is far easier to foresee the impact of technological trajectories than to decide what specific products or processes will be "winners" within a given technological trajectory [28].

Consortia of this type can serve a number of purposes. First, and perhaps most important, they can highlight the developing technologies which will be important in international com-

petitions for the next generation of products in one or more sectors. As noted above, this need not involve picking a winning process or product, but rather can be centered on reducing risks by work on generic technologies thus delineating the direction which product designers need to follow, and the constraints they need to recognize, for successful developments.

If consortia are to be effective, it is critical that they be organized to increase, rather than decrease, downstream competition among the companies involved. The Japanese were able to achieve this increased downstream competition by insistence on "shadow projects" within the participating companies. A shadow project is a group of engineers who remain within the home company but follow the work of the consortium actively. The word "actively" implies that the home group contributes to the idea generation within the consortium, visit the consortium relatively frequently, and also replicate critical experiments and processes at the home company facilities. This active following creates two conditions that are essential to later transfer of the technology: 1) there is a group at the home company with deep hands-on expertise; and 2) the home group remains integrated in the company network and can therefore act as in house champions for the developing technology, thus securing sufficient priority to activate appropriate projects in a timely manner.

Because of rapid product and process innovation, much gap filling needs to be transitory in nature. Some projects should start and end at predetermined dates to preclude their perpetuation beyond a useful cutoff time. Gap filling depends on current information and thus is especially dependent on intelligent surveying. This surveying is in turn dependent on the networks which facilitate prompt flow of information.

▼ *The Interrelationship of Functions*

Perhaps nothing is more evident from the preceding discussion than the interrelationship and interdependence of the four functions. Thus, any effort to formulate a technology policy without addressing each of the functions will likely lead to a crippled policy and program.

Unavoidable Policy Decisions

Advanced nations cannot avoid the many policy decisions which together create a more (or less) favorable climate for innovations. This climate strongly affects the health of technological enterprises operating within the nation's borders. Thus, a technology policy will exist in any advanced country. The relevant question, therefore is "How will policies be created?" At one extreme, policies can be created implicitly through a series of *ad hoc* indirect responses to various crises and by many decision makers whose primary concerns are not commercial technology. At the other extreme, policies can be formulated in a direct, self-conscious way which provides opportunity for explicit formation of consensus on the issues which affect the national interest among the interests involved.

Over the past few decades, the U.S. has largely employed the implicit, *ad hoc* mode in the commercial technology sector. The Japanese have primarily employed the explicit mode. This comparison speaks strongly for the value of explicit policy formation.

Similarly, over the post World War II period the U.S. has benefited from direct, well developed technology policies in the areas of defense, medicine, and agriculture. These policies produced not only U.S. technological leadership in the three sectors but these sectors also contributed most of the large trade surplus generating products during the 1980's. Commercial technologies, by comparison, experienced poorer performance. This comparison argues strongly for the value of direct, self-conscious formulation of technology policies.

An adequate technology policy needs to serve four interrelated functions: climate setting; surveying; coordinating; and gap filling. A lack of clear delineation of these functions in the past tended to confuse the debates needed to reach appropriate decisions about government support of technologies in the commercial sector. Inasmuch as some of these functions are unavoidable, and others can be carried out using a wide range of instruments, clear delineation is important.

Exercise of the four functions of technology policy by the U.S. government requires action of two fundamental kinds. First, attaining and maintaining competitive commercial technologies must become a national consensus, high priority goal. This will require that the President and the Congress, by both word and deed, impress on the nation the centrality of competitive technologies to a high national standard of living and thus to the achievement of many other national goals.

Second, the government needs to establish and maintain a sophisticated institutional capacity for surveying the global situation and trends in technology. It then needs to use the knowledge generated through surveying to facilitate consensual decisions about technology policies in many other parts of government, and to issue appropriate decisions regarding coordinating and gap-filling actions.

References

[1] Council on Competitiveness, *Gaining New Ground: Technological Priorities for America's Future*, 900 17th Street, N.W. Washington, DC, 1991.

[2] Executive Office of the President, *U. S. Technology Policy*, Washington, D.C.: Office of Science and Technology Policy, Executive Office of the President, Sept. 1990.

[3] Robert Kuttner, *The End of Laissez-Faire: National Purpose and the Global Economy After the Cold War.* New York, NY: Knopff, 1991.

[4] S. J. Kline, "A numerical index for the complexity of systems: The concept and some implications," in *Proc. Conf. Responsible Leadership in Computing*, 1990.

[5] J.E. Aubert and Andrikopoulos-Iliopoulos, "Evaluation of scientific and technological programmes and policies: A selection of current experiences in OECD countries," *Science, Technol., Indus. Rev.*, p. 163, Dec. 1989.

[6] Office of Science and Technology, *State Technology Programs in the United States.* St. Paul, MN: Minnesota Dept. of Trade and Economic Development, 1988.

[7] Michael Porter, *The Competitive Advantage of Nations.* New York, NY: The Free Press, 1990.

[8] Lester C. Thurow, *The Zero-Sum Solution: Building a World-Class American Economy.* New York, NY: Simon and Schuster, 1985.

[9] Stephen S. Cohen and John Zysman, *Manufacturing Matters*, New York: Basic, 1987.

[10] Michael L. Dertouzos *et al.*, *Made In America: Report of the MIT Commission on Productivity.* Cambridge, MA: M.I.T., 1989.

[11] J.P. Womack, D.T. Jones, and D. Roos, *The Machine That Changed the World.* New York, NY: Rawson and Assocs., 1990..

[12] The National Critical Technologies Panel, *Report of the National Critical Technologies Panel.* Washington, DC: Office of Science and Technology Policy, Mar. 1991.

[13] Philip A. Roussel, Kamal W. Saad, and Tamara J. Erickson, *Third Generation R & D: The Link to Corporate Strategy.* Boston, MA: Harvard Bus. School Press, 1991.

[14] Robert B. Reich, *The Work of Nations: Preparing Ourselves for 21st-Century Capitalism.* New York, NY: Knopff, 1991.

[15] Philip Barkan, "Strategic and tactical benefits of simultaneous engineering," *Design Management Journal,* 1991.

[16] Philip Barkan, "Productivity in the process of product development," in *Managing the DFM Process,* G. L. Sussman, Ed. New York, NY: Oxford Univ. Press, 1991.

[17] S.J. Kline, "Research, invention, innovation and production: Models and reality," Rep. INN-1, Dept. of Mechanical Eng., Stanford Univ., Mar. 1985, revs. Sept. 1985, 1987, 1989; also *Res. Manag.*, July-Aug. 1985 and *Science, Technol. & Soc. Newsl.*, Sept. 1985.

[18] S.J. Kline, "Innovation styles in Japan and the United States: Cultural bases and implications for competitiveness" (1989 ASME Thurston Lecture), Rep. INN-3B, Dept. Mech. Eng., Stanford Univ., Stanford, CA; also *Chemtech*, July 1991.

[19] S.J. Kline, "Models of innovation and their policy consequences," in *Proc. NISTEP (Japan) Int. Conf. Sci. Technol. Res.*, H. Inose, M. Kawasaki, F. Kodama, Eds., Shimoda City, Japan, 1991, pp. 125-140; also Rep. INN-4, Dept. Mechanical Eng., Stanford Univ., Stanford, CA.

[20] S.J. Kline and N. Rosenberg, "An overview of innovation," in *The Positive Sum Game*, R. Landau and N. Rosenberg, Eds. Washington, DC: National Acad. Press, pp. 275-305.

[21] Fumio Kodama, "Direct and indirect channels for transforming scientific knowledge into technological innovations," in *Transforming Scientific Ideas into Innovations: Science Policies in the United States and Japan,* B. Bartoka and S. Okamora, Eds. Japan Society for the Promotion of Science, 1985, pp. 198-204.

[22]Leon W. Zelby, "Impact of government funding on engineering colleges," *Bridge of Eta Kappa Nu,* vol 74, no. 4, pp. 6-9, Aug. 1978.

[23] Chalmers Johnson, MITI and the Japanese Miracle: The Growth of Industrial Policy, 1925-1970. Stanford, CA: Stanford Univ. Press, 1982.

[24] Fumio Kodama, "Policy innovation at MITI," *Japan Echo* vol. 11, no. 2, pp. 66-69, 1984.

[25] Don E. Kash, *Perpetual Innovation: the New World of Competition.* New York, NY: Basic, 1989.

[26] Fumio Kodama, "The erosion of American technological competitiveness — An interview of Richard R. Nelson," *Sci. & Technol. Japan*, pp. 89-94, June 1990.

[27] "The Government's guiding hand: An interview with ex-DARPA Director Craig Fields," Technol. Rev., Feb.-Mar. 1991, pp. 35-40.

[28] Simon Ramo, *The Business of Science.* New York, NY: Hill and Wang, 1988.

JOHN E. ULLMANN

The Image Bank/S. Wilkinson/1999

Task for Engineers — Resuscitating the U.S. Rail System

"The U.S. [passenger] rail system is almost as bad as East Germany's."
—Eberhard Jaensch of the German Federal Railways [15].

During the 1992 Presidential campaign, a new high-speed rail system was prominent in the infrastructure projects proposed by Bill Clinton to restore the country to economic health. The details were left vague and little has happened since, beyond proposed increases in prior budgets that provide few resources for major projects and, at this writing, may not survive the current budget crisis.

Building a new network the way it should be built would not only create jobs, important as that is, but would revive a whole equipment industry, necessary as part of both technical and manufacturing renewal. A national project of this sort would also help to redirect some of the technical and scientific resources now used by the U.S. military. Although the problems of conversion are difficult and little has been done so far [13], public projects of this sort are essential to boost the economy at this time of military cutbacks. Given the preemption of many consumer markets by imports in important technical areas, the low level of consumer confidence, and widespread waves of layoffs, little help can be expected from an unaided private sector.

Even more important in the longer run is that shifting more passengers and freight to rail would reduce automobile and truck congestion, air pollution, the steady pulverization of the roads, and further incursions on the land and on communities. Railroads have an almost four to one energy advantage per ton-mile [6], [12, pp. 183-195], and in relation to their traffic capacity, take up a fraction of the space of modern highways. A more recent estimate by the Austrian Federal Railways [1, p. 32] gives rail an energy advantage per passenger-mile of 1 to 3.5 over cars and of 1 to 8.7 per ton-mile of freight for trucks. As a footnote to this, when the Swiss were reluctantly persuaded to build a road tunnel

The author is Professor of Management and Quantitative Methods at Hofstra University, 134 Hofstra University, Hempstead NY 11550-1090.

Reprinted from *IEEE Technology and Society Magazine,* Vol. 12, No. 4, pp. 13–20, Winter 1993.

under the St. Gotthard Pass, they found that the energy required to run its ventilation system alone exceeded the total used by the present busy railroad over the pass.

To make any sense, however, and reap the potential energy advantages, a new system would have to be electrified, rather than run by diesels. All major rail systems in the world except in the United States (and Canada) have electrified their main lines and major electrification projects are underway, as in China, where this was not done already. Except for a couple of small gaps likely to be closed before long, it is now possible to travel under electric traction from Seville in Spain to Vladivostok on the Russian Pacific Coast and from Narvik in Norway to Sicily.

The reason has always been that electricity can be obtained in many ways, including renewable sources and cogeneration, i.e., making it from industrial and other heat now wasted. For instance, in the electrifications in the Alpine countries such as Switzerland and Austria, hydroelectric plants furnished the needed power. As long as the mountains stood and the snow melted in Spring, there was a basic transportation system largely immune from interruptions in the fuel supply.

On the other hand, cars, trucks, and diesel locomotives rely on oil alone. Reducing the demand for oil in the United States would do much for U.S. security and balance of payments and also tie in with what must be done anyway to improve energy use. Various steps, drastic in their context, are already being taken to alleviate automobile pollution in such heavily affected areas as Southern California.

For a time, there was interest in turbo-electric propulsion and France especially introduced turbo-electric units in its system and exported the train sets that now run between New York and Albany. However, the high fuel consumption, greater pollution, and poorer acceleration of turbine sets caused them to be considered and rejected for the European high speed lines which are all electric.

Following the energy crisis that began in 1973, conservation efforts, especially by industry, were enough to wean the country from Gulf oil and almost from North African oil as well. However, conservation and efficiency standards for energy-using equipment were downgraded in the 1980s and funds for energy conservation and the development of alternative (especially renewable) energy sources dried up. As a result, U.S. dependence on imports generally and on OPEC and on Gulf oil in particular, have risen to what can only be called dangerous proportions. By 1990, U.S. imports of petroleum were at 45% of consumption; no oil now comes to the United States from Iran and Libya but in that year, Iraq still accounted for 8.7% of U.S. imports and Saudi Arabia for 20.2% [14, pp. 563, 572].

To create an effective new system, passenger and freight services would be needed, both of them integrated and standardized in a national plan. It is absurd to pose the future of railroads as a complete dichotomy between superspeed passenger lines on one hand and slow freights meandering over decrepit trackage on the other. One would hope that the latter category would steadily shrink, but a unified system means that long distance passenger service, commuter lines including airport access, and freight would move at optimal speeds and provide maximum convenience for their users.

This objective — an integrated system, electrified for maximum energy saving and security — defines the details of what should be national policy as well as the tasks that American engineers must face in tackling the job. If a new system is to fulfill its purposes, engineers will have to become part of an industrial rescue operation — not to say resurrection — that has few parallels even in our current sorry industrial state.

Problematic Heritage

There is not now any national plan for any major resuscitation of the U.S. rail system. Essentially, the railroads are the survivors — leftovers is a better term for some of them — of what had been a network of about 250 000 route miles at its height in 1920. It then included a by now almost entirely vanished system of electric interurban trolley lines of some 50 000 miles which were concentrated in the "corridors" that are now the main and almost sole focus of interest in passenger service. At present, about 150 000 route miles in various states of repair or disrepair, remain operable.

This cutback is not an entire calamity. In the heyday of the industry, there was much duplication, which was due in part to incorrect estimates of the location and extent of economic development. Thus it was said of the Carson & Colorado Railroad, a star-crossed, albeit scenic, line that once ran from Nevada down the Owens Valley to California, that "it had been built three hundred miles too long or three hundred years too soon."

In many parts of the country, competing lines literally ran within sight of each other, like the now abandoned Pacific extension of the Milwaukee Road next to the Northern Pacific, or the New York Central (now Conrail) and Nickel Plate (now Norfolk-Southern) which still run a few hundred yards from each other from Buffalo to Toledo. Not even the most hardened railroad fan or civic booster would have considered it necessary to have a choice of seven railroads between New York and Scranton, Pennsylvania, but that was once the case.

The exuberance that produced such excess, however, did not necessarily mean good service or attention to the marketing aspects of transportation. Rather, corporate identity and one-upmanship often took precedence over travelers' convenience and so the railroads foisted impossible transfers on them, including those between distant rail stations. Eventually, there was a move to shared "union" stations in several communities and some resulted in splendid edifices, but originally, each railroad had to have its own.

Incredibly, the same mistake was made when New York's old Idlewild, now John F. Kennedy, airport was built. Each airline then existing had to have its own terminal, as if they would each outlast the Pyramids. This now greatly complicates any plans for rail access which is badly needed; the point is worth mentioning because several of the by now almost universal rail lines to European airports offer ready connections to the new high speed networks. As matters stand, locating airline terminals at JFK from the lists on overhead road signs is a difficult exercise in speed reading and rapid reaction.

Elsewhere in the New York area, many passengers were dumped on the Jersey shore in what were, as late as the 1950s, six stations serving nine different railroads. To get to New York, there were ferries which then left the passengers on the West Side piers, far from the subways. The Hudson Tubes, a subway which is now the PATH system, offered some connections to Manhattan, but the line was deliberately built with tiny tunnels so that no single railroad that acquired it could use it for direct access. All this required inconvenient and expensive transfers and no doubt was a major factor in the abandonment of much of the rail service then offered.

The New York region now has a big job on its hands trying to integrate its systems through direct connections and modernization. It will soon be running more New Jersey trains into Penn Station (on the West side of midtown Manhattan), in addition to its present load of all of Amtrak, including upstate New York State service, and the Long Island Railroad. However, Penn Station's once great edifice was torn down with the Pennsylvania Railroad's last gasp and the trackage is likely to prove inadequate for its load before long. Such alternatives as completing the connection of the Long Island Railroad to the now underused Grand Central Terminal on the East side, for which an East River tunnel was built but now stands empty, are indefinitely on hold.

What happened elsewhere is best described as civic vandalism. In at least a dozen major metropolitan areas, transit systems were torn out as late as the 1950s when explosive suburban growth could have given them new markets, provided only that they had been upgraded here and there and extended to serve new population centers. The end of the Pacific Electric system in Los Angeles, once at over 1100 route miles the largest of its kind in the world, is perhaps the most calamitous example. [10, pp. 109-127].

It is distressing to compare current U.S. activities with what is happening in other countries. Even a cursory examination of such trade journals as the International Edition of the Railway Gazette shows that while U.S. activities are limited to a few painful and expensive resuscitations of "light rail" (the new name for trolley cars), or upgrading of diesel-operated commuter lines on trackage used for freight or virtually abandoned before, many communities and regions in Europe and Asia are putting ambitious new systems in place.

Beyond the above passenger-related blunders, there has been a degree of mismanagement of markets, finances, operations and technology in many of the mainly freight-carrying railroads that borders on a death wish. Chapter and verse on this are beyond our scope here; such debacles as the Penn-Central alone spawned notable business disaster books [2], [8]. Some of the lines judged success only by how many people they fired and how much track they ripped up. Comparing today's rail maps with those of bygone days shows in many places an almost willful determination to break up secondary routes that could today serve as freight lines, leaving the better ones to carry passengers. One result of the latter practice was to complicate enormously the task of getting freight across the Middle West in the great floods of 1993; many secondary through lines that might have been used in the past no longer exist.

Railroads have an almost four to one energy advantage per ton mile.

It is necessary to become aware of these misjudgments which involved not only private, but civic and governmental shortsightedness, incompetence, and venality on a grand scale. As Winston Churchill once said, "the use of recriminating about the past is to enforce effective action at the present." Beyond resolving to do better this time around, a first step in the recovery process we now need is the simple use of a good map so we can rescue and really fix up usable rights of way and limit the need to start over completely.

Routes, Operations, and Markets

In 1980, the Federal Railroad Administration was beginning to think in terms of overall systems and prepared to make more detailed plans at a later time. In April 1980, I had published a short article on a possible system [11] and was then invited to Washington to discuss the FRA's own ideas. I had entitled my article "The I.D.E.R. (The Interstate and Defense Electric Railroad)", since a similar designation had been used in the 1950s for the new highway network and a defense-related (and, as it turned out, illusory) fig leaf had proved helpful in getting that job through Congress. However, this time nothing helped; the election of Ronald Reagan soon put an end to any of the FRA's plans and its political leadership was, of course, purged.

By that time, however, the FRA had at least identified 30 000 miles of Class A mains, meaning those with the highest traffic density, another 15 000 miles that could become A mains with improvements and 42 000 B-mains — lesser lines, but still with good potential. Taking the A's, the potential A's and part of the B trackage would suggest a total of about 60 000 miles of actual or potential high traffic density lines; those should be the core of the new system.

An important element of a new passenger system in the United States should be car trains; these are now the main new initiative in Europe beyond expanding the existing high speed lines. France has taken the lead in planning the TGV Routier service that will be a high speed version of Amtrak's Auto Train, and will carry trucks as well. Long trips in Europe are fraught with delays (including those at borders) worse than usually found in the United States. Still, it is frustrating to take long and fatiguing car trips in the U.S., while a railroad line, weed grown and neglected, lies idle beside the Interstate.

Nothing resembling such plans is now in the hopper in the U.S. The projections in Clinton's embattled budgets envisioned $646 million for high speed rail or magnetic levitation (maglev — of which more later) from 1994 to 1997, plus $258 in 1997, to be added to the $725 million for other rail transportation projects. The intention was to provide $13 billion over five years. As the budget battles went on in Spring 1993, it became clear that not only were these really far too modest amounts in peril, but the very survival of Amtrak was being called into question [6].

The 1993 budget contained a High Speed Ground Transportation Demonstration Program which will fund studies on such issues as the economics of high speed rail, ride quality, and improved switch design. One can but wonder, however, how many wheels would be reinvented in studies of conditions and technologies amply proven out abroad.

If it were done correctly, feeder lines of the basic system outlined above would continue as they are. This is important; mistaken railroad planners the world over always seem to believe that "pruning" the system of its branches improves it. Yes, taken by themselves, many branches lose money, but they supply customers to the main lines. By analogy, the lumber industry would no doubt prefer it if trees consisted only of trunks, but that is not nature's way either.

For railroads, these rather elementary facts suggest what should be one of the main marketing principles of transportation systems, but is still vengefully ignored in all too many cases: If people have to drive far to pick up a passenger line, or have a truck go far out of its way to a rail container terminal, only to have people and goods deposited at the other end far from their ultimate destinations, they will conclude that they might as well drive or run the truck all the way. In other words, what is needed is a network, not a disembodied line here and there that serves only a limited area and needs a special trip to reach it, as if it were a tourist attraction.

A close corollary of this is that one must absolutely limit the number of transfers, changes and other disturbances to a journey. People carry things, briefcases, the week's groceries, small children, etc. or they want to rest during their journey; that in itself is a reason for using a good public transport system. Even a transfer "across the platform" then becomes a nuisance, potentially damaging to the market, especially since this almost always means waiting for the connection by both vehicles.

Current discussion has focused on the "corridor" projects, i.e., service for clusters or stretches of communities that were once described as megalopolises and where rail journeys of up to about 400 miles would be highly competitive with air travel if they could be speeded up substantially beyond their current slow pace. While the best European and Japanese rail lines serve such conurbations and distances between population centers in the United States tend to be larger, there are many comparable corridor opportunities in the Eastern half of the United States, in the Middle West and also along the West Coast.

The Intermodal Surface Transportation Efficiency Act of 1991 therefore provided $30 million - a trifling amount, but a start - for high speed corridor studies and these are the focus of attention in the above-mentioned Clinton budgets. Five proposals of more than a dozen submitted have been accepted for such further analysis, but this is a piecemeal approach at best. The areas served are limited and there is no plan at the moment to link them to each other. For instance, one accepted proposal is Detroit-Chicago-St.Louis-Milwaukee, but why not include Minneapolis-St.Paul, or Kansas City, or Toledo-Cleveland-Buffalo where it could not only join an improved Empire State service into New York, but serve the Toronto corridor as well? Two other approved studies are Oakland-Sacramento-Los Angeles-San Diego and Vancouver-Eugene, Oregon, but why not link them and thus establish a whole West coast network? Such logical expansions alone would suggest that the corridor approach should not serve as an end in itself.

To be sure, the connections would not have to be special high speed passenger lines. There is not much population from Eugene, Oregon, to the Bay area, nor, say, from Kansas City to Salt Lake City between which only the short Colorado Springs-Denver-Cheyenne corridor is a major population center.

Note, however, that the successful European high speed systems do not rely only on new special trackage. For example, the Paris-Lyon TGV service is on new track starting in the Paris suburbs, but its service to Lausanne runs on upgraded regular track east of Dijon, as do the Riviera trains south of Lyon. It is similar with the Direttissima high speed line in Italy which includes one segment (Bologna-Florence) going back to Mussolini's public works and new construction only for two thirds of the line from Florence to Rome. That had to be straightened out, having originally been built in the 19th century by a British contractor who was paid by the kilometer, with predictably convoluted results.

The new passenger systems would need track upgrading (including concrete ties - a lot of wooden ones are still used in the U.S.), getting rid of grade crossings, and fencing in the right of way. In fact, some of the freight and dual use trackage would have to be revamped in a similar manner, as was done in Europe. Experience shows that gradients don't slow down electrified trains as much as they do diesels, which is another reason for electrifying the new lines. It is curves that have been shown to be the biggest obstacle to high speeds.

In late 1992, Amtrak started testing a Swedish train that moves its wheels (in effect, steers) and tilts so as to go faster around curves. It is to be used on the New York -Boston line which is scheduled to be totally electrified, rather than the present absurdity of using diesel traction east of New Haven. The new train is mainly needed so that high speeds can be maintained along the Connecticut coast where there are huge speed-killing horseshoe curves around some of the major estuaries, the worst one probably being the one in New London. The new train will help, one hopes, but straightening the curves is still the best answer; it should be used wherever remotely feasible. As noted later, however, this raises major cost problems.

Several major freight railroads now run profitably. They could run much faster and improve their market share by using the simplest of all methods to do this: It is to run fast, short, frequent trains, instead of waiting to assemble the six-locomotive two-hundred-car monsters they now often prefer. Those need constant and expensive switching and reshuffling because, except for specials like unit coal trains, not many cars go to the same destinations. Short and frequent freight trains are standard procedure in Europe.

To make any sense, a new system would have to be electrified.

To attract customers, the passenger lines with the best potential would have to match the best speeds, say 180-200 mi/h, of the new European lines. This is much higher than the current U.S. maximum of 125 mi/h, permissible only on stretches between Washington and New York; the rest of Amtrak is restricted to about 100 mi/h or less, with average speeds way below that. This raises important questions beyond electrification with respect to the technologies to be used.

Technology

Inevitably, major plans such as the foregoing, invite attention to drastic new methods, to ways of starting over entirely. However, in the kind of system proposed here, a radical departure beyond the great improvements that have been made in older technologies is not necessary and highly problematic. The obvious candidate for something new is maglev which has attracted much public comment as a possible end run around perceived limitations of conventional railroads. Maglev was invented in the United States and much has been made of the fact that the most ambitious development work in it is now going on in Germany and Japan rather than in the U.S. Maglev is therefore often presented as yet another technology invented in the U.S. and taken up by others, when commercial development of products was one of the industrial glories of the United States in better days.

There are, of course, more than ample reasons to be concerned over such takeovers; they are an appalling consequence of the neglect of U.S. technical resources through military preemption [12, pp. 3-26]. In fact, electric traction and rail equipment as a whole are among the worst such examples, as will be discussed later. However, some realism is in order for maglev. First, its technical problems are anything but solved. Its trains are very noisy at 250-300 mi/h, their power consumption is a substantial multiple of that of high speed conventional trains and even its most ardent advocates would have to concede that this is not a technology ready to roll.

For one thing, the most advanced development work in the United States in which the Grumman Aircraft Corporation plays a leading role, focuses on the use of high temperature superconducting technology. While promising much better efficiency than the current German and Japanese proposals, it needs very much more work in itself. Little would be gained, given the urgent need to move with new transportation concepts, by relying on a technology probably as much as two decades in the future.

At least equally seriously, a new and incompatible system like maglev would run counter to the marketing considerations set forth in the last section. The maglev project furthest along, in Orlando, Florida, is now on hold, because Disney Epcot Center decided to put its terminal a mile from the nearest hotel, using "shuttle buses" which, of course, are exactly what one should avoid if at all possible. For any maglev lines that would aspire to something more than Disney-type attractions, changes and long trips to the system would be required in most cases; even if little time is lost as a result, it would not take much to wipe out any conceivable speed advantages from maglev.

Furthermore, for the foreseeable future, capital will be short even for what we can and should do now, let alone building an entirely new single purpose system whose advantage would only be marginal. And "single purpose" it is: There is no realistic possibility of dual use; though theoretically there might be applications, maglev is not now being proposed as a freight carrier, nor for lines with frequent stops.

In some places it might even skim the most profitable parts of an already troubled service sector as well as duplicating capital facilities. Some years ago, this was avoided in another such case, when eminent domain for coal slurry pipelines was blocked in Congress; they would have taken much of the badly needed coal traffic of the railroads. As matters stand, though, this is not necessary. Great savings in both first cost and construction time can be realized by working with what already largely exists. We need only improve its technology to the impressive extent that ample experience abroad has shown to be feasible, beginning with the Japanese Shinkansen trains almost thirty years ago.

We can start by just catching up with the technology of electric traction that has been developed in the last fifty years or so. Electric traction, of course, was largely invented by Americans. From early pioneers like Leo Daft and Frederick Vandepoele, to Thomas A. Edison, B.G. Lamme, George Westinghouse and above all Frank Sprague, the main movers in the field were Americans, with only Werner von Siemens standing out as a comparable foreign figure.

However, as mainline electrification largely fell victim to Depression-era disinvestment and deferred maintenance, and as trolley systems and the interurbans vanished, interest in the field also waned. For practical purposes, there was little if any U.S. participation in the major traction developments, such as AC systems at commercial frequencies, AC traction motors, and thyristor controls. If anything was used, it was mainly imported.

This decline was reflected in what happened with electric locomotives. As the celebrated 1930s vintage GG1's on the New York-Washington line reached operational senility, General Electric built some replacements. They were enormous 80-ft units that not only failed to operate at their promised speeds but wrecked a particularly strongly designed roadbed when they attempted to do so. Some of them survive on New Jersey commuter trains where they run much below their originally intended speeds. Amtrak tried out another engine from Sweden's ASEA which was promptly dubbed "Mighty Mouse"; it was barely 50 ft long and did yeoman service.

Amtrak then got the Electromotive Division of General Motors (EMD) to manufacture the engines in the U.S. under license. It did so for a while, in what one might call Third World fashion, importing the high value-added parts like controls. After a while, EMD stopped making the engines and later ones were made by ASEA (since joined with Switzerland's Brown-Boveri to become ABB), in its plants, affiliates, and subcontractors in Sweden and Austria.

EMD, incidentally, was the cradle of the diesel locomotive in the United States and pioneered its manufacture as a virtually mass-produced item, but it has now been closed by General Motors. Its diesel locomotive business was transferred to a plant in London, Ont. Wages were somewhat lower there, but the main reason was reported to have been that Ontario has public health insurance and the United States does not and GM wanted to get out of paying for health care.

Meanwhile, matters were not much better with other equipment. In 1987, the Budd Company, the last U.S. maker of passenger cars, went out of business, and with it went the capability to bend stainless steel into passenger car shells, an expensive and highly skilled process that Budd had first developed. The result was dependence on non-U.S. suppliers for new rolling stock as well as motive power [5].

Reviving the Equipment Industry

Since there are no longer any independent U.S. manufacturers of the equipment likely to be needed, it will be necessary to rely on others who have at least some American facilities. Both ABB and Siemens do, and Bombardier, the Canadian maker of subway cars and other items, also with a plant in the United States, has a close relationship with France's Alsthom-Francorail. Among the Japanese makers, Sumitomo and Kawasaki also have worked with major American subcontractors. The American firm of Morrison-Knudsen which started as a rebuilder of old diesel locomotives and other equipment, plans to go into more of original manufacture, again in part with foreign partners [5].

Building a new high-speed rail network would create jobs and revive a whole equipment industry.

The problem with such arrangements is the perennial one of the kind and proportion of domestic content. The interests of American workers and engineers are not served if all they are allowed to do is the heavy castings and other rough stuff, leaving the high value items for imports. As noted, this happened in the past, but is not inevitable. It must be made possible for the United States to take back the manufacture of a lot of products that has been allowed to move abroad as the result of military preemption and managerial indifference. Otherwise, there is no way of creating the new manufacturing jobs to which President Clinton gave high priority in the employment agenda of his Administration. Clearly, action is now needed in the face of a tide of industrial layoffs that shows no sign of abating.

The industrial remedies must be closely integrated with the national project as a whole. The next two years should be a time of planning and preparation. First, the U.S. should train a cadre of engineers in the latest technology of high speed rail, probably best by sending them abroad for intensive study. Second, the two years should be used to organize production facilities that could do more of the work in the U.S., and eliminate possible bottlenecks by creating or recreating specialized facilities. The Federal Railroad Administration would also have to rise from the desuetude to which it was condemned in the 1980s and take the lead in planning and sponsoring research, heeding in the latter area the above caveats against reinventing the wheel.

Third, the time should be used to settle on the network and start on the upgrading of tracks, before going on to the electrification phase. In this step, it is also necessary to improve the productivity of heavy construction, a major task in its own right [12, pp. 195-198].

Questions and Obstacles

How will all this be financed? The present electrification of the 158 miles from New Haven to Boston is budgeted at $200 million, or about $1.3 million a mile. This does not allow for much track realignment which would be very costly along that particular stretch; avoiding wetlands and Indian burial grounds and building two very large bridges as part of the job would add a further $1.8 billion, or a total of 11.4 + 1.3 = $12.7 million a mile [9]. Such high costs would, however, be required in few places; the European systems avoid much of this additional cost by running more slowly in metropolitan areas.

Accordingly, a national average of $3 million a mile would be conservative; it would allow for electrification as well as some track improvement. A 60 000 mile system would then cost about $180 billion, spread over at least ten years, probably longer. Government startup money or loan guarantees may be needed, but railroads would do the maintenance. It would make sense for railroads to pay for the locomotives out of their equipment replacement budgets, or they could rent them from a national cooperative that could sell its securities, such as equipment trust certificates, to the public.

However, some recent developments have complicated the situation. Just at this time when electrification should be the focus of their technical and operational attention, railroads have gone on something of a buying spree for diesels. Notable is an order by the Burlington Northern for 350 new diesels costing $675 million. As one concession to modernity, AC traction motors are used, but this is still not the rule in new orders by other lines.

More directly related to our topic, Amtrak is buying a special new diesel locomotive type from General Electric which it has called, no doubt with a dash of fervent hope, "Genesis" [4]. It is largely built by Krupp in Germany and has some of the same features as Germany's high-speed Inter-City Express (ICE) trains, but it will run no faster than 80 or 90 mi/h on the tracks available on Amtrak's freight-hauling hosts; reasons such as openness of rights of ways, grade crossings, lack of cab signaling, etc., are cited. Grade crossing eliminations would have to be a central part of a new high speed system, however, even if major new track construction is to be limited.

The body of the new engines is assertedly designed for 150 mi/h but the present maximum speed is no more than 103 mi/h. The engines are also designed for eventual conversion to straight AC electrics, but that is clearly for another era, given the present evidently limited focus of rail managements and related government agencies; they espouse performance standards which, in Europe and Japan, date back a generation. *If anything significant happens at all* in the United States in the near future, it will aim for no more than 150 mi/h, be limited to a few lines, and will not be newly electrified, other than the New Haven-Boston stretch.

Basically, any new system would have to be treated as a capital investment that has to pay for itself out of operating savings. It is deplorable, albeit a subject beyond our present scope, that government budgets generally make no distinction between capital spending and operating expense, i.e., current consumption. This bedevils discussion of such topics in the context of the financial stringency referred to at the outset.

What are the obstacles beyond the financial ones? A major one is the NIMBY ("not in my back yard") problem. It has even handicapped railroads wanting to revive commuter passenger service on existing rights of way; Frobom [3] cites cases in New Jersey, California, Minnesota, and Ontario. They remind one of two Long Island Railroad branches that could be adapted to serve as airport access to JFK; one, in fact, was originally abandoned only with the proviso that it could be resuscitated for such a purpose. When this was later proposed, "community activists" blocked it. All this suggests problems ahead, but clearly, they can be minimized if existing roadbeds are used to the maximum extent possible, especially getting out of town. Frobom suggests that the opposition will be greatest in the suburban rings, which is a serious matter, given their sprawl. Unless equitable but speedy condemnation procedures are established, the outlook is difficult.

Any improvements in rail systems are, in fact, part of a global road versus rail struggle, which has so far favored road traffic and thus helped create probably the largest single source of wasted energy on earth. Trucks enjoy the political advantage of needing very much greater numbers of people than railroads to operate, maintain and service them. They don't do their task efficiently, of course, but the jobs exist here and now, without waiting for industrial recovery or development; it is an important factor where there is chronic high unemployment. Few governments seem to have the stomach to take on an ever-potent lobby of truckers, oil producers, and road services. Still, it will eventually become indispensable for them to do so and to persuade their citizens that it is the right thing to do; the environmental argument is already a strong one in European railroad promotion.

It may well be that hopes of getting on with the job in something like the above form, will once again end in disappointment. If this happens, we will once again have accepted the fallacy that in the new rail system, as in most infrastructure projects, the choice is between doing them and spending money on the one hand, and not doing them and saving money on the other. Rather, the choice is between different ways of paying for environmental and other societal impairment. If the transportation system is not organized more efficiently, the omission will be paid for in more car and truck accidents, higher repair bills, travel delays, respiratory ailments and, in the extreme, military action to assure U.S. oil supply. As the French poet Tristan Corbiere wrote in another context, "You don't want me as a dream? Then you'll have me as a nightmare!" Choosing between the two is still possible.

References

[1] Austrian Federal Railways (Oesterreichische Bundesbahnen), *Die Bahn der Zukunft*. Vienna: OeBB, 1992.

[2] J.R. Daughen and P. Binzen, *The Wreck of the Penn-Central*. New York: Mentor, 1971.

[3] A.H. Frobom, "High-speed trains vs. the NIMBYS," *Trains*, p. 122, Dec. 1992.

[4] B. Johnston, "Genesis: Amtrak's new breed," *Trains*, p. 37, Sept. 1993.

[5] D. Phillips, "Morrison Knudsen's big gamble," *Trains*, pp. 44, Oct. 1992.

[6] D. Phillips, "Don't bet the farm on high-speed rail or Amtrak," *Trains*, p. 12, Oct. 1993.

[7] L. Schipper and A.J. Lichtenberg, "Efficient energy use and well-being: The Swedish example," *Science*, p. 1003, Dec. 3, 1976.

[8] R. Sobel, *The Fallen Colossus*. New York: Weybright and Talley, 1977.

[9] S. Toy *et al.*, "Next Stop America?," *Bus. Week*, p. 106, Apr. 19, 1993.

[10] J.E. Ullmann, *The Suburban Economic Network*. New York: Praeger, 1977.

[11] J.E. Ullmann, "The I.D.E.R.," *Trains*, pp. 66, Apr. 1980.

[12] J.E. Ullmann, *The Prospects of American Industrial Recovery*. Westport, CT: Greenwood Press-Quorum Books, 1985.

[13] J.E. Ullmann, "Converting Defense facilities," *Society*, p. 25, May-June 1993.

[14] U.S. Department of Commerce, *Statistical Abstract of the United States*. Washington, DC: U.S. Govt. Printg. Office, 1992.

[15] Eberhard Jaensch, cited in *Bus. Week*, p. 107, Apr. 19, 1993.

T&S

Chapter 6

Risk and Product Liability

Risk Assessment and Communication

RISK assessment and communication are rich examples of the interrelation of engineering ethics and public policy. Questions related to risk assessment and risk management are arguably the most prominent issues encountered in engineering and public policy. During the past few decades, engineers have been embroiled in numerous controversies surrounding technological and environmental risks, including nuclear power, toxic chemicals, and transportation safety.

The duality of risk and safety, and the clear ethical issues involved in determining acceptable risk, are central to most treatments of engineering ethics (1). Less obvious to engineers are the ethical issues involved in determining or measuring risk (2) and the ethical issues raised by differences in risk perception between technical experts and ordinary citizens.

For example, although many engineers believe that public risk perceptions are irrational, many humanists and social scientists have noted that these views are indeed rational, albeit a different rationality (known as social or cultural rationality) from the purely technical rationality employed by engineers and technical experts (see Table 3). As I argue in my article included in the background reading, such perceptual differences, which have been widely studied and characterized by behavioral and social scientists, reflect real limitations of expertise. These limitations result in an ethical obligation for engineers to listen to and attempt to understand the concerns of nonexperts, as well as to incorporate such concerns into their consideration of risk. Others have made similar arguments in favor of incorporating lay risk perception in risk assessment activities, based on the principle of informed consent (3). These arguments imply that, in order to succeed, efforts at risk communication should be two-way, with risk information being exchanged between experts and nonexperts. Such an approach is opposed to the traditional, one-way model of risk communication in which the expert merely informs or educates the lay public as to the facts of the situation.

This argument also extends to the importance of considering both quantitative methods and information, favored by engineers and other experts, and qualitative methods and information, utilized by some social scientists and the lay public. Fischhoff and Merz (5) have noted, for example:

> Quantitative understanding is essential if people are to realize what risks they are taking, decide whether those risks are justified by the accompanying benefits, and confer informed consent for bearing them. Qualitative understanding is essential to using products in ways that achieve minimal risk levels, to recognizing when things go wrong, and to responding to surprises. (p. 160)

Even if one holds to the traditional model of risk communication in which the expert is the primary communicator and the nonexpert the primary receiver of risk information, some ethical questions remain to be resolved, including the purpose of the risk communication, the motivation of the risk communicator, and the issue of whether the risk message is overt or covert (5,6).

Although researchers in the field of engineering and public policy often incorporate such considerations in their work, many engineers and engineering societies that address risk issues cling to the traditional model of risk communication, belittle and devalue lay perceptions of risk, and characterize any approaches to risk other than their own as irrational and irrelevant to public policy. For example, a leader of the Committee on Man and Radiation (COMAR) of the Institute of Electrical and Electronics Engineers (IEEE) maintains that all views on the health

TABLE 3 TECHNICAL AND CULTURAL RATIONALITY: SELECTED FACTORS (Adapted from 4)

Technical Rationality	Cultural or Social Rationality
Trust in scientific evidence	Trust in political culture
Appeal to expertise	Appeal to peer groups and traditions
Narrow, reductionist boundaries of analysis	Broad boundaries of analysis
Risks depersonalized	Risks personalized
Statistical risk emphasized	Impacts on family and community emphasized
Appeal to consistency and universality	Focus on particularity

impacts of electromagnetic radiation other than that of COMAR are simply incorrect and are spawned by "electrophobia," an irrational fear of electrotechnology. His indictment goes beyond members of the lay public, extending even to other IEEE members and staff who do not accept COMAR's position (7). Similarly, a civil engineering educator (8) recently argued that

> extreme environmentalism, akin to a religion, is causing the expenditure of massive amounts of limited resources of both money and attention to solve relatively unimportant problems. ... The public perception of environmental problems is far removed from that of the scientific consensus. Environmental policy is now based more on emotion and debate than on facts and rational calculation. ... A moderate, rather than fearful, reaction to environmental concerns is urged. Engineers should promote quantitative solutions to environmental problems, and should appreciate the economics of pollution control and risk reduction. (p. 79)

In addition to skirting serious ethical questions, such posturing on the part of engineers and engineering societies is often counterproductive, contributing to a negative public image of engineers (9) and obscuring important risk information that might fall outside the bounds of expert risk models (5,10).

Even engineers with a strong concern for public safety in their personal and professional roles can retreat to the traditional model of risk communication in their public role. Consequently, a good place to begin persuading engineers to take public risk perception more seriously is through a framework for engineering ethics that is conducive to better integration of the personal, professional, and public roles (see the Introduction to Part III).

Product Liability

The recent interest in tort liability reform is a particularly vivid example of how issues of risk can cause conflicts between public policy and engineering ethics. The 104th Congress passed legislation that would severely limit the effect of product liability litigation by placing a cap on punitive damages and enacting stricter requirements for holding manufacturers liable. Although President Clinton, as expected, vetoed the bill, product liability is almost certain to be a reoccurring issue in future congressional sessions (11).

The proponents of product liability reform have argued that the current system unjustly rewards plaintiffs and stifles technological innovation, resulting in a lack of competitiveness on the part of U.S. manufacturers and decreased product safety. On the other hand, supporters of the current system point out that although there might be a need for minor refinements, the system generally works as intended in discouraging the manufacture of defective products and compensating people injured by such defects (12).

Some observers view the debate over product liability reform as a classic industry versus consumer confrontation. A *New York Times* editorial (13), for example, described the proposed legislation as "The Anti-Consumer Act of 1996." Engineers and engineering societies, on the other hand, have tended to side with the proponents of product liability reform. For example, a vice-president of engineering of a major U.S. automobile company has argued that product liability restricts engineering practice by inhibiting innovation, discouraging critical evaluation of safety features, and preventing implementation of new or improved designs (14).

The position statement on product liability tort reform of the United States Activities Board of the IEEE (15), issued in 1998, calls for stringent limits on product liability, including holding the manufacturer blameless when existing standards are met, when adequate warnings are provided, or when the product is misused or altered by the user. Other engineering societies, such as the American Society of Mechanical Engineers (16), have also actively supported product liability reform.

Although many engineers are supportive of changes in the product liability system, few have considered the effect that decreasing the impact of product liability would have on engineering ethics, most notably on engineers seeking to alert their employers to design defects that might jeopardize safety. Indeed, it is dismaying how little attention the engineering community has paid to the ethical implications of product liability. For example, a major 1994 study of product liability and innovation by the National Academy of Engineering, while considering such issues as corporate practice, insurance, regulation, and the role of scientific and technical information in the courtroom, contained no ethical analysis at all, with the exception of a single chapter by Fischhoff and Merz (5) which, in addressing public risk perceptions, raised the issues discussed in the preceding section of this chapter.

Even the ethics literature is equivocal on the issue of product liability (17). For example, in De George's well-known essay on the Pinto case in which Ford engineers and executives knowingly produced a vehicle with poor crash-worthiness (18), he advocates stronger regulation and fines and imprisonment for corporate officials when market mechanisms fail to achieve desired levels of safety. De George gives only passing notice to the role of product liability litigation in this regard.

The evidence appears to be mixed concerning whether or not product liability rewards result in improvements in product safety (12), but the role of product liability litigation in creating an environment wherein engineers with safety concerns are given a hearing by their managers is certainly worthy of consideration. As Ladd (see background article in Chapter 3) and others have argued, corporations are not moral agents. The sole goal of a corporation is to generate profits. In order to influence a corporation's behavior, then, we need to make it in its economic interest to do the right thing. Product liability litigation would seem one mechanism for achieving this goal. At the very least, the connection between the threat of product liability suits and the ability of engineers to raise safety concerns is a question that ought to be scrutinized by those concerned about engineering ethics.

Another aspect of product liability reform that merits further attention is the effect sweeping changes might have in weakening the legal doctrine of informed consent, which evolved from tort law (17), and its role in providing support for the ethical principle of informed consent. As noted previously, informed consent is often the basis for arguments concerning engineering responsibility for understanding and accommodating the public's views on risk.

Finally, it should be noted that the overriding concerns here are not limited to product liability but also to other forms of tort liability of interest to the engineering profession, including malpractice issues and liabilities resulting from environmental risks and accidents in public facilities (12). As liability issues grow in importance in the public policy arena, engineers will need ways of reconciling their public positions with their personal and professional ethics.

The Bjork-Shiley Heart Valve Case

This case is one of many in a growing catalogue of product liability cases involving biomedical devices. Like the Hyatt case (see Chapter 3), it illustrates the often-complicated interplay between ethical and legal issues. This artificial heart valve, which was manufactured by a company subsequently bought by industry giant Pfizer, Inc., had a structural failure that caused the death of more than 400 recipients. Evidence suggests that the manufacturer was not forthcoming with information about the flaw and indeed experimented with fixes in subsequent commercial versions of the valve. Lawsuits have included claims by victims of actual heart valve failures, or their survivors, as well as by people who currently have the defective valves in place. In an interesting analysis (included as a case study in Chapter 9), John H. Fielder argues that the failure rate of the valve is not unusual for this kind of device. Rather, he finds the manufacturer guilty of an ethical lapse in failing to be forthright about the flaws in the valve—a lapse, he argues, that caused the public to lose confidence in the product. The case is thus an effective vehicle for examining the role of engineering ethics in risk assessment and product liability regarding such issues as informed consent.

Background Reading

In 1986, philosopher *K. S. Shrader-Frechette* contributed an article, "The Conceptual Risks of Risk Assessment," in which she challenges the conventional wisdom that the methods of risk assessment are value-free. Moreover, she points out some ethical issues in determining the acceptability of risk.

According to Shrader-Frechette, the value judgments in risk assessment methods include measuring risk solely in terms of potential fatalities, to the neglect of injury, financial losses, and cultural losses. Other value judgments arise because the various methods of risk identification, such as animal research and epidemiology, all have limitations that require evaluative judgments on the part of the risk assessors. Risk estimation also involves evaluative judgments due to uncertainties in dose-response relationships, the population at risk, and the dose actually received by the at-risk population. Lastly, the author argues that the prevalent methods of risk evaluation (i.e., determining whether or not a risk is acceptable) also include evaluative judgments such as ignoring risk distribution or assuming that past levels of acceptable risk should determine acceptable risks for new technologies or hazards. Many of the evaluative judgments required in these three stages of risk assessment, she argues, derive from the problem of not being able to adequately determine causality. To buttress her claims, the author provides a brief review of a controversial risk assessment of various energy technologies performed by Herbert Inhaber in the late 1970s.

Beyond these methodological matters, Shrader-Frechette identifies three ethical dilemmas that risk assessors typically must face. The first, the contributor's dilemma, involves the additive or synergistic impacts of many small risks. The second, the threshold dilemma, refers to setting an acceptable level of average risk, when the average may be inappropriate for a given individual or group. The third, the consent dilemma, refers to the difficulty in securing true informed consent, especially in circumstances where those exposed to the risk, such as workers in low-income areas, are not in a situation where consent can be freely given.

In closing, Shrader-Frechette suggests two mechanisms for dealing with these dilemmas. In the first, risk cost-benefit analysis would be "ethically weighted" (i.e., alternative analyses would be conducted on the basis of clearly stated ethical systems and priorities). Second, she suggests that adversarial risk assessments be conducted by different groups utilizing differing ethical and methodological frameworks, including participation by both experts and members of the public.

In my 1994 article, "Ethical Risk Assessment: Valuing Public Perceptions," I note that as modern technology becomes more complex the assessment of environmental and safety risks also becomes more complex, lending itself to an increasing array of moral issues and dilemmas. By virtue of their significant role in the process of technological development, engineers are inevitably involved, directly or indirectly, in risk assessment and risk management, and hence must be prepared to wrestle with the accompanying moral issues.

The evolution of engineering codes of ethics reflects a growing acknowledgment on the part of professional engineering societies that engineers have both a professional and a moral responsibility for the public safety, health, and welfare. In addition, the engineer's responsibility for the public safety and welfare extends to informing the public about technology, its applications and its consequences in a more general sense. Moreover, their responsibility to the public requires that engineers themselves be informed with respect to nontechnical perspectives on technology, including a sensitivity to and respect for the differences between expert and nonexpert perception of risk. Risk communication efforts will not go very far if engineers persist in dismissing public perceptions of risk, responding to oversimplified notions of what the public wants, and utilizing quantitative methods to camouflage their own value judgments. There is thus both a practical necessity and an ethical imperative for engineers to become actively engaged in risk communication efforts that incorporate an understanding of and respect for public perception of risk.

This effort calls for nothing less than a transformation of the engineering culture, including an integration of technical concepts with concepts drawn from the humanities and social sciences. A meaningful transformation in the engineering culture will require substantive institutional changes, especially in schools of engineering, and the development of "moral imagination" by individual engineers.

Case for Discussion

In his 1992 article, "Health Risk Valuations Based on Public Consent," *E. S. Cassedy*, an electrical engineering professor, presents a case study of valuing health risks of electric power generation, transmission, and distribution by obtaining input from members of the public on their willingness to pay for reducing such risks. The valuation includes existing risks from nuclear power and coal and risks that have more recently gained attention, such as the effects of electromagnetic fields from power lines and household wiring. The benefits of this approach, Cassedy argues, include public participation in the decision process and a measure of informed consent. It also avoids the problems encountered in conventional risk cost-benefit analysis relating to valuing human life and injury.

In evaluating current public health risk with respect to ambient air quality standards (primarily in conjunction with coal burning), Cassedy finds inadequate the current standards that are based on establishing a threshold value of pollutant concentration—that is, a level below which it is assumed no harmful effect will occur. These standards are subject to criticism on a number of grounds, including questions relating to distribution of risks, informed consent, and the necessity to place an economic value on human life. Using a study of ozone pollution in Los Angeles as an example, Cassedy suggests that these problems could be eliminated by incorporating the public's willingness to pay for reduction in the risk from these pollutants as a "benefit" in the standard risk cost-benefit analysis technique.

Cassedy reaches similar conclusions regarding nuclear radiation standards for routine operations of nuclear power plants. In this case he is even more concerned about the threshold approach due to the controversy in the scientific community over whether or not there is an actual threshold and, if not, what the nature of the dose-response relationship is at low levels of exposure. Again, Cassedy suggests, a willingness-to-pay approach would avoid these dilemmas. Of particular importance in the nuclear case, this method would include attaching a value to the public's perceived risk of nuclear power, as evidenced by their willingness to pay to reduce such risks (in the form of higher electric rates). Although the evidence of health impacts from electromagnetic fields is still inconclusive, Cassedy suggests the willingness-to-pay approach would be useful here as well, should evidence develop that such fields do pose a health problem. In this case, as in the others, the cost would be considerable. It therefore becomes very important to get as accurate an assessment of the benefits as possible, including the public's willingness to pay for the reduction of such risks.

References

[1] Flores, A., ed. 1989. *Ethics and risk management in engineering.* Lanham, Md.: University Press of America.

[2] Mayo, D. G. and Hollander, R. D., eds. 1991. *Acceptable evidence: science and values in risk management.* New York: Oxford University Press.

[3] Martin, M. W., and Schinzinger, R. 1989. *Ethics in engineering.* 2nd ed. New York: McGraw-Hill.

[4] Plough, A., and Krimsky, S. 1987. The emergence of risk communication studies: social and political context. *Science, Technology and Human Values* 12 (3&4): 4–10.

[5] Fischhoff, B., and Merz, J. F. 1994. The inconvenient public: Behavioral research approaches to reducing product liability risks. *Product liability and innovation*, ed. J. R. Hunziker and T. O. Jones, 159–189. Washington, D.C.: National Academy Press.

[6] Morgan, M. G., and Lave, L. 1990. Ethical considerations in risk communication practice and research. *Risk Analysis* 10: 355–358.
[7] Osepchuk, J. M. 1996. COMAR after 25 years: still a challenge! *IEEE Engineering in Medicine and Biology* 15 (3): 120–125.
[8] Chadderton, R. A. 1994. Should engineers counteract environmental extremism? *Journal of Professional Issues in Engineering Education and Practice* 121: 79–84.
[9] Vesilind, P. A. 1993. Why do engineers wear black hats? *Journal of Professional Issues in Engineering Education and Practice* 119: 1–7.
[10] Slovic, P. 1987. Perception of risk. *Science* 236: 280–285.
[11] Lewis, N. A. 1996. Clinton vetoes bill to curb awards in product liability suits. *New York Times* (May 3).
[12] Hunziker, J. R. and Jones, T. O., eds. 1994. *Product liability and innovation.* Washington, D.C.: National Academy Press.
[13] New York Times (editorial). 1996. The Anti-Consumer Act of 1996. *New York Times* (March 21).
[14] Castaing, F. J. 1994. The effects of product liability on automotive engineering practice. In *Product liability and innovation*, ed. J. R. Hunziker and T. O. Jones, 77–81. Washington, D.C.: National Academy Press.
[15] IEEE United States Activities Board. 1998. Tort Law and Product Liability Reform (position statement). Washington, D.C. (Available from World Wide Web site, http://www.ieee.org/usab/FORUM/POSITIONS/liability.html.)
[16] ASME International. 1996. Letter to the House and Senate Conferees on Product Liability Bills. (Available from World Wide Web site, http://www.asme.org/gric/96-01.html.)
[17] Rabins, M. J., Harris, E., and Pritchard, M. 1992. Engineering design: literature on social responsibility versus legal liability. (Available from World Wide Web site, http://ethics.tamu.edu/ethics/essays/design.htm.)
[18] De George, R. T. 1981. Ethical responsibilities of engineers in large organizations: The Pinto case. *Business and Professional Ethics Journal* 1: 1–14.

FEATURE ARTICLE

The Conceptual Risks of Risk Assessment

K. S. SHRADER-FRECHETTE

Abstract—Many, if not most, risk assessors allege that their methods of risk identification and risk estimation are wholly objective. Contrary to this view, it can be shown that a variety of value judgments are inherent in classical risk assessment methods. In addition to these methodological threats to objectivity, there are at least three ethical problems which require risk assessors or policymakers to make normative decisions. After outlining these ethical and methodological difficulties, the essay closes with two suggestions for improving risk assessment and for rendering its evaluative components explicit.

INTRODUCTION

Widely touted as a "developing science" [1], the new discipline of risk assessment or risk analysis is currently being used by engineers, statisticians, epidemiologists, economists, and policymakers, both to identify technological threats to our health and safety and to evaluate their acceptability. The goal of quantitative risk assessment is to deliver society from the twin extremes of dangerous, unrestrained technological development and ignorant, paranoid opposition to all new technologies. Instead, it offers a middle ground, an analytic framework within which thousands of risks can be comparatively evaluated—from those posed by liquefied natural gas facilities and nuclear power plants to those arising because of new contraceptives and food additives.

In employing a comparative framework within which diverse risks may be evaluated, assessors aim to give policymakers a rational basis for decisions about health and safety. They provide schemes showing how to save the most lives for the available dollars and how to spend funds so as to reduce the greatest risks first. In theory, there are at least two main assets of risk assessment. First, it could enable society to maximize the benefits of government expenditures for health and safety. Second, it could promote equity and consistency in allocation of funds among safety programs. As one commentator put it, risk assessment provides a framework for asking "why OSHA intends, in a set of proposed regulations on coke-oven emissions, to protect the lives of steelworkers at [a projected cost of] $5 million each, while a national Pap smear screening program that would save women's lives at [a projected cost of] less than $100,000 each has gone unfunded" [2]. In sum, risk assessment could in theory respond to our reasonable demands for efficiency and equity in reducing risks. In practice, however, the technique is beset with a number of methodoligical and political problems.

The greatest liability of quantified risk assessment, according to its critics, is that it attempts to reduce qualitatively diverse risks to mere mathematical probabilities, numbers of fatalities, and dollar costs. They also charge that use of this tool threatens to remove health and safety from democratic social control by the public and by parties bearing societal risks, and to place control instead in the hands of experts who contend that many of society's fears about technological hazards are irrational.

Regardless of whether opponents of risk analysis are correct in their criticisms of its political integrity, it nevertheless presents other risks as well. These conceptual risks are important precisely because they are often not recognized by practitioners of risk analysis. Because they are not, risk assessment conclusions sometimes err in relying on a number of crucial but hidden value judgments. The two sorts of value judgments to be discussed here arise in connection with difficulties over risk methdology and over ethically acceptable risk. After outlining typical situations in which these two types of value judgments arise, the essay closes with two suggestions for amending risk assessment methods, and hence for beginning to address its major methodological and ethical problems.

VALUE JUDGMENTS AND RISK METHODOLOGY

"Risk" is generally defined as a compound measure of the perceived probability and magnitude of adverse effect, and is often expressed in terms of average annual probability of fatality. One of the greatest value judgments in risk assessment arises precisely from interpreting risk in terms of fatalities, and ignoring injuries, financial losses, and carcinogenic, mutagenic, and teratogenic harms. Perhaps one of the greatest risks of risk assessment is that cultural, sociopolitical, and philosophical risks are typically not taken into account. Computer and communications technology, nuclear weapons, commercial nuclear fission, and many other technological developments pose great risks to civil liberties because of a variety of privacy, security, and public safety considerations. Each of the three stages of risk assessment—risk identification, risk estimation, and risk evaluation—is characterized by a number of methods, techniques, and assumptions fraught with conceptual difficulties [4]. Consider some of the most important evaluative problems associated with typical methods used in each of the three stages of risk assessment.

A number of authors divide risk assessment methods into three stages: risk identification, risk estimation, and risk evaluation. (Although not a part of risk assessment,

Reprinted from *IEEE Technology and Society Magazine*, Vol. 5, No. 2, pp. 4-11, June 1986.

"risk management" refers to the various political, regulatory, and technological strategies for controlling risks once they are assessed.) The standard view, according to Starr, Whipple, Okrent, Maxey, Cohen, Lee, and other widely published risk assessors, is that expert methods employed in the first two stages—risk identification and risk estimation—are objective and neutral, whereas the methods employed in the last stage, risk evaluation, are somewhat subjective and evaluative [3].

In claiming that risk identification and risk estimation are objective and value-free, Starr, Whipple, and others overlook a variety of unavoidable evaluative judgments that arise in connection with these processes.

Risk Identification

In the case of carcinogenic risks arising from particular chemicals, for example, the risk identification state is particularly problematic. It is difficult to identify risks because each of the five commonly used methods of identification has serious defects whose presence requires the assessor to make evaluative judgments.

The use of case clusters, which look for adverse effects to appear in a particular place, is helpful only when specific hazards cause unique diseases, and when the population at risk is known in some detail. Since often it is not known whether these conditions are satisfied, especially in cases involving new risks, assessors must interpret the situation as best they can.

Comparison of compounds in terms of structural toxicology, a second method of chemical carcinogenic risk identification, is likewise problematic. It reveals only that a chemical compound has, for example, a molecular structure similar to that of a known carcinogen. Yet, in using this method, assessors sometines assume that this similarity of structure is sufficient to classify a substance as a putative carcinogen.

Use of a third identifying method, mutagenicity assays, raises the issue of another normative judgment. This method is compromised to the extent that it rests on the assumption that most chemical carcinogens are mutagens; experience has shown that mutagencity assays are rarely sufficient to support the conclusion that a particular mutagen is also carcinogenic.

A fourth method of risk identification, use of long-term animal bioassays, is likewise limited, in that it depends on the questionable inference that results from animal experiments are applicable to humans.

A fifth class of methods, biostatistical epidemiological studies, are much more sophisticated versions of case clusters. They aim to show a statistical association between exposure to an agent and the subsequent manifestation of a disease. The obvious flaw in this method is that it is often difficult to accumulate the relevant evidence, particularly if exposure is at low dosage, or if the effects are delayed, as is often the case with cancers having latency periods of up to 40 years. In the absence of complete data and long years of testing, assessors are forced to interpret and to extrapolate from limited data. Most substances are moreover not even screened using formal epidemiological methods. This is because, apart from other sources of hazards, there are roughly 60,000 chemicals used in various manufacturing processes, and at least 1000 new ones are added each year [5]. Deciding which of the chemicals to screen, by any or all of the above methods, is perhaps one of the greatest evaluative judgments in risk identification.

Risk Estimation

At the second stage of risk assessment, risk estimation, one is concerned primarily with determining three things: a dose-response relationship for a particular hazard, the population at risk, and the dose it receives from the hazard. Dose-response methods are conceptually problematic because they typically require extrapolation from high to low doses and from animals to humans. The problems associated with such extrapolation are well known, as when health physicists, given data points only for higher-level exposures, try to extrapolate a dose-response curve for low-level radiation exposure. Environmentalists, industry representatives, and government bodies have each extrapolated a different dose-response curve for low-level effects of radiation. On the basis of their differing value judgments, they have concluded that low-level effects of radiation are, respectively, very dangerous, not dangerous at all, and moderately dangerous [6].

Estimating the population at risk and the dose received is just as problematic for at least two reasons. Actual measurements of particular doses—e.g., of a chemical—cannot be made in all situations; as a consequence, a mathematical model of assumed exposure must be used. Such estimates are also problematic because seemingly unimportant pathways of exposure can assume great significance owing to biomagnification due to food-chain and synergistic effects. Moreover, it is rare that a substance is uniformly distributed across pathways or across time. For all these reasons, the assessor is forced to make a number of interpretative judgments.

Risk Evaluation

At the third stage of risk assessment, risk evaluation, analysts typically determine whether a risk is acceptable or tolerable to society. They employ at least four different methods: risk cost-benefit analysis (RCBA), revealed preferences, expressed preferences, and natural standards.

RCBA is widely used and consists simply of converting the risks, costs, and benefits associated with a particular project to monetary terms and then aggregating each of them in order to determine whether the risks and costs outweigh the benefits [7]. The great advantage of RCBA is that it allows assessors to calculate how to save the greatest number of lives for the fewest dollars. Its most obvious deficiency is that simply adding up risks, costs, and benefits ignores who gets what, or the equity of risk distribution. Assessors using RCBA are thus forced to make the highly evaluative ethical judgment that the *magnitude* of risks, costs, and benefits is more important than their *distribution*, and that distribution need not be taken into account at all.

The second method of risk evaluation, revealed preferences, consists of making inductive inferences about acceptable risk on the basis of the levels of risk which existed in the past [8]. The most worrisome assumption in this method is that past societal risk levels reveal correct or desirable risk levels for the present. This assumption requires one to judge that, with regard to risk, the present ought to be like the past, and that what risk was accepted in the past ought to have been accepted. Both of these assumptions involve one in highly evaluative judgments, since past levels of risk may be indefensible in certain respects, and since our ethical obligations regarding present risk may be greater than those regarding past risk.

The third method of risk evaluation, expressed preferences, consists of using psychometric surveys to determine the acceptability of particular risks [9]. It is built on the questionable assumption that the preferences people express via instruments such as surveys provide reliable indicators as to acceptable risks. Obviously, however, preferences are not always authentic indicators of welfare: some persons have irrational fears and other persons are too ignorant to realize a serious risk. This method also requires the assessor to make a number of evaluative judgments whenever he encounters inconsistencies in survey preferences or failure of the responses to correspond with actual behavior regarding risk.

The fourth method of risk evaluation, natural standards, uses geological and biological criteria to determine risk levels current during the evolution of the species [10]. This method is based on the assumption that if a particular level of risk was present in the past, then that same level is acceptable at present. Again, the obvious problem is that because a risk "is" present naturally does not mean that we "ought" to allow it to remain, especially if it can be avoided, or especially if no overarching benefits arise from allowing the risk to remain high. Moreover, since the method is based on a "natural standard" for each different hazard, it totally fails to take account of synergistic or cumulative effects which could be many orders of magnitude greater than the actual "standard level" of exposure. In the case of radiation, for example, where the effects of exposure are cumulative, one's annual exposures could be within the standard level, although the cumulative effect of 40 years of standard exposures could be many times greater than the effect of annual exposure at the standard level. Likewise, one might be exposed to standard levels of radiation, toxic chemicals like vinyl chloride, and hazardous substances like asbestos; although the level of exposure to each of these three agents might be within the "natural standard" limits, their combined effect in terms of carcinogenesis could be greater than that of any one of them alone. For all these reasons, the method of natural standards embodies a highly evaluative notion of actual risk.

All the assumptions, extrapolations, and inferences built into each of the methods in the three stages of risk assessment appear to arise because nearly every situation of risk identification, estimation, or evaluation is empirically underdetermined. That is, either complete data cannot be obtained, or, where they can be, they are arrived at only by means of highly evaluative assumptions, inferences, extrapolations, and simplifications on the basis of what is observed. Because most situations of risk assessment are empirically underdetermined, assessors are forced to make some value judgments so as to interpret available data. At root, these assumptions and inferences often reduce to the problem of determining causality. This is difficult, since causes are not seen; they are inferred from the recurrence of symptons and effects. Cancers don't wear tags saying they were caused by their host's smoking, use of oral contraceptives, intake of diet drinks, or inhalation of the emissions from the chemical plant next door. Moreover, we know statistically how much of one agent causes cancer, but we don't know in individual cases whether a given agent caused a particular cancer; this we can only infer.

A classic example of the difficulty of establishing causality of harm, given a certain probability of risk, occurred during the fifties and sixties, when U.S. servicemen were exposed to above-ground nuclear weapons tests in the western United States and South Pacific. Many of them were within five miles of ground zero for as many as 23 separate explosions, or were marched to within 300 yards of ground zero immediately after detonation. Yet of the half a million soldiers so exposed, only ten have won benefits when they or their survivors claimed that their deaths and injuries were caused by the fallout [11]. True, good risk assessors can use statistical methods to determine when there is a significant rise in particular deaths or injuries. In the end, however, all they have to go on is a correlation between an effect and an imputed cause, and never a strict proof of causality, even in experiments with strict controls. As a consequence, there are always some data lacking, some need for inference, extrapolation, and simplification. This means that even the best situations are empirically underdetermined and force one into the realm of doing philosophy under the guise of doing science.

INHABER'S RISK ASSESSMENT

Sometimes the philosophy underlying a particular risk assessment is done well; other times, the methdological assumptions are so suspect on logical and epistemological grounds that the resulting risk estimates are of questionable value. One famous example of such a risk assessment is that of Herbert Inhaber. His study was commissioned by the Canadian Atomic Energy Control Board and then summarized in *Science*. The major conclusions of his study, which estimated the relative risks of alternative energy technologies, were that the risk from conventional energy systems, such as nuclear and coal power, is less than that from nonconventional systems, such as solar and wind energy; and that noncatastrophic risks, such as those from solar energy, are greater than catastrophic risks, such as those from nuclear energy [12].

How did Inhaber arrive at his surprising conclusions? He made some highly questionable evaluative assumptions. In estimating the risk posed by particular energy technologies, he assumed that all electricity was of utility-grid quality, i.e., able to be fed into a power grid [13]. This means that the low-risk benefits of solar space heating and hot-water heating, neither of which can be fed into a power grid, are ignored. Indeed, the wide variety of low-temperature forms of solar energy, which could supply *40 percent of all U.S. energy needs* at competitive prices and at little risk, were ignored [14].

Another assumption central to Inhaber's risk estimates is that all nonconventional energy technologies have coal back-ups [15]. This means that 89 percent of the risk attributed to solar thermal electricity comes not from the solar technology (especially construction of components), but from the coal backup, which he classifies as a solar risk! Moreover, Inhaber assumes that nuclear fission requires no backup [16], even though these plants require a down time of approximately 33 percent per year for check-ups, refueling, and repairs.

In the area of risk evaluation, Inhaber's assumptions are just as questionable. When he aggregates and compares all lost work days for all energy technologies, he ignores the fact that lost work days are more or less severe, depending on the nature of the accident causing them and whether or not they are sequential. In Inhaber's scheme, a lost work day due to cancer or the acute effects of radiation sickness is undifferentiated from that due to a sprained ankle [17]. Yet obviously the cancer could result in premature death, and the radiation exposure could result in mutagenic effects on one's children. Neither is comparable to a sprained ankle.

Inhaber made a similarly questionable assumption in evaluating the severity of risks. Unlike other risk assessors, he totally ignored the distinction between catastrophic/non-catastrophic risks, and assumed that 1000 construction workers' each falling off a roof and dying in separate accidents was no different than 1000 workers' dying because of a catastrophic accident in a nuclear fuel fabrication plant [18]. Numerous risk assessors typically make a distinction between catastrophic and non-catastrophic accidents. They suggest, for example, that because of the increased societal disruption that such an accident incites, n lives lost in a single catastrophic accident should be assessed as a loss of n^2 lives [19]. Their reasoning is that catastrophic accidents generally promote more fear and aversion than do non-catastrophic accidents; that they arise from situations in which potential victims exercise less control over their fate than they do in other risk situations; and that catastrophic accidents do not merely kill separate individuals in different and unrelated events, but also often wipe out whole social units—communities and neighborhoods.

Regardless of whether or not this n^2 interpretation is reasonable, the point remains that Inhaber made numerous evaluative assumptions, such as that the distinction between catastrophic and non-catastrophic accidents could be ignored, and that while so doing, he passed off his results as purely objective risk assessment. Indeed, were one to trace Inhaber's methods step by step, it would be clear that virtually every assumption he makes in estimating and evaluating alternative risks has the effect of increasing his alleged conventional risks and decreasing his alleged nonconventional risks [18].

ETHICAL JUDGMENTS AND RISK EVALUATION

Even if risk assessment situations were not empirically underdetermined for all the reasons already surveyed, and even if persons such as Inhaber were to avoid their questionable means of aggregating and estimating risks, subjective values would unavoidably shape the risk assessment process. This is because value judgments are often the only ways to resolve some of the ethical dilemmas facing risk assessors. Consider three of the most prominent such difficulties: I call them the Contributor's Dilemma, the Threshold Dilemma, and the Consent Dilemma.

The Contributor's Dilemma

The Contributor's Dilemma is as follows: citizens are subject to numerous small risks—e.g., to certain carcinogens—each of which is allegedly acceptable, yet all of which are cummulatively clearly unacceptable. Each of the numerous carcinogens to which we are exposed is alleged to be acceptable because it is below the threshold at which some statistically significant increase in harm occurs. Yet statistically speaking, 25 to 30 percent of us are going to die from cancers, 90 percent of which have been estimated to be environmentally induced and hence theoretically preventable [20]. It is reasonable to assume that many of the cancers are caused by the aggregation of numerous exposures to carcinogens, no one of which alone is alleged to be harmful.

The Contributor's Dilemma is problematic for both synergistic and non-synergistic risks. In the case of synergistic risks, assessors are forced both to assume (in the case of aggregate risks, e.g., of cancer of all types) and not to assume (in the case of individual risks contributing to the genesis of any cancer) that the whole risk faced by an individual is greater than the sum of the parts of that risk. In the case of non-synergistic risks, even when the sum of the parts of the risk equals the whole risk, one still faces the Contributor's Dilemma. This is because a risk may be called "negligible," but, together with thousands of other "negligible" risks, it may be substantial. Risk assessors who condone sub-threshold risks but condemn the deaths caused by the aggregate of these subthreshold risks are something like the bandits who eat the tribesmen's lunches in the famous story of Jonathan Glover:

"Suppose a village contains 100 unharmed tribesmen eating their lunch. 100 hungry armed bandits descend on the village and each bandit at gun-point takes one tribesman's lunch and eats it. The bandits then go off, each one having done a discriminable amount of harm to a single tribesman. Next week, the bandits are tempted to do the same thing again, but are troubled by new-found doubts about the morality of such a raid. Their doubts are put to rest by one of their number [a government risk assessor] . . . They then raid the village, tie up the tribesmen, and look at their lunches. As expected, each bowl of food contains 100 baked beans . . . Instead of each bandit eating a single plateful as last week, each [of the 100 bandits] takes one bean from each [of the 100] plate[s]. They leave after eating all the beans, pleased to have done no harm, as each has done no more than sub-threshold harm to each person" [21].

The obvious question raised by this example is how a risk assessor can say both that sub-threshold exposures are harmless, as the data indicate, and yet that the sum, or contribution, of these exposures causes great harm. It appears that risk assessors need to amend their theory regarding additive or synergistic risks like cancer.

The Threshold Dilemma

The Threshold Dilemma poses many of the same problems as the Contributor's Dilemma. It consists of the fact that society must declare some threshold for the acceptability of a given risk (e.g., that it would cause less than a 10^{-6} increase in one's average annual probability of fatality), since a zero-risk society is impossible. Yet no threshold standard is able to provide equal protection from harm to all citizens. Choosing the 10^{-6} standard appears eminently reasonable, both because society must attempt to reduce larger risks first, and because 10^{-6} is often said to be the natural hazards mortality rate [22]. This choice poses a dilemma, however, because it is a standard based on *average* annual probability of fatality.

Because this 10^{-6} threshold seems on the average acceptable does not mean that it is acceptable to each individual. Most civil rights, for example, are not accorded on the basis of the *average* needs of persons, but on the basis of *individual* rights. For instance, we do not accord constitutionally guaranteed civil rights to public education on the basis of average characteristics of students. If we did, then retarded and gifted children would have rights only to an education at the average level. Instead, we say that according "equal" civil rights to education means according "comparable education," given each student's aptitudes and needs. That is why the state can provide special schools for both the retarded and the gifted.

This example from the field of education raises an interesting question for risk assessment: if civil rights to education are accorded on the basis of individual, not average, characteristics, then why are civil rights to equal protection from risks not accorded on the basis of individual, rather than average, characteristics? Why is a 10^{-6} average threshold accepted for everyone, without compensation, when adopting it poses risks higher than 10^{-6} for the elderly, for children, for persons with previous exposures to carcinogens, for those with allergies, for persons who must lead sedentary lives, and for the poor? Blacks, for example, appear to face higher risks from air pollution than do whites, even though they share the same "average" exposure [23]. Although average air pollution levels are typically said to be the same for blacks and whites in a given community, blacks tend to live in areas with higher pollution concentration levels, and whites tend to live in areas with lower levels—facts hidden by the community *averages* for levels of air pollution.

The Consent Dilemma

A third problem faced by assessors who must estimate and evaluate risks is the Consent Dilemma. It arises from the facts that imposition of certain risks is legitimate only after consent is obtained from the affected parties, and that those parties genuinely able to give legitimate consent are precisely those who are likely never to do so.

Probably the best example of the Consent Dilemma arises in workplace situations. Here there is an alleged compensating wage differential, noted by both economists and risk assessors. According to the theory behind the compensating wage differential, all things being equal, the riskier an occupation is, the higher the wage required to compensate the worker for bearing the risk will be [24]. Moreover, imposition of these higher workplace risks is legitimate apparently only after the worker consents, with knowledge of the risks involved, to perform the work for the agreed-upon wage. But the dilemma arises once one considers who is most likely to give legitimate informed consent. It is a person who is well educated and adequately informed about the risk, especially its long-term and probabilistic effects. It is a person who is not forced, under dire financial constraints, to take a job which he knows is likely to harm him. Yet sociological data reveal that, as education and income rise, persons are less willing to take risky jobs, and that those who do so are primarily those who are poorly educated or financially strapped [25]. This means that the very set of persons least able to give free, informed consent to workplace risks are precisely those who most often are said to give consent.

If this observation about workplace risk is accurate, then medical experimentation may have something to teach us about risk assessment. We know that the promise of early release for a prisoner who consents to risky medical experimentation provides a highly coercive context which could jeopardize his legitimate consent. So also high wages for a desperate worker who consents to take a risky job provides a highly coercive context which could jeopardize his legitimate consent. What is the way out of this dilemma?

SKETCHES OF TWO SOLUTIONS

Although there is no space here to develop extensive arguments for the best ways to avoid some of the most detrimental effects of the three dilemmas just outlined, there are two possible solutions. One is to use weighted RCBA, placing additional ethical costs on each of the

RCBA parameters so as to counteract the consequences of the dilemmas noted earlier. Thus, for example, when we faced the Contributor's Dilemma, we could place heavier costs on allegedly acceptable, small, individual risks which combine to form unacceptable aggregate risks.

There are a number of reasons for using ethically weighted RCBA. One is based on the fact that there is no necessary connection between Pareto Optimality, the central concept of RCBA, and socially desirable policy, although there is a great tendency to assume that RCBA by itself reveals socially desirable policy. Using a weighting scheme would help avoid this erroneous tendency. It would also aid democratic decision making in the sense that the ethical aspects of policy analysis would not be left merely to the vagaries of the political process. Instead, they could be clarified through alternative, ethically weighted RCBA's. Ethically weighted RCBA's would also aid public decision making in that values could be brought into the policy making process explicitly and at an early stage, rather than later, when RCBA conclusions were already determined [26].

Several economists have proposed that policymakers use alternative systems of ethical weights for RCBA. Two of the best known suggestions in this regard are that of Weisbrod and that of the United Nations Industrial Development Organization [27]. Both of these weighting schemes are problematic. Weisbrod's approach is questionable because it is built on the assumption that efficiency and distribution are the only two criteria to be used for weighting; the UNIDO approach is questionable because it allows practitioners to totally disregard political constraints and democratic opinions. Both approaches fail to discuss the ethical dimensions of the weighting problem in any comprehensive or sophisticated way.

Attempting to correct this flaw, economist Allen Kneese (working with Ben-David and Schulze) developed an alternative weighting scheme. On the Kneese account, the public should choose which ethical system, of the many available, it wishes to implement in public policy, and it should reweight RCBA by means of this ethical system. Kneese, *et al.* represent each alternative ethical system as a general, transitive criterion for individual or social behavior. For example, their criterion for representing an ethical system of Benthamite utilitarianism is that one ought to maximize the sum of the cardinal utilities of all individuals in society. In proposing that such a general criterion be used to represent an ethical system, Kneese, *et al.* reject the idea that each ethical system be represented by a list of rules treated as a set of mathematically specified constraints on the outcomes of a particular RCBA [28].

The obvious problem with this approach is that in employing simple criteria, Kneese, *et al.* fail to capture the complexity of ethical systems. In particular, they are unable to represent a priority weighting of different ethical claims within the same ethical system. This is a significant omission, since in Rawls' system, for example, there are at least two central principles of justice which he believes ought to be lexicographically ordered, so that the claims of the first are satisfied first. The first claim specifies guaranteeing equal rights to everyone under certain conditions, and the second specifies that social and economic inequalities ought to be arranged so as to provide benefits to the least advantaged and equality of opportunity to all. Since most ethical systems appear to demand a lexicographically ordered list of rules (e.g., recognize all constitutional rights of a certain kind, *then* work to maximize welfare), the Kneese framework is inadequate to represent the ethical nuances of typical risk policy situations. For this reason, it would be desirable to use a list of transitive, lexicographically ordered rules to represent the priority weighting of different ethical claims within a given ethical system, and then to prepare alternative RCBA's, each weighted in terms of a different ethical system, among which the public could decide. (A lexicographic ordering is one in which the attributes of alternatives are ordered by importance, and the alternative is chosen with the best value on the most important attribute. If there are two alternatives of equal value, then this procedure is repeated for those alternatives in terms of the next attribute, and so on until a unique alternative emerges or until all alternatives have been considered.) The mathematics behind such a proposal has been described elsewhere; although original, it relies on a theorem of Debreu, a recent Nobelist in economics [29].

The great advantage of a lexicographically ordered system of weights is that it could render RCBA virtually immune to the classical objection against it: that is is utilitarian and collapses a social choice into one dimension. Since this new weighting scheme provides for lexicographically ordered rules, it provides for social choices to be made in terms of many ethical dimensions, not all commensurable in terms of one dimension. No other weighting system proposed has this advantage.

The second solution for mitigating the effects of the three risk assessment dilemmas is to use an adversary system of assessment. This amounts to requiring that a number of different risk assessments be performed for the same project, and that each assessment employ alternative ethical and methodological assumptions as the basis of its calculations. Once these alternative assessments were completed, then policymakers and the public could debate their merits and observe how alternative methodological and ethical assumptions generate alternative risk estimates and evaluations. Such a strategy would allow the public and policymakers to choose not only which risks they want, but which philosophies (ethics and methodologies) to use in identifying, estimating, and evaluating those risks.

The suggestion to employ a system of adversary assessment has much in common with the recent proposals to establish "science courts" [30]. The greatest deficiency of the "science court" proposal, however, is that it provides for no citizen participation. It requires that controversial factual issues central to questions involving science/technology/risk be decided by an allegedly impartial court of scientists. Numerous commentators have charged both that allegedly factual issues always involve some partially evaluative assumptions and inferences, and that scientists alone ought not be allowed to formulate public policy, especially when that policy affects the public [31].

My proposal for adversary assessment attempts to take account of these objections to the "science court." It stipulates that alternative risk assessments be prepared by experts working from different ethical and methodological perspectives, and that the final decision be made by both the experts and public representatives. Described in detail elsewhere [32], this proposal amounts to requiring, for all issues involving analysis of technological risks, that adversary assessments be performed. To a limited degree, this has already been accomplished on an experimental basis: in Cambridge, Massachusetts, in San Diego, California, and in Ann Arbor, Michigan, city councils have formed boards comprised of both citizens and scientists/engineers. The purpose of these boards has been to assess alternative evaluations of risk. (For example, in the Cambridge case, the board assessed the safety procedures required by the U.S. National Institute of Health for recombinant DNA research.) Each board operated by means of adversary proceedings complete with cross-examination from opposing sides. Citizen panelists on each board were also required to educate themselves with respect to the merits of alternative risk assessments, each with different ethical and methodological assumptions [33].

Using ethically weighted assessment and adversary assessment entails spelling out our philosophical assumptions as well as our conclusions about risk. These steps would take us closer to the public policy ideal of Albert Einstein, who urged that all important issues be publicly debated in "the marketplace of ideas."

Admittedly both proposals face a number of objections, many of which focus on the desirability of allowing the public so much voice in the highly technical business of evaluating alternative risk assessment methods and ethical assumptions. Although there is no time to discuss these objections here, there are reasonable responses which can meet each of them [34]. The cores of two such responses are brief enough to be formulated here. One is that, if anyone doubts the appropriateness of adversary proceedings in which laymen play a role together with scientists and engineers, then he should also doubt the congressional system of committee hearings which, in many cases, has been enormously successful despite the fact that congressmen are rarely experts on the issues they help decide. The second response was eloquently framed by Thomas Jefferson:

> "I know of no safe depositor of the ultimate powers of the society but the people themselves; and if we think them not enlightened enough to exercise their control with a wholesome discretion, the remedy is not to take it from them, but to inform their discretion" [35].

ACKNOWLEDGEMENT

The author is indebted to S. H. Unger of the Computer Science Department of Columbia University and to an anonymous referee for a number of helpful suggestions for strengthening this essay. Whatever flaws remain are the sole responsibility of the author.

REFERENCES

[1] S. Levine, "Panel: Use of Risk Assessment...," in *Symposium/Workshop...Risk Assessment and Governmental Decision Making,* The Mitre Corporation, Ed., McLean, Virginia: Mitre Corporation, 1979, p. 634.

[2] F. Hapgood, "Risk-Benefit Analysis: Putting a Price on Life," *The Atlantic,* Vol. 243, p. 38, Jan. 1979.

[3] See W. Hafele, "Benefit-Risk Tradeoffs in Nuclear Power Generation," in *Energy and the Environment,* H. Ashley, R. Rudman, and C. Whipple, Eds., New York: Pergamon, 1976, p. 181; D. Okrent and C. Whipple, "Approach to Societal Risk Acceptance Criteria and Risk Management," U.S. Department of Commerce, Washington, D.C., Report No. PB-271264, pp. 1-9, 1977. For a discussion of the value judgments implicit in RCBA, see Carl Barus, "On Costs, Benefits, and Malefits in Technology Assessment," *IEEE Technology and Society Magazine,* Vol. 1, p. 3, Mar. 1982. For discussion of some of the social and theoretical values of risk assessment, see R. J. Bogumil, "Limitations of Probabilistic Risk Assessment," *IEEE Technology and Society Magazine,* Vol. 1, p. 24, Sept. 1982.

[4] For discussion of the methodological problems arising at each of the three stages of risk assessment, see K. S. Shrader-Frechette, *Risk Analysis and Scientific Method,* Boston: Reidel, pp. 15–51, 1985; hereafter cited as: Shrader-Frechette, *RASM.* See also F. Press, *Risk Assessment in the Federal Government,* Washington, D.C.: National Academy Press, 1983; and L. Lave, Ed., *Quantitative Risk Assessment in Regulation,* Washington, D.C.: Brookings, 1982.

[5] Shrader-Frechette, *RASM,* p. 20

[6] K. S. Shrader-Frechette, *Nuclear Power and Public Policy,* Boston: Reidel, 2nd ed., pp. 25–27, 1983; hereafter cited as: *NPPP.*

[7] See E. Mishan, *Cost-Benefit Analysis,* New York: Praeger, 1976.

[8] See C. Starr, "General Philosophy of Risk-Benefit Analysis," in *Energy and the Environment,* H. Ashley, R. Rudman, and C. Whipple, Eds., New York: Pergamon, pp. 6 ff., 1976.

[9] See B. Fishchollf, *et al., Acceptable Risk,* New Rochelle, New York: Cambridge University Press, 1981; hereafter cited as: Fischhoff, *AR.*

[10] Fischhoff, *AR,* pp. 87-88.

[11] Shrader-Frechette, *NPPP,* pp. 99-100.

[12] H. Inhaber, "Risk with Energy from Conventional and Nonconventional Sources," *Science,* pp. 718-723, Feb. 23, 1979; hereafter cited as: Inhaber, "Risk." See also Inhaber, *Risk of Energy Production,* Atomic Energy Control Board, Ottawa, Canada, Report AECB-1119, 1978.

[13] Inhaber, *Risk,* p. 718.

[14] L. S. Johns, *et al., Application of Solar Technology to Today's Energy Needs,* U.S. Office of Technology Assessment, Washington, D.C., vol. 1 (of 2 vols.), 1978, p. 3: "On-site solar devices could be made competitive in markets representing over 40 percent of U.S. energy demand by the mid-1980's, although the output of solar equipment installed by this date is unlikely to be able to meet more than a small fraction of this potential market. Existing Federal programs controlling fuel prices and subsidizing nonsolar energy sources have created a situation where, without compensating subsidies, solar energy is uniquely disadvantaged." The OTA staff goes on to say that high-temperature solar power (solar thermal electric) is currently expensive, whereas low-temperature solar uses, which comprise 40 percent of total U.S. energy needs, are currently competitive economically with existing alternatives (pp. 13-14), even in cities such as Boston, Albuquerque, and Omaha (pp. 31 ff.).

[15] Inhaber, *Risk,* p. 721.

[16] Inhaber, *Risk,* p. 721.

[17] Inhaber, *Risk,* pp. 721-722.

[18] Inhaber, *Risk,* pp. 721-722. For further criticism of the Inhaber Report, see J. Herbert, C. Swanson, and P. Reddy, "A Risky Business", *Environment,* pp. 28-33, July/Aug. 1979; and J.

Holdren, K. Smith, and G. Morris, "Energy: Calculating the Risks," *Science,* pp. 564-568, 1979.

[19] See, for example, W. Rowe, *An Anatomy of Risk,* New York: Wiley, p. 290, 1977; and B. Fischhoff, P. Slovic, and S. Lichtenstein, "Facts and Fears," in *Societal Risk Assessment,* R. Schwing and W. Albers (Eds.), p. 208, New York: Plenum, 1980. Moreover, as S. H. Unger has pointed out (private communication), even though the average number of deaths may be the same for two different risks to which the same population is subjected, it is not clear that both risks are equally desirable or undesirable. For an analysis of this and related problems, see Shrader-Frechette, *RASM,* Chapter 6, especially pp. 170-171. See also Chapters 3 and 5 for discussion of problems with determination of harms from average numbers of fatalities and for analysis of difficulties involved with using risk and harm calculations based on expected utillity.

[20] J. C. Lashof, *et al., Assessment of Technologies for Determining Cancer Risks from the Environment,* Health and Life Sciences Div. of the U.S. Office of Technology Assessment, Washington, D.C., June, 1981, p. 3: Cancer "strikes one out of four Americans," and the number is increasing; "60 to 90 percent of cancer is associated with the environment and therefore is theoretically preventable." Admittedly, as S. H. Unger points out, many of these cancers are not *in practice* preventable, as they are caused by factors such as natural background radiation. For discussion of the synergistic effects of carcinogens, see, pp. 6 ff. of the volume by Lashof *et al.,* and also R. Baker, *Pesticide Usage and its Impact on the Aquatic Environment in the Southeast,* Washington, D.C.: U.S. Environmental Protection Agency, pp. 2-3, 187-191, 1972.

[21] Quoted in Derek Parfit, *Reasons and Persons,* Oxford: Clarendon Press, 1984, p. 511. I am still grateful to S. H. Unger for pointing out the need to clarify the fact that synergistic and nonsynergistic risks both involve the Contributor's Dilemma.

[22] The figure of 10^{-6} is typically taken as a cut-off point below which risks as assumed to be negligible. Some assessors claim that the natural hazards mortality rate is 10^{-6}; in this regard see, for example, C. Starr, *Current Issues in Energy,* New York: Pergamon, 1979, pp. 14 ff.

[23] See, A. Freeman, "Income Distribution and Environmental Quality," in *Pollution, Resources, and the Environment,* A. Enthoven and A. Freeman, Eds., New York: Norton, 1973, p. 101.

[24] See. W. Viscusi, *Risk by Choice,* Cambridge: Harvard University Press, 1983, pp. 38 ff.

[25] E. Eckholm, "Unhealthy Jobs", *Environment,* pp. 31-33, August/September 1977. See also D. Berman, *Death on the Job,* London: Monthly Review Press, 1978.

[26] See K. S. Shrader-Frechette, *Science Policy, Ethics, and Economic Methodology,* Boston: Reidel, 1984, pp. 261-265; hereafter cited as: *SP.*

[27] B. A. Weisbrod, "Income Redistribution Effects and Benefit-Cost Analysis," in *Problems in Public Expenditure Analysis,* S. B. Chase, Ed., Washington, D.C.: Brookings, 1968. See also P. Dasgupta, A. Marglin, and A. Sen. *Guidelines for Project Evaluation,* New Yori: UNIDO, 1972.

[28] For relevant discussion, see A. Kneese, S. Ben-David, and W. Schultze, "The Ethical Foundations of Benefit-Cost Analysis", in *Energy and the Future,* D. MacLean and P. Brown, Eds., Totowa, N.J.: Rowman and Littlefield, 1982, pp. 59-74.

[29] Shrader-Frechette, *SP,* Chapter 8.

[30] A. Kantrowitz, "The Science Court Experiment," *Science,* pp. 653-656, Aug. 20, 1976.

[31] See Shrader-Frechette, *SP,* pp. 290-293.

[32] For an account of an adversary system of risk assessment, see Shrader-Frechette, *SP,* Chapter 9.

[33] See K. G. Nichols and the OECD for Science, Technology, and Industry, *Technology on Trial,* Paris: Organization for Economic Cooperation and Development, 1979, pp. 99-100.

[34 Shrader-Frechette, *SP,* pp. 294-312.

[35] Cited in D. Bazelon, "Risk and Responsibility," *Science,* p. 280, 1979.

JOSEPH R. HERKERT

Ethical Risk Assessment: Valuing Public Perceptions

Engineers are confronted with an array of moral issues and dilemmas as the complexity of modern technology results in equally complex efforts to assess the accompanying environmental and safety risks. The ethical responsibilities of engineers and the need for workable solutions to technological controversies dictate that engineers be able to discuss technological risk with the public. The openness required for meaningful risk communication, however, is often at odds with the prevailing engineering culture.

It was argued in [1] that for meaningful communication to occur between experts and the lay public on issues related to technological risk, substantive attitude changes are necessary on the part of experts, including engineers and other technical specialists, with respect to the relevance of public perception of risk. While that paper included a case study of risk communication in connection with "inherently safe" nuclear reactors, no attempt was made to relate the analysis to the ethical responsibilities of engineers or to the norms of engineering practice.

The purpose here is to examine the connections between engineering ethics, risk communication, and the engineering culture. First moral issues in risk assessment are reviewed and the ethical responsibilities of engineers with respect to risk assessment and risk communication are discussed. The conventional model of risk communication, which holds that only experts possess relevant risk information, is then critiqued, and the findings of social scientists and humanists with respect to the dual importance of expert and public risk information are reviewed. Following a discussion of the prevailing engineering culture, particularly as it relates to the problems involved in risk communication, some suggestions are made for transforming the engineering culture in a manner conducive to more meaningful discussion of risk.

Moral Issues in Risk Assessment

As modern technology and the assessment of environmental and safety risks become more complex, an increasing array of moral issues and dilemmas is created, including [2]:

1) the overlapping of political or normative judgments with scientific judgments;

2) the potential for manipulation of risk assessments to legitimize deployment of risky technologies;

3) the determination of the "value of life" used in risk assessment;

4) the imposition of risks upon others, particularly those less empowered (e.g. developing nations); and

5) the distribution of risks and consequences across geopolitical, cultural, or generational boundaries.

These points are illustrated by events such as the 1984 chemical leak in Bhopal, India, the 1986 explosion of the space shuttle, Challenger, and the 1986 accident at the Chernobyl nuclear power plant.

An earlier version of this paper was presented at The International Symposium on Technology and Society 1993, Washington, DC, October, 1993. *J.R. Herkert is at 312 March Street, #4, Easton, PA 18042. He is Vice President of the IEEE Society on Social Implications of Technology.*

Reprinted from *IEEE Technology and Society Magazine,* Vol. 13, No. 1, pp. 4–10, Spring 1994.

The overlapping of political or normative judgments with scientific judgments follows inevitably from the use of risk assessment in determining the acceptability of risks. The now-infamous remark of one Morton Thiokol executive to another during an engineering teleconference on the eve of the Challenger launch — "Take off your engineering hat and put on your management hat" — is a stark reminder both of how technical judgments regarding the amount of risk can become intertwined with value judgments regarding the acceptability of risk, and how risk decisions ostensibly made on technical grounds can be manipulated to camouflage economic or political reasons for imposing risks on others [3].

The "risk assessment" performed by the National Aeronautics and Space Administration (NASA) prior to the Challenger accident further illustrates how risk assessment can be misused to legitimize a risky technology. Although, at the time, no formal risk assessment techniques were employed by NASA in the space shuttle program, the agency estimated the risk to be very small when contributing to a study of the risk of flying a plutonium payload. And while the risk estimates were supposedly based upon "engineering judgment," the actual risk estimates of the working engineers were three orders of magnitude higher. [4]

Yet another example of the misuse of risk analysis is presented by the claims of engineers and others who promote a nuclear power renaissance based upon a new generation of "inherently safe" reactors [1]. Recognizing that the industry's current risk estimates for core-melt accidents are unacceptably high for any scenario that projects large-scale deployment of nuclear power, some experts have projected that the risks of future reactors will be as much as one hundred times lower than existing reactors. Such *ad hoc* risk analysis, while reflecting the experts' own notion of what the public will find acceptable, does not necessarily address the level of safety that is actually obtainable, or for that matter, publicly acceptable. The Three Mile Island and Chernobyl accidents have cast doubt upon the reliability of risk assessments pertaining to the current generation of reactors. Moreover, the new generation of reactors has yet to be demonstrated in a manner that even approaches commercial application. Highly optimistic projections of future safety improvements thus amount to not much more than wishful thinking, even when cloaked in the "respectability" of numerical estimates and accompanied by claims of inherent safety.

Determining the "value of life" is among the thorniest problems in risk assessment. Frequently, lives are valued based upon criteria such as economic worth or expected earnings, which translates into "life is cheap" in poorer neighborhoods or less-developed nations. This issue is clearly illustrated by the Bhopal case. Life in Bhopal was implicitly valued less than life in the United States — the safety equipment and emergency preparedness at the Bhopal plant were far less adequate than those at a similar facility

operated by Union Carbide in Institute, West Virginia [5]. The low valuation of life in Bhopal was made explicit in the legal arguments made by Carbide regarding the extent of the company's liability and in the actual settlement. Although nearly 3000 people died as a result of the Bhopal accident, the settlement was less than half the settlement agreed to in the case of the Exxon Valdez oil spill.

The Bhopal case also highlights the moral issues involved in the imposition of risk upon others, particularly those less empowered, as in the case of multinational corporations "exporting" risk to developing countries by building hazardous facilities abroad, by selling products in developing countries that are banned in the West on health or safety grounds, or by paying developing countries' governments for the disposal of hazardous waste.

While the direct impacts of the Bhopal catastrophe by and large were contained within the local community, the Chernobyl accident illustrates how such impacts are not always limited by geopolitical, cultural, or generational boundaries [2]. Significant amounts of radioactive fallout from Chernobyl were deposited throughout Europe and the then Soviet Union. Ironically, perhaps the most serious damage was done to a people who have very little to do with complex contemporary technologies such as nuclear power. The Lapps of northern Scandinavia, whose culture revolves around the herding of reindeer, were devastated when much of their herds had to be destroyed due to ingestion of radioactive cesium that had been deposited on the lichen that made up the reindeer's diet. Radioactive fallout from Chernobyl also raises the possibility that future generations will be adversely impacted by the shortcomings of today's technologies.

The points mentioned so far by and large deal with the value judgments made in the use of risk assessments, or put another way, in determining acceptable risk. A number of these examples also illustrate, however, that contrary to the conventional wisdom within the technical community that the process of risk assessment is entirely objective, value judgments can and often do enter into the process. The risk evidence in the Challenger case was perceived differently by the Thiokol engineers, who relied to a great extent on past experience and their engineering judgement, than by the NASA and Thiokol managers, who insisted that there was no hard and fast data that justified scrubbing the launch. The uncertainties involved in evaluating the risks of nuclear power are valued differently by technical specialists, who argue that "inherent safety" reduces risk and uncertainty, and a public skeptical of such claims as a result of previous accidents and near-misses. Determining the "acceptability of evidence of risk," then, can also involve moral questions [6].

In summary, determination of the acceptability of risk and of risk evidence is a value laden process, accompanied by a wide array of moral issues and dilemmas. By virtue of their significant role in the process of technological development, engineers are inevitably involved, directly or indirectly, in risk assessment and risk management and, hence, must be prepared to examine the values these judgements incorporate and wrestle with the accompanying moral issues [7].

Engineering Ethics and Risk Assessment

Although engineering ethics codes have existed for a century, it is only during the past thirty years that such codes have been revised to include explicitly the notion that engineers have a responsibility for protecting the public safety, health, and welfare. The first "fundamental canon" of the code of ethics of the National Society of Professional Engineers (NSPE), for example, holds that engineers shall:

Hold paramount the safety, health and welfare of the public in the performance of their professional duties.

Likewise, the first pledge contained in the current code of the Institute of Electrical and Electronics Engineers (IEEE) states that IEEE members agree:

to accept responsibility in making engineering decisions consistent with the safety, health, and welfare of the public, and to disclose promptly factors that might endanger the public or the environment.

The evolution of such codes reflects a growing acknowledgment on the part of professional engineering societies that engineers have both a professional and a moral responsibility for the public safety and welfare [8]. In this respect, ethics codes have also become more consistent with traditional moral theories in both the utilitarian and rights/duties traditions [7].

The moral issues raised by risk assessment have significant implications regarding the professional responsibilities of engineers. Since, as we have seen, the scope of such issues is not limited to technical matters, their ultimate resolution, in addition to technical expertise, also requires the judgments of experts in management and public policy [7]. In addition, since most technological risks are borne by the public, and since societal values are relevant in the evaluation of risk, the judgments of nonexperts are also legitimate components of risk assessments.

The concept that the public should have an active role in the assessment of risk is consistent with the IEEE code of ethics, the fifth pledge of which states that IEEE members agree:

...to improve the understanding of technology, its appropriate application, and potential consequences....

Improving understanding of technology, however, is not a one-way process. In order to fully understand technology and its consequences, engineers must be informed with respect to nontechnical perspectives on technology [8]. Indeed, as I have argued elsewhere [1], for successful communication about risk to take place between engineers and the public, engineers must develop a sensitivity to and respect for the differences between expert and nonexpert perception of risk. Unfortunately, such sensitivity is not widely apparent among engineers, even among those who have addressed the problem of risk communication.

Two Models of Risk Communication (1)

The conventional view of risk communication, held by many engineers, is that risk communication need consist merely of "educating" the public to endorse expert judgment concerning which risks are acceptable and which are not. Under this model of risk communication, the experts have a corner on the truth of the matter; the only problem is to see that the public is properly informed of the experts' views. Nonexperts (and even experts in other fields) are seen as incapable of understanding the problem, and of making rational risk decisions. The role assigned the public under this model of risk communication is a passive one: listen and learn from the experts. Success is measured by how well the message gets across, and whether or not the public accepts the judgments of the experts [9], [10].

This view of risk communication is so imbued in the engineering culture that it is readily endorsed by most engineers and technologists who turn their attention to addressing the problem of risk communication. Markert, for example, has argued [11]:

Since only a few persons truly understand new technologies, most of us are dependent on those few "experts" to present us with all the facts regarding safety. Prudent control and management of risk demands a high level of technology awareness unclouded by emotional biases. It appears that the most effective risk management decisions will be made only when the risks are clearly communicated.

The conventional model of risk communication is also exemplified by the risk messages contained in discussions of advanced, "inherently safe" nuclear reactor designs. By implying that accidents are impossible, the term "inherently safe" epitomizes the conventional model of risk communication with the experts assuring the public that all is well. Implicit in its use is the notion that the public wants zero risk, and, therefore, the solution is to give them "inherent safety." Although increased public acceptance of nuclear power is a stated goal of the planners of the new nuclear power era, one wonders whether the public would ever accept a claim of inherent safety by an industry whose credibility has been so heavily damaged by the failures of the first era [12].

In contrast to the conventional model, an emerging view of risk communication recognizes that nonexperts are also in possession of relevant risk information, thus necessitating an exchange of information between experts and the public if effective communication about risk is to occur [10]. Under this broader definition of risk communication, success is measured by the level of increase in understanding of risk problems by all of the involved parties [9]. This two-way, interactive model of risk communication follows from research in the behavioral and social sciences which has indicated that while experts and the public view risk in fundamentally different ways, each has something valuable to offer to the understanding of risk.

To date, most of the work on the perception of risk has been done by psychologists who have determined that people employ mental strategies, known as heuristics, as aids in decision making in the face of uncertainty. While the use of heuristics is essential to avoiding a life frozen with indecision, use of these mental strategies also introduces systematic biases in the way we evaluate risks. Such factors as difficulties in judging probabilities, sensational media coverage, and personal experiences often lead us to underestimate or overestimate risks [13]. The well-known gambler's fallacy (the longer I play a game of chance, the more likely I am to win) is an example of a heuristic which would cause one to underestimate risk. Television coverage of an airline crash and a close relative suffering from a rare disease are events which could lead to overestimation of risk. Significantly, experts can also fall victim to the same sort of biases, particularly when they are in the realm of applying intuition as opposed to dealing with available data [13].

Using a technique known as the psychometric paradigm, psychologists have also determined that the concept of risk has a different meaning to experts, who usually focus on the probability of fatality from a given activity, than it does to members of the lay public, who also tend to factor in qualitative characteristics of hazards such as catastrophic potential, lack of control, and delayed harm [13]. Psychometric research

The conventional model holds that only experts possess relevant risk information.

has also indicated that the risks that are least understood and most dreaded by the public — such as those posed by nuclear power and other potentially catastrophic technologies — often have a very high "signal potential" regarding the indirect impacts of environmental pollution or technological accidents [13]. Many engineers, for example, view the Three Mile Island nuclear accident, in which there were no apparent deaths, as testament to the safety of nuclear power. The impacts of the accident, however, have rippled throughout the industry and the economy. Had more attention been paid by industry and government experts to the public's perception of the risks involved, the accident and its costly indirect impacts may have been prevented.

Some psychologists and risk analysts have viewed these findings as justification for the conventional model of risk communication. They view the public perception of risk as irrational, when compared to the rational judgment of experts. In order to close this gap, the experts' information must be transferred to the public [10]. Slovic [13], on the other hand, argues that psychometric research implies a broader conception of the risk communication process:

...there is wisdom as well as error in public attitudes and perceptions. Lay people sometimes lack certain information about hazards. However, their basic conceptualization of risk is much richer than that of the experts and reflects legitimate concerns that are typically omitted from expert risk assessments. As a result, risk communication and risk management efforts are destined to fail unless they are structured as a two-way process. Each side, expert and public, has something valid to contribute. Each side must respect the insights and intelligence of the other.

Slovic's statement highlights the fact that there are practical reasons for engineers to become better informed regarding the public's attitudes toward risk. The broader understanding we have about how people evaluate risk, and how they respond to the continued development of technology in the aftermath of accidents, the more likely we are to prevent accidents with costly direct and secondary impacts. The impacts of the cases previously cited — Three Mile Island, Chernobyl, Bhopal, and Challenger — could have been reduced or eliminated entirely if the warnings contained in lay risk perceptions had been heeded as valid supplements to expert risk judgements. By reducing complexity of system design, improving operating procedures, siting facilities in a more prudent manner, and leveling with the public about the uncertainties contained in quantitative risk models, the proponents of technological development will benefit from a safer operating record and far greater public support.

Moreover, engineers have a *moral* obligation to acknowledge the dual importance of expert and public risk information. We not only expect professionals to be experts in their field; we also expect them to know their limitations. Engineers thus have a professional obligation to recognize and acknowledge the limits of scientific rationality and the value of public perception of risk [14].

Substantive attitude changes are necessary on the part of experts.

Risk Communication and the Engineering Culture

There is both a practical necessity and an ethical imperative for engineers to become actively engaged in risk communication efforts that incorporate an understanding of and respect for public perception of risk. This calls for nothing less than a transformation of the engineering culture.

The prevailing engineering culture is readily recognized from both inside and out. Engineers are no-nonsense problem solvers, guided by scientific rationality and an eye for invention. Efficiency and practicality are the buzzwords. Emotional bias and ungrounded action are anathema. Give them a problem to solve, specify the boundary conditions, and let them go at it free of external influence (and responsibility). If problems should arise beyond the workbench or factory floor, these are better left to management or (heaven forbid) to politicians.

These attributes of the engineering culture are succinctly summarized by Lichter's "core principles" of engineering [15]:

1) A concern for the efficiency of practical means;

2) A commitment to concrete problem-solving, constrained to some necessary degree by time and available resources;

3) The pursuit of optimal technological solutions based on scientific principles and/or tested technical norms and standards;

4) The pursuit of creative and innovative designs;

5) And the development of new tools for the accomplishment of each of these.

Given the preoccupation of engineering with efficiency and merely technical solutions, to the neglect of all other criteria, it is not surprising that the conventional view of risk communication, which devalues lay perception of risk, is reinforced by the current engineering culture.

For example, perhaps the biggest obstacle in persuading engineers to listen more carefully to the concerns of the public is the overwhelming bias toward quantification in their evaluation of risk [1]. As the Challenger and "inherently safe" reactor examples illustrate, merely assigning numbers to the decision parameters in a technological problem doesn't change the fact that value choices are often made by experts at the expense of the people affected by technology. Closing the gap between expert and public perception of risk thus implies closing the gap between the quantitative and the qualitative. And this, in turn, implies changing the engineering culture so that questions of enduring human value are at its core [15], even when such questions can only be dealt with qualitatively.

In other words, a complete understanding of risk assessment and the problems of risk communication is not possible without an integration of technical concepts with concepts drawn from the humanities and social sciences. In particular, meaningful participation in risk communication efforts by engineers requires significant attitude changes on their part with respect to the relationship between expert and public [1]. For example, engineers should value, on an equal footing with scientific and technical rationality, other ways of knowing and expressing. Engineers should also have a commitment to *listening* to others, and to incorporating alternative points of view into the search for technical solutions to problems. Such attitude changes cannot be expected until knowledge gained through study of the humanities and social sciences is relocated from the margin of engineering to its essence [15].

If a meaningful transformation in the engineering culture is to occur, significant change must take place within engineering institutions. The most important of these institutions are the places where engineers first become immersed in the culture — the engineering schools.

Transforming the Engineering Culture

Several options are available for remaking the engineering culture through transformation of engineering education. A number of educators have recently called for increased emphasis in the engineering curriculum upon the relationship between technology and social, political, economic, ethical, and legal concerns [16]-[19]. Most agree that the typical practice of selection of a few elective courses in the humanities and social sciences does not adequately facilitate the making of connections between engineering and other modes of inquiry.

A solution that does not rely on isolated humanities and social science courses for emphasizing the relationships between technology and society is the development of engineering courses which incorporate the study of these relationships [20], [21]. For example, in order to prepare engineering students for meaningful communication with the public regarding risk, such topics as the role of engineering in society, engineering ethics, safety, risk assessment, and the professional responsibilities of engineers should become standard fare within the engineering curriculum, along with increased emphasis on communication skills [16].

Another option worth pursuing is increased interdisciplinary education for engineers [22]. In recent years, a new field has emerged in the area of Science, Technology, and Society (STS) studies. Although much of the recent interest in STS by the engineering community has been aimed at increasing the "technological literacy" of liberal arts students, the STS field has a strong tradition of integrating technical perspectives with perspectives on technology drawn from the humanities and social sciences for the benefit of both engineering and liberal arts students [23]. Indeed, the STS model is ideally suited for the development of courses that include material from a variety of perspectives on such topics as technological catastrophes, ethical responsibility, risk assessment, and risk communication [12].

Innovative approaches to teaching engineering students, such as collaborative learning [24], are also called for. Collaborative learning includes various processes by which students engage in peer learning in semi-autonomous small groups. The benefits of collaborative learning include improved student performance and enthusiasm for learning, development of communication skills, and greater student appreciation of the importance of judgment and collaboration in solving real-world problems such as those relating to safety concerns and the ethical responsibilities of engineers. In an interdisciplinary setting, collaborative learning has the added benefit of exposing engineering students to the concerns and biases of peers who lack technical training, but who nonetheless have a stake in the outcome of engineering decisions, thus enabling the difficulties of risk communication to be modeled in the classroom.

In addition to engineering schools, other institutions must contribute to a transformation of the engineering culture. The professional engineering societies can play a critical role in creating an engineering culture more conducive to effective risk communication by:

1) taking a lead role in promoting the educational reforms previously discussed;

2) providing stronger support and protection for engineers who apply ethical principles in the exercise of their professional judgment [8], [15]; and

3) initiating efforts, such as forums and journals, that recognize the dual importance of public attitudes toward risk and expert judgment [14].

Educational materials such as the video *Gilbane Gold*, produced by the National Institute for Engineering Ethics founded by NSPE, and "How to Be a Good Engineer," a guide to engineering ethics presentations available from IEEE's United States Activities Board, are examples of how professional societies can facilitate meaningful discussion of risk and other ethical issues confronting engineers. Significant changes in the engineering culture will also require the cooperation of their corporate employers [15]. As the cases mentioned in this paper illustrate, shortcomings in the conventional model of risk communication have created many problems both for corporations and for the advancement of technology; acknowledging and accepting the importance of the public's role in risk decisions would be beneficial to corporations as well as their engineering employees.

Finally, the role of the individual engineer in transforming the engineering culture should not be overlooked. As with other ethical issues, engineers need to approach the problem of risk communication with what Werhane [25] calls "moral imagination":

> *...to try to change one's habits, or ask oneself weird questions, to try in some way to place oneself in a different perspective so as to regard events from another point of view.*

Valuing other ways of knowing and expressing, as well as listening to the views of others, are essential prerequisites if engineers are to become effective risk communicators [1]. While it may be difficult to imagine an engineering culture in which this is possible, all engineers have a professional and moral responsibility to do so, and to work toward a transformation to such a culture.

Acknowledgments

I am indebted to Steve Unger for his friendship, support, and encouragement and for the inspiration he provides me and other engineers through his persistent work on engineering ethics within both the IEEE and the scholarly community. Portions of this text draw on material (especially [1]), including work with a former student, Rachel M. Rankin. This work has benefitted from many helpful comments on earlier drafts by Norman Balabanian and two anonymous reviewers.

References

[1] J.R. Herkert, "Risk communication and the technical expert," in *Natural and Technological Disasters: Causes, Effects and Preventive Measures*, S.K. Majumdar *et al.*, Eds. Easton, PA: Pennsylvania Academy of Science, 1992, pp. 400-411.

[2] R.M. Rankin and J.R. Herkert, "Moral issues and dilemmas in high risk technology," in *Natural and Technological Disasters: Causes, Effects and Preventive Measures*, S.K. Majumdar *et al.*, Eds. Easton, PA: Pennsylvania Academy of Science, 1992, pp. 484-493.

[3] J.R. Herkert, "Management's hat trick: Misuse of 'engineering judgment' in the Challenger incident," *J. Bus. Ethics*, vol. 10, pp. 617-620, 1991.

[4] R. Feynmann, *What Do You Care What Other People Think?* New York: Norton, 1988.

[5] B.I. Castleman and P. Purkavastha, "The Bhopal disaster as a case study in double standards," in *The Export of Hazard: Transnational Corporations and Environmental Control Issues*, J. Ives, Ed. Boston: Routledge & Kegan, 1985.

[6] D.G. Mayo and R.D. Hollander, Eds., *Acceptable Evidence: Science and Values in Risk Management*. New York: Oxford Univ. Press, 1991.

[7] A. Flores, "Value issues in risk management," in *Ethics and Risk Management in Engineering*, A. Flores, Ed. Lanham, MD: University Press of America, 1989, pp. 1-15.

[8] S.H. Unger, *Controlling Technology: Ethics and the Responsible Engineer*. New York: Holt, Rinehart and Winston, 1982.

[9] National Research Council, Committee on Risk Perception and Communication, *Improving Risk Communication*. Washington, DC.: National Academy Press, 1989.

[10] A. Plough and S. Krimsky, "The emergence of risk communication studies: Social and political context," *Sci., Technol., Human Values*, vol. 12, pp. 4-10, 1987.

[11] L.R. Markert, "Living amidst high risk technologies — Making tradeoffs," in *Proc. Conf. A Delicate Balance: Technics, Culture and Consequences*, 1989, pp.185-189.

[12] J.R. Herkert, "High-risk technology and technological literacy," *Bull. Sci., Technol., Soc.*, vol. 7, pp. 730-737, 1987.

[13] P. Slovic, "Perception of risk," *Science*, vol. 236, pp. 280-285, 1987.

[14] R.D. Hollander, "Expert claims and social decisions: Science, politics, and responsibility," in *Acceptable Evidence: Science and Values in Risk Management*, D.G. Mayo and R.D. Hollander, Eds. New York: Oxford Univ. Press, 1991, pp. 160-173.

[15] B.D. Lichter, "Safety and the culture of engineering," in *Ethics and Risk Management in Engineering*, A. Flores, Ed. Lanham, MD: University Press of America, 1989, pp. 211-221.

[16] J.R. Herkert and B.V. Viscomi, "Introducing professionalism and ethics in engineering curriculum," *J. Professional Issues in Engineering Education and Practice*, vol. 117, no. 4, pp. 383-388, Oct. 1991.

[17] E. Wenk, Jr., "Social, economic, and political change: portents for reform in engineering curricula," *Eng. Ed.*, vol. 79, no. 2, pp. 99-102, 1988.

[18] W.H. Vanderburg, "Engineering, technology, and the university," *IEEE Technol. Soc. Mag.*, vol. 6, no. 4, pp. 5-11, 1987.

[19] V.M. Karbhari, "Quality undergraduate engineering education — A critical perspective," *J. Prof. Issues Eng.*, vol. 115, no. 3, pp. 241-251, 1989.

[20] D.A. Andersen, "Liberalizing aspects of a civil engineering education," *J. Prof. Issues Eng.*, vol. 112, no. 3, pp. 190-193, 1986.

[21] B.M. Olds and R.L. Miller, "Integrating the two literacies: Humanities in the engineering curriculum," *Bull. Sci., Technol. Soc.*, vol. 7, nos. 5 & 6, pp. 592-597, 1987.

[22] J.R. Herkert, "Science, technology, and society education for engineers," *IEEE Technol. Soc. Mag.*, vol. 9, no. 3, pp. 22-26, 1990.

[23] S.H. Cutcliffe, "Technology studies and the liberal arts at Lehigh University," *Bull. Sci., Technol., Soc.*, vol. 7, nos. 1 & 2, pp. 42-48, 1987.

[24] J.R. Herkert, "Safety, ethics and legal aspects of engineering: a collaborative learning approach," in *1991 ASEE Annu. Conf. Proc.*. Washington, D.C.: Amer. Soc. Eng. Ed., pp. 896-898.

[25] P.H. Werhane, "Engineers and management: The challenge of the Challenger incident," *J. Bus. Ethics*, vol. 10, pp. 605-616, 1991.

T&S

E. S. CASSEDY

Health Risk Valuations Based on Public Consent

Health risks due to electric power generation are examined with regard to the worth of their possible reduction. The risks include those already known to exist due to air pollution and nuclear radiation and the newer potential risks of electromagnetic fields at power frequencies. The risks considered here are those for the public at large. It is shown that the question of: "how safe is safe enough?" can be answered by having members of the population give valuations of the worth of proposed reductions in any one of these risks. In soliciting their responses, the people are given scientific information on the nature of the hazards involved and their chances of being affected. Thus, the process is termed one of "informed consent," preferable to having the valuation judgment imposed technocratically. The process also allows a cost benefit comparison without the moral dilemma of valuing human life and suffering.

The "Willingness-to-Pay" Approach

The benefits of technology have been manifest historically in the development of Western societies and in America in particular. It is only in recent years, however, that the hazards and costs of technologies have been widely perceived. Foremost among the technologies so viewed are those for the production and distribution of the commodity of energy, with electric energy being a salient example.

With electric energy generation, for instance, we can weigh the benefits of energy in its useful form against the hazards and costs of pollution, radioactivity, catastrophic accidents and possible climate change. With electric power transmission and distribution we are now learning of possible health hazards due to extremely-low-frequency (ELF) fields, which we all may have to weigh back against the long-ingrained expectation of electricity at the flick of a switch.

In this article, we will focus our attention on hazards to public health due to the generation, transmission and distribution of electricity. As such, we will consider the impacts of the conversion of fossil and nuclear fuels in the generation process as a historic concern that continue today to be subjects of controversy. These concerns serve as the background, showing how public policy regulated these health hazards and considered their reduction. We will then look into the possible extension of such public health policies to the new highly controversial area of potential hazards due to exposure to electric and magnetic fields at power-line frequencies.

More specifically, we will consider the existing standards for protecting public health due to air pollution as directed toward power plant emissions, and to ionizing radiation as directed toward the operation of nuclear power plants. While air quality standards cover pollutants from sources other than power plants, such as "mobile" (automotive) sources, some are pollutants (e.g., SO_2) from "fixed" fossil-fired sources, such as electric power plants and also industrial heat production. Likewise, we will consider the radiation standards as they apply to the normal operation of nuclear power plants and the (uranium-based) fuel cycle which supplies them.

In examining these existing standards, we will be interested in the criteria [1], [2] that were used in their original formulation and subsequent revisions, as well as the scientific theory and data that were applied. In all cases of such public-health standards, we face questions of limiting exposure of the public at large to

The author is with the School of Electrical Engineering and Computer Science, Polytechnic University, 6 Metro Tech Center , Brooklyn, NY 11201.

Reprinted from *IEEE Technology and Society Magazine*, Vol. 11, No. 4, pp. 7–16, Winter 1992/1993.

hazards which are largely of an involuntary nature. The policy dilemma in all these cases is: How low should such an exposure limit be? In some cases, such as air-quality standards, *threshold values* have been established below which policy makers have asserted that no measurable health effects can be discerned. In other cases, such as carcinogenic effects of (ionizing) radiation, there is evidence that *no threshold* exists and effects decrease to zero only as the exposure (dosage) decreases to zero [3, and references therein]. In either of the two cases, however, the question arises of how low the exposure limit should be.

Zero risk from a technological hazard has been recognized as unattainable (aside from a reversion to a nontechnological society) without diverting resources from reduction of other risks or other public-welfare programs. Policies appealing to the zero-risk notion have been enunciated, or advocated as qualitative guides for standards, which state that exposures (e.g., radiation dosage) should be kept "as low as is reasonably achievable" (ALARA) [4]. Similarly, intended policies have required the use of "the best available control technology" for controlling public risks from energy production.

If we move toward doctrines of limited yet nonzero risk or exposure, several concepts have been laid down in the past, either by policy makers or in the courts. The notion of "acceptable risk" [5] has been used to guide the allowable limits to probabilities of catastrophic accidents and are applicable cases of public exposure as we are considering here. Even more to the issue of exposure dosage is the "significant risk" doctrine laid down by the U.S. Supreme Court [6], [7] in a case of occupational exposure to chemical carcinogens, which clearly implies some sort of threshold dosage. This court ruling rests on the earlier legal "de minimus" doctrine which implies the converse, since it applies to a dosage level below which exposure can be safely ignored.

A currently emerging approach to determining how stringent standards should be (or "how safe is safe enough?") is *cost-benefit* analysis. When applied to risk reduction it is termed risk-cost-benefit analysis (RCBA) [1]. Cost-benefit analysis had its origin in the public-works projects of the U.S. Corps of Engineers in the 1930s and had some use in other agencies in the decades that followed. The balancing of benefits versus costs has grown up in environmental law starting in the 1970s. It has been limited, however, by legislation enabling the various environmental regulations. Often, the requirement for weighing benefits against costs results from litigation; but just as often, the courts have ruled in terms of "health risks" alone, without reference to the balancing of the benefits of risk reductions [7]. In 1981 the process was given additional impetus with Executive Order 12291 which required budgetary justification for all government expenditures, including health, environmental and safety regulations activities [7].

Cost-benefit analysis is straightforward in concept when applied to purely material costs and benefits, such as reducing flood damages with which the Corps of Engineers were concerned. It deals with costs of damages ex ante (before the fact) as estimated by policy makers and planners, not damages ex post facto, as they would be dealt with in the courts. It got problematic when attempts were made to apply it to health and safety regulation in the workplace or for the public at large. When a more stringent regulation (e.g., a lower allowable exposure dosage) is imposed on an industrial operation (e.g., power generation) the added *costs* are still straightforward in concept to calculate. Additional equipment is often needed and less efficient, more costly operation

Setting a dollar measurement for human suffering and death is difficult, if not downright immoral.

commonly results. The *benefits*, however, must measure the value of reductions in illness and injury or lost lives which result. Since the measure of benefits must be balanced against the costs, this logically requires *dollar* measures of human suffering and death, or "life and limb." Such a quantification is considered difficult by most and downright immoral by others [1], [2], [8], [10].

New ideas developed recently have led to a transformation of the RCBA balance problem [1], [11] which puts the quantitative comparison of benefit to cost on a basis that can be defended in terms of commonly held ethics and social philosophy. It attempts to measure the value of less risk, not the value of the lives saved. This new basis, now used in some government policy analyses [12, and references therein], [13], is called the "willingness to pay" (WTP) basis. It is based on the "expressed preference" means of valuation [11], [14] by an ordinary member of the public at large of a reduction in his/her level of risk. It builds, to a certain extent, on earlier psychological research on "perceived risk" and how people weigh their own chances and impacts [15], [16]. The perceived risks are incorporated, theoretically, by individuals in a "pricing," so to speak, of the value to them of a reduction of risk, treating the reduction as a nonmarket commodity. Such a measure of worth is called an *economic valuation.*

The WTP approach to benefits valuation is not only an innovative construct of behavioral science but has also several advantages for policy making in a democratic society. First, its basis, in principle, is with the individuals in the public at risk. As such, it tends to lower possible technocratic biases due to the projection of the values of the expert analyst on the outcome [2], [17]. As such, it also places an overlay of *consent* for risks assumed by the public, which is a gain for policy analysis in the view of researchers in the philosophy of technology [10].

Current Public-Health Risks

▼ Air Quality Standards

The unregulated burning of fossil fuels left an indelible mark on the industrial countries in the late 19th and early 20th centuries. Similar grim scenes have recently come to light from contemporary Eastern Europe. Whereas such public health statistics were not widely kept in 19th century Europe and America, reports of air pollution episodes, with high "excess mortality" figures attributed directly to high pollution levels, continued into the first half of this century. London, England, and Donora, Pennsylvania, are two of the better known locales for these episodes which were associated with atmospheric inversions [18]. Mounting public pressure, culminating in Earth Day in 1970, led to the Clean Air Act (CAA) and other environmental legislation. The impact of the CAA (1970) on the energy industries was to limit air emission from fossil-fired sources, such as power plants and industrial processing, so that the ambient air quality would be protected from pollutants [19]-[22, and references therein]. The CAA (and its subsequent amendments [21]) have set standards, called the National Ambient Air Quality Standards (NAAQS), which specify allowable concentrations of the various pollutants, either in terms of volume concentrations (e.g., micrograms per cubic meter) or relative concentrations (e.g., parts per million (PPM)).

There are two types of NAAQS standards: Primary Standards, designed to protect the public health; and Secondary Standards to limit effects on vegetation, materials, visibility, and other non-health related aspects of public welfare. The Primary AAQ Standards will be our focus here since they deal with risks to public health. There are six pollutants that are covered: particulate matter, sulphur dioxide (SO_2), nitrogen dioxide (NO_2), carbon monoxide (CO), ozone (0_3) and lead (Pb). Of these, the first three are most closely associated with fixed, fossil-fired (principally coal) sources as opposed to mobile (automotive) sources. High concentrations of particulates, SO_2 or NO_2 are known to have adverse health effects [18] such as: respiratory impairment and lung damage of otherwise healthy persons, aggravation of the conditions of persons suffering from cardiovascular and respiratory ailments, and, often, a lowering of bodily defenses against carcinogenesis and infections. Especially vulnerable are the very young or the very old, and people suffering from asthma, emphysema, influenza or cardiovascular conditions.

The approach taken to NAAQ standards was to establish a threshold value of pollutant concentration below which was a "lowest best judgment estimate" for health effects [18]-[20], [23], based on epidemiology. Objections have been raised in the past that these thresholds do not take adequate account of the more vulnerable individuals in the populations at risk [19], [20], [24], even though the CAA manifestly requires protection for "sensitive population groups" [19], [20]. There are also questions with respect to the spatial distribution depending on winds and weather. Objections have also been expressed more recently about the adequacy of statistical data and the differing responses of susceptible individuals to the various pollutants [25].

The present NAAQ primary standards [26], affecting the (fixed) energy-related sources are as follows:

▼ *Particulates:* a) Total Suspended Particulates (TSP) less than 75 $\mu g/m^3$ long-term average, and 260 $\mu g/m^3$ short term. The lowest best judgment estimate for health effects was 100 $\mu g/m^3$ long term and 70-250 $\mu g/m^3$ short term.

b) In 1987 the TSP standard was replaced with a standard focused on particulate matter less than 10 μm in diameter (PM_{10}), which is likely to be more responsible for adverse health effects. The new standard is 50 $\mu g/m^3$ annual average, and 150 $\mu g/m^3$ 24 h average, for the density of small particles.

c) Long-term averages of TSP and PM_{10} concentrations nationwide are within the standard and have a downward trend. However, about 27 million people live in localities where the PM_{10} standards are still not attained [26].

▼ *Sulfur Dioxide (SO_2):* The standard is

0.030 PPM or 80 μg/m^3 on an annual average, and 0.140 PPM or 365 μg/m^3 on a 24 h average. National trends for SO_2 are running less than 0.010 PPM and going slowly down. The lowest best judgment effects estimates were: 91 μg/m^3 long-term and 300-400 μg/m^3 short term. The thresholds were challenged by the Clean Air Scientific Advisory Committee in 1988, but the EPA did not change them [24]. Reduction of SO_2 emissions is also a high priority for the reduction of acid rain, and further reductions were mandated in the 1990 revisions of the CAA [21]. Only about 100 000 people nationwide live in nonattainment localities for SO_2 [26], as defined by the existing threshold standards.

▼ *Nitrogen Dioxide (NO_2):* The standard is 0.053 or 100 μg/m^3 on an annual average, with 141 μg/m^3 the long-term best judgment estimate. Current national trends are running about 0.023 PPM and are going down slowly. However, over eight million people nationwide live in NO_2 nonattainment localities [26].

It remains controversial whether these thresholds are sufficiently low. From a philosophic view point, a threshold that doesn't protect certain groups in society is a denial of *distributive justice* [1]-[27] — a basic tenet of Western democracy. Others would object to the setting of such a threshold to risk by experts without any form of *consent* from members of the population at risk [28]. Furthermore, any attempt by the experts to make an economic calculation, based on the dollar values of lives saved or suffering avoided, quickly runs into moral objections [9]. An irresolvable dilemma for air quality policy therefore seemed to exist - at least until several years ago.

The WTP approach, as mentioned in the introduction, has been applied to air pollution studies as a way out of these dilemmas. Krupnick and Portney [12] reported on a benefit-cost assessment of reducing ambient air pollution for the Los Angeles air basin. While the pollutant of major interest is ozone (O_3), due principally to automotive sources, the method is applicable to any AQ study [22]. In the survey, a sampling of the population at risk is asked to put a dollar value on a specified increment of reduction of their annual mortality risk due to the proposed pollution control measures. Similarly, valuations are solicited on reduced morbidity (respiratory illness and symptoms). To these valuations medical costs can be added, as well as lost income due to respiratory illness or distress, estimated before the fact. If the major pollution sources were fossil-fired power plants, the WTP estimates could be related directly to electricity rates.

Estimates of benefits, per increment of reduction of pollution levels, can be tied back to the pollution controls through "dose response" functions [12], [18]. These functions, for air-quality studies, are fitted to data correlating morbidity (respiratory illness) and mortality (premature deaths) with "dosage" (air pollution concentrations for specified time durations). Thus, knowing the size of the regional population at risk and having a dose-response curve, the total morbidity and mortality benefits of a specified reduction in pollution concentrations, over specified time intervals, can be estimated for that region. Then, using the WTP valuation methodology, Krupnick and Portney came up with total annual health benefits of $3 billion, aggregated for the residents of the South Coast region of Los Angeles. This was an ex ante estimate for bringing ambient ozone concentrations within the "attainment" levels required by the NAAQS.

The costs which must be borne to attain reductions in air pollution, while more straightforward in concept, are nonetheless complicated. In the case of ozone reduction in the Los Angeles basin [12], the needed reduction in O_3 levels had to be achieved mainly through reductions in volatile organic compounds (VOCs), since ozone is generated by photochemical reactions with airborne nitric oxides and other components under particular atmospheric conditions. A 35% reduction of VOC has been estimated to be needed, meaning a reduction of 4 million tons emitted annually for the region. Further reductions in NO_2 emissions from power plants will also contribute to O_3 reductions because of NO_2 involvement in the photochemical reactions [12], [18]. The control costs for reducing VOC and companion pollutants involves over 100 types of emission controls, including those on manufacturing of rubber products, use of solvents, auto refinishing

The willingness-to-pay approach allows a cost benefit comparison without the moral dilemma of valuing human life and suffering.

and even backyard barbecues, in addition to automotive emissions. For the latter, all the region's vehicles must be converted to low-emission operation. The aggregate control costs for the region are estimated at about $13 billion, or about $2700 per household [12]. While these estimates by Krupnick and Portney were disputed by others in the field [25], there was no implication that the benefit-valuation approach was not feasible.

While this example has not been one concerned principally with (electric or industrial) energy production, it does illustrate the application of the presently used RCBA methods for health effects of air pollution. For energy-production cases, the WTP estimation of health benefits would not differ in method from those used in the Los Angeles basin case. The control costs, of course, would differ. Most obviously, the control of the main power plant emissions (particulates, SO_2 and NO_2) would be required. Costs of precipitators and fuel-gas scrubbers have long been established [1], [22]. Newer combustion technologies exist also, such as fluidized-bed combustion, which have cost-reduction benefits [1], [22], in addition to benefits of reduction of SO_2 and NO_2 emissions.

▼ Nuclear Radiation Standards

Risks from nuclear radiation are the most dramatically etched in the public mind in the form of nuclear-plant accidents or nuclear weapons. Fears of the effects of invisible yet penetrating radiation are deeply engrained in the human psyche [29], [30]. The regulatory bodies, however, are faced with the setting and reviewing of standards for limiting public exposure to radiation, even for low-level exposures. Here, we will focus on the regulations designed to limit the dosage of members of the public due to the routine operation of nuclear plants and other parts of the uranium fuel cycle. We limit ourselves to routine operations, rather than (abnormal) accident conditions, to fit the theme of the paper - public exposure standards of ongoing risks associated with electric-power supply.

Nuclear radiation is classed as "ionizing radiation" because each atomic particle (e.g., alpha or beta particles) or radiation quantum (e.g., gamma rays) has sufficient energy to break chemical bonds. As such, it has the capacity to directly alter living cells and cause biological injury [3]. High-level radiation exposure, such as that experienced by victims of nuclear weapons or serious nuclear accidents, kills cells directly, resulting in radiation sickness and/or prompt deaths. Lower levels of exposure have delayed stochastic effects which can only be determined on an epidemiological basis, such as genetic effects and carcinogenesis [3].

The units of radiation dosage (exposure) are related to the energy absorbed per unit of mass of biological tissue and include a "quality factor" of biological effectiveness (depending on the radiation type, e.g., gamma rays, neutrons, or alpha particles). The SI dose unit is the "sievert" (Sv) which corresponds to one joule of energy absorbed per kilogram mass of tissue. A similar unit of common usage is the rem (*roentgen equivalent man,* named after the discoverer of x-rays since the quality factor is referenced to the biological effects of x-ray radiation) and 100 rem = 1 Sv. Dosages above 0.5 Sv (50 rem) are in the high level range. For a scaling at the low end, annual average dosage of the U.S. population due to natural causes (including cosmic rays and terrestrial sources except radon) are about one mSv (100 mrem) and about 0.5 mSv (50 mrem) for medical purposes.

Exposure standards for ionizing radiation of the public are set nationally by the U.S. EPA [31] and the U.S. NRC [4]. Both are given in terms of allowable annual dosages and both have undergone changes over the years. The EPA sets 0.25 mSv (25 mrem) as the annual limit of whole-body dose for the public. The NRC, in 1990, lowered its allowable limit from 0.5 mSv to 0.1 mSv (10 mrem) for public exposure under normal (non-accident) operating conditions for nuclear power plants. Of concern here are the routine or "planned" discharges of radioactive gases and water, which occur in normal operation of the nuclear reactor and its auxiliary apparatus.

Federal guidelines for radiation protection of the public will reportedly [4] be formulated by an EPA Task Group, which has representation from the NRC. The matter of a lower "reference level" of exposure, more clearly defining a threshold, will presumably be considered in view of recent reevaluations of radiation effects [3], In the meantime, however, the NRC, under pressure from the nuclear industry, formulated regulatory rules including a policy of "Below Regulatory Concern" (BRC) [4]. This policy clearly follows the "de minimus" [7] principle and represents a definitive threshold by the Commission below which they will not concern themselves. The BRC threshold level is being debated by intervenors and the NRC staff, with the lowest level proposed being 1 mrem per year, but 10 mrem is favored by several commentors. These levels appear to have application mostly to radioactive waste disposal rather than nuclear power plant operation. There is no indication of seeking any form of public consent or cost-benefit norm by the NRC in the criteria for the BRC policy.

Allowable low-level radiation dosages have been debated for decades. Major inputs for this policy have been the series of studies called "The Biological Effects of Atomic Radiation" (BEIR) started in 1956 by the National Research Council of the National Academy of Sciences. The most recent of this series BEIR V [32] again raises doubts about the use of thresholds in risk assessment for ionizing radiation. In particular, assertions have been renewed (having their beginnings in the 1940s) that *no threshold* exists for carcinogenic or teratogenic (noncancerous abnormal growth) effects due to ionizing radiation and fears that the previous estimates of excess mortalities due to radiation-induced cancers were too low [3]. The findings of other radiation advisory groups, nonprofit groups, and international commissions [3] reflect an uncertain concern over threshold and non-threshold standards as well.

If there is no threshold for these effects of radiation, then the question is raised as to what the dose-response function must be, as the dose is considered down toward the zero level. The BEIR V report interprets data from nuclear-bomb tests and medical radiology records to indicate that the risk of thyroid cancer is a linear function of dose [3]. (The slope of this linear function is estimated to be 7.5 fatal thyroid cancers per sievert over the lifetimes of 10 000 persons.) The model for leukemia is thought to be linear quadratic. There is a fear, furthermore, that the *rate* of increase in the dose-response curves is still higher for children with thyroid cancers and possibly also with leukemia.

The application of existing standards implies expected numbers of excess fatalities in the population at risk, if the rates of the (zero-threshold) dose-response functions of the BEIR V study are used. In the case of thyroid cancers, an allowed level of 10 mrem/yr (0.1 mSv/yr) would imply several excess fatalities in the lifetimes of the roughly one million population served by a base-load nuclear power plant (800 MW capacity), whereas the 1 mrem/yr (0.01 mSv/yr) NRC de minimus level would imply less than a single fatality over the lifetimes of that same population. Similar incremental reductions in expected fatalities, and morbidity as well, might be found for other types of cancers. All such consequences of long-term stochastic consequences of a reduction of exposure, however, are subject to large uncertainties and thus these estimates can only be taken as indicative.

If the establishment of the probabilistic risks due to nuclear radiation in the form of threshold or dose-response functions is difficult, the economic valuation of those risks is even more so. It took decades of controversy over nuclear power (and weapons) before the behavioral phenomena in *perceived risks* [15], [33], [34, and references therein] were recognized. What was vehemently denounced by nuclear advocates as the "irrational fears" of the public, when the advocates presented the low probabilities of the occurrence of accidents, was belatedly recognized as the public weighing the *consequences* of the accidents with fear and loathing [29], [30]. The weighing of such consequences, it turned out, could be determined on a systematic scientific basis, being reproducibly measured psychometrically [35], [36], and thus placing these subjective responses on a quantifiable basis [15], [16]. This being the case, it would appear that the benefits valuation of reducing radiation risks, as emotionally charged as they may be, might be susceptible to cost-benefit analysis [37], [38].

As mentioned in our introduction, a weighing of risks, or the benefits of reducing those risks, can be incorporated in principle into an analysis of the balance of benefits and costs [13], [14], [22]. An approach is the WTP method, as was illustrated in the case of AQ benefits, where members of the public at risk give their "expressed preferences" to a survey or questionnaire. In the case of standards for low-level radiation, the valuations of members of the public would be ex ante on potential reductions of permitted dosages of ionizing radiation to the public.

The estimation of the costs of further controls on emitted radiation from nuclear power plants and nuclear fuel-cycle operations, like the air-quality case, would be complicated in practice but straightforward in principle. Both capital investment costs and operational costs could be expected to increase. These costs would vary according to reactor type and particular design [22], [39]. More thorough purging of radioactive leakage in containment structures and improved processing of (radioactive) noble gases resulting from fission in fuel rods might be required. Longer storage of such radioactive gases would likely be required to allow for more radioactive decay before release into the atmosphere. More complete filtering of particulates (carrying radio nucleides) from hot-air chimneys would also likely be required. Additional training of operating personnel might be a cost-effective means of lowering emissions at some facilities, also.

Potential Public-Health Risks of Electric and Magnetic Fields

Electric and magnetic fields are a physical presence wherever electricity is transmitted or distributed. Electromagnetic fields surround any set of electric current carrying conductors. Intense fields of both types are found around transmission lines and lesser fields around primary and secondary distribution lines. But these fields drop off in strength rapidly with distance and it is possible for the public to avoid continual immersion in them.

Not so for the fields inside of homes and places of work, where members of the public must spend major fractions of their lifetimes. Inside of any modern (wired) building exists an ambient low-level background of electric and magnetic fields. Electric fields are quite low (10 V/m or less) for unshielded conductor pairs in service wiring and can be reduced to values several orders of magnitude lower by conducting shielding. Magnetic fields however, cannot be effectively shielded from open spaces and have therefore an ambient presence in the interiors of buildings and homes, which cannot be practicably reduced lower than the level provided by the field cancellation effects of conductor pairs. Ambient ac magnetic fields inside a home or office run 1-2 mG (0.1-0.2 μT) at locations not near appliances (TVs, electric ranges, hair dryers, etc.).

Until recently, there was no concern over the effects of these fields. After all, on a common sense basis, the public has been exposed to them for over a century with no obvious ill effects. Furthermore, a near universal belief had been shared by electrical engineers and physicists that the physical effects of such weak fields could not possibly cause harm to people. After all, it was typically reasoned, the electric fields are very much smaller than those experienced with electrostatic shocks on carpeting during the winter and the magnetic fields are over a hundred times smaller than the earth's (ambient) magnetic field.

The approach builds on research on "perceived risk" and how people weigh their own chances and impacts.

Within the past two decades, however, concern for the biological effects of nonionizing electromagnetic radiation arose with the introduction of microwave ovens [40]. Microwave and radio-frequency radiation and fields had not previously been matters of great public concern unless they were of high intensity, such as found near a powerful transmitter. Disquieting reports about health effects of low-level radiation came out of research in this country and Eastern Europe [40] and public worry was heightened by some expose journalism on microwave radiation [41]. As a result of the public debate, microwave safety standards have been revised [42] and are still under study.

When reports followed in the late 1970s of the possible hazards of electric and magnetic fields at power-line frequencies (60 Hz), the public tended to lump the two effects together. They are, in fact, quite different. Here, we will only deal with the effects of electric and magnetic fields which are oscillating at the very much lower powerline frequencies. Nonetheless, the histories of the two forms of electromagnetic fields are similar and the association in the public's mind is likely to persist.

High-voltage, high-power transmission lines are long recognized as having strong electric and magnetic fields, with associated hazards. Transmission-line rights of way have long had required widths intended to limit the field strengths to which people might be exposed [43], [44]. The electric fields were historically those of concern and existing practice in the 1970s was to limit exposures to the 1-2 kV/m range. These limits and transmission corridor practices were challenged in 1974, however, in a project for the New York Power Pool to construct a very-high voltage (765 kV) transmission line [45]. This case led to other legal challenges, and the public awareness of possible hazards due to electric power fields grew, a concern which would later carry into low-level fields.

Concern over the possible effects of *low-level* 60 Hz (power-frequency) fields followed the findings of a pioneering epidemiological study carried out by Wertheimer and Leeper (1979) [46] who found a disturbing correlation of the incidence of childhood cancers to electric wiring configurations in and around homes. They posited that children who had grown up in homes where the constant presence of ac magnetic fields was stronger than the usual ambient value were more susceptible to leukemia, lymphomas, and nervous-system tumors.

Reactions to these findings ranged from alarm in the public to disdain in the scientific and technical communities. Public alarm was heightened by another exposé [47]. At the other extreme was disbelief and disparagement by engineers and physicists of the earlier studies, some vehement and persisting to the present [48]. Further epidemiological studies were conducted throughout the 1980s [49]. The Wertheimer-Leeper results on Denver children were substantially confirmed, thus causing pause among all but the most vehement critics. Of the other epidemiological studies, some confirmed a link of magnetic fields at 50 and 60 Hz frequencies to cancer and others did not. One study even claimed to have established a dose-effect curve.

The studies included both residential and occupational (long-time) exposures and involved adults as well as children. Whereas the results did not conclusively prove the link, it became generally accepted that further studies, both clinical and epidemiological, were needed [50], [51]. As a result, federal and private funding for electromagnetic field (EMF) research is expected to reach over $15 million per year [52].

The biological mechanisms [49] that could explain why magnetic fields almost two orders of magnitude smaller than the earth's (steady) field could be harmful centers on the frequency range of the power lines — called the extremely-low-frequency (ELF) range. Magnetic fields at these frequencies are virtually the same inside the body as outside. They are little absorbed by the body. This suggests, of course, that gross heating effects from these weak but penetrating fields cannot account for biological effects. Current densities induced internally by ELF magnetic fields at domestic (home) levels are below endogenous physiological levels, and so the effects are evidently not due to gross electric perturbance.

Theories have been advanced [49] to explain EM bio-interactions, developing some deeper understanding, but to date none have wide predictive ability. Among these theories are: cell-ion-transport models, ion-cyclotron resonance and many-body cooperative mechanisms at the molecular level. Such mechanisms may affect cell communication or long-range biological coherence which are known to be involved in control of growth and immune reactions and which operate well below the level of thermal fluctuation. Currently, laboratory studies on animals, tissues (animal and human), and organic chemicals at high field strengths are being carried out to help elucidate the possible mechanisms of these interactions, including mitosis of DNA, gene mutations, chromosome aberrations, calcium efflux (in cellular processes), hormone and enzyme responses, and immunological effects. The major thrust of these studies is to find biological processes relevant to carcinogenesis. Some of the experimental results show inconsistent effects, or effects under special conditions, and others have negative results, even though all are conducted at strong field strengths. Nonetheless, there is hope of learning more about either cancer producing or cancer promoting interactions. The research program has, clearly, not yet reached the stage of predictive models that could be used to design epidemiological experiments.

If a causal relationship is established between ELF field exposure and cancer, then, of course, health standards will have to be formulated. The same kinds of questions for these effects can be expected to arise as have arisen for air-quality and (ionizing) radiation standards.

- Will there be thresholds established or will the bio mechanisms suggest a zero threshold, as they have for carcinogenicity due to ionizing radiation?
- If there is a threshold, will it account for the more vulnerable members of the population, since early epidemiology suggests that children are possibly more at risk?
- How will the threshold level, or allowable exposure for a nonthreshold mechanism be established? Through a balancing of benefits against costs for establishing an allowable exposure or by means of "health-related" risk analysis?

The costs of controlling field-strength exposure of the public could be enormous. Further changes in transmission-line rights of way or conductor configurations may be required. Substations may have to be reconfigured or moved. Primary lines and distribution transformers may have to be reconfigured. And finally, wiring in homes and places of work may have to be redone for a significant fraction of the building stock. Cumulative costs to the nation's electric utilities could run into tens of billions of dollars [53]. These costs would, of course, be passed on to the ratepayers, under the public utility laws of the various states, resulting in stiff increases in electric bills. Added to this, for the ordinary homeowner or commercial building owner, might be the possibly heavy capital outlays for rewiring.

The question of "how safe is safe enough?" can be answered by the population at risk.

If benefits were to be weighed against such costs for reducing ELF field exposures, they could be done in a very real and concrete willingness-to-pay analysis. Members of the public exposed to these fields would have to weigh for themselves, and their families, what they are willing to pay for reduction of such risks, which are at the present (uncontrolled status) sufficiently rare in the population that they must be uncovered by (mortality or morbidity) statistics. It might be assumed that the members of the population (at risk) might conclude that their chances of contracting cancer from this cause are so small that they could ignore them. On the other hand, these same members weighing the consequences of the event, rare as it might be, may well be willing to pay much more than sheer probabilities might indicate, thus paralleling the case of nuclear risks. In the case of the potential risks of ELF fields, the consequences are cancer, frequently fatal, with children appearing possibly more at risk, both factors making for a heavy weighing of the consequences by members of the population exposed.

How Safe is Safe Enough?

We have reviewed the existing and prospective standards for protecting the public health against the ambient hazards associated with electricity production. These hazards include air pollution, ionizing radiation, and electromagnetic fields, to which individual members of the public are exposed to one degree or another, in an ambience of exposure to the effects of these hazards and on a largely involuntary basis. Studies have shown that members of populations at risk are capable of making valuations of the worth to them of reduction of such risks, if they are appropriately informed as to the probabilities and possible consequences of the effects of the hazards. Such valuations of risk reduction should ideally be carried out *after* the expert assessment of the probabilities of death, injury, or illness. The role of the expert can thus be limited to applying the epidemiological models and biological causes studies involved in the (probabilistic) risk assessment, without intruding on the perceived worth of a reduction in that risk. This expert knowledge could then be used for informing the members of the population at risk, not for making the final judgment for them of: How safe is safe enough. At the same time, it places an element of *consent* by those involuntarily at risk,into the balancing of costs versus benefits in setting levels of hazardous exposure.

At the present time, the use of any such benefits valuations in public health and safety policy is infrequent and fragmented. The application of cost-benefit analysis of any sort is anything but uniform among the regulatory agencies. This is due in part to the legislative history. For older legislation, prior to 1981, there was no uniform benefits versus costs evaluation. While Executive Order 12291 attempted to conform regulation throughout the government, many of the regulations, including air quality and radiation standards, have not been reformulated using cost-benefit studies, much less willingness-to-pay valuations. This implies, of course, that the allocation of funding, limited currently even more than ever, cannot be made on any basis of rational comparisons.

Presently, policy guidance for standards relies on a bewildering mix of criteria from earlier legislation and practices. Some of it is simply "risk-based," meaning that it depends on expert scientific judgments of what are acceptably low probabilities of death, illness, or injury among the population at risk. (This is termed "risk assessment" in the agencies and much of the literature.) Sometimes the legislation, such as regulating toxic substances, will even state that the risk-based standard will be set without regard for cost, thereby disallowing either allocation of resources for health and safety or consent by the population at risk. Often agency policies have to conform to court decisions (e.g., de minimum judgments) made ex post facto to legislation.

It is clear at this time that the public health and safety regulations pertaining to electric utilities could be beneficially reexamined with these questions of cost-benefit valuation in mind. The existing ambient air quality standards and the radiation standards, as they affect the public at large, are outstanding examples of regulations for such reviews. To date, exploratory studies have been made, for instance on the NAAQs [54], but not carried further toward implementation in the standards. The potential hazards of electric and magnetic fields due to electricity distribution will have to be addressed if the cause-effect relationships with cancer are established. These standards, carrying many of the same perceptual aspects as radiation hazards, would be enhanced by expressed-preference valuations as well. If the relationship to cancer is proven, the perception of the risks and consequences of exposure will carry many of the same deep-seated emotions as that for ionizing radiation. If children are shown to be more at risk, then this is bound to deepen fears in the public. In such a case, a valuation of the benefits of lower exposure to EM fields by the population at risk would seem even more to be mandated.

Acknowledgment

The author is indebted to several staff members at the U.S. EPA and the U.S. NRC for guiding him to references and clarifying agency policies. Their cooperation is to be commended as a model of open government. Any criticisms in this paper of regulatory policies is entirely that of the author and should not be attributed in any way to these civil servants. The author is also indebted to his colleagues Leo Birenbaum and Walter Kiszenick for assistance in locating references and information.

References

[1] E.S. Cassedy and P.Z. Grossman, *Introduction to Energy - Resources, Technology, and Society*. Cambridge Univ. Press, 1990. ch. 10.
[2] F. Fischer and J. Forester, *Confronting Values in Policy Analysis. the Politics of Criteria*. Newbury Park, CA: Sage Pubs., 1987.
[3] A. Upton, "Health effects of low-level ionizing radiation," *Physics Today*, Aug. 1991.
[4] "Standards for Protection Against Radiation: Final Rule," 10-CFR, Part 20, U.S. Nuclear Regulatory Commission, Washington, DC, 1991.
[5] W.W. Lowrance, *Of Acceptable Risk*, W.Kaufman, Inc., Los Altos, CA, 1976.
[6] J.D. Graham *et al.*, *In Search of Safety: Chemicals and Cancer Risk*. Cambridge, MA: Harvard Univ. Press, 1988.
[7] P.F. Ricci and L.S. Molton, "Risk and benefit in environmental law," *Science*, vol. 214, no. 4, pp. 1096-1100, Dec. 1991.
[8] K.S. Shrader-Frachette, *Science Policy, Ethics, and Economic Methodology*. Dordrecht, Holland: D. Reidel, 1983.
[9] S.E. Rhodes, Ed., *Valuing Life: Public Policy Dilemmas*. Boulder, CO: Westview, 1980.
[10] D. MacLean, *Values at Risk*. Totowa, NJ: Rowman & Allanheld, 1986.
[11] E.J. Mishan, *Cost-Benefit Analysis*. New York: Praeger, 1976.
[12] A.J. Krupnick and P.R. Portney, "Controlling urban air pollution: A cost benefit assessment," *Science*, vol. 252, pp. 522-528, Apr. 26, 1991.
[13] E.M. Gramlich, *Cost-Benefit Analysis of Government Programs*. Englewood Cliffs, NJ: Prentice Hall, 1981.
[14] B. Fischhoff and L.A. Cox, "Conceptual framework for regulatory benefits assessment," in *Benefits Assessment: the State of the Art*, J.D. Bentkover *et al.*, Eds. Dordrecht, The Netherlands: D. Reidel Publishing Co., 1986.
[15] B. Frischhoff *et al.*, "Defining Risk," *Policy Sciences*, vol. 17, pp. 123-139, 1984.
[16] C. Hohenemer *et al.*, "The Nature of Technological Hazard," *Science*, vol. 220, pp. 378-384, Apr. 1983.
[17] J. Byrne, "Policy science and the administrative state: the political economy of cost-benefit analysis," in F. Fischer and J. Forester, *Confronting Values in Policy Analysis. the Politics of Criteria*. Newbury Park, CA: Sage Pubs., 1987; and J. Byrne and S. M. Hoffman, "Nuclear optimum & the technological imperative...," *Bull. STS*, vol. 11, pp. 63-77, 1991.
[18] R. Wilson *et al.*, *Health Effects of Fossil Fuel Burning - Assessment and Mitigation*, Cambridge, MA: Ballinger, 1980.
[19] E.H. Haskell, *The Politics of Clean Air*. New York: Praeger, 1982.
[20] C.O. Jones, *Clean Air - the Policies and Politics of Pollution Control*. Pittsburgh, PA: Univ. of Pittsburgh Press, 1975.
[21] The latest amendment to the CAA was in 1990: Committee on Energy and Commerce, U.S. House of Representatives HR 3030, U.S. Government Printing Office, Washington, DC, May 1990.
[22] R.L. Ottinger, *The Environmental Costs of Electricity*. New York, NY: Oceana Press, 1990.
[23] National Research Council (National Academy of Sciences), "Implications of environmental regulations for energy production and consumption," vol. VI, Analytical Studies for the U.S. Environmental Protection Agency, Washington, DC, 1977.
[24] "News and Comments," *Science*, vol. 236, p. 389, Apr. 22, 1988.
[25] F.W. Lipfert and S.C. Morris, "Air pollution benefit-cost assessment," *Science* (Letters), vol. 253, pp. 606-609, Aug. 9, 1991.
[26] U.S. EPA, "National Air Quality and Emissions Trends Report, 1989" Rep. EPA 450/4-91-003b, Feb. 1991.
[27] H.B. Leonard and R.J. Zeckhauser, "Cost benefit analysis applied to risks: Its philosophy and legitimacy," in D. MacLean, *Values at Risk*. Totowa, NJ: Rowman & Allanheld, 1986.
[28] B. Fischhoff, "Cognitive and institutional barriers to informed consent," M. Gibson, Ed., *To Breathe Freely*. Rowan & Allenheld, Totowa, NJ, 1985.
[29] R.J. Lifton, "Nuclear energy and the wisdom of the body," *Bull. Atomic Scientists*, Sept. 1976, pp. 16-20.
[30] S. Weart, *Nuclear Fear - A History of Images*. Cambridge, MA: Harvard Univ. Press, 1988.
[31] "Environmental radiation protection standards for nuclear power operations," 40-CFR, Part 190, U.S. Environmental Protection Agency, Washington, DC (source: 42-CFR-2860, Jan. 1977).
[32] National Research Council, Committee on Biological Effects of Ionizing Radiations (BEIR V), *Health Effects of Exposure to Low Levels of Ionizing Radiation*. Washington, DC: Nat. Academy Press, 1990.
[33] R. Wilson and E. Crouch, *Risk-Benefit Analysis*, Cambridge, MA: Ballinger, 1982.
[34] W.R. Fruedenburg, "Perceived risk, real risk: Social science and the art of probabilistic risk assessment," *Science*, vol. 242, pp. 44-49, Oct. 7, 1988.
[35] S.S. Stevens, *Psychophysics: Introduction to its Perceptual, Neural and Social Aspects*. New York: Wiley, 1975.
[36] E. Galanter, "The direct measurement of utility and subjective probability," *Amer. J. Psychol.*, vol. 75, pp. 208-220, 1962.
[37] R.J. Zeckhauser and W.K. Viscusi, "Risk within Reason," *Science*, vol. 248, pp. 559-564, May 4, 1990.
[38] J.A. Bradbury, "The policy implications of differing concepts of risks," in *Science, Technology & Human Values*. Beverly Hills, CA: Sage, 1989.
[39] J.R. LaMarsh, *Introduction to Nuclear Engineering*, 2nd ed. Reading, MA: Addison-Wesley, 1983.
[40] N.H. Steneck, *The Microwave Debate*. Cambridge, MA: M.I.T. Press, 1984.
[41] P. Brodeur, *The Zapping of America: Microwaves, Their Deadly Risk and the Cover Up*. New York, NY: W. W. Norton, 1977.
[42] "American National Standard Safety Levels with Respect to Human Exposure to Radio Frequency Electromagnetic Fields, 300 kHz to 100 GHz," American National Standards Institute, ANSI C95.1 -1982, published by IEEE, NY, 1982.
[43] National Electrical Safety Code, American National Standards Institute (ANSI), Published by the Institute of Electrical & Electronic Engineers (IEEE), New York, 1990.
[44] E.L. Carstenson, *Biological Effects of Transmission Line Fields*. New York, NY: Elsevier, 1988.
[45] "EMF and Human Health" *EPRI Journal*, Oct./Nov. 1987.
[46] N. Wertheimer and E. Leeper, "Electrical wiring configurations and childhood cancers," *Amer. J. Epidemiol.*, vol. 109, pp. 273-284, 1979.
[47] P. Brodeur, *Currents of Death: Power Lines, Computer Terminals and the Attempt to Cover Up their Threat to Our Health*. New York, NY: Simon & Schuster, 1989.
[48] R. Adair, *Physics Today*, p. 103, Dec. 1991.
[49] "Evaluations of the Potential Carcinogenicity of Electromagnetic Fields," Review Draft, Office of Research and Development, U.S. Environmental Protection Agency, EPA/600/6-90/005B, Oct. 1990, Washington, DC, and "EMF Cancer Report" (draft), Science Advisory Board, Radiation Advisory Committee, Non-ionizing Electric and Magnetic Fields Subcommittee, U.S. EPA, Dec. 20, 1991.
[50] News & Comment, "Is there an EMF-cancer connection?" *Science*, vol. 249, pp. 1096-1097, Sept. 7, 1990.
[51] Power Industry News, "Current status of research on power-frequency electric and magnetic fields," *IEEE Power Eng. Rev.*, Aug. 1991, reprinted from *Electra*, no. 135, Apr. 1991.
[52] Science Scope, "Voltage rising in EMF research," *Science*, vol. 254, Dec. 6, 1991, p. 1443.
[53] H.K. Florig, "Containing the costs of the EMF problem," *Science*, vol. 257, p. 468-469, July 24, 1992.
[54] "Regulatory Impact Analysis on the National Ambient Air Quality Standards for Sulfur Oxides," Draft Rep. (for public comment), Environmental Protection Agency, Research Triangle Park, NC, Mar. 1988.

T&S

Chapter 7

Engineering and Sustainable Development

THE concept of sustainable development has become a major public policy issue worldwide, and not just within the engineering community. Following the Brundtland Commission report in 1987, which defined sustainable development as "development that meets the needs of the present without compromising the ability of future generations to meet their own needs" (1), the concept attracted considerable attention within the international community and on national agendas. The 1992 UN Conference on Environment and Development in Rio de Janeiro produced a blueprint for global sustainable development, Agenda 21, which resulted in the establishment of dozens of national commissions (2), including the U.S. President's Council on Sustainable Development (3).

The engineering community reacted with the establishment of the World Engineering Partnership for Sustainable Development (WEPSD) in 1992 (4) and issuance of position statements by committees formed within the traditional engineering organizations, including the American Society of Civil Engineers (ASCE) (5,6) and the Institute of Electrical and Electronics Engineers (IEEE) (7), as well as the American Association of Engineering Societies (AAES) (5).

Sustainable Development

Sustainable development theory emerged from the field of ecological economics. According to this theory, sustainable development involves three realms: ecological, economic, and social. The overall objective of a sustainable society is to achieve sustainability in economic, ecological, and social systems (see the background article in this chapter by Alex Farrell).

An alternative, but consistent, way of characterizing development is to think of three distinct systems—biological, economic, and social—each with its own set of goals. Sustainable development, then, consists of maximizing the achievement of these goals across the three systems, subject to inevitable tradeoffs and priority setting in any given time or place (8). As noted by the International Institute for Environment and Development, typical goals for each system might include (8)

- biological [ecological] system goals: genetic diversity; resilience; biological productivity
- economic system goals: increasing production of goods and services; satisfying basic needs or reducing poverty; improving equity
- social system goals: cultural diversity; social justice; gender equality; participation

This three-prong approach is consistently advocated in the sustainable development literature.

Despite some criticisms of the term, the concept of sustainable development maintains considerable currency in a number of circles, including engineering. In some cases engineering societies have even proclaimed sustainable development to be an ethical responsibility (9).

Engineering, Policy and Sustainable Development

Public policy's success in promoting sustainable development is dependent on achieving all three objectives of a sustainable society. However, despite proclamations that engineers have an ethical responsibility to endorse the principles of sustainable development, questions of just distribution and other questions of equity (such as risk distribution) are often left off the table. Indeed, almost all the effort of engineers and engineering organizations on the issue of sustainable development focuses on striking a balance between economic development and environmental protection. Although these efforts are commendable, they are limited by their failure to come to grips with the third essential element of sustainable development—the social objective.

The social dimension of the sustainable development problem is often either ignored entirely or relegated to a secondary status. It is common, for example, for engineers writing on the topic to refer to development that is "environmentally and economically sustainable" (10) or to view sustainable development as an attempt to balance environmental quality and economic growth (11).

This approach to environmental problems, which is often manifest through such techniques as cost-benefit analysis, is firmly entrenched in the engineering culture. However, as Harris, Pritchard, and Rabins (12) have pointed out, this approach, though consistent with utilitarian ethics, may be inadequate when addressing questions of just distribution and hence the social dimension of sustainable development. Such questions generally require consideration of ethical frameworks grounded in rights or duties, factors rarely considered in cost-benefit methods.

Social objectives, if made explicit at all by engineers, are often considered only to the extent that they impact on the economic or ecological objectives, as in the following statement by a prominent member of the WEPSD (10): "Factors in the project that pertain to economics, culture, environment, or technology are all interdependent. This interdependence requires that the projects be examined in terms of their impacts on the ecosystem" (p. 218).

The former IEEE Environment, Health, and Safety Committee (EHSC), now known as the Technical Committee on Electronics and the Environment, was charged with tending to sustainable development. The committee explicitly rejected the social goal by downplaying the concept of sustainable development because it is "somewhat value-laden ... implying for some people, for example, redistribution of wealth or a need to restrict current consumption." The EHSC instead embraced the concept of "industrial ecology," which limits its focus to "industrial and economic systems and their linkages with fundamental natural systems" (7, p. 2).

Some engineers have even argued for a preeminent position for engineering that clearly amounts to a technocratic vision of sustainable development. A founder of the WEPSD goes so far as to portray engineers as the best arbiters of *all* knowledge that needs to be brought to bear on the problem (4): "It is important for engineers to provide the implementing interface between science, society, and the decision-making bodies (public and private) to ensure that the most appropriate knowledge in natural, technological, and social sciences is implemented to meet the needs of both present and future generations" (p. 238).

With respect to public policy, then, the involvement of many engineers and engineering organizations in promoting sustainable development appears to be characterized by a technocratic and/or self-serving attitude with limited concern for social ethics.

In contrast to the technological fixes which the engineering community advocates for the challenges posed by sustainable development, ethicists have argued from both the perspectives of environmental ethics and development ethics that sustainable development is at heart an ethical issue (13). Kothari, for example, has stated (14):

> The shift to sustainable development is primarily an ethical shift. It is not a technological fix, nor a matter of financial investment. It is a shift in values such that nature is valued in itself and for its life support function, not merely for how it can be converted into resources and commodities to feed the engine of economic growth. Respect for nature's diversity, and the responsibility to conserve that diversity, define sustainable development as an ethical ideal. Out of an ethics of respect for nature's diversity flows a respect for the diversity of cultures and livelihoods, the basis not only of sustainability, but also of justice and equity. (p. 35)

It would seem, then, that engineering ethics ought to have much to contribute to the understanding of the engineer's public role with respect to sustainable development (see the Introduction to Part III). This is especially true in connection with the social objective, where insights drawn from environmental ethics and development ethics can be integrated with engineering conceptions of sustainable development.

Engineering Ethics and Sustainable Development

Some engineering codes of ethics have attempted to integrate environmental and social equity concepts.

Until recently, only two major U.S.-based engineering societies, the ASCE and the IEEE, even mentioned the environment in their codes of ethics. The former ASCE code, adopted in 1977, contained a "Guideline to Practice" committing engineers to "improving the environment." The first article of the current IEEE code, adopted in 1990, pledges engineers "to disclose promptly factors that might endanger the public or the environment." As Harris et al. point out (12), however, these codes provide only limited support for environmental principles beyond the impact of the environment on human health.

More recently, efforts have been aimed at incorporating sustainable development principles in engineering codes of ethics. Here again, however, social dimensions of sustainable development have been secondary to environmental issues.

For example, although the World Federation of Engineering Organizations (WFEO), one of the founding organizations of the WEPSD, promotes social and cultural sensitivity by engineers, it does so within the framework of a

Code of Environmental Ethics for Engineers. The WFEO code, adopted in 1987, invokes the notion of sustainable development. In Article 4 it clearly expresses concern for both ecosystems and socioeconomic systems, and in Article 3 it notes the importance of "social equity and the local system of values." However, the WFEO code emphasizes the natural environment, as reflected in the title and the encouragement in Article 4 for engineers "to select the best alternative for an environmentally sound and sustainable development." Although it might be argued that the WFEO uses the term "environment" to signify both the natural and social environments, such use would not reflect how "environment" is generally understood in engineering literature and in engineering practice.

An addition to the ASCE Code of Ethics, first proposed in 1983 in the form of an additional "Fundamental Canon" beyond the seven existing canons, and subsequently reworded, called for engineers to "sustain the world's resources and protect the natural and cultural environment," where the phrase "cultural environment" apparently referred to the built environment. Even the inclusion of this limited notion of sustainable development in the ASCE code generated considerable controversy for nearly two decades (15). Recent attempts to add responsibility for consideration of environmental impact to the code of ethics of the American Society of Mechanical Engineers encountered similar resistance but eventually succeeded (16).

Nevertheless, in 1996, the ASCE adopted changes in the code that made explicit the engineer's commitment to the environment and also introduced the notion of sustainable development to the code. In the new ASCE code, the first Fundamental Principle pledges engineers to "using their knowledge and skill for the enhancement of human welfare and the environment." Although the controversial eighth Canon was not enacted, the first Canon was amended in order to require engineers to "strive to comply with the principles of sustainable development in the performance of their professional duties." In addition, revisions were incorporated in the Guidelines to Practice Under the Fundamental Canons of Ethics under Canons One and Three, which require engineers to adhere to the principles of sustainable development, disclose instances where such principles are not adhered to, and be active in civic affairs and in educating the public in connection with sustainable development.

Nowhere, however, does the new ASCE code endeavor to define the principles of sustainable development. Moreover, the "Guidelines" strongly imply that the principles of sustainable development extend only to protection of the environment. Section 1.e, for example, refers to "protection of the environment through the practice of sustainable development," whereas section 1.f states that "engineers should be committed to improving the environment by adherence to the principles of sustainable development."

Despite progress in incorporating environmental language in certain engineering codes of ethics, as in the case of policy statements by engineers and engineering societies, the codes appear to come up short in their treatment of the social goals of sustainable development.

Background Reading

In his 1996 article "Sustainability and the Design of Knowledge Tools," *Alex Farrell* provides an introduction to the theory of sustainable development and an overview of analytical techniques that might assist engineers and others in developing policies to foster sustainable development. Farrell's goal is to move beyond the often-stated definition of sustainable development requiring concern for the needs of future generations to "operationalize sustainability" so that it might be rigorously analyzed and pursued.

Farrell begins with a critical overview of the conventional approach to economic development, which values everything in monetary terms and ignores the natural environment, thus rendering it an inadequate tool for dealing with sustainability issues. He then moves to a discussion of ecological economics, which was developed in order to integrate concern for the physical environment with economic theory. The author then makes a critical distinction between growth and development, growth representing a quantitative increase in scale and development representing qualitative improvements. Hence, in a world with finite limits, the term *sustainable development* is appropriate even if unlimited growth is not feasible.

Farrell then notes that a sustainable society must have three objectives: economic, ecological, and social. Other factors typically downplayed in conventional economic approaches are the scale of human activities with respect to the environment and the irreplaceability of certain environmental assets, such as the earth's protective ozone layer, known as "critical natural capital." Farrell argues that knowledge tools designed to operationalize sustainable development must account for critical natural capital. Sustainability problems that would benefit from new and better knowledge tools include completeness (i.e., tools not solely focused on economic factors to the neglect of environmental and social factors), measurement of natural capital and more adequate definitions of critical natural capital, increased understanding of the biosphere's carrying capacity, and identification of institutional forms required for attaining the social objectives of sustainability.

Farrell mentions five knowledge tools that show promise. The first two, cost-benefit analysis and national accounting methods, are useful conventional techniques but have limitations that must be acknowledged and corrected if possible (e.g., by including environmental and human welfare measures in national accounting systems). Three new knowledge tools that Farrell finds promising are standards and principles for sustainability, natural capital theory, and sustainability indicators (i.e., indicators used to measure progress toward the ecological, economic, and social objectives of sustainability).

Sharon Beder in her 1994 article critiques "The Role of Technology in Sustainable Development." Beder begins by noting that environmental impact is a function of population, consumption, and technology, and that world leaders concerned about sustainable development have yet to reach agreement on controlling population and consumption. Consequently, the burden falls on technology to achieve reduction in the environmental impact of economic development. Following this premise, policies to promote sustainable development have heretofore focused on changing the impacts of economic growth, rather than limiting it, through the application of technological fixes such as recycling, waste reduction, and pollution control.

Beder, citing the work of Barry Commoner, maintains that piecemeal, end-of-the-pipe technological approaches have not been designed with environmental protection in mind. Newer approaches that seek to develop "clean technologies" offer a promising alternative. However, Beder cautions that previous attempts to redesign technology, such as the appropriate technology movement of the 1970s, have failed to produce major technological changes.

The reasons technological alternatives are not adopted, Beder argues, go beyond simple economic explanations. Other explanatory factors include an engineering culture that favors the status quo, opposition to radical technological change by corporations that have vested interests in the status quo or that are resistant to the accompanying social changes that would be required, government inaction in the face of business lobbies, public opinion favoring private enterprise over government intervention, and failure by the promoters of the alternative technologies to recognize the importance of these social and political factors.

Beder concludes that technological fixes alone are not likely to result in the achievement of sustainable development, arguing that "radical technological innovation" is required. Moreover, even cleaner technologies may be inadequate. In her view, radical changes in values, institutions and cultures will also be necessary, including patterns of population growth and consumption.

Case for Discussion

A case study of sustainable development is presented in the 1992 article by *Steven B. Young* and *Willem H. Vanderburg* entitled "A Materials Life Cycle Framework for Preventive Engineering." Young and Vanderburg define preventive engineering as an approach used "to achieve the best possible compatibility between technology and its contexts." Similar to other preventive approaches to pollution, the materials life cycle approach advocated by the authors considers material flows through the entire life cycle of a product. In response to calls for sustainability and various pressures from government regulations, public opinion, and economic imperatives, industry is being forced to reconsider traditional approaches to environmental issues including materials usage.

The authors are among those who advocate a change from end-of-pipe pollution control to pollution prevention at the source or reduction of the waste stream throughout a product's life cycle. Such approaches not only foster sustainable development, but also eliminate uncertainties in the business climate, are cost-effective, and encourage technological innovation. The preventive materials life cycle approach proposed by the authors distinguishes between the product cycle and materials life cycle, the latter being concerned with the materials streams of all of a product's constituent materials. They describe a simplified model of a generic product as including four stages—resource extraction, manufacturing, use, and disposal—each of which relies on energy, material, and transportation, and has feedback loops between the various stages.

Among the process pollution prevention strategies identified by the authors are changes in technology, improvements in operating procedures, recycling within the plant, changes in input materials, and redesign of end products. The last category, they contend, is the most novel and requires the most fundamental change. Noting that products are generally designed in the reverse order of their materials flow (i.e., from the product design working back to the constituent materials), the authors advocate reduction and recycling as two approaches to prevention. Reduction implies reducing toxic materials in the production process, conserving energy and materials, and so on. These environmental criteria can be linked to the design process through such approaches as design-for-quality. Young and Vanderburg's proposed design for recycling strategy includes four steps: selection of recyclable materials, avoidance of complex designs, design for convenient disassembly, and component labeling. In conclusion, the authors caution that current methods of life cycle cost assessment have a number of limitations that require critical analysis before using them as evaluative tools for the preventive approach they propose.

References

[1] World Commission on Environment and Development. 1987. *Our common future* (The Brundtland Report). Oxford: Oxford University Press.

[2] French, H. F. 1995. Forging a new global partnership. In *State of the world 1995*, ed. L. R. Brown et al., 170–189. New York: W. W. Norton.

[3] President's Council on Sustainable Development, World Wide Web site: http://www.whitehouse.gov/PCSD/.

[4] Carroll, W. J. 1993. World engineering partnership for sustainable development. *Journal of Professional Issues in Engineering Education and Practice* 119: 238–240.

[5] Grant, A. A. 1995. The ethics of sustainability: An engineering perspective. *Renewable Resources Journal* (Spring): 23–25.

[6] Baetz, B. W., and Korol, R. M. 1995. Evaluating technical alternatives on basis of sustainability. *Journal of Professional Issues in Engineering Education and Practice* 121: 102–107.
[7] IEEE TAB Environment Health and Safety Committee. 1995. *White paper on sustainable development and industrial ecology* (Available from World Wide Web site, http://www.ieee.org/tab/ehswp.html.)
[8] Holmberg, J., and Sandbrook, R. 1992. Sustainable development: what is to be done? In *Making development sustainable.* ed. J. Holmberg, 19–38. Washington, D.C.: Island Press.
[9] AAES. 1994. Statement of the American Association of Engineering Societies on the Role of the Engineer in Sustainable Development. In *The Role of Engineering in Sustainable Development*, Washington, D.C.: AAES.
[10] Hatch, H. J. 1993. Relevant engineering in 21st century. *Journal of Professional Issues in Engineering Education and Practice* 119: 216–219.
[11] Prendergast, J. 1993. Engineering sustainable development. *Civil Engineering* (October): 39–42.
[12] Harris, C. E., Jr., Pritchard, M. S., and Rabins, M. J. 1995. *Engineering ethics.* Belmont, Calif.: Wadsworth Publishing Company.
[13] Engel, J. R., and Engel, J. G., eds. 1990. *Ethics of environment and development.* Tucson: University of Arizona Press.
[14] Kothari, R. 1990. Environment, technology and ethics. In *Ethics of environment and development*, 27–35. Tucson: University of Arizona Press.
[15] Vesilind, P. A. 1995. Evolution of the American Society of Civil Engineers Code of Ethics. *Journal of Professional Issues in Engineering Education and Practice* 121: 4–10.
[16] Rabins, M. J. 1997–1998. Personal correspondence.

Sustainability and the Design of Knowledge Tools

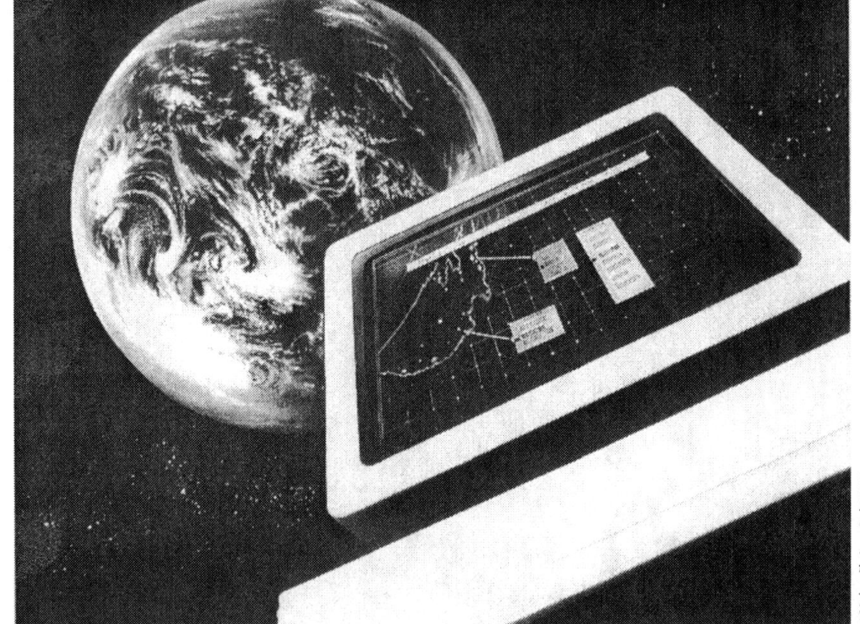
©J-F Podevin/The Image Bank/1999

The concept of sustainable development has gained widespread support over the last decade, with many organizations adopting it as an important goal. The motivation behind people's interest in sustainable development is the belief that current human activities may degrade the environment and cause serious negative consequences for human populations [1]. These concerns are especially understandable when we consider that the current practices of industrialized countries will in the future be used to support an ever-growing population.

Sustainable development as a concept is, however, vague; supporters disagree as to what it means, while critics deny that it has any meaning at all [2],[3]. A general definition is this: Sustainable development requires that "the needs of the present generation are met without compromising the ability of future generations to meet their own needs" [1]. While appealing, and even noble, this definition is imprecise — leaving the concept of sustainable development like an empty bottle into which one may pour the wine of one's own views.

What needs to be present in any useful discussion of sustainability is a rigorous application of both relevant theory and technological expertise. Theory and technological expertise are examples of knowledge tools — implements which can be used to organize and give meaning to

Alex Farrell is with the Wharton Risk Management and Decision Processes Center, University of Pennsylvania, 1323 Steinberg-Dietrich Hall, Philadelphia, PA 19104-6366. Email: <farrell@dolphin.upenn.edu>. *This article is based on a paper presented at the Conference "Knowledge Tools for a Sustainable Civilization," Toronto, June 1995.*

Reprinted from *IEEE Technology and Society Magazine,* Vol. 15, No. 4, pp. 11–20, Winter 1996/1997.

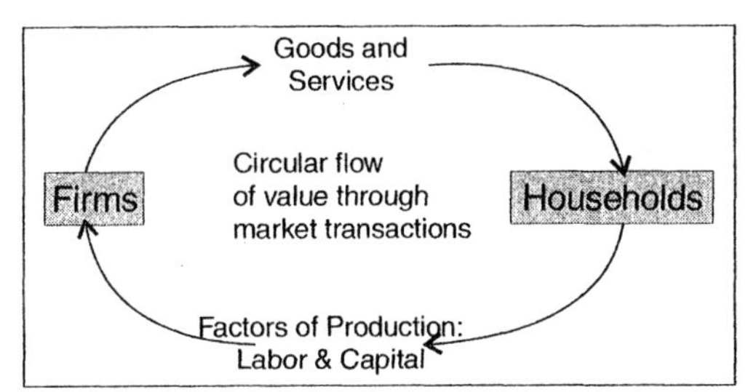

Fig. 1. The worldview of neoclassical economics.

data, create and analyze concepts, and synthesize the two (data and concepts) to aid in decision making. This process can be called "operationalizing" sustainability. In this article, I hope to offer some insights which may help in the design of knowledge tools for a sustainable civilization.

Worldviews Old and New

The theory most relevant to the concept of sustainable development is found in the emerging field of ecological economics. Ecological economics incorporates many neo-classical ideas, so it is appropriate to start with the conventional economic worldview. The foundation of neo-classical economics is an isolated, circular exchange of value between firms and households, as shown in Fig. 1 [4]. The physical environment does not enter into this worldview at all, even though issues such as resource use and pollution control are major topics in economics and a large literature on environmental economics exists [5]. Many of the issues in environmental economics have to do with "externalities," indicative of how removed the environment is from conventional economic thinking.

Economists have used this worldview to consider sustainability at some length, although significant difficulties in doing so have been noted by many authors [1],[6]. Particularly relevant here are the difficulties of the neo-classical economic approach apparent in comments by Solow, who notes that the central concept of sustainability is the preservation of the capacity to produce economic well-being [7]. Solow claims that productive capacity includes nonrenewable resources, the stock of plant and equipment, inventory of technical knowledge, and the general level of education. He feels that Net National Product (NNP), adjusted for environmental degradation, is the appropriate measure of sustainability.[1] NNP is a standard knowledge tool developed through neo-classical economics, and its chief problem, shared with many other knowledge tools, is that it uses a single unit of measurement: dollars. Given the worldview in Fig. 1 this makes some sense; since goods, services, labor, and capital are all exchanged in markets and can generally substitute for one another, total transactions are measured in a single unit. Solow, however, acknowledges that there is some difficulty in applying such a knowledge tool to environmental issues:

"The claim that a feature of the environment is irreplaceable, that is, not open to substitution by something equivalent but different, can be contested in any particular case, but no doubt is sometimes true. The calculus of trade-off does not apply. . . we are going to have to keep depending on physical and other special indicators in order to judge the economy's performance with regard to the use of environmental resources", [7, pp. 171- 172].

This statement, though correct in recognizing the inadequacy of conventional economic knowledge tools when it comes to the use of environmental resources, fails to provide guidance on what "special indicators" are needed. This may be because such indicators are outside of the worldview of conventional economics. To solve this problem, one may look at the worldview which underlies the field of ecological economics.

The probing research on sustainability in the ecological economics literature has its roots in the work of Nicholas Georgescu-Roegen and Kenneth Boulding [8], [9]. Georgescu-Roegen first rigorously applied thermodynamic concepts, entropy in particular, to economic systems, and Boulding carried these ideas forward with his notion of a "spaceship economy," in which the earth is the spaceship. Thus in the field of ecological economics the economy has *physical* dimensions which are essential to its nature, in addition to its role as a mechanism for exchange between firms and households. The human economy is embedded in, and relies upon, the biosphere for energy, matter, and ecological life-support services, as seen in Fig. 2. The biosphere is made up of many overlapping ecosystems, each of which contains multiple biological and nonliving systems, which creates a complex web of simultaneous interactions at many scales with many different dynamics. In this view, the human economy is an open subsystem of the biosphere, which itself is materially closed but open to energy transfers, as indicated. The interaction of the two creates a complex nonlinear system with many types of feedback, and which probably has several different equilibrium points.

In Fig. 2 the interactions between the biosphere and the human economy are shown as M, E, and L, representing matter, energy, and eco-

[1] NNP is measured in dollars and is equal to Gross Domestic Product less the depreciation of manufactured capital.

logical life-support services, respectively. These interactions can be classified into three different functions, which are at minimum valuable, and may be irreplaceable. First, natural resources provide the actual matter and energy used in production; these are the *M* and *E* inputs on the left and can be called ecological goods. Second, the biosphere serves as a sink to absorb and recycle (i.e., assimilate) many waste streams that current practices create; these are the *M* and *E* outputs of the human economy on the right and are a type of ecological service. Third, the biosphere provides a large set of generalized life-support services, represented by *L* [10]. Human-made capital requires these inputs to be produced and operated: the human economy also has people in it, who need a number of biologic inputs and services. Thus diverse inputs, including iron ore, good soil quality, energy supplies, clean air, and protection from harmful ultraviolet radiation (a service of the ozone layer), are needed. It is unlikely that there exists a single-dimension measurement that adequately captures the nature of these inputs.[2] As Solow indicates, "physical and other special indicators" will be needed, and one of the objectives of ecological economics is to develop knowledge tools to understand and use such special indicators.

With this worldview as background, we can define some basic terms used to discuss sustainability. These terms are in the common vocabulary, but they are used here in a more technical way.

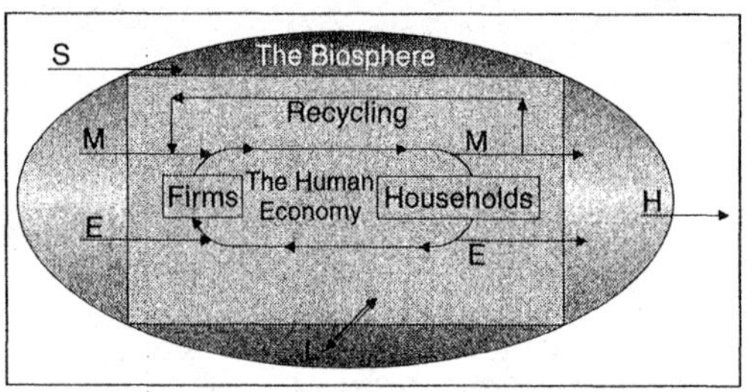

Fig. 2. The worldview of ecological economics. S-solar energy, M-matter, E-energy, H-heat, L-ecological life-support systems. Area within rectangle represents human-made capital. Area within oval represents natural capital. (Adapted from [4].)

▾ Growth and Development

The terms "growth" and "development" are sources of great confusion, but Herman Daly has drawn a clear distinction between the two:

"To *grow* means to 'increase naturally in size by the addition of material through assimilation or accretion.' To *develop* means 'to expand or realize the potentialities of; bring gradually to a fuller, greater, or better state.' In short, growth is quantitative increase in physical scale, while development is qualitative improvement or unfolding of potentialities. An economy can grow without developing, or develop without growing, or do both, or neither. Since the human economy is a subsystem of a finite global ecosystem which does not grow, even though it does develop, it is clear that growth of the economy cannot be sustainable over long periods of time. The term sustainable growth is rejected as a bad oxymoron. The term sustainable development is much more apt" [12].

Human aspirations clearly go beyond economic growth, and many observers feel that issues such as literacy, aesthetics, fulfillment in the workplace, and political freedom are important components of development (see Table II) [13]-[16]. A similar idea has been advocated by the United Nations, with added emphasis on the fact that sustainability is grounded in concerns about both intra-generational equity: "It would be odd if we were deeply concerned for the well-being of future generations while ignoring the plight of the poor today" [17].

▾ Sustainability and Sustainable Development

The word "sustain" means "to support, hold, or bear up from below; to keep up or keep going; to supply with the necessities of life" [18]. It is particularly important in this context to understand what is being supported (human life) and what is doing the supporting (the biosphere and a host of human institutions). At the core of concerns about sustainability is the idea that this support for human life — ecological goods and services *and* the products of industry *and* the benefits of a good society — must be provided continuously into the future.[3]

Focusing on the environment, examples of ecological goods include lumber, fish, and petroleum, which are available in limited supply, even if they are renewable. Assimilative capacity includes the ability of the atmosphere to absorb industrial pollutants like sulfur dioxide, and the ability of water supplies to clean up household sewage. The assimilative capacity of

[2]All single-dimensional measurement systems are susceptible to this criticism, whether the unit of measure is utility-based (dollars) or a thermodynamic value such as entropy, an idea which seems in line with Georgescu-Roegen's later work [11]. However, measurements such as entropy have their basis in observable physical processes, not on human preferences, and thus have provided very valuable insights into sustainability.

[3]This is an especially difficult issue where nonrenewable goods are concerned, and has received considerable attention from natural resource economists. One of the principal ideas to emerge is that some of the profits created through the use of nonrenewable resources must be reinvested in replacements [19].

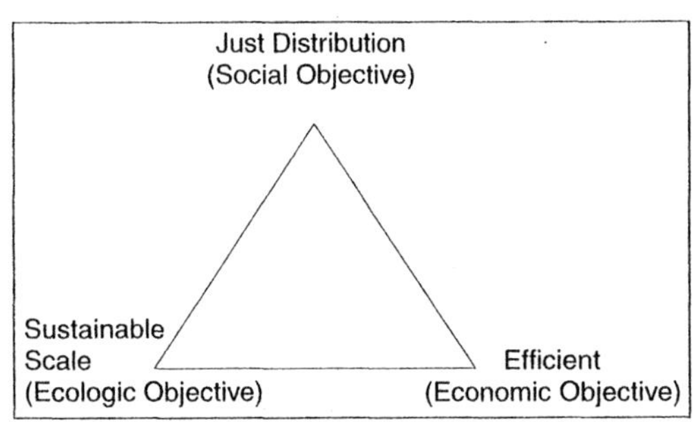

Fig. 3. The objectives of a sustainable society. (Adapted from [10].)

the biosphere is limited, and instances of its overloading (i.e., excess pollution) are among the most widely acknowledged of environmental problems. Ecological life-support services include ultraviolet radiation noted earlier, clean air and water, and so forth. Conflict between these three functions often exists; utilizing too much of one function (overlogging, for instance) diminishes the capacity of the biosphere to provide the others (soil erosion increases, water and air quality decrease, and fish catch decreases). This has occurred in both North America, where the effects are relatively minor, and in Bangladesh, with more serious consequences. This points toward a maximum level at which ecological goods and services can be provided, called the carrying capacity of the biosphere [20]-[22].

If something is to be "sustainable" with regards to human life, all three supporting functions of the biosphere need to be provided indefinitely; or put another way, human demands on the biosphere must remain within carrying capacity. Thus, sustainability is a state in which the relationship between human activities and larger, dynamic ecological systems allows for human potential to be realized to the maximum extent possible; but in which the effects of human activities remain within bounds, so as not to destroy ecological systems and thus lose their supporting functions [23]. Sustainable development is the process which tends to make human economic systems, social systems, and institutional arrangements achieve sustainability. The appropriate knowledge tools can help us identify those actions which are part of the process of sustainable development.

The social, economic, and ecological objectives of a sustainable society are often described with the use of the triangle shown in Fig. 3. Brief definitions of these objectives are given in Table I [10],[24].

▾ Scale

Scale refers to the volume and form of human-commanded ecological goods and services. Ecological economists note that the size of the human economy has grown large in comparison to the environment, as indicated in Fig. 2. There is no outer boundary to the human economy in the neo-classical worldview, so economics considers scale issues secondarily or ignores them, and thus does not provide a framework capable of answering some of the fundamental questions that arise in considering sustainability.

▾ Natural and Human-Made Capital

Ecological economists distinguish between "natural capital," "manufactured capital," and "human capital" (the latter two together forming the human-made capital in Fig. 2) [25].[4] Manufactured capital is the produced means of production — assets generated through economic activity. Natural capital is made up of renewable resources and nonrenewable resources — assets found in nature which provide ecological goods and services. Human capital refers to factors that 1) provide human societies with the skills and adaptations to utilize and modify the environment, 2) allow for the orderly control and development of the human economy, and 3) help achieve goals that the environment and the economy do not offer, such as political rights, a system of values, and so forth. The many institutions in society (such as laws and norms, levels of education, forms of governance, etc.) form a key part of human capital, and changing current institutions is perhaps the most critical and difficult problem in implementing sustainable development [26]-[28]. Clearly, knowledge tools are part of human-made capital.

▾ Limits to Substitution

An important insight in ecological economics is that *the many components of natural and human-made capital cannot always be substituted for one another perfectly, and are often complements.* While not an entirely new idea, ecological economists tend to focus on this issue rather than treat it as an unusual exception. There are many cases where natural and man-made capital can substitute for one another, but to base decision making on a framework where all forms of capital (and labor as well) are assumed to be perfectly substitutable seems imprudent. A key issue is to identify when substitution is possible, and when it is not. Examples of natural capital for which there are no substitutes include the ozone layer, productive soil, and biodiversity. This idea of non-substitutability has been captured most succinctly in the concept of "critical natural capital" which are environmental assets from which flow ecological services that

[4]Some researchers add a fourth type of capital in order to differentiate between human capital and cultural (also known as ethical or moral) capital.

Current national accounting methods exclude the value of leisure time, non-market environmental services, and ecological assets.

are indispensable for continued human existence and for which no replacement is feasible [4].

Consider the ozone layer, a feature of the earth which predates humankind and provides a service which no available technology can provide: protection from harmful ultraviolet radiation. Thus the ozone layer is a natural capital asset which displays the characteristic of non-substitutability. Some industrial chemicals (particulay chlobons, plete, or thin, the ozone layer when they are released [29]. Thinning of the ozone layer is believed to lead to increased skin cancers, eye cataracts, and human immune system suppression, among other effects [30]. If the results of ozone depletion were limited to just these adverse health and economic effects, using CFCs might be acceptable to society if the benefits of doing so were large enough. Here Solow's "calculus of trade-offs" *does* apply, because society might be willing to accept gains in one area along with losses in another.

However, further ozone layer depletion could seriously upset some ecosystems and have calamitous results for society. At this point, the critical nature of the ozone layer becomes apparent — the consequences of a significantly degraded ozone layer are difficult to estimate, but they could well make current consumption and population levels impossible to continue. In this case, the idea of trade-offs seems inappropriate because the potential outcome of continued ozone layer depletion is not of the same magnitude or type as the potential gains of using CFCs. Thus the ozone layer is a critical natural capital asset.

It is difficult to see how one might use neo-classical economics to think about this problem. Knowledge tools which use money as the single unit of measurement would end up asking questions like, "What is the value of the ozone layer?" It is hard to imagine what would be a meaningful answer. Similarly, it's not clear how a knowledge tool based on thermodynamics would be applied to this example. Thus the need for knowledge tools which are compatible with critical natural capital becomes apparent if sustainability is to be operationalized.

The Design of Knowledge Tools

The ideas being explored in ecological economics are far from complete, but the preceding section suggests a few of the necessary characteristics of knowledge, and knowledge tools, that a sustainable society needs.

▾ *Completeness*

The essential nature of human activities as embedded in and dependent on the biosphere (which forms a complex nonlinear system with many forms of feedback) suggests that a wide range of knowledge, and powerful knowledge tools, are required. Efforts at understanding sustainability which are too simple or incomplete may be useful, but are not likely to be sufficient to operationalize sustainability.

Neo-classical economics contains a powerful set of knowledge tools which have provided invaluable insights in the past and will continue to be used widely in a sustainable society. Proposals for implementing sustainable development that do not use markets to meet material needs as much as possible are likely to fail, because markets are by far the best method to meet the economic objective of a sustainable civilization — efficient allocation. However,

Table I
Definition of the Objectives of Sustainable Development

Objective	Definition
Efficient Allocation	A division of resource flows among alternative uses that conforms with consumer preferences weighted by ability to pay.
Just Distribution	A division of resource flows among different people, including future generations, that is fair or limited in inequality.
Sustainable Scale	A physical volume and form of energy and material flowing through the environment that does not erode carrying capacity.

adapted from [4].

Table II
Sustainability Indicator Sets (not all indicators shown)

Name	Ecological	Economic	Social
Altieri and Masera [59]	crops displaced, erosion, human health effects	capital costs, cash flow, skill level	acceptance, ownership issues, participation
Bergstrom [60]	NO_x, green space, garbage, water quality	cost/benefit ratios for environmental criteria	beauty
Chi [61]	air/water/soil quality, deforestation, waste	basic needs	equality, rights, health, education, population
Hill [62]	forest stock, erosion, biodiversity	project NPV	inter-generational equity, empowerment, variance of effort
International Institute for Sustainable Development [72]	BOD, air emissions, energy used, environmental accidents		compliance, environmental reporting
Keeney and McDaniels [63]	environmental impact	capital and energy cost, regional economic development	industry control of technology choice
Meier and Munasinghe [64]	CO_2, acid rain, thermal plume, biodiversity, soil permeability, exposure to air pollution	electricity demand, capital and operating costs	local employment, electrification
Minnesota Milestones [65]	wildlife populations, soil productivity, emissions	farm profitability, full-cost prices, cost-effective regulation	cultural diversity, viable rural communities, cluster development
Lake Superior Binational Program [66]	wildlife and habitat measures, human health	human health and welfare	discussion of values in public forums, land form
Shell Canada, Ltd. [67]	solid waste, energy use, air and water emissions, water use		legal compliance, reporting, policy development
Sustainable Seattle (100 indicators) [68]	wild salmon, air, energy, and water use	affordable housing, economic security, children in poverty	adult literacy, voting participation, population

definitions of sustainability which are little more than a simple notion of cost-effective pollution control are likely to prove inadequate because they do not address ecological goods and services very well. Further, many concerns about sustainability are ethical and political in nature, indicating that narrow economic or technological perspectives are insufficient.

▾ *Measuring Natural Capital and Defining Critical Natural Capital*

One area in which knowledge tools are critically needed is the measurement of natural capital, which presents both conceptual and practical difficulties. First, it is not clear what kind of knowledge is created by aggregating together the dollar values of very different environmental assets into a single index [31]. For instance, if a nation's forests and fisheries were combined into a natural capital account, what would a constant level of that natural capital account represent? The same amount of forested land and of fish? More of one and less of the other? Less of each, but higher prices for both of them?

Second, because the unit of measurement itself (dollars) is tied to the valuation of capital goods, the problem of measuring the *quantity* of capital independently of its *price* arises. That is, if the price of an ecological good (say, lumber) goes up, the value of the assets which provide that resource (forests) goes up, even though the size of the assets (number of trees) is unchanged. Even determining the *size* of many natural assets, such as biodiversity level or fisheries stocks, is difficult [32],[33]. Moreover, for many nonrenewable resources, the size of the stock (measured in mass or volume) depends fundamentally on price. For instance, current prices are needed to define the *size* of economically recoverable petroleum reserves; as oil prices increase, size of reserves, measured in barrels, also increases because it becomes economically feasible to produce from marginal fields. (Thus, the *value* of petroleum reserves goes up for two reasons: there are more barrels, and each barrel is worth more.) Lastly, current oil prices drive exploration efforts, which helps determine discoveries and thus future proven oil reserves.

The concept of limited substitutability between certain capital assets is among the primary interests of ecological economics, and underlies the idea of critical natural capital. However, there are many theoretical and practical challenges in applying the concept of nonsubstitutability which call for new, and better, knowledge tools. Probably the most important is defining the range in which nonsubstitution applies because it may be important only in a limited number of cases. If critical natural capital assets are maintained sufficiently well, it may be easier to devise and use knowledge tools relating to them. Thus, among the most important areas of research associated with critical natural capital assets are in determining *what* they are and *at what levels* they become critical.

▾ *Carrying Capacity*

Our understanding of how the biosphere functions is very incomplete, as demonstrated by environmental surprises such as the ozone hole, and recognized uncertainties about anthropogenic greenhouse effects. The increase in the scale of human activities has made us *de facto* managers of the environment, even though it is not clear how the environment operates, nor what the appropriate management objectives for the environment are. The concept of carrying capacity has been suggested as one way to develop appropriate measures of natural capital, but further research in this area is needed [34]. This is an area in which new knowledge tools are needed in order to better describe the productivity, organization, and resilience of the biosphere to ensure that it can continue to support human life [35],[36].

A related concern is that carrying capacity is frequently politicized in public debate. However, carrying capacity is partially scientific in nature; the ecosystems of the earth obey certain physical laws and can withstand perturbations of limited size. It is only our responses to environmental limitations that are political. Knowledge tools that can define carrying capacity better, measure it, and use the concept in a scientific manner as much as possible are needed. This would probably mean defining upper limits of the flows of ecological goods and services that are available over time. Achieving those flows, and allocating them to different groups in society, are the political problems, and are different than determining what those levels are. Other important issues related to carrying capacity include uncertainty and irreversibilities [37]. Finally, because human activities cannot be completely separated from a highly complex biosphere, knowledge tools needed for a sustainable civilization will need to be powerful and applied with ingenuity. The recent application of signal-processing techniques to analyze global climate change is an excellent example of this [38].

▾ *Institutional Capabilities*

A significant difficulty with much of the ecological economics approach is that a wide set of social concerns is reduced to the simple "equity in the division of resource flows between people" (from Table I). This understanding may be appropriate for many decisions, such as how to spread the costs of environmental programs through society. However, this more narrow view seems insufficient for many questions of a social or political nature about sustainability [39].

The nature of the objectives of a sustainable society leads to a fundamental question that is not yet well answered by the ecological economists: What sorts of social structures will be able to attain the non-economic objectives of a sustainable society? A pressing concern is that the breakdown of existing institutions, including nation-states, may be driven by environmental degradation, and that this breakdown has positive feedback (i.e., it will accelerate degradation at the same time). Equally important is the question: How can technology be used to slow and reverse these problems? Of course, this is a central concern of the engineering profession, but some important insights may come from induced institutional innovation theory, which has been developed by agricultural economists and economic historians [40],[41]. Achieving sustainability will require changes in political structures, business practices, accepted ethics, and many other areas, which suggests that it is necessary to look beyond the field of economics to fully appreciate the concept [27],[42],[43]. Knowledge tools which link cultural and political factors to the ecological and economic objectives of sustainability are especially needed.

Examples of Knowledge Tools

A wide variety of knowledge tools has been developed for structuring and evaluating complex decisions [44]. Below, two traditional knowledge tools are briefly evaluated in light of the necessary conditions given above, and three new ones are mentioned.

▾ *Cost/Benefit Analysis*

Cost/benefit analysis (CBA) calls for comparing the dollar value of the costs of a policy or project with the dollar value of the benefits. The technique is one of the most common analytic tools in the policy analysis suite, and many volumes have been written on it. This includes a vigorous debate on the applicability of the technique to environmental issues and sustainability [13],[45]. CBA is often faulted for relying on the monetary valuation of goods which are not sold on any market and thus for which prices are not available. Nonetheless, the technique can indicate the magnitude of the costs and benefits of environmental regulation, and the use of these data could greatly improve policy making, especially in developing countries [46],[47]. Of course, CBA uses a single unit of measurement and does not meet the completeness criteria, and is subject to criticism for those reasons as well. It is important to continue to use CBA, but to respect its limitations and acknowledge the uncertainties inherent in the technique [10],[48].[5]

▾ *National Accounting Methods*

National accounting produces measurements such as GDP, NNP, capital flows, and trade balances which denote the size, structure, and performance of whole economies. They can be adapted to help understand sustainability in policy analysis and are beginning to be applied in this way [33],[49],[50].

There are two main themes in this area. First, current national accounting methods do not accurately represent the economy of a nation because many important issues such as the value of leisure time, non-market environmental services, and ecological assets are excluded [51]. The United Nations has published initial guidance on how environmental accounting can be used with current methods [32].[6] The second main theme is that measuring the size of the economy accurately is not the real issue. What is truly important is increasing human welfare (i.e., development), for which total economic activity is a poor proxy. For instance, Daly argues that expenditures which are "defensive" in nature and have little relation to human development should not be counted in any measurement of development [14]. In this regard, a number of alternate indices have been suggested (two of the most recent being the Human Development Index and the Genuine Progress Index) but none are universally accepted [17],[52]-[54].

▾ *Standards and Principles*

Guiding standards and principles are important policy tools which have been suggested for use in attaining sustainablilty. They include concepts such as "no net biodiversity loss," or the stabilization of greenhouse gas emissions, and are sometimes called "safe minimum standards" [25].

Perhaps the most important of these standards is the "Precautionary Principle," which requires that when the ecological outcomes are uncertain, extra weight is given to avoiding possible negative outcomes. As noted above, many environmental issues important to sustainability contain significant uncertainty, so this conservative approach would make it less likely that a given project would be undertaken. The precautionary principle is distinguished from other environmental policy tools in that it is oriented forward in time towards unknown (and potentially unforseeable) environmental problems. Other techniques usually acknowledge only the subset of environmental problems which are known from past experience and which can be evaluated quantitatively. This forward orientation is in line with the motivation for considering sustainable development in the first place.

Costanza has suggested that a precautionary polluter-pays principle (4P) could be one of the fundamental building blocks of a sustainable public policy [55]. Add a rule for dealing with natural resource depletion, and another on international trade, and Costanza feels that the resulting policy set would have a high likelihood of assuring sustainability, suggesting the potential power of standards and principles.

▾ *Natural Capital Theory*

Natural capital theory is a formal approach to the concept mentioned above and proposes to model ecological assets (natural capital) which provide ecological goods or flows of ecological services that may be irreplaceable. This approach is essentially a re-extension of the concept of capital beyond the usual two-factor production function which has been developed in neoclassical economics [56],[57]. Natural capital theory has been best described by Victor [31] and Turner [25]. It has been used to test for sustainability in at least one instance, where several industrialized nations and one developing country (Costa Rica)

[5]Although deep ecologists and some others feel that CBA must be abandoned entirely, this is not the case among most ecological economists.

[6]These methods rely upon the ability to price nonmarket goods, and thus share some problems with CBA.

passed a test of "weak" sustainability, while several other developing nations failed [58]. It may be possible to adapt other traditional economic methods of public policy analysis to sustainable development using natural capital methods, if they are developed further. However, natural capital theory uses relatively simple production functions which have a single unit of measure, and so may be limited in its applicability.

▾ *Sustainability Indicators*

Interestingly, there is a fairly substantial number of organizations that are creating "special indicators" to describe sustainability, in the spirit of Solow's comments above. Sustainability indicators (which come in sets) are an attempt to define the concept by enumerating and measuring the attributes that a sustainable society, a sustainable community, or a sustainable energy system would have. A sustainability indicator set is taken to be a complete representation of the concept of sustainable development. A partial listing of a few of the existing sustainability indicator sets is given in Table II. Sustainability indicators can be used to operationalize (i.e., quantify) the concept. The indicators for each set in Table I have been separated into three categories (ecological, economic, and social) by the author, corresponding to the triangle in Fig. 3. Many of the indicator sets are divided in this way by their creators, and most of the indicators can be placed in one category without making many unjustifiable distinctions, thus increasing their structure and comprehensibility. Sustainability indicators are a promising, but still relatively crude form of knowledge tool, suggesting a research program with two major tasks — identifying appropriate indicators and then measuring them. It may be possible to resolve the problems of knowledge tools based on single units of measurement with the use of sustainability indicators.

Most sustainability indicator sets are used to establish numerical goals and analyze trends, much like economic indicators are used. Elsewhere, I have argued that it is possible to extend the use of sustainability indicator sets with multi-criteria decision processes to provide significant insight into the nature of sustainable development and to create a knowledge tool for public participation in the policy process [69]. Many observers have called for such efforts, including the President's Council on Sustainable Development [70]. A previous article in this magazine describes one methodology for doing so and provides a simple expository example [71].

Crucial Role for Engineers

Creating a sustainable civilization is one of the primary challenges facing humanity. Motivated by concern for resource depletion, anthropogenic alterations to natural systems upon which we depend, improvements in the quality of life in poorer countries today, and especially the fate of future generations, sustainability requires the careful and critical design of knowledge tools. These efforts must be guided by well founded theory, some of which can be provided by ecological economics, and some of which will come from scientists and engineers. However, efforts to provide a conceptual foundation for the concept of sustainability are far from complete, and this may be the most pressing need in the development of knowledge tools for a sustainable civilization. Working with both ecologists and economists, the engineering profession will play a crucial role in the development of such knowledge tools in order to better understand the complex nonlinear system which we call the biosphere, and in which we live.

Acknowledgment

The author thanks Clare Narrod, Roger Raufer, James Winebrake, and three anonymous reviewers for their invaluable comments. All views and remaining errors are strictly the fault of the author alone.

References

[1] The World Commission on Environment and Development, *Our Common Future (The Brundtland Report)*. New York: Oxford University Press, 1987.

[2] J. Pezzy, *Sustainable Development Concepts: An Economic Analysis*. Washington, DC: The World Bank, 1992.

[3] A. Heyes and C. Liston-Heyes, "Sustainable resource use: the search for meaning," *Energy Policy*, vol. 23, pp. 1-3, 1995.

[4] H. Daly, "Operationalizing Sustainable Development by Investing in Natural Capital," in *Investing in Natural Capital*, A. Jansson, M. Hammer, C. Folke, and R. Costanza, Eds. Washington, DC: Island Press, 1994, pp. 505.

[5] W. Oates, *The Economics of the Environment*. Cambridge: Cambridge University Press, 1992, pp. 608.

[6] M. Toman, J. Pezzy, and J. Krautkraemer, "Neoclassical Economic Growth Theory and "Sustainability," in *The Handbook of Environmental Economics, Blackwell Handbooks in Economics*, D. Bromley, Ed. Cambridge, MA: Blackwell, 1995, pp. 139-165.

[7] R. Solow, "An Almost Practical Step Towards Sustainability," *Resources Policy*, vol. 19, pp. 162-172, 1993.

[8] N. Georgescu-Roegen, *The Entropy Law and the Economic Process*. Cambridge, MA: Harvard University Press, 1971.

[9] K. Boulding, "The Economics of the coming spaceship Earth," in *Environmental Quality in a Growing Economy*, H. Jarrett, Ed. Baltimore, MD: Johns Hopkins University Press, 1966, pp. 3-14.

[10] M. Munasinghe, "Environmental Economics and Sustainable Development." The World Bank, 1993.

[11] N. Georgescu-Roegen, "The Entropy Law and the Economic Process in Retrospect," *Eastern Economic Journal*, vol. 12, pp. 3-25, 1986.

[12] H. Daly, "Toward some Operational Principles of Sustainable Development," *Ecological Economics*, vol. 2, pp. 1-6, 1990.

[13] M. Sagoff, *The Economy of the Earth*. Cambridge: Cambridge University Press, 1988.

[14] H. Daly and J. Cobb, *For the Common Good*. Boston, MA: Beacon Press, 1993.

[15] World Wildlife Fund, "Sustainable Use of Natural Resources," World Wildlife Fund, Gland, Switzerland, Aug. 1993.
[16] I. Serageldin, "Making Development Sustainable," in *Finance & Development*, Dec. 1993, pp. 6-10.
[17] U.N. Development Programme, *Human Development Report*. New York, NY: Cambridge University Press, 1994.
[18] American Heritage Dictionary of the English Language. New York, NY: Houghton Mifflin Company, 1992.
[19] J. Hartwick, "Intergenerational Equity and the Investing of Rents from Exhaustible Resources," *American Economic Review*, vol. 67, pp. 972-974, 1977.
[20] P. Vitousek, P. Erlich, and A. Erlich, "Human Appropriation of the Products of Photosynthesis," *BioScience*, vol. 36, pp. 368-373, 1986.
[21] J. Cohen, *How Many People Can the Earth Support?* New York, NY: W. W. Norton & Company, 1995.
[22] S. Postel, G. Daily, and P. Erlich, "Human Appropriation of Renewable Fresh Water," *Science*, vol. 271, pp. 785-788, 1996.
[23] R. Costanza, H. Daly, and J. Bartholomew, "Goals, Agenda and Policy Recommendations for Ecological Economics," in *Ecological Economics*, R. Costanza, Ed. New York, NY: Columbia University Press, 1991, pp. 1-21.
[24] H. Daly, "Allocation, distribution, and scale: towards an economics that is efficient, just, and sustainable," *Ecological Economics*, vol. 6, pp. 185-193, 1992.
[25] K. Turner, "Sustainability: Principles and Practise," in *Sustainable Environmental Management: Principles and Practice*, vol. 2, K. Turner, Ed. New York, NY: Belhaven Press, 1993.
[26] F. Berkes and C. Folke, "Investing in Cultural Capital for Sustainable Use of Natural Capital," in *Investing in Natural Capital*, A. Jansson, M. Hammer, C. Folke, and R. Costanza, Eds. Washington, DC: Island Press, 1994.
[27] R. Devon, "Sustainable Technology and the Social System," *IEEE Tech. and Soc. Mag.*, vol. 10, pp. 9-13, 1991.
[28] D. Pearce, R. K. Turner, T. O'Riordan, N. Adger, G. Atkinson, I. Brisson, K. Brown, R. Dubourg, S. Fankhauser, A. Jordan, D. Maddison, D. Moran, and J. Powell, *Blueprint 3: Measuring Sustainable Development*. London, UK: Earthscan, 1993.
[29] S. Rowland, "Chlorofluorocarbons and the Depletion of Stratospheric Ozone," *American Scientist*, vol. 77, pp. 36-45, 1989.
[30] National Academy of Sciences, *Ozone Depletion, Greenhouse Gases, and Climate Change*. Washington, DC: National Academy Press, 1989.
[31] P. Victor, "Indicators of sustainable development: some lessons from capital theory," *Ecological Economics*, vol. 4, pp. 191-213, 1991.
[32] U.N. Department for Economic and Social Information and Policy Analysis, *Integrated Environmental and Economic Accounting: Interim Version*, vol. F-61, Handbook of National Accounting ed. New York, NY: United Nations, 1993.
[33] R. Repetto, "Earth in the Balance Sheet," *Environment*, vol. 34, pp. 12-20, 43-44, 1992.
[34] J. Cohen, "Population growth and earth's human carrying capacity," *Science*, vol. 269, pp. 341-346, 1995.
[35] R. Costanza, "Toward an Operational Definition of Ecosystem Health," in *Ecosystem Health*, B. Haskell, B. Norton, and R. Costanza, Eds. Washington, DC: Island Press, 1992, pp. 239 - 253.
[36] M. Munasinghe and W. Shearer, "Defining and Measuring Sustainability: The Biogeophysical Foundations," H. G. d. Souza and I. Serageldin, Eds. Washington, DC: The World Bank, 1995, pp. 440.
[37] D. Ludwig, R. Hilborn, and C. Winters, "Uncertainty, resource exploitation, and conservation: lessons from history," *Science*, vol. 260, pp. 17-36, 1993.
[38] D. Thomson, "The seasons, global temperature, and precession," *Science*, vol. 268, pp. 59-77, 1995.
[39] T. Homer-Dixon, "Environmental scarcities and violent conflict: Evidence from cases," *International Security*, vol. 19, pp. 5-40, 1995.
[40] C. Runge, "Induced agricultural innovation and environmental quality: the case of groundwater regulation," *Land Economics*, vol. 63, pp. 249-258, 1987.
[41] V. Ruttan and Y. Hayami, "Toward a theory of induced innovation," *J. of Development Studies*, vol. 20, pp. 203-222, 1984.
[42] R. Lipschutz and K. Conca, "The State and Social Power in Global Environmental Politics," in *New Directions in World Politics*, H. Milner and J. Ruggie, Eds. New York, NY: Columbia University Press, 1993, pp. 363.
[43] C. Milbrath, Y. Downes, and K. Miller, "Sustainable living: A framework of an ecosystematically grounded political theory," *Environmental Politics*, vol. 3, pp. 421-444, 1994.
[44] P. Kleindorfer, H. Kunreuther, and P. Schoemaker, *Decision Sciences: An Integrative Perspective*. New York, NY: Cambridge University Press, 1993.
[45] OECD, "Environmental Policy Benefits: Monetary Valuation," OECD Publications Service, Paris 1989.
[46] E. Lutz and M. Munasinghe, "Integration of environmental concerns into economic analyses of projects and policies in an operational context," *Ecological Economics*, vol. 10, pp. 37-46, 1994.
[47] E. Barbier, "Valuing environmental functions: Tropical wetlands," *Land Economics*, vol. 70, pp. 155-173, 1994.
[48] M. Hazilla and R. Kopp, "Social cost of environmental quality regulations: A general equilibrium analysis," *J. of Political Economy*, vol. 98, pp. 853-873, 1990.
[49] P. Bartelmus, C. Stahmer, and J. v. Tongeren, "Integrated Environmental and Economic Accounting — Framework for an SNA Satellite System," *Review of Income and Wealth*, 1989.
[50] K.-G. Maler, "National Accounts and Environmental Resources," *Environmental and Resource Economics*, vol. 1, pp. 1-15, 1991.
[51] K. Arrow, B. Bolin, R. Costanza, *et al.*, "Economic growth, carrying capacity, and the environment," *Science*, vol. 268, pp. 520-521, 1995.
[52] C. Cobb, T. Halstead, and J. Rowe, "If the GDP is up, why is America down?", *The Atlantic Monthly*, vol. 276, pp. 59-78, 1995.
[53] M. McGillivrary, "The human development index: Yet another redundant composite development indicator?', *World Development*, vol. 19, pp. 1461-468, 1991.
[54] V. V. B. Rao, "Human development report 1990: Review and assessment," *World Development*, vol. 19, 1991.
[55] R. Costanza, "Three General Policies to Achieve Sustainability," in *Investing in Natural Capital: The Ecological Economics Approach to Sustainability*, A. Jansson, M. Hammer, C. Folke, and R. Costanza, Eds. Washington, DC: Island Press, 1994, pp. 392-407.
[56] P. Dasgupta and G. Heal, *Economic Theory and Exhaustible Resources*. Cambridge: Cambridge University Press, 1979.
[57] D. Pearce and R. K. Turner, *Economics of Natural Resources and the Environment*. Baltimore, MD: The Johns Hopkins University Press, 1991.
[58] D. Pearce and G. Atkinson, "Capital theory and the measurement of sustainable development: an indicator of "weak" sustainability," *Ecological Economics*, vol. 8, pp. 103-108, 1993.
[59] M. Altieri and O. Masera, "Sustainable rural development in Latin America: building from the bottom up," *Ecological Economics*, vol. 7, pp. 93-121, 1993.
[60] S. Bergstrom, "Value standards in sub-sustainable development: On the limits of ecological economics," *Ecological Economics*, vol. 7, pp. 1-18, 1993.
[61] C.-C. Chi, *Environment and Development in Developing Countries, with Case Studies of Kenya and Taiwan*. Buffalo, NY: State University of New York, 1992.
[62] D. Hill, "Evaluating Energy Alternatives in Rural Nepal: Saving Trees, Saving Time, or Saving Rupees?", in *Energy Management and Policy*. Philadelphia, PA: University of Pennsylvania, 1993.
[63] R. Keeney and T. McDaniels, "Value-focused thinking about strategic decision at BC Hydro," *Interfaces*, vol. 22, pp. 94-109, 1994.
[64] P. Meier and M. Munasinghe, *Incorporating Environmental Concerns into Power Sector Decisionmaking: A Case Study of Sri Lanka*, vol. 6. Washington, DC: The World Bank, 1994.
[65] Minnesota Planning, *Minnesota Milestones: a report card for the future*. St. Paul, MN: The State of Minnesota, 1992.
[66] Lake Superior Binational Program, "Ecosystem Principles and Objectives for Lake Superior," Lake Superior Binational Program, Thunder Ban, ON, Discussion paper Oct. 22, 1993.
[67] Shell Canada Limited, "Progress Toward Sustainable Development," Shell Canada Limited, Calgary, AB, Corporate Environmental Rep. 1992.
[68] Sustainable Seattle, "Indicators of Sustainable Community," Metrocenter YMCA, Seattle, WA 1993.
[69] A. Farrell, "Sustainability and Public Policy: Tools for Analysis," vol. 17th Annual Research Conference of the Association for Public Policy Analysis and Management. Washington, DC, 1995.
[70] PCSD, "President's Coun. on Sustainable Development Rep.," http://www.whitehouse.gov/WH/EOP/pcsd/Concil_report.h tml: World Wide Web, 1996.
[71] J. Herkert, A. Farrell, and J. Winebrake, "Technology choice in a sustainable development context," *IEEE Tech. & Soc. Mag.*, vol. 15, no. 2, 1996.
[72] International Institute of Sustainable Development, "Sourcebook on Sustainable Development," IISD, Winnepeg, MB, 1992.

T&S

The Role of Technology in Sustainable Development

The Image Bank/Tom Klare/1999

The inability of governments represented at the 1992 Earth Summit to reach a consensus on reducing either population growth or consumption, and the political need for the concept of sustainable development to accommodate economic growth, mean that the achievement of sustainable development will depend on our ability to reduce the environmental impact of resource use through technological change. This will require the redesign of our technological systems and not merely the application of technological fixes that are seldom satisfactory in the long term. Past attempts by the appropriate technology movement to affect such a redesign neglected the social dimensions of technological change. Modern advocates of sustainable development will similarly fail unless they recognize the need for fundamental social change and a shift in priorities.

Sharon Beder is senior lecturer in the Science and Technology Studies Department of the University of Wollongong, PO Box 1144, Wollongong, NSW 2500, Australia. She is the author of The Nature of Sustainable Development, Scribe, Victoria, 1993 (reviewed in T&S, Summer 1994).

Dependence on Technological Change

The concept of sustainable development has succeeded in gaining widespread support

Reprinted from *IEEE Technology and Society Magazine,* Vol. 13, No. 4, pp. 14–19, Winter 1994/1995.

amongst the world's decision makers and power brokers. In October 1987, the goal of sustainable development was endorsed by the governments of 100 nations in the U.N. General Assembly. The U.N. endorsement followed the completion of a report by the World Commission on Environment and Development [1], published as "Our Common Future," which defined sustainable development as "development that meets the needs of the present without compromising the ability of future generations to meet their own needs."

Sustainable development sets out to make necessary modifications that will enable normal economic activities to be sustainable into the future. At the same time it recognizes that serious and irreversible environmental degradation should be prevented because it could diminish the ability of the planet to sustain such activities [2].

At the heart of the debate over the potential effectiveness of sustainable development is the question of whether technological change can reduce the impact of economic development sufficiently to ensure other types of change will not be necessary. Changes in population growth and consumption levels seem to be off the agenda, since nations were unable to come to any agreement on these issues at the recent Earth Summit in Rio de Janeiro in June 1992, although population growth was the subject of an international conference in Cairo in September 1994. If environmental impact is a function of numbers of people (population), resource use per person (consumption), and environmental impact per unit of resource used (technology), this leaves technology as the remaining variable available for manipulation.

Sustainable development policies seek to change the nature of economic growth rather than limit it. They are premised on the belief that continual growth in a finite world is possible through the powers of technology, which will enable us to find new sources or provide alternatives if a particular resource appears to be running out. Otherwise, technology will help us use and reuse what we have left in the most efficient manner. The tools of sustainable development—economic instruments, legislative measures, and consumer pressures—are aimed at achieving technological changes such as recycling, waste minimization, substitution of materials, changed production processes, pollution control, and more efficient usage of resources.

The British Pearce Report [3] suggests that resource usage can be dealt with through recycling and minimizing wastage, and that the damage to the environment from disposing of wastes can be minimized in a similar way: "Recycling, product redesign, conservation, and low-waste technology can interrupt the flow of wastes to these resources, and that is perhaps the major feature of a sustainable development path of economic progress."

The Failure of Technological Fixes

Efforts to clean up the environment have tended to concentrate on "cleaning technologies" rather than "clean technologies"—that is, on technologies that are added to existing production processes to control and reduce pollution (end-of-pipe technologies and control devices). The alternative to end-of-pipe technologies is to adopt new "clean" technologies that alter production processes, inputs to the process and products themselves so that they are more environmentally benign. Clean technologies are preferable to end-of-pipe technologies because they avoid the need to extract and concentrate toxic material from the waste stream and deal with it.

It is suggested by Cramer and Zegveld [4] that process technologies should be used that require less water (for example, by alternative drying techniques), energy, and raw materials, and that reduce waste discharges (for example, by developing detection and separation machinery and process-integrated flue-gas cleaning and filter systems). Also, raw material inputs and processes can be changed so that, for instance, solvent-free inks and paints and heavy metal-free pigments are used. The end products can be redesigned to reduce environmental damage during both manufacture and use, and waste flows can be reused within the production process rather than dumped.

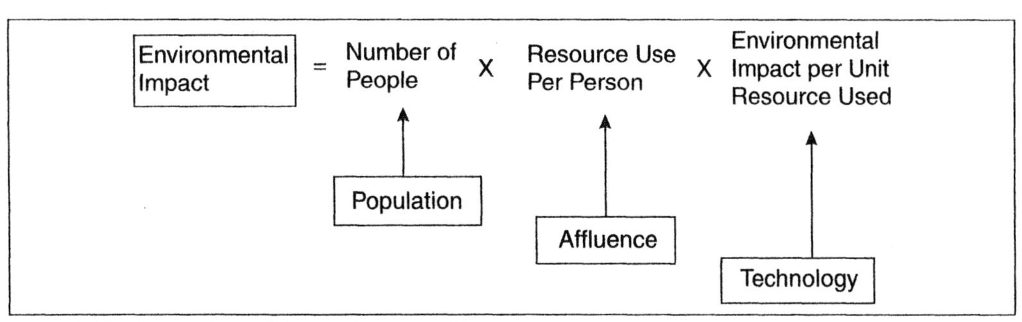

Fig. 1. The factors determining environmental impact.

The Organization for Economic Cooperation and Development, OECD [5], found that most investment in pollution control was being used for end-of-pipe technologies, with only 20 percent being used for cleaner production. Cleaner technologies are not always available and, even when they are, companies tend not to replace their old technologies until these have run their useful life. Also, companies prefer to keep to a minimum the organizational changes that need to be made; they like to play it safe when it comes to investment in pollution management.

The problem with measures such as end-of-pipe technologies is that they are technological fixes that do not address the cause of the problem. Such fixes can often cause other problems:

> A target for improving the efficiency of the combustion of fossil fuels is to convert all available carbon in the fuel into carbon dioxide. On the other hand, carbon dioxide is a major greenhouse gas. Moreover, our means of achieving better thermal efficiencies is by increasing the temperature of the combustion process. A result of increasing temperature, however, is that more oxides of nitrogen are formed from the air used in combustion. Oxides of nitrogen are an important element in the formation of photochemical smog. Thus, in the pursuit of more efficient energy usage, it is possible other potentially undesirable side-effects may arise.[6]

Barry Commoner [7] argues that a spiral of technical fixes occurs because of the failure to correct the fundamental flaw that technology is subject to in our society. He says that "if technology is indeed to blame for the environmental crisis, it might be wise to discover wherein its 'inventive genius' has failed us—and to correct that flaw—before entrusting our future survival to technology's faith in itself."

A common reaction to the litany of problems attributed to technologies is to argue that the problem is not so much in the technology but in how it is used or abused. Technologies themselves only become environmentally harmful if they are not applied with due sensitivity to the environment. Another reaction is to argue that technologies often have unexpected side effects or second-order consequences that were not originally designed into the technology. Pollution is one such side effect that is never intended by the designers of technology. However, Commoner does not accept these views, arguing that "these pollution problems arise not out of some minor inadequacies in the new technologies, but because of their very success in accomplishing their designed aims."

Commoner points out that plastics do not degrade in the environment because they were designed to be persistent; similarly, fertilizers were designed to add nitrogen to the soil, so it is not an accident that they add to the nitrogen reaching the waterways. Part of the problem, he argues, is that technologists make their aims too narrow; they seldom aim to protect the environment. He argues that technology can be successful in the ecosystem "if its aims are directed toward the system as a whole rather than at some apparently accessible part."

He gives sewerage technology as an example. He says that engineers designed their technology to overcome a specific problem: when raw sewage is dumped into rivers, it uses up too much of the river's oxygen supply as it decomposes. Modern secondary sewage treatment is designed to reduce the oxygen demand of the sewage. However, the treated sewage still contains nutrients which help algae to bloom; and when the algae die they also deplete the river of oxygen. Instead of this piecemeal solution, Commoner argues, engineers should look at the natural cycle and reincorporate the sewage into that cycle by returning it to the soil rather than putting it into the nearest waterway.

Commoner advocates a new type of technology that is designed with a full knowledge of ecology and the desire to fit with natural systems.

The Failure of Appropriate Technology

Attempts to invent and design different types of technology that fit with natural systems are not new. The appropriate technology movement which blossomed in the 1970s attempted to do just this. Appropriate technology has been defined as "technology tailored to fit the psychosocial and biophysical context prevailing in a particular location and period." [8] It was designed not to dominate nature but to be in harmony with it.

Appropriate technology involves attempting to ensure that technologies are fitted to the context of their use—both the biophysical context, which takes account of health, climate, biodiversity, and ecology, and the psychosocial context, which includes social institutions, politics, culture, economics, ethics, and the personal/spiritual needs of individuals.

One of the best-known early proponents and popularizers of appropriate technology was the British economist E.F. Schumacher [9], who talked about 'intermediate technology' in his book *Small is Beautiful: A Study of Economics as if People Mattered.* He was principally concerned with development in low-income countries, and recommended a technology that was aimed at helping the poor in these countries to do what they were already doing in a better way.

During the mid-1970s, the appropriate technology movement expanded from its initial focus on low-income countries to consider the problems in industrialized high-income countries. Advocates of appropriate technology were concerned about social as well as environmental problems.

Robin Clarke [10] differentiated between the appropriate technology response and the "technological fix" responses to environmental problems. For example, he characterized the technological-fix response to pollution as "solve pollution with pollution control technology"; the appropriate technology response, instead, would be to invent non-polluting technologies. Similarly, the technological-fix response to exploitation of natural resources was to use resources more cleverly; the appropriate technology response was to design technologies that used only renewable resources.

The appropriate technology movement has been going for more than twenty years in many countries, and today involves an extensive network of organizations, projects, and field experiments, and an identifiable literature of its own.

Despite this, its ideas have not transformed the pattern of technology choice exercised by mainstream society. Kelvin Willoughby [8], a U.S. scholar who has studied this movement, points out that it has "achieved a modestly impressive track record of successful projects which lend weight to the movement's claims" and become a significant international movement but still remains "a minority theme within technology policy and practice." As will be seen in the next section, this is because there are powerful vested interests supporting existing technologies as well as institutional and professional structures that have evolved alongside those technologies.

The Failure to Adopt Alternative Technologies

Not all technological options and alternatives are developed or explored. Although this is often because alternatives are more expensive or less economical, there are often other reasons, too. Even today, many firms are not implementing technologies aimed at waste reduction and minimization, despite their availability and probable cost savings. The reluctance of many engineers to take up alternative technologies can be explained partly in terms of technological paradigms. This is a term borrowed from Thomas Kuhn [11], who postulated in 1962 that science progresses through periods of "normal science," which operates within a scientific paradigm, interspersed with periods of "scientific revolutions."

Several writers have applied the concept of a paradigm to technological development. Edward Constant [12] argued that the routine work of engineers and technologists, which he called "normal technology," involves the "extension, articulation, or incremental development" of existing technologies. A technological paradigm or "tradition," Constant said, is subscribed to by engineers and technicians who share common educational and work experience backgrounds.

Giovanni Dosi [13] described a technological paradigm as an "outlook," a set of procedures, a definition of the "relevant problems and of the specific knowledge related to their solution." Such a paradigm, Dosi said, embodies strong prescriptions on which technological directions to follow and ensures that engineers and the organizations for which they work are blind to other technological possibilities. Richard Nelson and Sidney Winter [14] also observed that a technological paradigm or regime will define for the engineer what is feasible or at least worth attempting:

> The sense of potential, of constraints, and of not yet exploited opportunities, implicit in a regime focuses the attention of engineers on certain directions in which progress is possible, and provides strong guidance as to the tactics likely to be fruitful for probing in that direction.

As a result, technological development tends to follow certain directions, or trajectories, that are determined by the engineering profession and others (see Fig. 2). Ideas are developed if they fit the paradigm; otherwise, they tend to be ignored by the mainstream engineers, the bulk of the profession. An example is the development of sewerage engineering. The range of ways of treating sewage is limited by a sewage treatment paradigm that assumes that sewage will be delivered in pipes to centralized locations near waterways. Treatment is classified into three stages—primary, secondary, and tertiary, which build upon one another. The first stage is to remove some of the solids from the sewage; the second stage is to decompose the sewage; and the third stage either removes more solids or decomposes the sewage further. Any new technology will be thought of or developed only if it can fit within this system.

Generally, technological change is gradual and occurs within technological paradigms. Radical technological innovation is often opposed by firms because of the social changes that may need to accompany it—for example, changes to the work and skills of employees, to the way production is organized, and to the relationships between a firm and its clients and suppliers [4]. Dutch scholar Johan Schot [15] argues that radical technological change can occur only if the social context also changes.

Firms may also have vested interests in particular technological systems. According to McCully [16]:

> The reason that the USA is the most polluting nation in the world has little to do with a lack of energy-efficient technologies or renewable methods of producing electricity: it has a lot to do with the size of the country's oil, coal, and automobile industries and the influence they have on the political establishment. In the U.K., the public transport system is expensive, unreliable, and infrequent, not because the government cannot afford to improve it or does not know how, but because the vested interests behind public transport have negligible power compared to the influential road and car lobbies.

Because of the reluctance of governments to act against business interests, legislation and economic instruments are seldom tough enough to foster technological change of the type required for ecological sustainability. Although such regulation would probably strengthen business in the long run, businesspeople see strong government intervention as an infringement on their autonomy. Commoner [17] argues that businesspeople are supported in this because there remains a strong public conviction "that the decisions that determine what is produced and by what technological means ought to remain in private, corporate hands."

Langdon Winner [18] has argued that most people in the appropriate technology movement ignored the question of how they would get those who were committed to traditional technologies to accept the new appropriate technologies. They believed that if their technologies were seen to be better, not only in terms of their environmental benefits but also in terms of sound engineering, thrift, and profitability, they would be accepted.

Winner says that appropriate technology was generally seen as a way of effecting change without challenging the established power structure of Western societies. This allowed a sense of optimism that had been lacking in the ranks of those who were unhappy with the direction and values of the societies they lived in. They believed that, as the new technologies caught on, social change would follow:

> As successful grass-roots efforts spread, those involved in similar projects were expected to stay in touch with each other and begin forming little communities, slowly reshaping society through a growing aggregation of small-scale social and technical transformations. Radical social change would catch on like disposable diapers ... or some other popular consumer item.

Not all advocates of alternative technologies ignored the social and political dimensions, however. David Dickson [19] recognized the difficulties that would be encountered by those proposing radically different types of technology when he proposed the name "utopian technology." He said that he used the word "utopian" because an alternative technology could "only be successfully applied on a large scale once an alternative form of society had been created."

However, many of the advocates of appropriate technologies made no attempt to understand how modern technologies had been developed and why they had been accepted or why alternatives had been discarded. Winner claims that "by and large most of those active in the field were willing to proceed as if history and existing

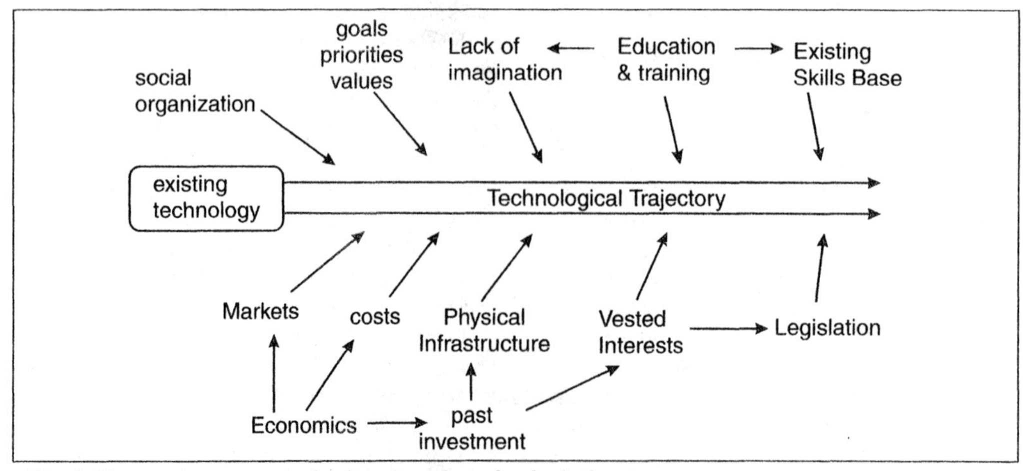

Fig. 2. Factors that constrain the direction of a technological trajectory.

institutional technical realities simply did not matter."

It is important not to put too much emphasis on technological factors without considering the social, political, and economic factors that can be crucial in the shaping and implementation of technologies. It seems that those pinning their hopes on technology to deliver to us a sustainable future may well be doing the same thing. Having the technological means to reduce pollution and to protect the environment does not mean that it will automatically be used.

Conclusions

Sustainable development relies on technological change to achieve its aims, but will governments take the tough steps that are required to force radical technological innovation rather than the technological fixes that have been evident to date? Such measures would require a long-term view and a preparedness to bear short-term economic costs while industry readjusts.

Even if cleaner technology can be implemented, will the reductions in pollution be enough? Cramer and Zegveld [4] argue that they will not if production continues to grow. Giving the example of their own country, where the purchasing power of the average person is expected to increase by 70 percent by the year 2010, they argue that "an incredible reduction in discharge levels and waste flows per product unit would have to be realized to achieve the aim of a sustainable society." They believe this is not realistic. On top of this, production would need to increase ten times if everyone in the world were to live at the same standard of living as those who live in affluent countries. They claim that the growth of both production and freely disposable income would have to be restricted if pollution levels are to be reduced.

It would appear that so long as sustainable development is restricted to minimal low-cost adjustments that do not require value changes, institutional changes, or any sort of radical cultural adjustment, the environment will continue to be degraded. Unless substantial change occurs, the present generation may not be able to pass on an equivalent stock of environmental goods to the next generation. "Firstly, the rates of loss of animal and plant species, arable land, water quality, tropical forests, and cultural heritage are especially serious. Secondly, and perhaps more widely recognized, is the fact that we will not pass on to future generations the ozone layer or global climate system that the current generation inherited. A third factor that contributes overwhelmingly to the anxieties about the first two is the prospective impact of continuing population growth and the environmental consequences if rising standards of material income around the world produce the same sorts of consumption patterns that are characteristic of the currently industrialized countries." [20]

Even if people put their faith in the ability of human ingenuity in the form of technology to be able to preserve their lifestyles and ensure an ever-increasing level of consumption for everyone, they cannot ignore the necessity to redesign our technological systems rather than continue to apply technological fixes that are seldom satisfactory in the long term. Technological optimism does not escape the need for fundamental social change and a shift in priorities. That was the mistake many in the Appropriate Technology Movement made. It takes more than the existence of appropriate or clean technologies to ensure their widespread adoption.

References

[1] World Commission on Environment and Development, 1990, *Our Common Future*, Australian edition, Oxford University Press, Melbourne.
[2] S. Beder, 1993, *The Nature of Sustainable Development*, Scribe, Newham, Australia.
[3] D. Pearce, A. Markandya, and E. Barbier, 1990, *Blueprint for a Green Economy*, Earthscan, London.
[4] J. Cramer and W.C.L. Zegveld, 1991, *The Future Role of Technology in Environmental Management*, Futures, 23(5), pp. 461-2.
[5] Organization for Economic Cooperation and Development, 1989, *Economic Instruments for Environmental Protection*, OECD, Paris.
[6] Ecologically Sustainable Development Working Groups, 1991, *Final Report—Manufacturing*, AGPS, Canberra.
[7] B. Commoner, 1972, *The Closing Circle: Nature Man & Technology*, Bantam Books, Toronto.
[8] K. Willoughby, 1990, *Technology Choice: A Critique of the Appropriate Technology Movement*, Westview Press, Boulder.
[9] E.F. Schumacher, 1974, *Small is Beautiful: A Study of Economics as if People Mattered*, Abacus, London.
[10] R. Clarke, 1974, "Technical Dilemmas and Social Responses," in *Man-Made Futures: Readings in Society, Technology and Design*, eds Nigel Cross et al., Hutchinson Educational, London.
[11] T.S. Kuhn, 1970, *The Structure of Scientific Revolution*, 2nd edn, University of Chicago Press, Chicago.
[12] E. Constant, 1984, "Communities and Hierarchies: Structure in the Practice of Science and Technology," in *The Nature of Technological Knowledge. Are Models of Scientific Change Relevant?*, ed. R. Laudan, D. Reidel Publishing Co, Holland, 1984.
[13] Giovanni Dosi, 1982, *Technological Paradigms and Technological Trajectories*, Research Policy, 11, pp. 147-162.
[14] R. Nelson and S. Winter, 1977, *In Search of Useful Theory of Innovation*, Research Policy, 6, pp. 36-76.
[15] J. Schot, 1992, *Constructive Technology Assessment and Technology Dynamics: The Case of Clean Technologies, Science, Technology, & Human Values*, 17(1), pp. 48-50.
[16] P. McCully, 1991, *The Case Against Climate Aid*, The Ecologist, 21(6), p. 250.
[17] B. Commoner, 1990, *Making Peace With the Planet*, Pantheon Books, New York.
[18] L. Winner, 1986, *The Whale and the Reactor: A Search for Limits in an Age of High Technology*, University of Chicago Press, Chicago.
[19] D. Dickson, 1974, *Alternative Technology and the Politics of Technical Change*, Fontana/Collins, Great Britain.
[20] Ecologically Sustainable Development Working Group Chairs, 1992, *Intersectoral Issues Report*, AGPS, Canberra.

T&S

STEVEN B. YOUNG AND WILLEM H. VANDERBURG

A Materials Life Cycle Framework for Preventive Engineering

Preventive engineering [1]-[3] uses information about how technology affects human life, society, and the natural ecology in order to adjust engineering methods and approaches to achieve the best possible compatibility between technology and its contexts. It constitutes a different engineering paradigm, aspects of which are emerging in industry and universities. One area of application, pollution prevention, considers the design and selection of processes, products, and materials to reduce negative impacts on the natural ecology. The materials life cycle approach is a conceptual framework with which to examine flows of materials from "cradle to grave," and a component of the product cycle of design, production, use and discard. These concepts are related to the demands placed on engineers resulting from the rapidly changing setting in which we practice.

An earlier version of this article was presented by S.B. Young at the IEEE-SSIT Conference "Preparing for a Sustainable Society," held June 21-22, 1991, in Toronto, Ontario, Canada. *S.B. Young is a Doctoral candidate in the Centre for Technology and Social Development and in the Department of Metallurgy and Materials Science, University of Toronto, Toronto, Ontario, Canada M5S 1A1. W.H. Vanderburg is Director of the Centre for Technology and Social Development, University of Toronto, Toronto, Ontario, Canada M5S 1A1.* *This work was supported by the Ontario Centre for Materials Research, Ontario Graduate Scholarships, and the Natural Sciences and Engineering Research Council of Canada.*

The Problem

Industry and engineers are functioning today in a changing climate being fueled by concerns about the sustainability of our way of life and worries about environmental damage. Popularized in 1987 by the report of the United Nations World Commission on Environment and Development [4], sustainable development is a concept which calls upon humanity to make our ways of doing things more compatible with the natural and social ecologies, protecting nature and advancing the quality of life for all people, particularly for those in the Third World.

Public concern in the industrialized nations has lead to government actions which are the primary environmental pressure for change. Operations are increasingly becoming regulated, requiring modifications of practices such as the installation of improved emission control equipment. These responses to public and group pressures are bound to continue and are expected to improve the effectiveness of environmental protection. Typical government actions will include recycling mandates on specific products and materials, additional emissions controls for manufacturing facilities, and local and regional remediation clean-up plans. As a result, the scope, intensity, and enforcement of regulations will continue to grow and, clearly, it will become increasingly expensive for polluters to comply with the laws. Already in 1986 in the United States the administration of and compliance to environmental laws was costing an estimated $10 million per page of government regulation [5].

Another pressure on industry is the greening of public values: the environment has now become a direct issue of commercial competition

Reprinted from *IEEE Technology and Society Magazine,* Vol. 11, No. 3, pp. 26–31, Fall 1992.

and marketing. Because the green image a company cultivates is linked directly to the real performance of its operations and products, companies are further encouraged to be environmentally responsible. Additional incentives to change include: the rising costs of waste disposal, liability insurance [6], worker health and safety protection, and concerns about the scarcity and vulnerability of materials and energy resources (see, for example, [7]).

Prevention: A Solution

As elsewhere, the old adage that "prevention is better than cure" also applies to pollution. Pollution prevention offers a route to an environmentally sustainable way of life, avoiding many of the environment-economy conflicts that hamper traditional strategies. Pollution prevention broadly encompasses technological and social activities which result in the nongeneration or reduction of waste, or its toxicity, over the entire life cycle of products and activities. It requires a new way of thinking, away from after-the-fact focusing on waste at the "end-of-pipe," to thinking instead about fully integrating clean and preventive criteria within the design, development, production, and consumption of products and services. Thus pollution prevention initiatives protect nature and human health at the most fundamental level, before they are threatened. [2], [3], [8], [9].

For business, perhaps the most unsettling consequence of the environmental pressures mentioned above is the climate of uncertainty, in which long-term decision making is difficult and risky. A pollution prevention strategy promotes economic stability and competitiveness [10] because the costs and hazards of managing pollution and dealing with regulations are reduced or avoided altogether. This is especially true for unforeseen future costs arising from new legislation, as well as the immediate costs facing a polluter. Furthermore, an effective prevention strategy is fully integrated into the day-to-day engineering and business activities of a firm and thus encourages technological innovation and development [5], [8]. 3M corporation is one notable example of a business progressing in this direction; over the last 15 years, the company has earned over $500 million in first-year savings under their *Pollution Prevention Pays* Program [11].

The Materials Life Cycle Approach

To address the concept of pollution prevention from the perspective of materials engineering, we distinguish between two levels of analysis: the product cycle includes the general development, production, use and discard of a product; whereas, the materials life cycle focuses on the life cycles of the constituent materials included in the product (Fig. 1). For example, it is clear that numerous materials are required to manufacture an automobile; they are first selected and designed into the car, then the metal, polymer, and glass are acquisitioned and processed into the physical product. The materials life cycle approach has been used before, but by applying it in a preventive context, new insights and applications emerge.

Materials form an integral foundation for technological systems of production and service. The initial designs and choices of materials determine the performance of products, their manufacturing requirements, and consequently their ecological impacts. It follows that, if well-informed and responsible decisions can be made about the selection, properties and processing characteristics of industrial materials, negative environmental impacts can be reduced, helping to make our way of life more sustainable with basic technological changes.

The materials life cycle approach is a useful conceptual tool with which to examine patterns of production and consumption (Fig. 2). Many stages and interrelations can be identified for complex industrial systems in terms of material and energy. A simplified industrial representation applied to a generic product may be visualized in four stages:

1) resources are *extracted*, refined, and processed into useful materials,

2) engineering materials are *manufactured* into parts and products, which are distributed,

3) finished products are consumed, *used*, and maintained, and

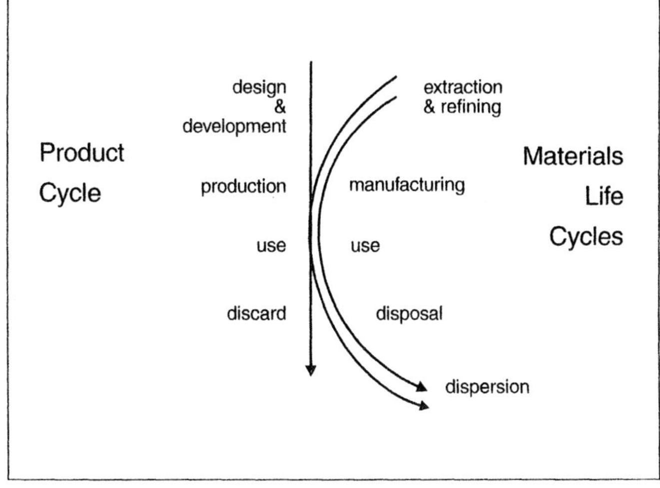

Fig. 1: Relationship of the product cycle to materials life cycles.

4) products with no remaining value are discarded, the resulting waste material is *disposed* and often dispersed.

Energy, material and transportation are necessary inputs at all these stages; and feedbacks and loops, of course, make up part of the cycle. Two quite different examples are illustrative. It is common for "prompt" plastic scrap that is produced during molding operations to be quickly and cleanly recycled in-plant directly back into production. This is done for reasons of cost effectiveness and material efficiency rather than any environmental imperative. Second, post-consumer waste collection and recycling programs are becoming a popular means to deal with the discard end of the product cycle; however, these programs have led to few new closed recycling loops and have not yet demonstrated a diminished societal reliance on virgin resources or significant pollution reduction [12]. It may be noted that these curb-side collection and recycling programs are largely end-of-pipe in character in that they are usually added onto existing waste disposal infrastructures, without adjustments being made to provide suitable markets for secondary materials.

For large scale production systems involving a number of materials cycles, the possibility of preventing waste by closing materials loops has been proposed as the concept of industrial metabolism or industrial ecosystem [13]-[15]. Input and output flows of materials and energy would be connected within and across industries and industrial sectors. The idea is directly analogous to nature, where cyclic ecological processes which have evolved over time ensure there is no waste or are nested into other cycles which turn the "waste" into essential inputs. To help achieve this, industrial parks would be designed to be "industrial ecosystems" as much as possible to achieve clean production and manufacturing.

Beyond the industrial parks, as materials and products travel along their life cycles, wastes are also discharged at each other stage; moreover, there is a general tendency for materials to disperse with time, both as wastes and as products. To examine these patterns we will draw on the previously introduced distinction between the life cycles of products and their constituent processes. Thus wastes due to processes of extraction and manufacturing can be distinguished from wastes due to product use and discard. In a materials systems sense, the extraction and manufacturing stages of the materials life cycle are characterized by processes of acquisition, combination, and separation, wherein the portion of material deemed valuable is concentrated in favor of waste emissions. The product use stage is preceded by widespread distribution of a product and is followed by dispersive disposal practices like landfilling and incineration. Product-design waste prevention can be further subdivided into strategies, firstly, for material and waste reduction and, secondly, plans to optimize recycling. From these general observations, a useful framework emerges with which to look for prevention strategies from a materials perspective.

▼ *Process pollution prevention strategies*

For industrial processes, manufacturing, and production, preventive techniques can be listed under five broad categories for source reduction of wastes [5]:

1) changes to production technology and equipment,

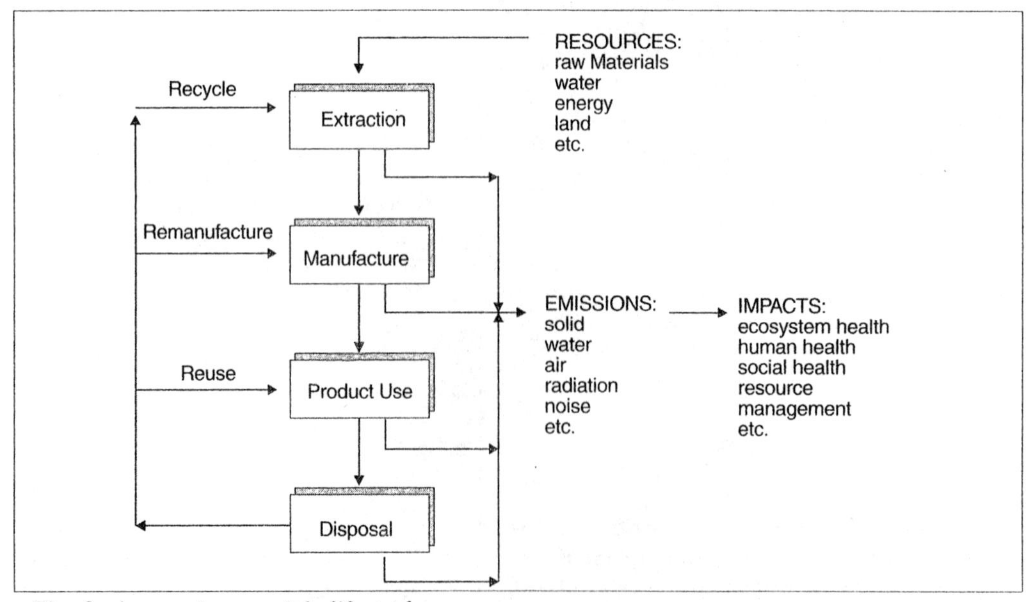

Fig. 2: A generic materials life cycle.

2) improvements to operations and procedures,
3) recycling of wastes within the plant,
4) changes to input materials, and
5) redesign or reformulation of end products.

The first three categories are relatively conventional routes to cleaner production, such as improving the efficiency of processes and closing loops to avoid emissions. A common example of changing input materials, the fourth category, is to substitute a more benign alternative for a toxic organic-based solvent. The last category is more fundamental in character, since redesigning and reformulating products to prevent wastes demands a complete reworking of the processes involved over the entire product cycle of the system, from the initial conceptual stages of design, through to production, distribution, use, and discard.

▼ Product design strategies for pollution prevention

It is generally accepted that about 80 to 90% of total life cycle costs are committed by the final design of a product, before production or construction actually begin [16], [17]. Similarly, wastes resulting from product creation, use, and disposal are largely determined by original design. Currently, however, disposal options and environmental impact are not common initial design criteria.

The design process necessary to meet an identified need is often described in three stages [18]: at conceptual design the product is conceived and outlined to meet its application; at embodiment design it is formulated more thoroughly; and at detailed design the final production requirements and specifications are spelled out. If these stages are compared directly to the materials life cycle (see Fig. 2), it is apparent that the design process runs conceptually counter-current to the materials flow. That is, products are conceived and embodied from the bottom of the product cycle (excluding discard) to the top (see Fig. 1) and in the reverse order from which materials are chronologically selected, created, used, and disposed. It follows that, if the ultimate disposal of products and materials is to be seriously considered, this should be done at the earliest stages of original conceptual design. With this in mind, two general directions for product design will be outlined in order of their preventive priority: reduction and recycling.

▼ *Reduction*: For the current discussion, design strategies for waste reduction encompass considerations which reduce the quantity or quality of harmful or toxic implications: for example, substituting benign alternatives for more polluting materials or reducing the amount of matter and energy required, either for manufacturing or during product use. A number of fairly obvious conservation principles can be broadly applied. On the product use side, it is routine engineering practice to optimize product function, efficiency and performance, for example by designing an engine for fuel efficiency. The matter and energy embodied in a product during manufacture can be directly examined by considering, for example, its degree of structural overdesign. Also temporal characteristics may be considered by examining durability, repairability, reusability, and the potential for multiple uses of a product.

A pollution prevention strategy promotes economic stability and competitiveness because management, clean up, and regulation costs are avoided.

From these basic considerations it follows that the routes to greater environmental conservation for product design tend to be very specific to the individual application, location, and service of interest. To bring pollution prevention into design, it is proposed to link environmental criteria to design criteria by drawing upon integrated design strategies like design-for-quality and concurrent engineering [17], [19]. These techniques promote improved product development by designing preventively for the entire manufacturing cycle; it is therefore quite natural to link integrated design strategies to the materials life cycle approach.

▼ *Recycling:* After value-added products have exceeded their useful lives, their remaining benefit is limited to the value of their material constitution. Consequently, much of the original engineering invested in a product is lost after it is deemed "useless," with the exception of the materials engineering. Recycling design strategies broadly refer to efforts to recover and reuse the material resources in product waste [20]. Energy-from-waste is sometimes considered as recycling but is not discussed here.

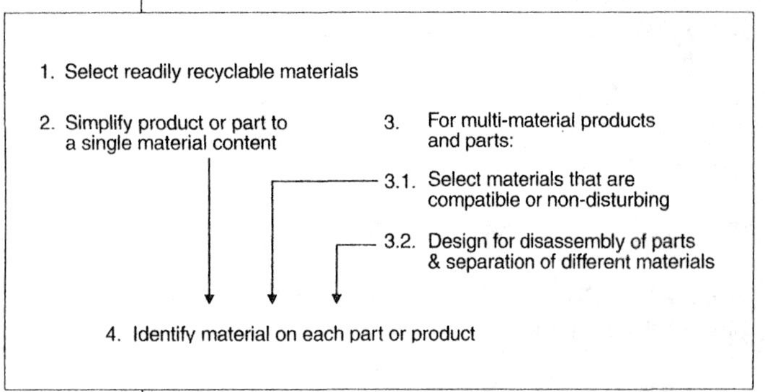

Fig. 3: Product design strategy for material recycling.

In a product design strategy for material recovery, four steps are proposed to design for recyclability (Fig. 3). The first principle is simply to select materials which are recyclable; however, in many ways this is a far more difficult task than it appears. For example, the definition of "recyclable" is often unduly stretched to include materials that are technically but neither practically nor economically recyclable. Factors determining recyclability include purity, availability, cost of recovery, recycled material properties, and potential markets for recycled material. Significantly, it is clear that these factors are largely extrinsic to the composition and structure of the material itself. Consequently the meaningful recyclability of a material is largely dependent on design and life cycle considerations other than material properties.

The second principle is to design products with the lowest possible material complexity. For example, a bottle whose lid, body, label and base all consist of the same thermoplastic is more amenable to recycling than a container consisting of a mixture of metal and incompatible plastics. The third principle, for more complex products like automobiles, is to design with disassembly in mind, so that parts and materials can be separated after use into groups that are compatible for recycling. Finally, to facilitate the correct processing of separated materials, labeling is necessary on each material component to identify composition, properties, potential contamination, etc. All in all, this strategy addresses concerns that valuable materials, which could continue to be used, will suffer irreversible dispersion and contribute to the expensive and overburdened waste management infrastructure.

Life Cycle Assessment

We have not yet explicitly dealt with the issue of how to decide between different alternatives within a preventive approach. Life cycle assessment (LCA) is being developed as one possible methodology for this and other related tasks [21]; however, as a tool, LCAs are limited in that they analyze only the environmental costs of an individual product without considering the larger preventive framework.

LCA is currently being applied to competitive products and packaging that are increasingly being scrutinized and criticized for their environmental soundness [22]. Disposable versus cloth diapers, paperboard versus polystyrene foam "burger boxes," and glass refillable versus plastic beverage containers are typical subjects of debate. As a methodological framework for LCA, three assessment stages have been proposed [21]:

1) *Inventory* the resources required and the wastes generated using a life cycle model.
2) Analyze *impacts* on the natural, human and social environments.
3) Study options to *improve* the product by reducing its negative environmental impacts.

To date, only the inventory stage of the LCA methodology has been developed in a critical but limited fashion; it is possible that the impact and improvement stages will be detailed in the future. Despite their limited status, life cycle assessments are increasingly being used and abused as a basis for product marketing and environmental-labeling schemes; therefore, critical examination of the quality, meaningfulness and applicability of these assessments is needed. Two issues are briefly outlined here.

The boundary conditions for a LCA are critical to its execution and can significantly alter the findings; in fact, system definition often seems to predetermine the comprehensiveness and outcome of a study. For example, if disposable single-use beverage containers are to be examined, "disposability" may be contained within the terms of reference and nondisposable reusable cups may not be included for analysis. A manufacturer of disposable cups would obviously prefer not to release a study that might damage its reputation or revenue.

Secondly, it is a complicated and challenging task to convert *quantitative* accounts of energy and material inputs and outputs, once they have been inventoried, into useful and often *qualitative* assessments of environmental impacts, like ecosystem damage and human health threat. In fact, most commercially sponsored life cycle assessments to date have not made the quantitative to qualitative transition necessary for meaningful conclusions to be drawn. To find and collect all the information required is daunting; to perform a useful interdisciplinary assessment and interpretation of collected information is very complex. Furthermore, like cost-benefit analysis, risk-benefit analysis and technology assessment approaches, it is likely that life cycle assessment will also be insufficient to resolve the fundamental value conflicts that arise with complex, interrelated and interdisciplinary problems. Each of these methods deals with only one technological undertaking at a time, thus excluding the analysis of positive and negative synergistic impacts.

A More Sustainable Future

The path to a more sustainable future is a complex mixture of technological, environmental, and social policies and actions. Preventive strategies offer the advantage of promoting innovation and economic benefits, while synergistically protecting human and natural health. By analyzing how industrial systems generate waste and disperse useful material, the materials life cycle approach can be used to indicate preventive measures to limit negative impacts. At the extraction and manufacturing phases, processes can be modified to prevent pollution through a variety of available source reduction techniques. Similarly for products, if life cycle criteria like disposal are considered within conceptual design strategies, resource requirements can be reduced and valuable materials can be diverted from the waste stream. Life cycle assessment is emerging to help judge products and to indicate priorities for change but, like other piecemeal approaches, will remain a limited tool unable to resolve fundamental value conflicts. Given a broad conceptual framework, life cycle evaluation and design present a way for preventive environmental criteria to be integrated with conventional technological practices, and consequently, for engineers to contribute effectively towards creating a more sustainable way of life for modern society.

References

[1] W.H. Vanderburg, "The globalization of technology and the need to transform engineering," in *Proc. Seventh Canad. Conf. Eng. Educ.*, Toronto, ON, 1990, pp. 690-699.

[2] W.H. Vanderburg, "Preventive engineering as a response to the globalization of technology," in *Proc. Seventh Canad. Conf. Eng. Educ.*, Toronto, ON, pp. 700-708, 1990.

[3] W.H. Vanderburg, "Sustainable development and the practice of engineering," in *Preparing for a Sustainable Society, IEEE/SSIT Conf. Proc.*, Toronto, ON, 1991, p. 9.

[4] World Commission on Environment and Development, *Our Common Future*. Oxford: Oxford Univ. Press, 1987.

[5] Office of Technology Assessment, Congress of the United States of America, *Serious Reduction of Hazardous Waste: For Pollution Prevention and Industrial Efficiency*. Washington, DC: US Government Printing Office, 1986.

[6] R.L. Brown, "Environmental liability insurance: An economic incentive for responsible corporate action," *Alternatives*, vol. 18, no. 1, pp. 18-25, 1991.

[7] L. Kempfer, "Strategic metals — back to basics: Elemental superalloy recycling," *Materials Eng.*, vol. 108, no. 7, pp. 30-32, 1991.

[8] J.S. Hirschhorn, "Cutting production of hazardous waste," *Technol. Rev.*, pp. 52-61, Apr. 1988.

[9] J.C. van Weenen, "Waste prevention: Theory and practice," Ph.D. thesis, Technical University of Delft, Delft, The Netherlands, 1990.

[10] M.E. Porter, "America's green strategy," *Scientific Amer.*, p. 168, Apr. 1990.

[11] 3M Environmental Engineering and Pollution Control Department, *3M's Pollution Prevention Pays*. St. Paul, MN: 3M Corp, 1990.

[12] V.W. Maclaren, "Waste management: Current crisis and future challenge," in *Resource Management and Development*, B. Mitchell, Ed. Toronto, ON: Oxford Univ. Press, 1991, ch. 2.

[13] The United Nations University & the International Federation of Institutes for Advanced Study, "Industrial metabolism: Restructuring for sustainable development," Human Dimensions of Global Change Workshop Rep. Maastricht, The Netherlands: IFIAS and UNU, Oct. 1989.

[14] J.H. Ausubel, H.E. Sladovich, Eds., *Technology and Environment*. Washington, DC: National Acad. Press, 1989.

[15] R.A. Frosch and N.E. Gallopoulos, "Strategies for manufacturing," *Scientific Amer.*, vol. 261, no. 3, pp. 144-152, 1989.

[16] W.J. Fabrycky and B.S. Blanchard, *Life-Cycle Cost and Economic Analysis*. Englewood Cliffs, NJ: Prentice Hall, 1991; or W.J. Fabrycky, "Designing for the life cycle," *Mechanical Eng.*, pp. 72-74, Jan. 1987.

[17] D.A. Gatenby and G. Foo, "Design for X: Key to competitive, profitable markets," *AT&T Tech. J.*, vol. 69, no. 3, pp. 2-13, May/June 1990.

[18] G. Pahl, W. Beitz, *Engineering Design*. London: The Design Council & Berlin: Springer Verlag, 1984.

[19] C. Overby, "Design for the entire life cycle: A new paradigm?," in *ASEE Annual Conf. Proc.* 1990. pp. 552-563.

[20] M.E. Henstock, *Design for Recyclability*. London: Institute of Metals & The Materials Forum, 1988.

[21] SETAC, *A Technical Framework for Life-Cycle Assessment*. Washington, DC: Society of Environmental Toxicology and Chemistry, Jan. 1991.

[22] National Association for Plastic Container Recovery, *Comparative Energy and Environmental Impacts for Soft Drink Delivery Systems*. Prairie Village, KN: Franklin Associates, Ltd, 1989. T&S

Chapter 8

Engineering in a Global Context

ENGINEERING and its products are not constrained by national borders. As noted in the case study of U.S. technology policy by Kline and Kash included in Chapter 5, global economic competitiveness should be a strong priority in formulating a national technology policy. The end of the cold war and the continuing emergence of developing nations have provided new markets for technological innovations. And many, if not most, corporations are now global or at least multinational in scope.

The issue of sustainable development introduced in Chapter 7 provides one window on the global context of the world economy, as well as on the corporate role in shaping and responding to it. A corporate response to the challenges posed by the 1992 Rio Conference on Environment and Development came from two organizations, the Business Council for Sustainable Development in Geneva and the World Industry Council for the Environment, an initiative of the International Chamber of Commerce in Paris. The two organizations merged in 1995 to form the World Business Council for Sustainable Development (WBCSD). The WBCSD is comprised of more than 120 international companies, in over 30 countries and more than 20 major industrial sectors (1).

An analytical framework for sustainable development, similar to that discussed in Chapter 7 but explicitly aimed at the business community, has been proposed by Hart (2) and others. In this model, the "global economy" is seen to consist of three components:

- the market economy
- the survival economy
- nature's economy

The need for sustainable development has been demonstrated in many ways, but the most striking illustrations focus on the developing world. For example, the explosive growth of urban areas in the emerging and developing countries poses a major threat to a sustainable future. In 1950, only Greater London and New York City exceeded 10 million in population. By 1994, 14 cities had a population of 10 million or more, only 4 of these in the developed countries. It is projected that by 2015 there will be 27 cities with a population in excess of 10 million, 22 of these in developing countries (3). Population growth of this magnitude in underdeveloped urban areas puts considerable strain on economic, ecological, and social systems.

Corporate Policy and Sustainable Development

The views of business organizations involved in promoting sustainable development closely parallel those of the engineering organizations, but with even more emphasis placed on the role of developing nations and other global issues. The overall aims of the WBCSD are to "develop closer co-operation between business, government and all other organizations concerned with environment and sustainable development... (and) to encourage high standards of environmental management in business itself." "Global Outreach" is one of four specific objectives of the WBCSD:

> BUSINESS LEADERSHIP... to be the leading business advocate on issues connected with the environment and sustainable development;
> POLICY DEVELOPMENT... to participate in policy development in order to create a framework that allows business to contribute effectively to sustainable development;
> BEST PRACTICE... to demonstrate progress in environmental and resource management in business and to share leading-edge practices among our members;
> GLOBAL OUTREACH ... to contribute to a sustainable future for developing nations and nations in transition.

The current "focus areas" of the WBCSD include such topics as trade and environment, sustainable production and consumption, financial markets, and eco-efficiency (1).

Despite the fact that the business perspective, like the engineering view, seems skewed toward the ecological and economic dimensions of sustainable development, Stuart L. Hart of the University of Michigan's Corporate Environmental Management Program argues (2) for a preeminent role for business in achieving a sustainable society:

> The roots of the problem—explosive population growth and rapid economic development in the emerging economies—are political and social issues that exceed the mandate and the capabilities of any corporation. At the same time, corporations are the only organizations with the resources, the technology, the global reach, and, ultimately, the motivation to achieve sustainability. (p. 67)

Hart comes to this conclusion (2) based on his examination of the qualitative relationship, first discussed by Paul Erlich, Barry Commoner, and others, showing that the total impact of human activity on the environment is a function of population, affluence (which might be viewed as a surrogate for resource use per capita, i.e., consumption), and environmental impact per unit of resource use (which is a result of the technology employed) (see background article by Beder in Chapter 7).

According to Hart, population control is difficult, if not impossible, to achieve, and there is a need for continued growth in consumption due to the development needs in emerging and developing nations. Improved technology is thus the only viable option for achieving sustainable development. And, says Hart, while "population and consumption may be societal issues, technology is the business of business" (p. 71).

This view that business should lead the sustainable development charge on the basis of its technological expertise is in marked contrast to the claims of engineering leaders. As discussed in Chapter 7, they feel that engineers and engineering organizations should be the key players in promoting sustainable development due to *their* mastery of technology. Of course, most engineers work in corporations, and many engineering leaders are captains of industry, so these views might only seem divergent from a standpoint external to the corporate walls.

To Hart's credit, he has outlined "Strategies for a Sustainable World" that go a long way toward addressing the ecological and environmental goals and that even begin to address the social goal of sustainable development. Hart advocates a progression in strategy from "pollution prevention" to "product stewardship" to "clean technology." Pollution prevention, a switch from the traditional approach of pollution control, is epitomized by 3M's highly touted and successful 3P (Pollution Prevention Pays) Program started in 1975 (4). In many multinational corporations, Hart states, a transition is occurring from pollution prevention programs toward product stewardship, a concept that incorporates processes such as Design for Environment that focus on the environmental impact of the full life cycle of a product. The ultimate strategy Hart advocates is changing to clean technology in order to create a sustainable technology base. An example of this approach is Monsanto's highly publicized shift from bulk chemicals to biotechnology.

Although Monsanto's ambitious focus on biotechnology begs a number of ethical questions outside the scope of this chapter, it is interesting to note that the company has organized seven "sustainability teams" in three categories: tools for better decision making; meeting of global sustainability needs (including addressing global hunger and water needs); and materials and training for both internal and external use (5).

Taking these concepts a step further, Hart's guidelines for Building Sustainable Business Strategies convey sensitivity to some of the social aspects of sustainable development in a global context (2):

- Develop clean products and technology
- Lower material and energy consumption
- Reduce pollution burdens
- Ensure sustainable use of nature's economy
- Replenish depleted resources
- Foster village-based business relationships
- Build the skills of the poor and the dispossessed

Hart's program falls short of encompassing the social goal of sustainable development in two very important ways. In the first place, by eschewing efforts to control population and consumption, Hart downplays the huge disparity between the developed and developing world. For example, he cites statistics indicating that the average American consumes 17 times as much as the average Mexican (emerging economy) and 100 times more than the average Ethiopian (developing economy) (2). Thus, achieving global sustainability will in all likelihood entail some limits to consumption in the developed world. Subtler is the fact that Hart ignores the social context of the development of clean technology. In other words, as Beder warns (see background article in Chapter 7), sustainable development requires more than the mere application of technological fixes to environmental problems.

Other public policy issues in the global arena extend beyond the notion of sustainable development. (Some of these are discussed in the background reading and case studies of this chapter.) Among the most important of these issues are economic exploitation of the Third World, export of environmental and safety hazards to less developed countries, technology transfer, respect for cultural diversity, and war and conflict resolution. Like sustainable development, however, these policy issues are also marked by important ethical questions that engineers working in the global economy need to be aware of.

Background Reading

Philosopher *Thomas Donaldson* considers "The Ethics of Global Risk" in his 1986 essay. Donaldson is concerned about the ethics of corporations based in the industrialized countries imposing technological risks on less developed nations, especially the underprivileged of those nations. Such risks include establishing hazardous industrial facilities (such as the one that resulted in the deadly chemical leak in Bhopal, India) and marketing products abroad that are banned for sale in the corporation's home nation. In such cases, environmental, health, and safety standards are either nonexistent or much lower in the other nation.

Donaldson begins with the twin assertions that cultural variables can make both ethical analysis and risk assessment problematic. Such factors as willingness to accept economic development/risk tradeoffs may differ among nations. Moreover, conventional risk assessments can pose such ethical problems as inequitable risk/benefit distribution in developing countries and valuation of life in solely economic terms (which makes life very cheap in the developing world). Donaldson suggests a helpful distinction between justifications for imposing risks made on cultural grounds and those made in the name of economic development for the nation being exposed to the risk.

In the latter case, he claims, the best moral approach is to "put ourselves in the shoes of the foreigner." If we would willingly accept the risks in question were our society in a similar state of economic and social development, then Donaldson believes it would be morally permissible to impose risks on others that we would not currently tolerate. When cultural factors are the justification, however, Donaldson believes that the home nation's views should be respected unless some universally accepted human right is being violated or there is reason to believe the common citizens of the other nation are not voluntarily accepting the risks.

Donaldson distinguishes his approach from the paternalistic "White Man's Burden" rationale by noting that what he proposes is not to regulate others but to regulate the actions of our own government and companies based within our nation. Donaldson also justifies intervention under the conditions outlined above based on the fact that most underdeveloped nations are not democratic and are usually unable to make informed decisions about technological risk. Although he supports efforts to improve these conditions in the developing world, he believes that the developed nations must accept moral responsibility for their own actions.

Cases for Discussion

John Byrne, Constantine Hadjilambrinos, and *Subodh Wagle* present an important case study with global repercussions in their 1984 article "Distributing Costs of Global Climate Change." Global warming may be occurring as a result of emissions of certain greenhouse gases, most notably carbon emissions from the burning of fossil fuels. The authors begin by critiquing three conventional approaches to climate change that have been developed by economists from industrialized nations.

The "no regrets" approach, they argue, places too high a premium on economic growth and too much faith in technological fixes to solve the problems caused by climate change. The "insurance approach" differs slightly in that it favors taking some steps to mitigate the effects of climate change while further scientific research is conducted to narrow the uncertainties. Both of these approaches, the authors maintain, are focused on economic growth and entirely neglect questions of distributive justice. The third approach, which advocates pollutant markets, seeks to determine the optimum level of greenhouse emissions and then turns the resolution over to market economics through devices such as carbon taxes or emission trading permits. This approach, too, fails to equitably account for the dominant position of the industrialized nations. Ethical problems common to all of these approaches include valuing human life on the basis of earnings capacity, assigning no intrinsic value to the environment, and devaluing sinks for carbon emissions (such as forests). Moreover, all these approaches put a disproportionate burden on the developing world, while giving the industrialized nations a free pass for their past carbon emissions that have led to the problem.

Byrne et al. are also critical of some proposed approaches that purport to acknowledge equity issues explicitly, including allocating emission rights on the basis of land area, in proportion to current emission levels, in proportion to population, and on a per capita basis taking into account past emissions. Although these approaches are an improvement, none of them focuses on achieving a sustainable limit to human-induced climate change. The authors therefore propose a sustainability-based equity principle that would allocate the biosphere's ability to absorb greenhouse gases on a per capita basis. This calculation, admittedly difficult, would become the basis of a tradable system of emission permits. One means of resolving the resulting imbalance in allowable emissions between the industrialized and developing nations would be to trade the financial debt of the developing nations for the "debt" accumulated by the industrial nations by excessive past emissions contributing to climate change.

Godfred Frempong discusses "Developing Telecommunications in Ghana" in a 1996 article. Frempong begins by noting the important relationship between telecommunications and socioeconomic development, followed by a brief review of the major telecommunications infrastructure initiatives in Ghana over the past two decades.

Despite these initiatives, the availability and quality of telecommunications services in Ghana remain quite low. Although Ghana's overall economic development has been higher than average for the region, as late as 1990, telephone service was available at a rate of only 0.30 per 100 people, with the vast majority of this service being available in urban areas. The penetration of telex services is also below par for the region, although telex is rapidly giving way to fax as a means of data transmission. As measured by such indicators of quality of service as call completion rates, Ghana, with an average call completion rate of about 50 percent, falls well short of the targets the International Telecommunications Union has set for developing nations. Not surprisingly, Frempong also finds investment rates in telecommunications to be lower in Ghana than in other countries in the region.

Frempong identifies a number of problems hampering Ghana's telecommunications efforts: inadequate funding, obsolete equipment, lack of qualified system operators, poor maintenance procedures, poor worker pay and morale, and lack of overall infrastructure in the less developed regions of the country. Frempong notes that the poor state of telecommunications in Ghana has serious implications for socioeconomic development. These include difficulties in attracting industrial and commercial activity due to lack of adequate telecommunications access and high costs for service; lack of support in rural areas for the agrarian economy on which the nation relies; and impediments to the country's socioeconomic development plans, which call for decentralization of government services.

Frempong concludes with a number of lessons learned from the Ghanian experience in telecommunications development. He believes that the publicly funded authority in Ghana is not up to the task of rapid telecommunications development, and he favors opening up the market to more competition from private companies, as is the case in cellular communications and paging. As a corollary, he notes that developing countries such as Ghana cannot be the sole source of investment in their telecommunications systems; successful development also requires private investment.

References

[1] World Business Council for Sustainable Development, World Wide Web site: http://www.wbcsd.ch/.

[2] Hart, S. L. 1997. Beyond greening: strategies for a sustainable world. *Harvard Business Review* (January–February): 66–76.

[3] Barnett, C. C., and Lulofs, F. 1996. *Sustainable cities.* Research Triangle Park, N.C.: Research Triangle Institute.

[4] Pollution prevention pays, World Wide Web site: http://www.mmm.com/profile/envt/3p.html.

[5] Magretta, J. 1997. Growth through global sustainability: An interview with Monsanto's CEO, Robert B. Shapiro. *Harvard Business Review* (January–February): 79–88.

FEATURE ARTICLE

The Ethics of Global Risk

THOMAS DONALDSON

In India, The Philippines, Nigeria, and elsewhere, technology is superimposed upon ancient cultures. Not long ago in Bhopal, India, the devastating potential of technology's hazards in a non-technological culture was brought home with awesome pain—over 2,000 dead and 200,000 injured. My aim here is to inquire about the justice of practices, like those in Bhopal, that subject foreign citizens to technological risks higher than those faced by either home country citizens or more favored foreign citizens. What moral obligations underlie, what extra-national responsibilities should inform the behavior of global actors such as Union Carbide and the United States? The question not only intrigues us, it demands an answer on the behalf of those who have been harmed or who are presently at risk. Yet it appears disturbingly clear that the question as framed eludes answers because we possess no viable scheme for applying traditional moral concepts to the twilight created by the juxtaposition of differing legal and cultural traditions.

THE BHOPAL TRAGEDY

Let us begin by sketching key elements of the disaster in Bhopal, India. Bhopal is by no means unique in the history of chemical catastrophes [1], but it is striking for the enormity of its scale and, more importantly, for the lessons it teaches [2].

Although the entire story remains to be told, blame for the disaster is likely to be spread through a complex constellation of persons and acts. In the preceding year, cost-cutting measures had severely weakened safety control. The refrigeration unit designed to cool the methyl isocyanate had been broken for some time, and more than a score of crucial safety devices specified in the safety handbook prepared by Union Carbide in the United States were conspicuously absent. The training, habits, and attitudes of Indian employees were lax and naive. Safety procedures specified in the book were routinely circumvented by technicians who, lacking adequate training, went on with their work blissfully ignorant of the dangers lurking behind their daily routines. In responding to the disaster, employees showed bad judgment and bad training: upon first learning of the initial leak, the officer in charge opted to think about it over tea. Outside the plant, government regulatory authorities and city officials were entirely at a loss whether to inspect and regulate the plant on an ongoing basis, or to respond appropriately to a disaster once it occurred.

Finally, Union Carbide itself, although it held a majority of its subsidiary's stock and had accepted responsibility for all major economic and safety decisions, had failed to maintain an adequate system of safety accountability, and consequently to exercise appropriate control over its subsidiary.

Yet Bhopal is not only a story about tragedy and human frailty, it is also a story about injustice. For the people who died and suffered were not citizens of the nation whose corporation held ultimate responsibility. To make matters worse, the people who suffered most were the slum dwellers, the poorest of the Indian poor, who had pitched their huts literally next to the walls of the Carbide plant.

CULTURAL VARIABLES AND RISK ANALYSIS

Cultural variables muddy moral analysis. Whereas in the context of our own culture we can estimate with some assurance the value of goods sacrificed or put at risk by undertaking a given act or policy, when dealing with a foreign culture our intuitions are opaque. Our extra-cultural vision may be sufficiently clear to allow us to understand a tradeoff between risk and productivity—between the dollar benefits of an increased gross national product on the one hand, and the dollar cost of the medical care necessary to accommodate higher levels of risk—but our vision is blurred when attempting to assess more culturally dependent tradeoffs. However difficult for us to understand, a citizen of Zimbabwe, Africa, may be willing to trade off a few marginal dollars in per capita gross national product for the unquantifiable benefit of his or her nation's economic independence from earlier colonial powers. For the same reason, he or she may even be willing to trade off a fraction of a percentage point in the nation's infant mortality rate. Similarly, a citizen of Pakistan may be more eager to preserve his or her country's Moslem heritage, a heritage with strict sexual differentiation in the division of labor, than to increase the country's economic welfare by integrating women into the workplace.

Cultural variables can also derail traditional methods of risk analysis. This is especially true of the tendency of most methods to focus on dollars and bodies at the expense of social and cultural criteria, a tendency which, while reprehensible in domestic contexts, becomes pernicious when the difference between two countries' social and cultural habits is striking. Consider the twin issues of the distributing and pricing of risk. It is well known that the techniques of cost-benefit analysis are often mute regarding issues of distributive justice; that is, they tend to bypass questions of the fairness of a practice from the

Thomas Donaldson with Loyola University of Chicago.

Reprinted from *IEEE Technology and Society Magazine,* Vol. 5, No. 2, pp. 17–21, June 1986.

perspective of its relative impact on social subclasses, such as the poor, the infirm, or the members of a minority ethnic group. Such silence is less neglectful in the context of a national legal system whose rules have as a central function the protection of individual rights [3]. In the context of international transactions, however, where one country's laws do not (and, in an important sense, cannot) regulate activities in another country, the silence is morally corrupt. Certainly pesticide risks to field workers must be weighed against the desperate need of a poor country for greater food production; when that development is carried entirely on the backs of the poor, when the life expectancy of the field worker is cut by a decade or more while the life expectancy of the urban elite *increases* by a decade, then considerations of fairness should trump cost-benefit considerations offered in the name of general welfare.

The common and sometimes criticized distinction in risk analysis between voluntary and involuntary assumptions of risks is of little help. If we are uneasy about the assumption that the decision of a lower class worker in the U.S. to take a high risk job is "voluntary," despite that worker's limited technological sophistication and pressing financial needs, then surely we must reject the label of "voluntary" when applied to the starving, shoeless laborer in Bangladesh who agrees to work in a pesticide-infected field.

Finally, the tendency of cost-benefit analysis to tie costs to market prices can distort risk tradeoffs in less developed countries. The dominant assumption of most risk analysis, and of cost-benefit analysis in particular—that risks must be balanced against costs—means that, in the instance of life-threatening risks, human life must be assigned a price. Despite the apparent barbarity of the very concept, defenders point out that most of us are willing to assume non-zero risks to our life for the sake of reducing cost and frequently do so when we, say, buy a smaller car or accept a higher paying but riskier job [4]. But while assigning a price to human life may have beneficial consequences against the backdrop of a single developed country—i.e., it may help policy makers better allocate scarce safety-promoting resources—in the Third World it can unfairly relativize human worth. Since the market price of a life is tied to the capacity of a person to generate income, and since, in most parts of the Third World, the absence of a capital infrastructure limits the average individual's productive capacity, it follows that in the Third World a human life will be assigned a lower price.

If cultural variables confound risk analysis, then how can such analysis address international problems? One tempting solution must be abandoned, namely, reliance on international market pressures for acceptable risk distribution. What the market does unsuccessfully in a national context, it fails to do utterly in an international context. As Charles Perrow has pointed out [5], even in developed countries "there is no impersonal fair market that rewards those that risk their lives with higher wages." The "jumpers" or "glow boys" in the nuclear industry, temporary workers "who dash into a radioactive area to make repairs, will be hired for two or three weeks' work, at only six dollars an hour . . . Textile workers are not compensated for brown lung disease, nor are chemical plant workers compensated for cancer showing up ten or twenty years after exposure" [6].

The average level of unemployment in the Third World today exceeds forty percent, a figure that has frustrated the application of classical economic principles to the international economy on a score of issues. With full employment, market forces will—all other things being equal—encourage workers to make tradeoffs between job opportunities using safety as a variable. But with massive unemployment, market forces in Third World countries drive the unemployed to the jobs they are lucky enough to land, regardless of safety. Does some criterion exist, itself not bound by culture or nation, that can give objectivity to intercultural assessments of risk distribution?

It will help to provide concrete contexts for this problem. Consider first the selling of domestically banned goods abroad [7], a practice which appears to involve double standards [8]. In 1979 the U.S. Congress passed legislation amending the Export Administration Act which gave the President broad control over exports [9]. But on February 17, 1981, just 36 days after the signing of the order, newly elected President Reagan revoked it [10]. He furthermore called for a repeal on the export restrictions affecting unapproved drugs and pharmaceutical products. (Banned pharmaceuticals, in contrast to other banned goods, have been subject to export restrictions for over 40 years.) In defense of the Reagan initiative, drug manufacturers in the United States appealed to differing cultural variables. For example, a spokesman for the American division of Ciba-Geigy Pharmaceuticals justified relaxing restrictions on the sale of its Entero-Vioform, a drug he agrees has been associated with blindness and paralysis, on the basis of culture-specific cost-benefit analysis. "The government of India," he pointed out, "has requested Ciba-Geigy to continue producing the drug because it treats a dysentery problem that can be life-threatening" [11].

Let us consider a second case, this time involving the world's worst pollution. A small triangle of land near Sao Paulo, Brazil, known as Cubatao, has more reported cases of cancer, stillbirths, and deformed babies than anywhere else in the world [12]. Factories, and especially petrochemical plants, dominate the landscape where about 100,000 people live and work. Cubatao's air is considered unfit on a record number of days, and its rainfall contains the highest level of pollutants recorded anywhere. In 1983, 100 slum dwellers living alongside a gasoline duct were killed when the duct caught fire. The town was constructed during the heyday of the so-called "Brazilian miracle," a time when right-wing military rulers maintained pro-business labor laws, stable political conditions, and some of the highest profit margins in the world: conditions that allowed them to attract an enormous influx of foreign investment. Even today, with the Brazilian miracle fallen into disrepute, substantial foreign investment remains. Cubatao's 111 plants are owned by 23 foreign and Brazilian companies.

According to Marlise Simons of *The New York Times,* "Squatters have built rows of shacks above a vast underground grid of ducts and pipes that carry flammable, corrosive, and explosive materials. Trucks lumber alongside, loaded with poison which has spilled in past accidents. . . 'But we need the work,' one man said. 'We have nowhere else to go' " [13].

Many people try to justify international risk inequities like those in the preceding two cases by appealing to a nation's special needs, e.g., for economic development or the elimination of a particular problem. Lower safety, pollution, and import standards are explicitly maintained by some countries in order to achieve special ends. In Brazil, for example, lax standards of pollution enforcement are justified in the name of Brazil's desperate need for greater productivity, and the claim has a persuasive edge in a country where malnutrition is sufficiently widespread that by some estimates one in every five Brazilian children will suffer permanent brain damage [14]. It seems morally arrogant to suppose that acts undertaken by representatives of developed nations that encourage or tolerate lower standards abroad are impermissible simply for that reason. On the other hand, the convenient relativism of some corporate and government officials, which excuses anything in the name of cultural difference, seems equally suspect.

Light is shed on this issue by drawing a distinction between those instances in which, from the standpoint of the foreign country, the reason for tolerating the "lower" norms refers to the country's relative level of economic development, and those in which the reason for tolerating them is related to inherent cultural beliefs, e.g., in religion or tradition. When an instance falls under the former classification, a different moral analysis is required than when it falls under the latter. Here it makes sense to do what, for cultural reasons, cannot be done in the later instance (where inherent cultural beliefs intrude), namely, put ourselves in the shoes of the foreigner. To be more specific, it makes sense to consider ourselves and our culture at a level of economic development relevantly similar to that of the other country. And if, having done this, we find that *under such hypothetically altered social circumstances* we ourselves would accept the lower standards, then it is permissible to adopt the standards that appear inferior.

What lies behind the thought experiment is an age-old philosophical insight, namely, that when considering the universality of moral principles like must be compared to like, and cases must be evaluated in terms of morally relevant similarities. Hence, when considering the acceptability of practices abroad, the moralist must not err by applying wholesale the principles relevant to his or her own nation, but instead must inquire what those principles would imply under the relevantly altered circumstances of the foreign nation.

Let us be more specific. We can "test" the practices of shipping banned products abroad and of operating multinational branch facilities in Cubatao by a thought experiment wherein we ask whether our own moral intuitions would find such practices acceptable were we in a state of social development relevantly similar to that of the country in question. This test works in such cases because the values that presumably prompt the lower standards in foreign countries are ones we share, i.e., economic and medical well-being. For example, it makes sense to ask whether we in the United States would find levels of pollution equal to those existing in Cubatao justified here for the sake of economic progress, were we at Brazil's present level of economic development. If we answer yes, then we may conclude that it is permissible for U.S. multinationals to adopt the lower pollution standards existing in Cubatao. If not, then the practice is not permissible. (I suspect, by the way, that we would *not* find Cubatao's pollution permissible.)

The same test is appropriate in the case of banned products. Were we at a hypothetically lower state of economic development similar to that of Ghana or Colombia, would we allow Tris-Treated Sleepwear (sleepwear treated with a fire-retardant known to be highly carcinogenic and hence banned from the United States market) to be bought and sold? Probably not. Yet, lest one think that the test always returns negative results, consider the case of India's special request for the drug Entero-Vioform. Dysentery, a widespread and virulent health problem in India, is often associated with undeveloped societies because of their lack of modern systems of food handling and sanitation. It may well be that, if we imagine ourselves in a relevantly similar social situation, the tradeoffs between the risks of the drug's side effects and the widespread dysentery that would occur without the drug would favor the import of Entero-Vioform, despite its properly being banned in developed countries.

As a practical matter of moral psychology, some acts of rational empathy are easier than others. This is reflected in the distinction between lower standards, that all justified in terms of relative economic development and those that are not. It is relatively easier for us to empathize with the need for economic development in a poor nation, since economic well-being is an almost universally shared value, than it is to empathize with the need for a purer form of Muslim government, or for a more African, less European social system. The general principle governing rational empathy seems to restrict it to situations wherein the fundamental values which motivate the decision making of those with whom we seek to empathize are values that we share. In instances where we fail to share the moral values that prompt lower safety standards, the test of rational empathy is inappropriate, for reasons already stated. Here the final appeal can only be to a floor of universal rights, with the presumption in favor of permitting the lower standards unless doing so violates a basic right. Unable to appeal to the values that must ultimately underlie social welfare tradeoffs, we must presume the validity of the foreign government's stance, except in the instance where a universal human right is at stake, or where we doubt the voluntary acceptance of the lower norms by rank-and-file citizens. Appealing to rights in such cases has special validity, because rights are, by definition, moral concepts that specify moral minimums and prescribe, as it were, the lowest common denominator of permissibility [15].

LIMITS ON THE IMPOSITION OF RISK

The preceding analysis has shown, then, that there are firm limits to the extent to which we can impose risks on persons in foreign countries, even when such risks fall within existing moral and legal guidelines operative in the foreign country.

But what about the "White Man's Burden"? Isn't it morally arrogant of us to prescribe levels of risk for other sovereign nations? Yes, of course, it is, but we are not talking here about regulating the actions of foreigners; we are talking about regulating *our* actions. We are talking about the standards *we* have a right to impose on corporations we control, or on products *we* export, not about standards that foreigners have a right to impose on their own corporations and exports. The point about the White Man's Burden also fails because, in the instance of most Third World countries, the agents who assume risks are surrogates, which is to say that they act on behalf of third parties to whom they are presumably responsible. Surrogate agency would be less damning were it true that both democracy and informed public opinion laid behind it, but in most Third World countries this is not the case. Most are far from democratic in the sense of democracy to which we are accustomed, and, even when democratic, possess a level of technological sophistication sufficiently low to rule out the possibility of rational risk assessment. In Bhopal (which happens to be a good-sized city), only one in a thousand households has a telephone. It is arrogant self-delusion for us to imagine that such people make rational decisions about exposing themselves to the risks of methyl isocyanate.

The idea of a society "choosing" to undertake risks when it lacks a sufficient political and technological infrastructure lies at the root of much unwitting technological imperialism. Again, consider Bhopal. Even the Indian employees of Union Carbide were unaware of methyl isocyanate's toxicity; most thought it was chiefly a skin/eye irritant, and almost all knew nothing of its lethalness.

Outside the plant, the Indian regulatory apparatus was woefully unequal to its task. A few weeks before the disaster, the Union Carbide Plant had been granted an "environmental clearance certificate" [16]. Enforcement was left not to the national government, but to the separate states. In Madhya Pradesh, the state in which Bhopal lies, 15 factory inspectors were given the task of regulating 8,000 plants, and those inspectors, who sometimes lacked even typewriters and telephones, were forced to use public buses and trains to get from factory to factory. The two inspectors responsible for the Bhopal area held degrees only in mechanical engineering, and knew little about chemical risks [17]. The regulators confessed that chemical safety was an issue they left up to the plant managers. It should be added that India is considerably more advanced technologically, with a better technological infrastructure, than most of its Third World counterparts.

Bhopal offers many lessons about what steps Third World countries must take in order to reduce irrational technological risks, among which are the institution of suitable zoning ordinances, better inspection and regulation of hazardous factories, and the acceptance of only those technologies that the local technological infrastructure is capable of handling. Similarly, there is little doubt that these same countries have unfulfilled responsibilities in other areas, including policies affecting the importation of banned products. Nicholas Ashford, for example, has offered a tidy list of recommendations: that such countries coordinate industrial development policy with environmental policy (frequently the Ministry of Industry does not confer with the Ministry of the Environment); that they develop a data base for the assessment of imported products' safety and their effects on productivity; and, finally, that they maintain a centralized purchasing control mechanism for choosing products or technology that will enter the country [18].

Realism, however, demands that we recognize the unlikelihood of such reforms in the near future. We cannot, as I hope this article has shown, justify our own irresponsibility by thrusting the moral burden on the shoulders of societies still adolescent in the age of technology.

NOTES AND REFERENCES

[1] In 1972 between 400 and 5,000 Iraqis were killed as a result of eating unlabeled, mercury-treated grain from the United States. In 1979, workers and livestock in Egypt were poisoned by the pesticide loptophos, and more recently, hundreds died and were injured in Mexico Coty as a result of a liquified natural gas explosion. Nicholas A. Ashford, "Control the Transfer of Technology," *The New York Times,* p. 2, Dec. 9, 1984.

[2] The information used to construct the following description comes largely from a series of four articles in *The New York Times,* Jan. 28, 30, 31, and February 3, 1985. These articles, as well as a preliminary article which appeared in the *The New York Times* on Dec. 9, 1984, shortly after the disaster, were written by Stuart Diamond. Mr. Diamond's account comes largely from interviews with workers, including Mr. Suman Dey, who was the senior officer on duty.

[3] The view that the fundamental function of a legal system is the protection of rights is articulated systematically by Ronald Dworkin in *Taking Rights Seriously,* Cambridge, Mass.: Harvard University Press, 1978.

[4] For an insightful account of the moral assumptions involved in risk analysis see Kristin S. Shrader-Frechette, *Risk Analysis and Scientific Method: Methodological and Ethical Problems with Evaluating Societal Hazards,* Hingham, MA: D. Reidel Publishing Co., 1985.

[5] Charles Perrow, *Normal Accidents,* New York: Basic Books, p. 68, 1984.

[6] *Ibid.,* p. 68.

[7] The problem, by the way, is not limited to the United States, since Europe exports even more hazardous products to developing countries than does the U.S. See Nicholas Ashford, "Control the Transfer of Technology," *The New York Times,* p. F2, Dec. 9, 1984.

[8] A 1979 United Nations resolution stressed the need to "exchange information on hazardous chemicals and unsafe pharmaceutical products that have been banned in their territories and to discourage, in consultation with importing countries, the exportation of such products. Quoted in "Products Unsafe at Home Are Still Unloaded Abroad," *The New York Times,* p. 22, August 22, 1982.

[9] With this as his basis, President Carter issued on January 15, 1981, an executive order that asked for a comprehensive approach to hazardous exports. The complex notification schemes for alerting foreign countries about hazards were to be coordinated and streamlined. An annual list of all products banned in the U.S. was to be compiled and made available, and government officials were empowered to seek international agreements on hazardous exports. Finally, the order required the creation of export controls on those "extremely hazardous substances" that constituted "a substantial threat to human health or safety or the environment." See "Control the Transfer of Technology," [7].

[10] Industry opposition, described as "massive" by Edward B. Cohen, Executive Director of the Carter Administration's Task Force on Hazardous Exports Policy, was probably what killed the Carter plan. See "Products Unsafe at Home," [8], p. 22.

[11] Quoted in "Products Unsafe at Home," [8], p.22.

[12] Most of the informaiton about Cubatao here described is from Marlise Simons, "Some Smell a Disaster in Brazil Industry Zone," *The New York Times,* p. 4, May 18, 1985.

[13] *Ibid.,* p. 4.

[14] See "Controlling Interest," a film produced and distributed by California Newsreel, California, 1977.

[15] In attempting to isolate the list of rights with true claim to cultural universality, we might, for example, consult international documents such as the U.N. Declaration.

[16] Surprisingly, the Union Carbide plant at Bhopal was considered almost a model for other plants. In constrast to a steel plant in the same state that had 25 fatalities in the past year, Union Carbide had in recent years only a single fatal accident.

[17] Robert Reinhold, "Disaster in Bhopal: Where Does Blame Lie?" *The New York Times,* p. 1, Jan. 31, 1985.

[18] Ashford, "Control the Transfer of Technology," [7], p. 2F.

JOHN BYRNE, CONSTANTINE HADJILAMBRINOS, & SUBODH WAGLE

Distributing Costs of Global Climate Change

Desertification in Mali.

The problem of global climate change links the issues of energy utilization, economic development, environmental degradation, and equity on a planetary scale. While questions concerning the scale and timing of the impact of continuously increasing emissions of greenhouse gases remain, the outcome of the Earth Summit conference in the summer of 1992 was that coordinated international action is necessary to begin addressing the problem [1]. While the debate over the type and extent of such actions continues and while some proposals based on principles of equity and fairness have been put forth [2], [3], a set of approaches has emerged which claim to objectively demonstrate that nothing or very little should be done to address this problem.

Presented at the International Symposium on Technology and Society 1993, Washington, DC, October 1993. *The authors are with the Center for Energy and Environmental Policy, University of Delaware, Newark, DE 19716.*

These approaches rely on standard economic theory and analytical methods to reach their conclusion that significant worldwide action is unnecessary. In our view, however, these approaches contain biases which place an exceptionally high value on maintaining the status quo of global patterns of resource consumption and distribution of wealth. When utilized to analyze various policy options, this bias results in a determination that equity considerations are too costly and may impede technological and economic progress [4]. As an alternative, we propose an approach based on a principle of equity in atmospheric resource utilization.

Economic Approaches to Policy Formulation

Policy formulations to address the problem of global climate change based on economic theory can be placed in three broad categories: the "no regrets," "insurance," and "pollutant market" approaches. While each of these approaches is

Reprinted from *IEEE Technology and Society Magazine*, Vol. 13, No. 1, pp. 17–24 & 32, Spring 1994.

distinguished by a particular combination of values, basic assumptions and arguments, several elements which are common to all three can be identified.

Perhaps their most important shared characteristics are that they have been developed by economists from industrialized nations, primarily the U.S., and were originally designed to evaluate response options of these nations in the event that international action was demanded. These characteristics are reflected in the basic values which shape the arguments and conclusions of these approaches.

▼ *No Regrets Approach*

Several economists who figure prominently in the global climate change policy debate, including Cristofaro [5], Nordhaus [6], and Shelling [4],[7], offer analyses which are typical of the "no regrets" approach. Two arguments distinguish this approach: the high value placed on economic growth and the great faith invested in the ability of as-yet-undeveloped technology to address humanity's major problems.

The "no regrets" approach proposes that societies respond to global climate change by comparing the costs and benefits which their economies will incur in responding to this problem with the costs and benefits of doing nothing. According to its proponents, only actions which result in benefit/cost ratio to a society that is greater than or equal to one are justified.

No restrictions are placed on the level of greenhouse gas (GHG) emissions so long as the net economic benefits outweigh the expected environmental costs. The latter are determined by assigning probabilities to the risks of global warming and multiplying them by damage estimates expressed in economic terms. Also, the scenarios used to evaluate different policy options include measures designed to mitigate, as well as to prevent, the possible negative effects of global climate change [5], [7]. This is significant because it places the burden of the argument on minimizing economic costs. Specifically, prevention can be justified under this approach only if it is cheaper than mitigation.

Not surprisingly, proponents of this approach generally conclude that no action to eliminate the possibility of global climate change should be taken unless it is justified by considerations that are extraneous to the issue of global climate change (such as productivity gains from advanced technologies that, incidentally, result in lower GHG emissions). The use of present values[1] in the cost-benefit analysis places greater emphasis on the costs of action, which will have to be undertaken early on, than on the benefits, which occur mostly in the distant future. Thus, the potential cost of adverse environmental effects, when combined with the uncertainty of whether and when they may occur, and then discounted into a present value, appears insignificant compared to the cost of prevention.

Adaptation to the effects of global climate change, rather than any concerted efforts at prevention, then, appears in this approach to be the most economically sound option. In an interview in the *New York Times* [8], Schelling cogently argues the "no regrets" position:

> *While climate has not changed rapidly in the last century, both the will and technological ability to adapt to radically different weather obviously has... While changes in rainfall, temperature and sea level could be dramatic, there is yet no reason to believe that the process would be completed too quickly to allow evolutionary responses — expanding irrigation, for example, or building dikes. The cost of growing food might conceivably rise by 20 percent, Mr. Schelling speculates. But this loss, he argues, is almost certain to be overwhelmed by a century's worth of improvements in seed strains and growing techniques... The quality of life in 100 years, he suspects, will depend as much or more on the endowment of technology and capital as on the percentage of carbon dioxide in the air. And if money to contain carbon emissions comes out of other investment, future civilizations could be the losers.*

Conducting similar analyses, and following the same reasoning, economists of the U.S. Department of Agriculture in a 1989 study concluded that the costs of decreased grain yields in industrialized countries would be less that one-tenth of one percent of GDP [8]. This is because losses in yields would be largely offset by higher world prices. Econometric models used by the U.S. Environmental Protection Agency indicate that global warming could actually increase agricultural income in the U.S., while technologies to protect coastal cities against sea-level rise — if such a phenomenon should occur — would probably cost far less than curtailments of greenhouse gas emissions [8],[9]. In a basic sense, the results of all of these analyses are the foregone conclusion of a method that assigns inherent value to economic growth and assumes that any serious environmental obstacle to growth can and will be overcome by new technology.

The "no regrets" approach, in its assumptions and its method of evaluation, anticipates but fails to justify the international distribution of costs and benefits that accompany a "no regrets" response. Global increases in food prices may

[1] The present value method discounts future costs and benefits to reflect their value in the present. The further cost or benefit is to be incurred in the future, the smaller its value in the present is, everything else being equal.

have little impact on the citizens of the U.S. and other industrialized nations but could have devastating effects on the citizens of developing nations. These nations may find it difficult or impossible to muster the resources necessary to protect themselves from a sea level rise, especially as some of them, particularly the small island and low-lying nations, face the possibility of complete inundation. The distributive perversity of "no regrets" is not accidental. The costs and benefits it assumes as data are the facts of a highly uneven global economy. It is logically inescapable that this data, when introduced into the "no regrets" method, will yield results that favor the interests of the industrialized nations over the developing countries.

▾ *Insurance Approach*

A second economic model, the "insurance" approach, seeks to determine the shares of resources which should be diverted from economic growth in order to "purchase" some measure of insurance to forestall catastrophic global climate change. This approach places great value in the scientific analysis of global change which, it assumes, can eventually determine with considerable certainty what the effects of increasing GHG concentrations will be on climate, oceans, species, etc. The aim of this "insurance policy" is to buy societies time until science can inform its actions [10]. Devoting resources to the scientific study of global warming, thus, constitutes a form of insurance against future risks. Such insurance can also be "purchased" through the expenditure of some resources to limit GHG emissions. The amount of resources to be spent on slowing the rate of GHG accumulation in the atmosphere is to be determined by an estimate of the time at which "adequate" information will become available and the likely severity of the environmental effects [10]. It is important to recognize that this approach assumes a first-order commitment to mitigation only. Prevention of global climate change is not justifiable at this time under the "insurance" reasoning of the problem.

The "insurance" approach shares with its "no regrets" counterpart the belief that economic costs and benefits should be the determining factors of social action. But its proponents are willing to concede that environmental consequences could go beyond what our economic systems are currently organized to evaluate. To grapple with this problem, science is summoned to predict the course of events accompanying global climate change. The "purchase" of "greenhouse insurance," i.e., some immediate action to lessen the effects of global climate change, is justified because scientific knowledge indicates that the process of global warming may have already started [11], but science does not yet possess the means to predict its course [10], [12].

According to proponents of this approach such as Manne and Richels [10] and Cline [12], any policy course designed to address this problem should be based on scientific knowledge and should be cost-effective. Manne and Richels spell out the reasoning behind this approach [6]:

> *It is straightforward to calculate the benefits from reducing scientific uncertainty. Better information leads to better decisions. Global 2100 allows us to calculate the aggregate economic impacts. The analysis leaves little doubt that there can be a big payoff to reducing climate-related uncertainties — something of the order of $100 billion for the U.S. alone... The near-term implications are clear. There is less need for precautionary emission cutbacks if we undertake a sustained commitment to reducing climate uncertainty and to developing new supply and conservation options. Better climate information reduces the need to hedge against a potentially hostile future. Improved supply and conservation technologies will enhance our ability to deal with such a future if it should occur.*

The "insurance" and "no regrets" approaches consider global warming of any magnitude as acceptable if it makes "economic sense." Both propose analytical methods which center around the evaluation of the aggregate monetary benefits and costs of various policy alternatives while virtually ignoring distributive effects. Neither approach considers the distribution of costs and benefits among social classes, nations, and regions of the earth. Indeed, the most fundamental difference between the two approaches is that "no regrets" proponents show a moderately greater reluctance to espouse immediate action to counter the threat of global climate change, thus placing complete faith in technological progress. The only other area of difference concerns the estimates of damage from global climate change. Insurance proponents use less conservative assumptions in estimating potential damages and also entertain the possibility that a technological fix may not be forthcoming [12]. However, these differences pale in comparison to their agreement on the core issues of economic primacy in valuation and the absence of any consideration of distributive effects.

▾ *Pollutant Market Approach*

Advocates of the third approach — "pollutant markets" — do not attempt to address the question of global climate change directly. Instead, they focus on the method of societal response. There is really no contest for economists on this question. They uniformly hold that market mechanisms are the most efficient way for achieving any reduction in greenhouse gas emissions. Once agreement has been reached on the permissible levels of such emissions, a free market is advocated to allocate emission rights. This is generally contemplated as taking the form of either a global carbon tax, or an international emissions trading regime [13], [14]. The rationale for these methods is illustrated by the following discussion by Cline [12]:

If the international community moves to limit global warming, it will be important to limit costs by applying efficient instruments. Most economists favor carbon taxes as the most efficient mechanism for reducing emissions... Most analyses show that a physical quota regime could approximate the efficiency advantages of a carbon tax if tradeable permits are allowed.

The pollutant market approach advocates that, if any action is warranted, it should be based on firm economic foundations. This approach is an articulation of the principle of maximization of economic efficiency and, as far as method is concerned, it is complementary with the "no regrets" and "insurance" approaches. Markets are advanced as simply an instrument for achieving an objective which is decided by some other method. This economic instrument, however, is not devoid of values. Like "no regrets" and "insurance," "pollutant markets" recognize no intrinsic value for the natural environment. The price of environmental stability is determined by the actors in the pollutant market, the "buyers" and "sellers" of pollution. The outcome of this approach is not difficult to guess. The "sellers" will be the industrialized nations of the world. As they already own most of the world's economic wealth, they possess the means for influencing prices to their advantage.

When no provision is made for addressing issues of equity, adherence to the principle of maximizing economic efficiency can provide the means for industrialized nations to dictate the terms of greenhouse policy to the developing nations. A more detailed analysis of the impact of the economic approaches on the on-going international policy debate illustrates their inherent problems.

Problems Inherent in Economic Approaches

The economic approaches reviewed here depend exclusively on the quantification of costs and benefits of various policy alternatives to select the most appropriate policy response. There are no overarching ethical principles upon which outright rejection of some costs can be based. Instead, the test of economic efficiency rules. The blindness of economic analysis to ethically objectionable positions is amply illustrated in proposals recently made by Lawrence H. Summers, chief economist of the World Bank, on the issue of the international trade and disposal of hazardous wastes [15]:

The measurement of the costs of health-impairing pollution depends on the foregone earnings from increased morbidity and mortality. From this point of view a given amount of health-impairing pollution should be done in the country with the lowest cost, which will be the country with the lowest wages. I think the economic logic of dumping a load of toxic waste in the lowest-wage country is impeccable and we should face up to that... I've always thought that under-populated countries in Africa are vastly under-polluted; their air quality is probably vastly inefficiently low...

There are a host of values and ideological predispositions in economic approaches to the global warming question. Four are highlighted below as among the most salient:

1) Human life is valued differentially, on the basis of wage-earning capacity. In pragmatic terms this means that people are valued on the basis of class, sex, ethnicity and country of residence.

2) The natural environment has no intrinsic value. Any amount of environmental degradation (for example, any level of global warming) is acceptable if the economic cost of avoiding such degradation is outweighed by the net benefits of economic growth that would have to be foregone.

3) CO_2 sinks such as the forests are devalued, their only inherent value being as lumber. The CO_2 absorption capacity of the biosphere is incidental from an economic point of view until and unless economic value is assigned to this capacity on the basis of, for example, a carbon tax.

4) Economic activity is intrinsically valued for its easily and directly measurable income-producing capacity.

Together, these predispositions of economic analysis determine a policy preference that prizes economic growth and sacrifices environmental capacity unless it can be monetized. Again, Lawrence Summers succinctly states the economic viewpoint [16]:

...[T]here is no intellectually legitimate case for abandoning accepted techniques of cost-benefit analysis in evaluating environmental investments... The argument that a moral obligation to future generations demands special treatment of environmental investments is fatuous. We can help our descendants as much by improving infrastructure as by preserving rain forests... as much by enlarging our scientific knowledge as by reducing carbon dioxide in the air.

By their nature, economic approaches to policy-making highly value the economic processes of industrialization and capital accumulation which have impressively served the needs and interests of the industrialized nations. On the other hand, these same processes have tended to recognize little of value in the natural environment *per se*, and have led to few — and some argue no — improvements in living conditions for the people of developing nations [17], [18].

The economic models under international discussion fail to take into account the fact that while the wealth "benefits" of environmental degradation are concentrated in a few nations of the world, its costs are widespread. In fact, the adverse effects of global climate change are likely to be borne disproportionately by developing nations because of their geographic location, as well as their poverty and lack of technological development, which make them especially vulnerable.

Yet, the market mechanisms which these models indicate as the most appropriate means for limiting the effects of this environmental threat may actually shift a disproportionate amount of the burden for environmental protection to developing nations. A carbon tax strategy, for example, in which only present or future emissions are taxed, would result in writing off the cost of greenhouse gases already concentrated in the atmosphere by the industrialized nations. Additionally, because CO_2 emission rates of industrial countries are declining, while those of developing nations are and will be increasing, the burden of tax payments could quickly fall upon developing nations. Without equity safeguards written into a policy, a carbon tax regime will require the developing nations to pay dearly for the opportunity to develop, while the industrialized countries, which thus far have been entirely responsible for the build-up of greenhouse gases in the atmosphere, will have received the opportunity to develop virtually cost-free in environmental terms.

A tradeable emission-permit regime for greenhouse gases could be designed on an equitable basis. Equity, however, would have to be demanded by some external force, as tradeable permits inherently promote only efficiency and not equity. Proponents of tradeable permits concede as much when they argue for a distribution of permits on the basis of "realism." This, of course, means that the status quo (with the industrialized nations emitting far greater amounts of greenhouse gases per capita) will not be disturbed, at least not for a significant period of time [12].

The principles, assumptions, arguments, and methods promoted by the economic approaches to policy making are also ill equipped to take into account the fact that "the environment may deliver nasty and irreversible surprises" [19]. While economic principles may deem that any degree of global climate change may be acceptable, it may be that a global average temperature rise of more than 4°C (7°F), which would create a warmer earth than at any time in the last 40 million years, may bring about the downfall of human civilization [20]. What meaning could there possibly be, in this instance, to Cline's calculations of the cost of a 10° to 18°C (18° to 32°F) global mean temperature rise as 28% of GDP "under the worst assumptions" [12]?

The favored position given to economic principles and analysis in the policy-making debate is justified by its proponents on the grounds that it offers an objective means for choosing among policy options. As has been shown, however, economic approaches promote a specific set of values. The use of economic analysis and methods to the exclusion of other social considerations can lead to the adoption of inequitable policies in the name of a false objectivity and rationality.

Equity-Based Approach

An equity-based approach to policy making is necessary, not only because it could produce more just policy options, but because, in all probability, it would produce the only tenable basis for policy adoption at the international level. It is likely that developing countries will simply refuse to participate in a policy-making process which they perceive to be biased against them. Only a policy which is perceived to be fair would have a reasonable chance to elicit the cooperation necessary for an international agreement on actions to avert the threat of global climate change.

The development of an equity-based approach was called for by the World Commission on Environment and Development [21], and again advocated during the 1992 Rio Conference on Environment and Development. In order, however, for a method to be developed which allows for the evaluation of various policy options on the basis of equity, it is necessary that an equity principle, as it applies to sustainable development in general, and the problem of global climate change in particular, be stated in concrete terms.

▼ *Equity Proposals for Reductions in Greenhouse Gas Emissions*

A variety of equity-based approaches to policy making at the international level have been proposed. Although these differ in the way in which the principle of equity is operationalized, they are in basic agreement that those who benefit from GHG-generating activities should bear a greater burden under policies to reduce the emission rates of these gases. Current equity-based approaches can be divided into four broad categories.

The first category proposes that GHG emission rights or quotas be allocated on the basis of the land area of countries [22]. The justification for this is that greater land area necessitates greater energy expenditures for movement of goods, people, etc. This distribution of emission rights, however, would discriminate against small nations and would reward those with large land masses and with sizable natural resource endowments.

The second category proposes a distribution of emission rights or quotas that is, at least to some extent, proportional to current emission levels [23]. Such a distribution would present the fewest problems of acceptance by industrialized nations and would cause few dislocations in the international economic system, but can hardly be justified on the basis of equitable treatment of developing country interests.

The third group of approaches proposes a distribution of emission rights or quotas on the basis of national populations [3], [24]. These approaches adhere to a principle of equality in the assignment of current and future emission rights. However, such equality will not reflect the disproportionate contributions of industrialized countries to the elevation of GHG concentrations over the last two centuries.

Finally, the fourth set of approaches proposes that GHG emission rights be allocated on a per capita basis, while taking into account the historical per capita emissions of different nations. Thus, nations which have, historically, produced more GHGs will be allocated lower per capita quotas [25]-[27]. This approach distributes the burden of GHG emission reductions in a way which reflects the responsibility for creating the problem of GHG accumulation in the atmosphere in the first place.

All of these approaches add some measure of equity over the economic models reviewed above. However, each of these approaches focus on emissions and emphasize actions to reduce current levels. Missing from these approaches is a consideration of a sustainable limit to anthropogenic emission rates. In fact, current equity proposals can lead to a significant increase in the amount of carbon in the atmosphere. Several anticipate a doubling of CO2 concentrations as inevitable [27].

▼ *Sustainability-Based Equity Principle*

A truly equitable policy approach would need to include a principle of CO_2 sustainability, defined as the worldwide organization of human activities such that emissions of this gas do not exceed its biospheric absorption. This is because solutions that ignore sustainability would inherently contain environmental threats that would pose disproportionate dangers for developing countries. Additionally, such an approach would need to adjust for the historical contributions of societies to the buildup of greenhouse gases.

One possible way of stating a sustainability-based equity principle in the context of global climate change is as follows: No human being is entitled to use the biosphere's carbon absorption capacity more intensively than another and, equivalently, no human being is entitled to store greater amounts of greenhouse gases than another. Operationalizing this statement requires that the biosphere's carbon absorption capacity be determined and allocated in an equitable manner. Several researchers such as Agarwal and Narain [28], and Mukherjee [29] have attempted this calculation. But considerable disagreement exists on whether and how such a calculation can be performed because of uncertainties concerning the relative effects of various greenhouse gases, their longevity in the atmosphere, nonlinear effects of natural absorption processes, etc. [29].

While we do not deny the complexities involved in this calculation, there are scientifically accepted estimates of the biosphere's current level of CO_2 absorption. The U.S. Environmental Protection Agency [30] and the World Resources Institute [31] indicate, with some confidence, that the biosphere is currently capable of absorbing between 14 and 17 billion tons of carbon dioxide per year. Dividing this number by the world population (about 5.2 billion in 1989) yields what we term a sustainable CO_2 emission rate of approximately 2.7 to 3.3 tons of CO_2 per person per year. A rough estimate of the inequity present in the status quo can be made on the basis of these numbers.[2]

▼ *Inequity of the Status Quo*

Fig. 1 shows that the industrialized countries of the world (including countries with centrally planned economies) have historically been responsible for most of the carbon dioxide emissions into the atmosphere (carbon dioxide is generated primarily from fossil fuel use). While in recent years developing countries have been contributing an increasing amount of these emissions (as a result of their economic development process), their contribution to the historical total amount of emissions remains minor.

[2]The CO^2 absorption capacity of the biosphere depends, to some extent, on the amount of this gas present in the atmosphere. As the amount of CO_2 in the atmosphere increases, the absorption capacity of the biosphere will also increase. This, however, is a dynamically unstable situation, for it is based on ever-increasing levels of atmospheric CO_2. For this reason, the present discussion is based on the estimation of a stable, sustainable absorption rate.

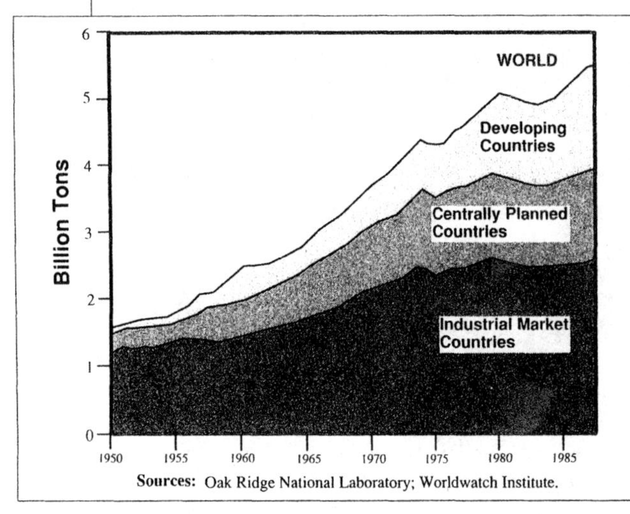

Fig. 1. Carbon emissions from fossil fuels: 1955-1988.

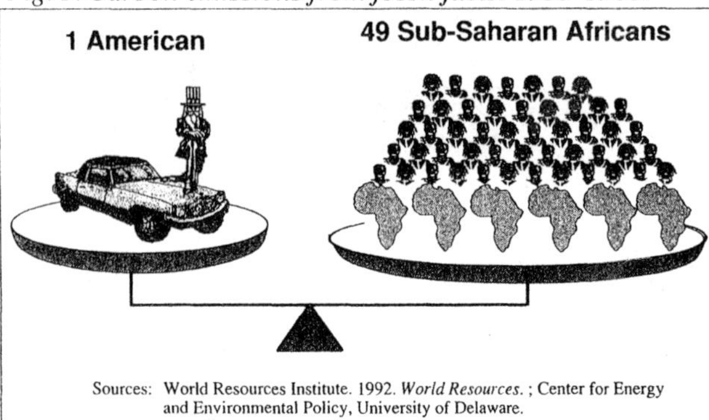

Fig. 2. Inequality of biosphere use.

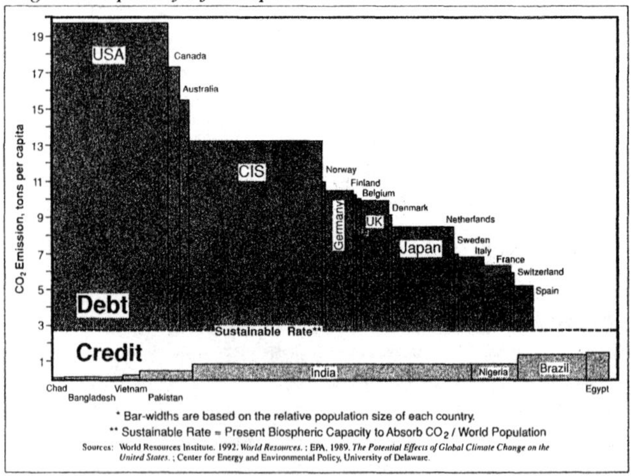

Fig. 3. CO_2 emissions for selected nations, 1989.

The sustainable CO_2 emission rate can be used to allocate a sustainable emission quota to each country of the world. The principle of equity would demand that allocations be based on each country's population. Such an equitable allocation indicates that almost none of the developing countries has yet exceeded their per capita quota, while all industrialized nations have exceeded theirs by large amounts. As an example, one U.S. citizen produces as much carbon dioxide annually as 49 residents of Sub-Saharan Africa (Fig. 2). Clearly, industrialized nations have appropriated more than their "fair share" of the biosphere's absorption capacity and use of the atmosphere to store their excess emissions and continue to do so.

As a basis for addressing this inequitable situation, the amount of the sustainable CO_2 emission quota which is not used by a nation can be thought of as "credit" which can be purchased by nations exceeding their own quota (Fig. 3). If industrialized countries could be induced to pay developing countries for the right to emit CO_2 beyond their sustainable quota, a system of tradeable emission permits would be established. If it is assumed that the price of a ton of CO_2 emissions would reflect the cost of avoiding a ton of such emissions (utilizing the avoided cost method for establishing value or price, routinely used in economic analysis), the total price tag for industrialized nations would range between \$0.3 and \$1.9 trillion per year. Such an amount would be adequate to retire the entire developing-nation debt within 44 weeks.[3] While it is improbable that the environmental debt of industrialized nations could actually be collected in this manner, its comparison with the economic debt of developing countries demonstrates two points:

- In terms of CO2 emissions in excess of the sustainable emissions rate, the industrialized nations are heavy borrowers from the developing nations. They are debtors to a much greater extent than developing nations are debtors in the strictly financial sense (see Fig. 3). Unlike industrialized nations, however, developing nations have almost no means at their disposal to collect on this debt.

- The comparison of CO_2 debt with financial debt suggests a means whereby the multidimensional nature of the equity problem may be addressed. The two types of debt could be traded, and the financial debt of developing to industrialized nations could be eliminated over some reasonable time period (probably several years), lifting a significant weight off the economies of these nations.

Perhaps as important, the calculation of the CO_2 debt of industrialized countries demonstrates the extraordinary level of dependence on

[3]The calculation of environmental debt is based on the cost of CO_2 emissions of OECD countries over their sustainable quota. The cost of CO_2 avoidance is based on estimates by Flavin [2] using energy efficiency technologies for the lower limit and combined-cycle combustion technologies as the upper limit. Excess CO_2 emissions were calculated by subtracting the product of the Sustainable CO_2 Emission Rate and the population of the OECD countries from their actual emission levels in 1989.

atmospheric inequality for the standard of living of these nations to be maintained. The dependence of industrialized economies on CO_2 emissions for the creation of wealth is so great that attaching almost any cost to these emissions would impose a significant burden on their economies. This is one reason why many economic analyses ([4]-[7]) suggest that very little should be done to address this problem. This points to a special challenge to an equity-based approach: any effort to develop realistic policies for addressing both the issue of global climate change and the inequity of the status quo would have to address the fundamental problem of transforming CO_2 emission-intensive industrial societies. Thus, both the analysis and the policy outcomes indicated by an equity-based approach must be quite different than those indicated by the economic approaches reviewed above.

The inequity of the status quo and the dependence of industrialized economies on carbon emissions for the creation of economic wealth should serve as motives for action rather than delay. Actions that begin to address these problems would clearly be better than no action at all [26].

▼ *Elements of an Equity-Based Approach*

One option for an equity-based policy is to adopt a regime of tradeable greenhouse-gas-emission permits as one of the means for managing the reduction of these gases. Such a system would ensure a high level of economic efficiency. The fundamental difference from similar methods proposed by existing economic approaches is that the initial allocation of "emission rights" under this regime must be based on the equitable distribution of the biosphere's CO_2 absorption capacity.

This difference has important ethical implications. First, it reflects and guarantees the equal rights of people regardless of where they live. Second, by being based on the biosphere's absorption capacity (the evaluation of which must become a priority for global climate change research), it recognizes the value of long-term environmental stability. Third, it avoids the reduction of the environment to the status of a commodity. Instead, commodity production is obliged to adjust to the social and environmental requirements of sustainability.

A sustainability-based equity approach would also provide considerable incentive to industrialized countries to pursue critically important energy and environmental policies such as: the increase of their own greenhouse sink capacity through, for example, extension of forested land; and the balancing of economic development and environmental goals through the pursuit of non-greenhouse-gas-producing energy options so that opportunities remain for material growth by developing nations. The pursuit of such an alternative would address the root problem of global climate change.

Equity Essential for Sustainable Solution

Popular economic approaches for addressing the threat of global climate change adhere to a distinct system of values which prizes capital accumulation, economic growth and quantifiable monetary costs and benefits, while discounting nonmonetary social and environmental values that are essential to identifying a long-term sustainable solution. As a result, policies proposed by prevailing economic approaches shift burdens unreasonably upon developing nations for the opportunity to develop.

A policy-making approach based on the principle of equity is necessary if the evolution of sustainable responses are to be made possible. In our view, the utilization of such an approach is the only means for securing just and viable international agreements and policies.

References

[1] United Nations, Agenda 21, *Rio Declaration*. New York: United Nations, 1992.
[2] Christopher Flavin, "Slowing global warming," in *State of the World 1990*, Lester R. Brown *et al.*, Eds. New York, Norton, 1990.
[3] *Princeton Protocol on Factors that Contribute to Global Warming*. Princeton, NJ: Woodrow Wilson School of Public and International Affairs, Fall 1988.
[4] Thomas C. Schelling, "Global environmental forces," in *Energy: Production, Consumption and Consequences*, John L. Helm, Ed. Washington D.C.: National Academy Press, 1990, pp. 75-84.
[5] Alex Cristofaro, "'No regrets' tax reform," in *Global Climate Change*, J.C. White, Ed. New York: Plenum Press, 1992, pp. 61-63.
[6] William D. Nordhaus, "To slow or not to slow: The economics of the greenhouse effect," *Econ. J.*, vol. 101, no. 6, pp. 920-937, July 1991.
[7] Thomas C. Shelling, "Economic responses to global warming: Prospects for cooperative approaches," in *Global Warming: Economic Policy Responses*, Rudiger Dornbusch and James M. Poterba, Eds. Cambridge, MA: M.I.T. Press, 1991, pp. 197-221.
[8] Peter Passel, "Curing the greenhouse effect could run into the trillions," *New York Times*, pp. 1, 18, Nov. 19, 1989.
[9] Environmental Protection Agency, *The Potential Effects Of Global Climate Change On the United States*, Rep. to Congress, EPA-230-05-89-050. Washington D.C.: United States Environmental Protection Agency, Dec. 1989.
[10] Alan S. Manne and Richard G. Richels, *Buying Greenhouse Insurance: The Economic Costs of Carbon Dioxide Emission Limits*. Cambridge, MA: M.I.T. Press, 1992.
[11] John T. Houghton, G. J. Jenkins, and J. J. Ephraums, Eds., *Climate Change: The IPCC Scientific Assessment, International Panel on Climate Change*. New York, NY: Cambridge Univ. Press, 1990.
[12] William R. Cline, *Global Warming: The Economic Stakes*. Washington D.C.: Institute for International Economics, May 1992.
[13] Daniel J. Dudek, "Energy and environmental policy: The role of markets," in *Global Climate Change*, J.C. White, Ed. New York, NY: Plenum Press, 1992, pp. 73-80.
[14] Marjorie Sun, "Emissions trading goes global," *Science*, vol. 247, pp. 520-521, Feb. 2, 1990.
[15] The Economist, "Let them eat pollution," *Economist*, Feb. 8, 1992.
[16] Lawrence H. Summers, "Economic focus," *Economist*, May 30, 1992, p. 65.

[17] Wolfgang Sachs, "On the archaeology of the development idea," *Ecologist*, vol. 20, no. 2, pp. 42-43, Mar./Apr. 1990.
[18] Vandana Shiva, *Staying Alive: Women, Ecology and Survival in India*. New Delhi, India: Kali for Women, 1992.
[19] The Economist, "Economic focus: The price of green," *Economist*, p. 87, May 9, 1992.
[20] John Bellamy Foster, "'Let them eat pollution': Capitalism and the world environment," *Monthly Rev.*, pp. 10-20, Jan. 1993.
[21] World Commission on Environment and Development, *Our Common Future*. Oxford & New York: Oxford Univ. Press, 1987.
[22] A.H. Westing, "Law of the air," *Environment*, vol. 31, no. 3, pp. 3-4, Mar 1989.
[23] R.E. Benedick, Ozone diplomacy: New directions in safeguarding the planet. Cambridge, MA: Harvard Univ. Press, 1991.
[24] Michael Grubb, "The greenhouse effect: Negotiating targets," Royal Institute of International Affairs, London, 1989.
[25] Kirk R. Smith, "Allocating responsibility for global warming: The natural debt index," *Ambio*, vol. 20, no. 2, pp. 95-96, Apr. 1991.
[26] Florentin Krause, Wilfrid Bach, and Jon Koomey, "Energy policy in the greenhouse, Volume one: From warming fate to warming limits," Int. Project for Sustainable Energy Paths, Final Rep. to the Dutch Ministry of Housing, Physical Planning and Environment, 1989.
[27] Barry D. Solomon and Dilip R. Ahuja, "International reductions of greenhouse-gas emissions," *Global Env. Change*, pp. 343-350, Dec. 1991.
[28] Anil Agarwal and Sunita Narain, *Global Warming in an Unequal World, A Case of Environmental Colonialism*. New Delhi, India: Centre for Science and Environment, 1991.
[29] Neela Mukherjee, "Greenhouse gas emissions and the allocation of responsibility," *Environment and Urbanization*, vol. 4, no. 1, pp. 89-98, Apr. 1992.
[30] Environmental Protection Agency, *The Potential Effects of Global Climate Change on the United States*. EPA 230-05-89-050, 1989.
[31] World Resources Institute, *World Resources 1990-1991: A Guide to the Global Environment*. New York, NY: Oxford Univ. Press, 1990.

T&S

GODFRED FREMPONG

Developing Telecommunications in Ghana

The contemporary global socioeconomic system is mainly organized through telecommunications. These systems have become the lifeline of the world's businesses and of its administrative, bureaucratic, and social organizations. They have moved from being a luxury to a strategic service in socioeconomic spheres, and should be seen especially in the developing countries as an essential part of the development process [1]; they have become synonymous with economic development. Increased economic activities generate more demand for telecommunications services, which has a positive multiplier effect on investments, exports, and growth.

Telecommunication development also assists social cohesion as it encourages interaction between communities. This is apt for developing countries like Ghana, which have poor rural and urban transportation systems and are plagued with ethnic conflicts. These inherently positive contributions to socioeconomic development create the necessity for every country (especially developing ones) to develop an effective telecommunications system. Ghana, recognizing the catalytic role telecommunications play in socioeconomic development, has over the past 15 years made some efforts to develop its telecommunications network.

In 1976, the Ghana Posts and Telecommunications (P&T) Corporation embarked on a First Telecommunications Project (FTP) which aimed at rehabilitating the telecommunications network in the country. This was followed by the Second Telecommunications Project (STP) in 1987. The main components of the STP included: 1) expansion of external plant and microwave transmission networks, 2) rehabilitation of an earth receiving satellite station, and 3) rehabilitation and expansion of the switch network in 13 urban and 26 rural areas and training of manpower [2].

The total amount of investment was about US$173 million obtained from multilateral sources. The STP commenced in 1987 and was originally scheduled to be completed in 1991 but had to be rescheduled to 1994. The rescheduling of the completion date was the outcome of delays in project design and implementation, and extensive procedures to be followed in the disbursement of funds. The STP on completion will increase the telephone national network by 52 000 lines and increase the telex network by 1500 lines.

The service provided by the P&T Corporation is to a large extent basic telephone service and to a lesser extent data transmissions using telex and facsimile. Mobile cellular communication and paging services are also available in the country. These services are provided by private operators who have been granted permission to operate under the liberalization policy in the telecommunications sector, which the Ghana government initiated in 1990.

The average telephone density remained at the level of 0.30 per 100 people for the period between 1985 and 1990, while the total number of telex lines within this period was 709 lines [3].

The author is a Research Officer at the Science & Technology Policy Research Institute, P.O. Box C 519, Accra, Ghana.

Reprinted from *IEEE Technology and Society Magazine,* Vol. 15, No. 1, pp. 23–31, Spring 1996.

Further, the quality of service provided by the P&T was far below the targets set by the International Telecommunications Union.

A number of problems have hindered improvements in telecommunications services in Ghana. Delays caused by those problems have implications for the socioeconomic development of the country as a whole.

Review of Services

Basic telecommunications services in Ghana, are far from being universally available. The quality of service is poor, as well as investment in it.

▾ *Types of Service*

A well-developed network provides different types of telecommunications services, particularly value-added services such as on-line services, video text, teleconferencing, etc., in addition to basic services. Such a network is well developed because it is able to incorporate modern services that have emanated from the technological convergence between telecommunications and computers.

In Ghana the main type of telecommunications services provided by the P&T are telephone and data transmission using telex and facsimile. Mobile cellular communication and paging services also exist in Accra and Tema, in addition to an on-line service which is used by the airlines in their operations.

International telecommunications service was expedited by the construction of an earth receiving satellite in the late 1970s, which was rehabilitated in 1988. Hitherto, Ghana was dependent upon a narrowband submarine cable for international telecommunications. The construction of the earth satellite station has enabled the P&T Corporation to provide an International Direct Dialing (IDD) facility to enable its subscribers to have quicker access to the outside world. The demand for IDD has been relatively higher in the Greater Accra Region than in other regions. Table I shows the number of IDD subscribers in the regions of Ghana from 1988 to 1990.

Table I
IDD Subscribers 1988-1990

Region	1988	1989	1990
Ashanti	6	6	6
Brong Ahafo	-	-	-
Central	-	-	-
Eastern	-	-	-
Greater Accra	526	906	1099
Northern	-	-	-
Upper East & West	-	-	-
Volta	1	1	1
Western	19	20	22
TOTAL	522	933	1128

Source: PORSPI, 1993, Telecommunications Assessment Report

From Table I, it can be observed that the total number of subscribers for IDD increased between 1988 and 1990, but the increase in demand was concentrated predominantly in the Greater Accra Region. For example, the number of subscribers in Accra with access to an IDD facility increased by 380 (72%) from 1988 to 1989, and by 193 (21%) in 1990. No increases were made in the other regions except in the Western region where there was a marginal increase of 1 and 2 subscribers in 1989 and 1990, respectively. The total number of 1128 subscribers in 1990 represented only 2.5% of the total 44 000 direct exchange lines (DELs) available in 1990.

In terms of independent radio communication systems, between 1980 and 1990, 67 institutions (both private and public) were granted licenses by the Ghana Frequency Registration and Control Board to operate radio systems. On the average, only seven institutions were granted licenses every year to operate a radio communication system for the 10-year period.

Mobile cellular communication is provided by two private companies in Ghana. As of August 1994, one of the companies had 3000 subscribers. The second company has just commenced operation.

▾ *Penetration*

Penetration of telecommunications service reflects the level of accessibility of service to the population of a particular country [4]. Indicators such as the growth rate of the main DELs and density of lines per capita can be used to show the level of availability of telephone and telex to the general populace.

Brown has criticized the use of main lines to show access to telephones in SSA [5]. According to Brown, most SSA countries' telecommunications systems have high loads per line. As a result, the number of telephone lines outstrips the installed capacity of the switching and transmission systems and does not normally portray the actual growth or accessibility to many others.

In Ghana, the basic characteristic of telecommunications penetration is that it is urban-based. According to a World Bank Report [6], about 75% of all DELs in Ghana are concentrated in the ten regional capitals which account for 10% of the country's population. Out of this, Accra alone has about 45% of the DELs with a density

Table II
Telephone Lines (DELs) per 100 Inhabitants

Country	Year					
	1985	1986	1987	1988	1989	1990
Ghana	0.30	0.29	0.30	0.29	0.35	0.30
Kenya	0.58	0.61	0.66	0.69	0.73	0.70
Senegal	0.34	0.38	0.41	0.42	0.51	0.60
Zambia	0.58	0.61	0.72	0.75	0.78	0.83
Zimbabwe	1.37	1.33	1.31	1.33	—	1.34

Source: ITU Yearbook of Common Carrier of Telecommunications Statistics, 1992

of 1.8 DELs per 100 people. The remaining 30% is shared among the other 9 regional capitals with a density of 1.4 DELs per 100 people. The bulk of the population (70%) who live in the rural areas shares the remaining 25% with a density of 0.07 DELs per 100 people [6]. This situation could have arisen because the rural economy is undeveloped. Therefore, the P&T finds it very uneconomic to provide service to rural areas or even repair the system in rural areas where the service previously existed.

The general penetration of telephones in Ghana is quite low as compared to other SSA countries. Table II shows the DELs in five selected African countries.

From Table II, it is seen that Ghana has the lowest telephone densities among the four other African countries listed. Zimbabwe has the highest density averaging 1.3 DELs per 100 people. The average telephone density in Ghana remained at 0.30. This is far below the African target of one telephone per 100 people. Though during this period (1985-1990) the Ghanaian economy and social development were above average, the subscriber densities of telephones of a number of African countries surpassed that of Ghana [7].

In terms of regional penetration in the country, telephone penetration is disproportionately concentrated in a few regions in the south of the country, mostly in Greater Accra and to some extent Ashanti, Western, and Eastern regions of the country. The Greater Accra Region accounts for 54.6% of the country's DELs, while the remaining nine regions collectively account for 45.4%. The regions having the fewest DELs are the Northern and Upper Regions.[1] They account for only 2.5% and 1.9%. respectively [9].

In addition to poor telephone penetration, the penetration of telex as a medium of data transmission in Ghana also lags behind the African countries represented in Table III.

As shown in the table, Ghana has the lowest penetration level of telex among the five selected countries. In 1990, Ghana had 709 telex lines while Zimbabwe had 2981 telex lines.

In terms of growth, Ghana has the highest growth rate of 15.5%, but this growth rate has had no significant impact on telex penetration. It is worth mentioning that with the advent of fax machines, the telex technology is becoming obsolete as improvements in technology have made fax cheaper.

Though it is very difficult to quantify the number of people operating fax machines in Ghana, it can be speculated that fax use in the country is on the rise. This situation has arisen because, as already mentioned, technological improvements have made fax cheaper and affordable. Also, it is fast in terms of transmission of written information.

▾ Quality of Service

Quality of service is a measure of the difficulties a user experiences when using the telecommunications system [8]. The quality of service is said to be high, when the user's terminal equipment has easy access to the telecommunications network without unreasonable delay. The quality of service is low or poor when the subscriber's terminal equipment cannot easily access the telecommunications network.

The quality of the service provided by a Post, Telegraph Telecommunications (PTT) authority demonstrates the level of technological development, performance of the PTT, and its ability to meet the needs of its customers. There are a number of indicators which could be used to measure quality of service. Call completion rates and average telephone breakdown time before repair are some of the indicators used. Call completion rate is the number of calls which are successful out of the total number of calls made. For example, if a subscriber makes ten calls and six of them were able to reach the called party, then the call completion rate is 60%. High call completion rates mean more of the subscriber telephone calls reach the intended destination and this should be close to or above the International Telecommunications Union (ITU) target of 90%. A low call completion rate signifies that

[1]The Upper Region embodies the Upper West and East regions of the country.

Table III
Size of Telex System: Subscriber Lines (in Thousands)

Country	Year						
	1985	1986	1987	1988	1989	1990	GR
Ghana	0.345	0.406	0.477	0.563	0.675	0.709	15.5
Kenya	2.190	2.420	2.531	2.531	2.322	2.375	1.6
Senegal	0.870	0.915	1.094	0.942	1.408	0.928	1.3
Zambia	1.476	1.478	1.708	2.520	2.770	2.875	14.3
Zimbabwe	2.065	2.102	2.480	2.635	2.801	2.981	7.62

*GR = Growth Rate
Source: Yearbook of Common Carrier of Telecommunications Statistics, 1992

few of the subscriber telephone calls reach the called party. Table IV depicts the quality of telephone service in Ghana from 1991 to 1992.

For call completion rates, as depicted by Table IV, the domestic call completion rate fell between 1991 and 1992. For example, the call completion rates for local and trunk calls in 1991 were 55% and 46%, respectively. However, in 1992, the call completion rates for the local and trunk calls fell to 54% and 43%, respectively. There was an improvement in international communication especially in outgoing calls in 1992. The same can be said about the average percentage of faulty DELs. There was a reduction of the average faulty DELs from 19.6% in 1991 to 6.9% in 1992. This means that there were few faults in the telephone network in 1992. However, the average breakdown time of telephone lines before repairs deteriorated in 1992 with an increase in days of breakdown time per line.

Weighed against the ITU targets, the quality of telecommunications service achieved by the P&T Corporation needs much improvement for users to enjoy efficient services.

▼ *Investment*

The telecommunications sector requires high fixed capital and investment to ensure its efficient functioning. The level of investment committed to telecommunications illustrates the commitment of the country to telecommunications development. Generally, Ghana's investment in telecommunications is below the ITU target of $4200 for SSA countries, but above the average of $1500 for developing countries.

Since the inception of the STP, the telecommunications sector has seen an injection of foreign capital investments with the aim of rehabilitating, modernizing, and expanding the existing facilities. Table V illustrates investment in telecommunications expressed as percentage of Gross Domestic Product (GDP).

From Table V, the average percentage of GDP invested in telecommunications in Ghana during the 5-year period represented was about 0.24%. The largest investment in telecommunications was in 1986, which marked the beginning of the STP. After that, there was a

Table IV
Quality of Telephone Services, 1991 and 1992

Indicator	Performance		ITU Goals
	Dec. 1991	Dec. 1992	
Call Completion Rates			
Local	55%	54%	90%
Trunk	46%	43%	90%
International			
-Outgoing	38%	51%	80%
-Incoming	23%	25%	80%
Average breakdown-time per line/year	19.4 days	30 days	5 days
Average % of faulty DELs	19.6%	6.93%	1%

Source: PORSPI, Telecommunications Assessment Report 1993.

Table V
Telecommunications Investment as Percentage of Gross Domestic Product (GDP)

Country	1985	1986	1987	1988	1989	1990	Average
Ghana	-	0.58	0.21	0.06	-	0.10	0.24
Kenya	0.64	0.38	0.86	0.62	0.78	12.42	0.79
Senegal	0.36	0.20	0.57	1.96	0.83	0.47	0.73
Zambia	2.24	3.18	1.43	0.01*	1.32	0.70	1.47
Zimbabwe	1.16	0.62	0.51	0.37	-	1.37	0.87

*This figure deviates from the normal range figures. It may be the result of wrong computation.
Source: ITU, *Yearbook of Common Carrier Statistics*, 1992.

Table VI
Total Telephone Expressed and Satisfied Demand in Ghana, 1985-1990

Year	Total Expressed Demand	Total Satisfied Demand (Installed Capacity)	Unsatisfied Demand
1985	67 067	38 153	28 914
1986	47 462	38 346	9116
1987	50 497	39 610	10 887
1988	52 433	40 442	11 991
1989	54 447	43 021	11 426
1990	56 143	44 243	11 900

Source: *Ghana Telecommunications Statistics 1992*
*Expressed demand = no. of lines + waiting list

remarkable fall in the percentage of GDP invested in the telecommunications sector.

In spite of the funds acquired from bilateral and multilateral sources to support the STP, the investment shown in telecommunications as a percentage of the GDP is lowest of the five African countries selected for comparison.

▾ *Demand for Services*

The importance of telecommunications in industrial and commercial activities, as well as in the governance of the country, has been identified by developing countries, including Ghana. However, such benefits have not fully been realized as demand for the services has outstripped supply.

Table VI illustrates how demand for telephone service has been met over the past 5 years. There was a drop of 19 605 (27.3%) in the expressed demand for the service in 1985 over 1986. However, there was no marked corresponding change in the installed capacity of the network (the installed capacity increased by 0.5%). It can be argued that the fall in the expressed demand could be due to administrative measures introduced by the P&T Corporation to reduce such high numbers and not actually providing the service to those who have applied for it.

One can further argue that there is high "demand" for the service. This is because the level of expressed demand may be distorted due to poor records keeping and "hidden demand" for the service.[2]

However, this view might not be wholly true. Under the STP, service in some of the main commercial towns and cities of Ghana improved. The improvement in the service was the result of the introduction of modern telecommunications technologies and the expansion of the network within these towns and cities. This has brought about considerable improvement in the quality of service and access to service in these areas to facilitate socioeconomic activities.

Problems Affecting Service Development

The availability of telecommunications services is far from universal in Ghana, because only the population in the urban areas has access to the services. In addition, the quality of service and the type of services emanating from the P&T Corporation cannot effectively support the emerging world political, social, and economic

[2]The term hidden demand is used to mean that some of the prospective applicants will become discouraged by the frustrations of those who have already applied and have been on the waiting list for a long time without getting the service, and thus will not bother to apply for service.

order where telecommunications plays a catalytic role. This situation has arisen because of a number of problems that have hindered the rapid development of the service in Ghana.

One of the main problems hindering telecommunications development in Ghana is the lack of adequate funding. As mentioned, the telecommunications sector requires high fixed capital and investment. The percentage of gross domestic product invested in telecommunications in Ghana, where the sector is still in its formative stage, is inadequate. This situation has arisen because sectors like agriculture, health, and education, which are equally critical as the telecommunications sector, are effectively competing for the scarce financial resources of the country. Though an amount of $173 million had been acquired under the STP from bilateral and multilateral sources, such an amount is inadequate to refurbish the whole telecommunications sector, which was near collapse before 1987. The insufficiency of the investment level under the STP is illustrated by the coverage of only 13 urban and 26 rural areas. However, the cause of the delay in implementing STP is not wholly due to the lack of availability of funds, but also to administrative bureaucracies.

Another problem facing telecommunications service development is the issue of obsolete equipment. Rapid technological changes are taking place in telecommunications, and this calls for timely replacement of equipment; however, this is not happening in the telecommunications sector in Ghana. Most of the switching equipment in the country before the STP had outlived its useful economic life. These switches were manufactured between 1957 and 1968 with a few manufactured in the 1980s [9]. These old switches with old transmission technologies (i.e., paired copper wires and lead sheathed paper insulated cables) are susceptible to the vagaries of the weather: the cables, especially the lead sheathed paper insulated cables, are often destroyed after heavy rains. Though the STP has introduced modern switching and transmission technologies, these modern technologies are limited to only 13 urban and 26 rural areas of the country. The majority of the areas not covered by the STP are still plagued with old equipment and this has resulted in the complete breakdown of the network in these areas.

Besides the problems with equipment, there is inadequacy of skilled personnel to operate the telecommunications sector in the country effectively. In 1993, the telecommunications section of the Ghana P&T Corporation had a total staff of 2272. Out of this total, 4.2% (95) were engineers and engineering technicians and 16.1% (365) were telecommunications technicians. This number is very inadequate to provide backup services to the sector to ensure its continuous operation.

Coupled with the problem of inadequate specialized staff are poor maintenance practices. This situation reduces the useful economic life of equipment, thereby hindering the recouping of the investments made in the sector. Poor maintenance practices also affect the quality of service enjoyed by subscribers. High numbers of faults per line, low call completion rates, and long periods of breakdown time before repairs are the result of obsolete equipment and a poor maintenance culture.

The poor work culture has been aggravated by a low level of remuneration and other incentives for the staff of the P&T Corporation. This affects staff morale and their ability to devote much of their attention and capabilities to their work. There is even the fear that with the advent of liberalization in the telecommunications sector, some of the P&T Corporation's skilled workers might resign and join the private operators who provide more attractive remunerations.

Inadequate funding and obsolete equipment inhibit telecommunications development in Ghana.

Another problem facing telecommunications development in the country is the disproportionate level of general development in the country where the main areas of industrial and commercial activities are concentrated in a few regions in the south. Telecommunications penetration has followed this pattern of development. Thus, there is high penetration of telecommunications services in Greater Accra, Ashanti, Western and Eastern regions of the country, with few services in the Northern and Upper Regions. The regions with high penetration of telecommunications services are the main commercial and industrial centers of the country. However, the use of telecommunications in the Northern and Upper regions is very necessary because of their distance from the national headquarters and the rest of the country. With this poor penetration rate in the northern part of the country, government officials and businessmen in these regions have to travel to the south to consult officials or to transact business. This has the attendant problem of delays as well as increasing operational costs of these institutions and individuals because of the general high cost of traveling in Ghana especially between the north and the south. Further, what can be achieved within minutes through the use of telecommunications can take days if the officials have to travel to the national headquarters or to the southern part of the country.

Implications for Socioeconomic Development

The problems hindering telecomunications development in Ghana, enumerated above, have overwhelmed attempts to refurbish and to some extent expand service. The implications of this situation are enormous.

As previously mentioned, as of 1990, there were only 1128 subscribers in the country who had access to IDD. This number is inadequate even in 1990 considering the increase in the volume of industrial and commercial activities due to the Economic Recovery Programme (ERP) being pursued by the government. The ERP has revamped old industries and attracted new foreign investments. The core of every industrial and commercial activity is to remain competitive and this can be expedited by a number of factors which include external telecommunications. This is because modern industrial and commercial activities require quicker access to international markets, research institutions, parent or subsidiary companies for vital information on prices, raw materials, spares, new production methods, etc.

One of the constricting factors hindering access to IDD, and dampening demand, is the initial deposit of five hundred thousand cedis required by the P&T Corporation before a customer is hooked onto the IDD facility. Only a few subscribers, i.e., big-time commercial and industrial concerns, can afford this initial deposit. The result is that numerous small-scale industries have been crowded out in the acquisition of IDD. This has affected their ability to contribute significantly towards national development as their competitiveness has been constrained by their inability to communicate easily with foreign companies outside the country.

The urban-based nature of service and the generally low penetration of telephones in the country affects the economic activities in the rural areas where the bulk of the country's wealth is generated. Ghana is predominantly an agrarian country. Through telecommunications, farmers can easily communicate with extension officers for technical advise, share experiences with their counterpart farmers in other areas, seek prices of agricultural products in the urban markets to enable them to sell their products at competitive prices to the middlemen, and generally seek information about government agricultural policies. But vital information, which could be obtained via telephones, has to be obtained through traveling to district, regional, or national headquarters, due to the absence of telecommunications services in rural areas.

According to Andersen Management International [7], there are high transaction costs in the agricultural sector of the country's economy due to the reduction in availability of telecommunications services over the past three decades. For example, the Ghana Cocoa Board urgently needs an increase in its telephone lines from 100 to 3000 in the rural areas to facilitate its commercial activities. However, the inability of the P&T Corporation to provide these additional lines has forced the Cocoa Board to set up and operate its own radio communications station which is using obsolete equipment, thereby making it expensive.

Modern industrial and commercial activities require quick access to international markets.

Invariably, increased productivity could be achieved in the cocoa industry and agriculture in general if adequate telecommunications facilities existed in the rural areas. This would facilitate extension services and provide easy access to the urban markets to aid farmers in marketing their produce.

In the socio-political sphere, a fair penetration pattern of telecommunications in the country could facilitate implementation of the county's decentralization policy. The country currently has 110 district capitals which have schools, hospitals, government departments, banks, etc. These institutions would perform creditably if they had access to efficient telecommunications services. In the absence of such services, personnel from these institutions have to travel to the regional or national headquarters when the need arises. Consequently, there are increased operational costs and delays in acquisition of vital information which will reduce their efficiency. In addition, the availability of telecommunications services in rural areas would facilitate the transmission and implementation of government policies in those areas. Emergency situations such as bush fires, epidemics, and social upheavals could easily be relayed to a district, regional, or national capital for prompt attention.

The poor quality and the lack of availability of telecommunications services put subscribers in Ghana at a disadvantage. Poor service quality means Ghanaians find difficulties in domestic as well as international communications. Long periods of faulty lines deprive users of the right of uninterrupted usage which is very important in commercial and industrial activities. High incidences of unsuccessful call completion rates associated with the sector in Ghana constrict the transmission of vital information at the most appropriate time.

Lessons and Observations

The state of telecommunications development in Ghana has a number of lessons for developing countries. The first lesson is that rapid telecommunications service development cannot be carried out by a publicly funded PTT authority. The P&T Corporation until 1990 had exclusive rights to the telecommunications sector. It was a network developer and service provider, and at the same time a regulator. The pace of telecommunications service development has not been rapid enough to reach the African target of one telephone for every hundred people.

In addition, the monopolistic situation hindered the introduction of other telecommunications services into the country. It was only when the sector was liberalized that services such as mobile cellular communication and paging were introduced by private operators. The liberalization has also led to the proliferation of terminal equipment of varied technological complexity to meet the needs of subscribers.

The other lesson is that governments of developing countries should not be the sole providers of investment for telecommunications development. The telecommunications sector requires high capital investment and experiences rapid technological changes. Developing countries alone cannot adequately provide such level of funding as well as keep pace with technological changes in the sector and also ensure high penetration rates owing to the poor state of their economies.

Ghana's experience confirms this. As discussed, Ghana's investment in telecommunications is below the ITU target for SSA countries. This has not provided the stimulus for rapid transformation which the sector presently requires.

The participation of the private sector relieves the government of some of its financial commitments to the sector and enables the injection of private investment, new managerial practices, and competition. The Ghana government introduced liberalization into the sector in 1990. Presently, two private companies are providing mobile cellular communication and one company is providing paging services. The rest are involved in the sale and installation of terminal equipment. The mobile cellular communication has introduced flexibility into the telecommunications sector by allowing users to be reached at any place and point in time. The numerous terminal equipment vendors have enabled subscribers to have access to terminal equipment of varying technological complexity to suit their particular needs. These services and the plethora

Small-scale industries are crowded out of international direct dialing access.

of terminal equipment would have taken a long time to enter the Ghanaian market under P&T Corporation's monopoly due to lack of funds.

Now that Ghana has opened its telecommunications sector to private operators, such liberalization should be perceived in two ways. First, it has served to facilitate the development of the telecommunications infrastructure and ensure its high penetration. Second, it has helped to meet the increasingly changing demands of the modern economic system. This should determine the role all the actors in the sector should play, i.e., the actors should be involved in both network development as well in the provision of service to ensure fair coverage of the country.

Facilitating Development

The poor state of the telecommunications services in the country needs much serious attention. It has placed Ghanaian business people and industrialists at a disadvantage by making them uncompetitive in the global market where telecommunications plays a cardinal role in ensuring competitiveness. It also affects the social and political organization of the country. Therefore, there is the need to make the introduction of liberalization into the sector blossom so as to facilitate development of telecommunications services in Ghana.

References

[1] H. Hudson, "Telecommunications and the developing world," *IEEE Commun. Mag.*, vol. 25, no. 10, pp. 28, Oct. 1987.
[2] W. Atubra, "The problems of network development," in *Technology Assessment of the Telecommunication Sector of Ghana*, PORSPI, Accra, 1993, p. 53.
[3] ITU, Common Carrier of Telecommunications Statistics, Geneva, 1992.
[4] G.K. Frempong, "Deregulation and privatisation of telecommunications: Implications for African countries," Masters thesis, Univ. of Lund, Lund, Sweden, 1993, p. 43.
[5] W. Brown, "Telecommunications network development in Africa," in *African Telecommunications Development (ATDC - 1990),* working paps., vol. 2, ITU, 1990, pp. 11.
[6] World Bank, *Ghana Second Telecoms Project*, 1988, p. 2.
[7] Andersen Management Int., "Report on the future development of the telecommunications sector in Ghana," 1993, p. ii.
[8] J. Graham, *Dictionary of Telecommunications*, rev ed., rev. by Sue Lowe. London, 1991.
[9] PORSPI/TUD, *Ghana Telecoms Statistics*, Accra, 1992.

T&S

Chapter 9

Technology and Health Care

ONE of the great technological achievements of the twentieth century is the advancement of health care, particularly in the industrialized nations. Such advances have made it possible for people to live longer, healthier lives and to enjoy quality of life well into advanced age. As Kline and Kash note in their background article in Chapter 5, health care policy has been one of the major success stories in the U.S. technology policy arena. Policy debates have continued in recent years, however, focusing on the need to contain health care costs while providing access to quality care to all Americans. Despite great advances in health care technology and continued refinement of the policy issues involved, the development of these technologies raises a number of important ethical questions. Engineers and professional engineering societies often support policies to develop new biomedical technologies without giving adequate attention to such ethical issues.

Health Care Technology and Ethics

The role of technology in health care costs is a key issue in today's policy debates (1). Although technology has made it possible to save and prolong lives, it has also raised questions concerning funding priorities and equitable access to health care. Are we spending too much on technologies that prolong life but not enough on prevention of disease (see background article by Cram, Wheeler, and Lessard)? For example, it is now possible, or will soon be possible, to produce artificial organs such as hearts, kidneys, and retinal implants. Some of the questions raised by these developments include: What are the potential benefits of artificial organs? How will artificial organs impact health care costs and access? How will artificial organs affect the structure and competitiveness of the U.S. medical device industry? How will the development of artificial organs affect the training of health care professionals? Will artificial organs be safe and reliable? How will artificial organs affect the product liability climate in the medical device industry? What impact will artificial organs have on existing patterns of organ donation? What are the religious, philosophical, and ethical implications of artificial organs (including, but not limited to, animal rights)? What laws, regulations, and incentives are in place for artificial organs? Are these measures adequate? Will artificial organs attract funding and attention away from preventive medicine strategies?

As daunting as such questions are, they pale in complexity in comparison to the ethical challenges posed by breakthroughs in genetic engineering. Humans are on the brink of achieving the means of redesigning human life. Beyond the obvious moral and religious questions raised are more practical questions regarding, for example, how information obtained from genetic testing will be used (see case study by Asch and Mennuti). The issue of the confidentiality of medical records extends beyond genetic information. As more and more of our medical records become computerized, privacy and confidentiality issues come to the fore (see Cushman's case study in Chapter 1). Other biomedical advances such as cloning, in vitro fertilization, and sperm and egg banks also raise difficult ethical questions.

On a level that is likely to impact individual engineers, the safety and reliability of biomedical devices is becoming more and more an ethical and legal concern, the latter in regards to product liability. In addition to the Bjork-Shiley Heart Valve case, discussed in Chapter 6 and in Fielder's case study at the end of this chapter, a number of other such cases have captured the public's attention.

The Dalkon Shield Case

Between 1971 and 1975, the A. H. Robins Company distributed over 4 million Dalkon Shield intrauterine devices in 80 countries, with false claims of efficacy and safety. In the United States alone, more than 2 million women were fitted by doctors who believed the misleading claims. To date, thousands of women have suffered serious damage caused by the Shield, from pelvic infection to sterility, miscarriage, and even death. Robins and its insurers have paid more than $340 million to litigants, and thousands of lawsuits are still pending (2).

Silicone Breast Implants

Over a million women in the United States have received silicone breast implants for cosmetic reasons, including women who have had mastectomies or other medical problems. During the 1990s, serious concerns arose regarding health impacts resulting from the deterioration and leakage of implants. The largest manufacturer of breast implants, Dow-Corning, has been found to have withheld from the public information concerning these impacts, as well as falsified data regarding the manufacture of the implants. Although the health effects of breast implants remain controversial, Dow Corning and its shareholder company, Dow Chemical, have been subject to extensive litigation (3).

Therac-25

Consumer products are not the only technologies subject to health and safety claims, as indicated by the Therac-25 case. Between 1985 and 1987, at least six patients were seriously injured or killed when they received massive overdoses of radiation from Therac-25, a radiation therapy machine. The overdoses were attributed to operator error, poor software engineering, overreliance on software for safety assurance, the manufacturer's inadequate response when accident reports began to surface, and unrealistic risk assessments (4).

As noted by Cram, Wheeler, and Lessard in the background reading, engineers should not limit their focus on biomedical technologies to the technological breakthroughs alone. In addition, they can and should become involved in policy debates and ethical deliberations regarding technologies that prolong, enhance and, potentially, alter human life.

Background Reading

"Ethical Issues of Life-Sustaining Technology" are discussed in a 1995 article by *Nicholas Cram, John Wheeler,* and *Charles S. Lessard.* According to these authors, technologies that sustain life pose several ethical dilemmas, including balancing the benefit to the patient versus the cost of the technology, inequities in access to life-sustaining technologies, and the administration of "futile care" in cases where patients are unconscious and unlikely ever to enjoy a basic quality of life. In analyzing these issues, the authors apparently adopt a utilitarian ethic: "an ethically sound medical technology provides the greatest possible benefit to the greatest number of people."

Bioengineers, they argue, can assist physicians in dealing with futile care decisions by providing information on the limitations and capabilities of medical technologies. These issues, however, are often complicated by risk avoidance strategies designed to limit legal liability and ambiguities in informed consent laws. Access issues arise owing to inequities among patient groups as well as geographical disadvantages, with most high-tech facilities located in large urban areas. Biomedical engineers have a role to play here by designing low-cost technologies, such as those that are noninvasive. Bioengineers can also help to dispel the prevailing thread of "technological favoritism" in medical care, which results from public opinion, physician choices, and policies of government, health care providers, and insurance companies, by ensuring their products are designed to meet patients' needs in a reliable manner. The authors note that some common alternatives utilized in futile care situations, such as advance directives (e.g., do not resuscitate orders and living wills) and hospice care, while potentially contributing to the patient's well-being, often do little by way of reducing costs.

Cram et al. outline three courses to deal with the ethical dilemmas posed by life-sustaining technologies: use community values to determine treatment protocols; perform outcomes research on various treatment regimes; and assess new technologies. The authors assert that all of these strategies would benefit from the active participation of bioengineers, and they challenge individuals and the bioengineering profession "to take a stand on ethical and economic issues related to technology development."

Cases for Discussion

David A. Asch and *Michael T. Mennuti* present a timely case study in their 1996 article entitled "Genetic Tests: Evolving Policy Questions." Advances in biotechnology, including the Human Genome Project, are making it increasingly possible to identify genetic diseases and disorders. The authors caution, however, that these developments also pose some vexing ethical dilemmas, particularly in cases where no treatments currently exist for the disorders that may be identified. The use and misuse of information from genetic tests is thus an important topic for consideration.

As genetic screening becomes more prevalent, more parties have a stake in the outcome of screening tests. These stakeholders include patients, physicians, genetic counselors, employers, insurance companies, and society at large. Broad social issues are at stake, including reproductive decisions. With respect to individual diagnoses, genetic screening offers many benefits, including detection of treatable genetic disease and predisposition to certain cancers. It also poses many ethical issues, often involving conflicts between the patient's right to privacy and the physician's duty to disclose. Some genetic disorders are untreatable, others may only affect future generations, and many may carry social stigma. In some cases, patients may not want to know the information provided by the tests, particularly if the condition is untreatable. Informed consent is also problematic in many cases of genetic screening owing to difficulties in communicating probabilities and error rates for the tests.

Reproductive planning can be enhanced by genetic screening, but it is inevitably tied to public debates over abortion and is susceptible to misuse in the form of eugenics, that is, selecting offspring on the basis of particular traits such as sex. The authors call for increased counseling to deal with difficult reproductive decisions arising from the new technology. Genetic testing by employers, though currently limited in practice, is also laced with ethical land mines relating to privacy, coercion of employees and unjust (and in some cases illegal) employment discrimination. Genetic screening could also affect insurance risk ratings, resulting, for example, in genetic disorders being labeled as preexisting conditions (i.e., uninsurable).

Asch and Mennuti conclude by advocating further education of stakeholders on the benefits and drawbacks of genetic screening, as well as promoting programs designed to protect confidentiality and avoid discrimination on the basis of genetic information. Policy in this arena is also linked to other, overarching health care issues such as the proper balance between prevention and treatment and access to health insurance. Most importantly, they argue, we need to maintain a deep respect for the importance of genetic diversity in society. Ultimately, responses to genetic screening must be based on human values.

Philosopher *John H. Fielder* in his 1995 article, "Defects and Deceptions—The Bjork-Shiley Heart Valve," presents an interesting case study of risk and ethics involving an area that is growing in importance in the United States—product liability. Fielder begins by noting that since no medical devices are risk free, commercial use is in effect an extension of clinical trials, and patients should be afforded the opportunity of informed consent. Although the failure rates of the Bjork-Shiley Convexo-Concave (C/C) heart valve may have been within generally acceptable limits, Fielder argues that the heart valve's manufacturer, Shiley, Inc. (as well as Shiley's subsequent owner, Pfizer), committed ethical lapses that eventually led to the withdrawal of the valve from the market.

Though considered an improvement in heart valve technology, the C/C valve exhibited defects during clinical trials and subsequent commercial use. The death rate of patients with defective valves is 2/3, with serious complications confronting many of the others. Through a process of trial and error, Shiley attempted to correct the problem with the valves, reissuing new models with letters of assurance to physicians that minimized the problems. When the failures continued, Shiley withdrew the valve under pressure from the Food and Drug Administration (FDA).

Accepting for the sake of argument that Shiley's controversial claim that the valve's benefits outweighed the risks of failure is true, Fielder believes that a utilitarian ethical analysis would find that the valve should have remained on the market. From a rights/duties ethical perspective, however, Fielder rejects Shiley's actions, since the patients receiving the modified valves were, in effect, being experimented on without their consent. Fielder reports that Shiley also misled the FDA with respect to the extent of the problems associated with the valves. Ultimately, Fielder believes, the valve was judged to be defective as much because of the manufacturer's defective ethics as because of the valve's structural defect. That is, he regards devices such as heart valves to be social as well as material artifacts.

References

[1] Foster, K. R., and Fielder, J. H., eds. 1996. Special issue on "Technology and the Revolution in Health Care Economics." *IEEE Technology and Society Magazine* 15 (3).

[2] Moore, P. 1994. Dalkon Shield (1970s–1980s). In *When technology fails: significant technological disasters, accidents and failures of the twentieth century*, ed. N. Schlager, 481–485. Detroit: Gale Research.

[3] Moore, P. 1994. Silicone-gel implants (1960s–1990s). In *When technology fails: significant technological disasters, accidents and failures of the twentieth century*, ed. N. Schlager, 475–480. Detroit: Gale Research.

[4] Leveson, N., and Turner, C. 1993. An investigation of the Therac-25 accidents. *Computer* (July): 18–41.

NICHOLAS CRAM, JOHN WHEELER, AND CHARLES S. LESSARD

Ethical Issues of Life-Sustaining Technology

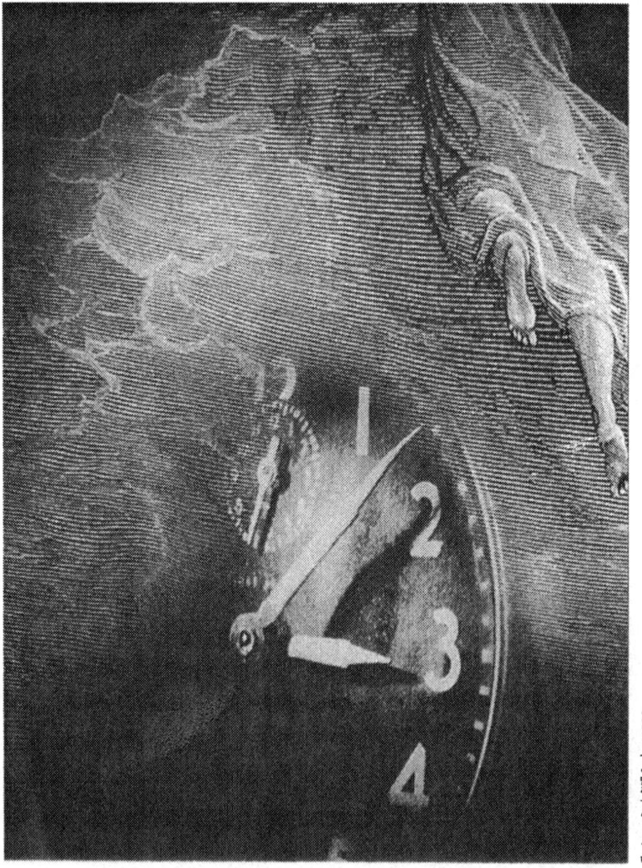

The Image Bank/Will Crocker/1999

Developments in medical technology have given physicians expanded means to sustain human life. In many instances, life-sustaining treatments are administered despite the fact that the patient is unlikely to benefit from the medical intervention. Because of "technology favoritism" in society, life-sustaining technologies influence the availability, financing, and use of existing technologies. Healthcare organizations are attempting to guide treatment decisions by providing physicians and patients with thorough information about the efficacy of technologies. Programs such as hospice care, advance directives, technology assessment, and outcomes research, are reducing the occurrence of "futile care."

Nicholas Cram and John Wheeler are students in the bioengineering program, Texas A&M University, College Station, TX 77843-3120. Charles S. Lessard is Associate Professor of Bioengineering, College of Engineering, Texas A&M University, College Station, TX 77843-3120.

Reprinted from *IEEE Technology and Society Magazine,* Vol. 14, No. 1, pp. 21–28, Spring 1995.

As developers of medical technology, biomedical engineers play a crucial role in ethical decisions concerning life-sustaining technologies. The curriculum for a degree in bioengineering includes courses in life sciences, as well as basic engineering, and bioengineers are found working directly in the health-care environment as clinical engineers, managing and acquiring medical technology, and as biomedical engineers involved in the design and evaluation stages of high-tech medical devices. Due to this unique combination of education and training, bioengineers can provide fresh insight and remedies to the health care versus technology debate. Biomedical engineers must provide not only clinically proven and reliable technologies, but also thorough information on the likelihood, duration, and degree of benefit that patients with specific conditions can expect to receive.

Life-sustaining treatments may be administered despite the fact that the patient is unlikely to benefit from the medical intervention.

Life-Sustaining Technologies

In 1903, William Einthoven devised the first electrocardiograph, giving medicine its first means to analyze physiological electrical phenomena related to chronic human disease. As society prepares to enter the twenty-first century, the availability of life-sustaining biomedical devices has given physicians imposing, and sometimes perplexing, choices in health care regimens. The existence of such devices as the "Jarvic Heart," defibrillator, pacemaker, renal dialyzer, and respirator, creates ethical conflicts between individual autonomy and the welfare of society.

The purpose of medical technology is to advance the general medical goals of alleviating suffering and restoring or maintaining the health of patients. Traditionally, the significance of a technology has depended on the value of the effect it produces in patients. Those technological developments that have the potential to prevent certain death enjoy widespread, although sometimes unwarranted, acceptance within society. Life-sustaining technologies have thrived especially in United States of America (U.S.), a society that demands the latest technological advancements available.

Life-sustaining technologies are defined as those drugs, medical devices, or procedures that can keep individuals alive who would otherwise die within a foreseeable, but usually uncertain, period. [13] Because these technologies postpone death, they present a multitude of ethical questions concerning the distribution of high-tech medical care. The likelihood, degree, and duration of benefit to the patient must outweigh the cost of the technology. In addition, life-sustaining technologies raise concerns about equality of access to medical services and unjust allocation of heath care resources. [10] An ethically sound medical technology provides the greatest possible benefit to the greatest number of people.

Aside from the intangible issues of ethics and morality, technology also affects the economics of dying, and has radically altered medically validated practice in the area of "futile care." The application of life-sustaining technology where there is little probability that the patient will recover to a state of full consciousness and enjoy a base level of "quality-of-life" can be considered "futile care." Americans' fascination with technology and demand for high-tech medical procedures will undoubtedly perpetuate the current problems.

Futile Care

For many individuals, e.g., very elderly patients, the administration of medical services provides little chance of prolonging life. In such instances, most people would agree that the administration of uncomfortable, nonbeneficial treatments is unreasonable and unethical. Unfortunately, health care professionals are reluctant to define this domain, known as "futile care," because the chance always exists for some patients to benefit from a questionable treatment. [13]

Physicians usually provide care until there is little doubt that the patient's condition will not improve. As a result, patients must endure unnecessary physical and emotional pain, and incur an additional financial burden. [4] The dilemma facing health care professionals is to define ethically sound criteria for determining who should receive life-sustaining medical treatments. [13]

In the current system, treatment decisions lie in the hands of physicians. Loosely defined societal and professional guidelines provide doctors with only limited assistance in determining treatment strategies. To determine the possible benefit of a treatment, physicians must compare life with the treatment to life without the treatment, and they must consider the possible physical, mental, and economic burdens derived from the application of the technology. Doctors must base their treatment decisions on professional judgment, experience, and their knowledge of the patient and the applied technology. [13]

Although it is clear that the decision of when to apply life-sustaining technology should not fall under the responsibility of the engineer, bioengineers can provide physicians and prospective patients with information concerning the limitations and expectations of applying high-tech medical devices. In the hospital environment clinical engineers could take responsibility for developing and maintaining data bases for technology assessment. These data bases could provide the structure for physician and patient in-services related to medical devices.

Legality

The basic premise of medical legal issues stems from the Hippocratic Corpus, in which the physician pledges nonmalfeasance, the avoidance of doing harm, and beneficence, the reduction of pain or suffering, improvement or restoration of health. [2] George Annas, an ethicist, describes the legal dilemma as "...an increasing trend to ask the courts whether life-sustaining treatment should be withheld from patients who are unable to (decide for) themselves." Judges are consulted for treatment decisions because only they can give physicians immunity for their actions. "In seeking this immunity, legal considerations quickly transcend ethical and medical judgments." [12]

Because the United States legal system is based on antecedent case law and common law, physicians face minimal realistic risk of criminal liability, regardless of the manner of use of the resuscitative devices. Physicians need not fear legal repercussions as long as treatment decisions are made "...in good faith and according to reasonable professional standards and judgment." Yet, risk-management decisions, intended to protect physicians from malpractice suits, occur frequently and inflict physical, emotional, and psychological burdens on patients and their families. [12] The patient or a legal guardian must have a voice in the treatment decision to protect both the patient's and the physician's interests. A common difficulty arises when the patient is incapacitated and no reasonable surrogate can be found to participate in the treatment decision. In such instances, physicians usually do everything in their power to prevent the death of the patient, even if the interventions will not ultimately succeed.

Informed-consent laws protect patients from having medical intervention forced upon them; however, their validity is often questionable due to the coercive atmosphere and desperate situations encountered in a hospital setting. [13] In addition, high-tech medicine has made it increasingly difficult for patients to understand information about the treatments and has also reduced the validity of informed-consent documents. [1]

Ethics of Access

The development of high-tech, life-sustaining treatments present an ethical dilemma concerning equality of access to medical technologies. Access to medical technology is restricted primarily in two ways. First, the high resource requirements of new technologies take revenues away from primary and preventive care for disadvantaged population groups. Second, high cost prevents the widespread distribution of medical technologies and results in geographically limited access. [1]

Hospital intensive-care units (ICU's) provide a perfect example of displacing health care revenues for the benefit of a small population. Studies have shown that 14 to 20% of hospital budgets are devoted to staff and equip ICU's, providing for only 2% of the patients served by hospitals. [13] Many professionals consider the research and development of life-sustaining technologies harmful to the health of society, because perfection of these technologies diverts revenues from the widespread needs of the public. For instance, perfection of a total artificial heart would add to the already massive burden of health-care costs. As with dialysis, artificial hearts would be government subsidized because of the crucial, life-saving benefit they provide. Denying such a life-saving treatment based on the expectation of payment would be ethically unjust and unacceptable. Government funding necessary to support total artificial heart devices would divert resources from other areas of research and treatment. [9]

The examples above imply that controlling the forces of technology, as in the case of government subsidies, payment stipulations of health insurance companies, and physician preferences, regulates access to treatment. These forces become more important as the life-saving capabilities of the device increase. Ethically, all population groups should have equal access to technology; however, when resources are limited, the technology must be distributed based on patient need. Organizations such as the Organ Procurement Transplant Network must be created to establish and maintain an ethical procedure for the distribution of the technology. [10]

Medical technology is controlled by many different entities, each having its own motivation and impact on the distribution of medical services. Physicians command the most direct control over technology. Because of differences in their personal views and beliefs, doctors' decisions will bias medical treatments for certain population groups. For instance, physicians may consider a patient's age as a factor in the aggressiveness of the treatment, even when it has little bearing on the probable outcome of the treatment. [13]

Physicians must work within the confines of hospital policies which sway treatment decisions for economic gain. Hospital policies provide or deny life-saving technologies based on a fiscal budget and will fund only those technologies it believes will be profitable. [13] Recent growth in the number of for-profit hospitals has contributed to the misuse of life-sustaining technologies. Since high-tech procedures bring in greater revenues, hospitals promote the use of new technologies, often underestimating their risks and exaggerating their benefits. [1]

On a societal scale, the availability of technology is determined by the society's decision of how to allocate health-care funds. Political agendas and social movements have a significant influence on the distribution of medical funding. Those ailments with the most devastating impact, or popular appeal, receive the largest portion of the available funding. [13]

Insurance agencies not only dictate the type of health care received by individuals, but also influence the direction of research and development in the medical industry. Insurance companies ration health care by limiting the expenses of beneficiaries and by basing coverage on patient need. Insurance companies have an immense effect on the distribution of new medical technologies because treatments are usually beyond the private financial means of individuals. Because of the high cost of medical care, the treatment distribution policies of insurance companies deny certain segments of the population access to medical technology. [15]

When designing medical devices, biomedical engineers must consider cost and applications which are noninvasive. A noninvasive application is a procedure which can be accomplished "without surgical invasion of the body." Noninvasive procedures can be provided at a lower cost and with less discomfort to the patient. Noninvasive procedures also tend to require fewer hospital personnel, which would lower the overall cost to the patient. These considerations of cost and modality in the design process provide biomedical engineers with a means to transform the available technologies, which in turn modifies the dictates of insurance companies and special government funding.

Ethics of Use

In addition to the information provided by the technology developer or original equipment manufacturer (OEM), physicians have the assistance of standards of practice to aid them in their treatment decisions. The Institutional Ethics Committee of the Joint Commission On Accreditation of Healthcare is only one of many organizations to publish treatment protocols [13]; however, the advice available to physicians suffers from "technology favoritism" and may result in an unethical treatment decision.

"Technology favoritism" describes Americans' popular worship of science, Americans' faith in the usefulness and benevolence of technology, the general appeal of instruments and gadgets, and the prestige awarded to the developers and operators of new technology. The environment of health care in the U.S. not only fosters, but demands the development of high-tech medical devices and procedures. Physician training has always taught that medical tradition "emphasizes giving the best care that is technically possible; the only legitimate and explicitly recognized constraint is the state of the art." [1] The policy of "technology favoritism" stems from the desires of different forces controlling technology. "Technology favoritism" satisfies patient demand for the most up-to-date technologies, physician preference for technology-intensive care, insurance reimbursement practices, and hospital desire for technology-based competition. These controlling forces of access to technology determine the equity of using high-tech medicine. [1]

Political, economic, and legal forces can alter the development and application of technology to focus less on the needs of the patient and more on the desires of the driving organization. [10] For example, hospitals use technology to attract patients and distinguished medical staff. Economics drives interhospital competition, resulting in the duplication of new technologies and decreased access to other, less costly, procedures. As the life-saving potential of a technology increases, devices become more susceptible to economic manipulation. [1] From an ethical standpoint, hospitals must be certain that the adoption of life-sustaining technologies serves the needs of the patients before the economic goals of the institution.

Several institutions, besides hospitals, contribute to the unjust adoption of medical technologies. The financing policies of the United States Government and the reimbursement policies of insurance companies create an environment that promotes early commercialization and extensive use of unproved technologies. Government support of underserved population groups has increased futile care practices by increasing the population with access to life-saving technologies. Government funding of medical research and development has also caused a proliferation of new technologies that have questionable utility. By reimbursing more generously for technology-intensive care, insurance companies have taken the importance of price out of heath-care decisions and created an environment favorable to futile care. A vicious cycle exists in which government financing and insurance policies boost investment in medical technologies and increase the number of technologically advanced, but unproved medical devices. [1]

Bioengineers control the development and assessment of medical technologies; therefore, they must recognize and address the two main factors contributing to the inappropriate use of technology, "technology favoritism" and evaluation consistently lagging adoption. Bioengineers can limit the unethical use of life-sustaining technologies by insuring that medical devices provide reliable and necessary results. As with all health professionals, bioengineers must guarantee that technological developments serve primarily patients' needs.

Ethics of Cost

In response to the American demand for high-tech health care, medical device manufacturer's have given physicians the ability to turn "inexpensive dying into prolonged living, usually through expensive means." [1] Unfortunately, the costs of most medical technologies are increasing. While in most industries new technologies reduce labor requirements, the opposite is true for the medical industry. New technologies usually increase labor intensity, skill requirements, the complexity of procedures, and formal training requirements. State-of-the-art procedures usually replace low-tech, inexpensive alternatives. [1]

The current state of America's health care industry has made cost controls a vital issue. Health care experts have noted that interventions in cases of imminent death are inherently wasteful, because they rarely cure or ameliorate disease or disability. Reducing expenditures for futile care may be a justifiable means of cutting health care spending and freeing valuable resources. [4] Still, the appropriateness of cost-based rationing of life-sustaining technologies must be determined by weighing the expected benefits against the expected financial, physical, and emotional burdens associated with the procedure. [1]

Inconsistency in the definition of technology has made determining its cost impact very difficult. Over the past two decades, 20 to 50% of the annual increase in health-care costs resulted from the adoption of new life-sustaining technologies. [1] Obviously, society needs less expensive, yet ethical alternatives to life-sustaining technologies. Bioengineers stand at the gateway to three key areas in the medical technology cycle: 1) conception and design, 2) evaluation and surveillance, and 3) product selection. Decisions and choices in these areas are pivotal for future changes in health-care and medical-device development.

The availability of life-sustaining biomedical devices has given physicians imposing, and sometimes perplexing, choices in health care.

Advance Directives

Advance directives limit the aggressiveness of medical intervention by providing the patient's written instructions concerning the extent of care that should be given to save the patient's life prior to the actual circumstances. Advance directives are prepared in similar manner to a "last will and testament." Do not resuscitate orders (DNR's) provide one example of an advance directive in which health care providers are asked not to perform cardiopulmonary resuscitation to prevent the patient's death. Living wills and durable powers of attorney, which allow the individual to specify the aggressiveness of medical treatment, clarify the wishes of the patient, and remove encumbrances from the physician, are also forms of advance directives. Unfortunately, durable powers of attorney usually fail to prevent futile care practices. In most instances, family members are more reluctant to refuse life-saving technologies than doctors or even the patients themselves. DNR orders also do not provide extensive cost savings for dying patients. In a study of more than 830 patients, the average cost of hospital stays for patients with

DNR orders was $62 594, while those patients without DNR orders paid a slightly lower average of $57 334. [4]

Many people rightfully argue that the intention of advance directives is to improve the physical and mental well-being of the patient rather than to reduce health care costs. However, two studies of more than 200 patients concluded that advance directives had little effect on the patient's well-being, health status, or type of medical treatment. The study showed that health-care professionals and family members are still very reluctant to give up hope for recovery from terminal conditions. [4]

Those technological developments that have the potential to prevent certain death enjoy widespread, although sometimes unwarranted, acceptance within society.

▾ *Hospice Care*

Hospice care provides another common alternative to life-sustaining technologies. Patients who choose hospice care refuse life-saving interventions and receive only palliative care in their own home. Savings range from 31 to 64% of health care costs in the last month of life; however, according to the National Hospice Study, the longer a patient stays in a hospice program the lower their cost savings. High-quality palliative care — pain medications and help with the activities of daily living — requires skilled and costly personnel. Over a long period, hospice costs reach the same proportions as those incurred from hospital stays. [4]

The amount of savings realized by frequent use of advance directives and hospice care are not as dynamic as some proponents suggest. Assuming that every patient provided advance directives that refused life-sustaining interventions, and chose hospice care instead of hospital stays, the estimated national savings would only reach 3.3% of health-care spending. Considering the $900 billion spent on health care in 1993, this would have amounted to a savings of only $30 billion.

Current estimates of savings are based on Medicare records, because they are the most extensive source of information available. The true savings could be greater if applied across a less selective population. However, refusing life-sustaining intervention does not reduce the amount of care, but only changes the type of care given at the end of life. The ethical merit of hospice care and advance directives lies in their respect for the patients wishes, their ability to reduce pain and suffering, and the compassionate and dignified care they provide at the end of life. [4] Perhaps if the public were better informed of the legal choices available to them, more people would initiate living wills and durable powers of attorney. Patient education and enlightenment may initially create legal obstacles for hospital administrators, but there is a moral obligation to intervene in these matters.

Solutions to the Ethical Dilemma

▾ *Treatment Protocols Based on Community Values*

While advance directives and hospice care offer some relief from the misuse of life-sustaining technologies, bioengineers can take steps to ensure that those patients who finish out their days in a hospital setting do not suffer from a barrage of inappropriate and ineffective technologies by providing educational inservices on technology management to nurses, physicians, and administrators. As experts in medical technology, bioengineers must position themselves to effectively manage medical technology with consideration of social, ethical, economic, and political factors. Physicians need guidelines on which to base their treatment decisions. A Denver, Colorado, program called GUIDE (*Guidelines for Use of Intensive Care in Denver*) is attempting to define those cases in which intervention is futile. GUIDE combines the professional opinions of health care administrators with the value judgments of members of the community to create specific criteria for limiting medical intervention. [6]

Many ethicists are concerned that proposals such as those embraced by GUIDE will infringe on patients' rights and have a dramatic effect on a very vulnerable population. Ethics experts consider the GUIDE project, and similar proposals in Oregon, to be an embryo of future health care reform. The quandary surrounding the issue of community-based treatment policies concerns the enforcement of rationed care. Such programs will be ethically unjust if wealthy individuals can "buy out" of the agreement. [6]

▼ *Outcomes Research*

A promising technique for determining the appropriateness of life-sustaining interventions is outcomes research. Outcomes research bases intervention decisions on data collected through previous applications of the technology in question. SUPPORT, a study conducted by the Program on the Care of Critically Ill Hospitalized Adults, is attempting to determine the effect of outcomes data on patient health and medical costs. SUPPORT combines research from six different hospitals in areas ranging from survival rate and functional status to family and psychological effects of life-sustaining intervention. The program hopes to show that outcomes data will help physicians and patients choose the medical intervention that meets patients' desires. [14]

Since the initiation of programs such as SUPPORT, the health-care industry has seen a major shift toward determining the efficacy and cost effectiveness of medical technologies though outcomes research. Many professional societies are now using outcomes data to create clinical standards that target specific technologies for patients with specific medical conditions. [13]

As outcomes research becomes more prevalent in the health-care industry, bioengineers working in clinical settings must manage and perform similar studies if the profession is to have an impact on medical technology. Through research, bioengineers can assist physicians and patients in determining the most appropriate technological intervention for a given circumstance. This places bioengineers in a unique and advantageous position to be an effective force in technology management.

▼ *Technology Assessment*

Outcomes data provides critical information for the application of existing technologies, but becomes ethically questionable when applied to new, unproved technologies. The value and effectiveness of new medical technologies must be determined through scientific methods of experimentation and clinical trials. Thorough technology assessment plays a crucial role in the prevention of undue pain, suffering, and expense resulting from inappropriate use of medical technology. [13]

Several different methods of technology assessment exist, the most rigorous method of which, randomized controlled clinical trials (RCCT's), uses double-blind assignment of large populations to study the effects of new technologies. In a Case Series, experimenters base the appropriateness of a particular technology on a collection of results from individual cases. Another type of technology assessment, consensus development, combines the opinions of several doctors into a written recommendation for uses of the technology. [7]

Rigorous technology assessment presents conflicts with society's eagerness for new medical technologies and hospitals' desire for profits and a competitive edge. In addition to the ethical difficulties associated with human experimentation, clinical evaluation of new technologies suffers from economic, social, and time demands. As a result of external pressures, assessments are rarely thorough and unbiased. [13]

Several professional organizations, including the American Medical Association, the American College of Physicians, and the Office of Technology Assessment (OTA), are attempting to improve the methods used to assess new medical technologies. [7] However, the efforts of these organizations can provide only guidelines for technology developers to follow. Bioengineers in the development stages of technology, and those in the area of clinical assessment, must abide by these ethical guidelines to insure the safety and autonomy of patients.

An ethically sound medical technology provides the greatest possible benefit to the greatest number of people.

Role of Bioengineers

As the developers of medical technology, bioengineers play a crucial role in the ethical decisions concerning life-sustaining technologies. Their design decisions and methods of technology assessment, affect not only the well-being of individual patients, but also society's attitude toward the value of life.

The development of any new life-sustaining technology has ethical ramifications on the availability, financing, and use of existing technologies. Because life-sustaining technologies target small segments of society, developers must weigh the technology's benefit against the ramifications of shifting finances away from programs that are more beneficial to the general population.

In order to have a positive influence on the current technology cycle, bioengineers must consider the long-range implications of an invention before placing it at the disposal of a technology-hungry society. American "technology favoritism" and the economic driving forces of hospital competition, insurance revenues, and government subsidies, create an environment that fosters misuse of life-sustaining technologies. Bioengineers must provide not only clinically proven and reliable technologies, but also thorough information on the likelihood, duration, and degree of benefit that patients with specific conditions can expect to receive. Physicians and patients need reliable information on the technology and treatment that will be used, as well as on alternative treatments in order to make justifiable ethically correct treatment decisions. For many individuals, advance directive and hospice care impose less of a financial, physical, and emotional burden than life-saving intervention.

Community-sponsored treatment protocols, technology assessment, and outcomes based research are keys to ethical-treatment decisions. By adopting standard assessment procedures, bioengineers can provide clear justification for the use of their technological developments.

Bioengineers play a hidden but crucial role in treatment decisions of doctors. Since bioengineers are responsible for not only the development and acquisition of medical technologies, but also device assessment and training of operators, they have a direct effect on the availability and reliability of new medical technologies. Physicians will base their treatment decisions on the performance information provided by the developer of the new technology. Therefore, technology assessment and design decisions have ethical ramifications that span a multitude of social and economic fronts.

With the current health care debate still unresolved, it would seem an opportune time for bioengineers to take a stand on ethical and economic issues related to technology development. Clinical engineers are seeking opportunities to define their role in the future of health care. Physician inservices, medical device selection, and technology assessment could provide a vehicle for defining the role of clinical engineers.

According to Millie Solomon, coordinator for the Decisions Near the End of Life program, " People are not afraid of death, they are afraid of being abandoned. They are afraid of losing control, and they are afraid of pain." [3] It is the duty of every health professional to fight for the life of the patient, but it is also in the best interest of medicine and society to respect the wishes and the dignity of the patient. There is a place for hospice care and advance directives in medical treatment regimens. The application of life-sustaining technology is not always appropriate. Making the decision as to when it is and is not appropriate to employ high-tech medical devices will always ultimately rest in the hands of the attending physician. Bioengineers can have an impact on the physician's perspective of life-sustaining technology by providing outcomes information and education on the use of various medical technologies.

The ethical considerations of the health-care industry place bioengineers in an ironic juxtaposition to most other engineering professions. While some engineers ask themselves "What is the lifespan of my device?," bioengineers must ask themselves "Can medical technology be limited in life-sustaining capabilities?"

References

[1] Irene H. Butter, "Premature adoption and routinization of medical technology: Illustrations from childbirth technology," *J. Soc. Issues*, vol. 49, no. 21, pp. 11-31, 1993.

[2] Joseph D. Bronzino, *Management of Medical Technology*. Boston: Butterworth-Heinemann, 1992.

[3] "Decisions' program educates staff on dealing with death," *Hospitals*, pp. 30-31, Feb. 5, 1991.

[4] Ezekiel J. Emanuell and Linda Emanuell, "The economics of dying: Illusion of cost savings at the end of life," *New England J. Med.*, pp. 540-543, Feb. 24, 1994.

[5] "Ethics in life-sustaining technologies," *Futurist*, p. 47, Jan.-Feb. 1988.

[6] Diane Ganelli, "One community looks for consensus on futile care." *Amer. Med. News*, p. 1+, Sept. 20, 1993.

[7] William Hendee, "Technology assessment in medicine: Methods, status, and trends," *Med. Progress through Technol.*, vol. 17, pp. 69-75, 1991.

[8] Robert L. Jayes, Jack E. Zimmerman, P. Douglas, Elizabeth A. Draper, William A. Knaus, "Do-not-resuscitate orders in intensive care units: Current practices and recent changes." *J. Amer. Med. Assoc.*, vol. 270, pp. 2213-2217, Nov. 10, 1993.

[9] Albert Jonsen, "The artificial heart's threat to others," *Hastings Center Rep.*, pp. 9-11, Feb. 1986.

[10] Eric Juengst, "Developing and delivering new medical technologies: Issues beyond access." *J. Soc. Issues*, vol. 49, no. 2, pp. 201-210, 1993.

[11] Helen Kapila and Nicholas Coni, "The application of modern diagnostic and therapeutic techniques to aged patients." *Aging and Soc.*, vol. 9, pp. 165-177, 1989.

[12] Marshall B. Kapp, "Law, medicine and the terminally ill: Humanizing the approach to risk management." *Health Care Manag. Rev.*, vol. 12, no. 2, pp. 37-42, 1987.

[13] Marshall B. Kapp, "Life-sustaining technologies: Value issues," *J. Soc. Issues*, vol. 49, no. 2, pp. 151-163, 1993.

[14] Mary T. Koska, "Can outcomes data help patients make end-of-life decisions?" *Hospitals*, June 5, pp. 42-43, 1991.

[15] David Orentlicher, "Rationing and the Americans with Disabilities Act," *J. Amer. Med. Assoc.*, vol. 27, pp. 300-314, Jan. 26, 1994.

T&S

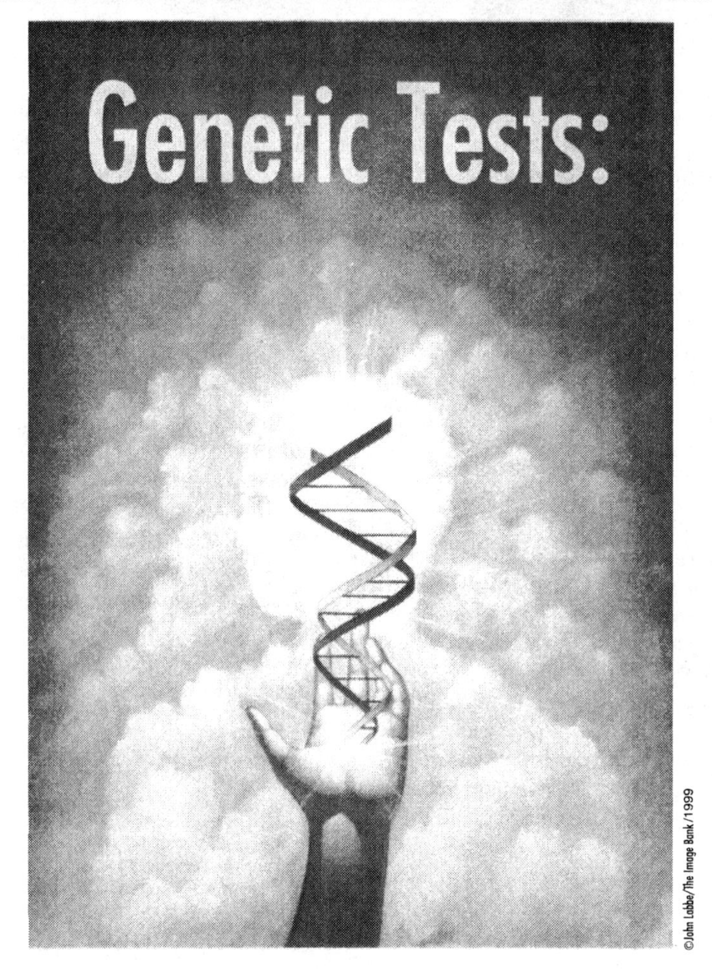

Evolving Policy Questions

Rapid advances in molecular biology are changing the world and the way we think of ourselves. As the Human Genome Project moves forward, we will identify more and more of the estimated 100 000 genes that make up our genome. This information will help us understand, identify, and eventually treat many genetic disorders and other diseases in which genetic predisposition plays an important role [1].

A central challenge posed by these advances is that the technology to diagnose genetic conditions and predispositions inevitably precedes the ability to intervene with specific treatments. During this window, genetic information has little therapeutic outlet and may find social outlets with desirable but also undesirable consequences. For example, the gene for cystic fibrosis was identified in 1989, making it possible in many cases to identify heterozygous carriers of this mutation who, though unaffected themselves, may bear children with cystic fibrosis. This discovery was an important medical advance, but it posed a challenge because there is no definitive treatment for this disorder. In such cases, genetic information may have undesirable social consequences [2]-[7].

Current or soon to be available genetic tests such as these raise the issue of potential uses and misuses of information. This is an issue of which both clinicians and the public need to be aware.

David A. Asch is a Department of Veterans Affairs Health Services Research and Development Research Associate at the Division of General Internal Medicine, University of Pennsylvania School of Medicine, 317 Ralston-Penn Center, 3615 Chestnut Street, Philadelphia, PA 19104-2676. Email: <asch@wharton.upenn.edu>. Michael T. Mennuti is Chairman of the Department of Obstetrics and Gynecology, University of Pennsylvania School of Medicine, Philadelphia, PA.

Reprinted from *IEEE Technology and Society Magazine,* Vol. 15, No. 4, pp. 4–10, Winter 1996/1997.

Screening and Stakeholders

As the genes for more diseases are identified, our concept of screening for genetic diseases is changing. The entire population is becoming the potential audience for genetic tests. Screening is no longer performed only in specialized medical settings by those experienced with interpreting and delivering genetic information, but in the offices of other health professionals, and in employment and insurance settings where safeguards for counseling and confidentiality cannot be assumed. This promises to make potentially valuable information available to more patients, but raises concerns about how this information will be transmitted and used. A report from the House Committee on Government Operations states that the complex problems that will result from our expanding knowledge of the human genome must be addressed "before inappropriate uses of genetic information become institutionalized" [2].

Different stakeholders will use genetic information differently. Patients, the most obvious and most legitimate stakeholders, have the most to benefit or lose from genetic information. They benefit from a genetic test when they learn they are not at risk for a familial condition or, if at risk, that they can minimize or eliminate that risk. They may lose if the information causes uncompensated anxiety or distress, or leads to social stigma or discrimination. Similar potential benefits and losses are provided by other medical tests that do not have a genetic basis, but there are important differences as well.

Genetic tests have special meaning because of their implications for the health prospects of relatives and future generations. Knowledge that an individual carries genetic mutations may be at once valuable and troublesome for relatives. Tests to detect infection with the human immunodeficiency virus raise similar issues: sexual contacts and children of a tested individual have a stake in the test results and may benefit or lose.

Physicians, genetic counselors, and other health professionals are stakeholders as well. Increasingly they will be asked to educate patients about the tests, and to help them make decisions about their use. Employers and insurers may also feel they have a stake in genetic information because it may help in their hiring practices or risk ratings.

Finally, advances in testing impact on social and religious values. Genetic testing often takes place in the prenatal setting and is linked with decisions regarding reproduction. Also, the ability to identify certain conditions and predispositions as genetically based will change the way these conditions are viewed. If we identify a partial genetic basis for phenomena that we previously thought had only a social basis—like criminal behavior—our views toward responsibility and punishment may change.

Some uses of genetic tests achieve legitimate social and personal goals. But often the distinction between desirable and undesirable uses of genetic tests depends on one's perspective and values.

Genetic Testing for Individual Diagnosis

The use of genetic tests for diagnosis of symptomatic patients is the most direct benefit of our expanding ability to probe the genome. Detection of two cystic fibrosis mutations in a child may establish the diagnosis even though the clinical findings are equivocal. Genetic tests are also beneficial when they identify predispositions that can be mitigated. Ataxia-telangiectasia, an autosomal recessive disorder, is associated with an increased sensitivity to ionizing radiation. The cancer rate for homozygotes for this trait is about 100 times greater than in the general population. Recent evidence suggests that heterozygotes also are more sensitive to radiation and have approximately a four-fold increased risk of certain cancers. The individual and public health implications of this finding are profound. Although the homozygous condition is rare, approximately 1.4% of the population is heterozygous for the gene. Such carriers, identified by DNA testing, will have a strong reason to minimize their exposure to diagnostic radiation. For heterozygous women the risk of mammography-induced breast cancers may exceed the benefits of early detection.

Individuals may also benefit from presymptomatic testing, although the benefits of such testing are not always clear. Huntington's disease, an autosomal dominant disorder, causes severe neurologic symptoms that do not develop until middle age. Individuals with an affected parent face a 50% chance of eventually manifesting the disease. There is a predictive test for Huntington's disease, but the disease is incurable. The test is a mixed blessing: learning that one has inherited the gene is disturbing and the information cannot be used in managing one's health, but it may be important in reproductive planning [3]. Moreover, the information has implications for children already born. For these reasons, many at risk do not want to be tested, and the mere availability of the test can cause anguish.

The recent discovery of the BRCA1 gene [4],[5], associated with breast and ovarian cancer, raises similar troubling issues for an even more common disease. About 1 in 200-400 women are believed to carry this gene. These women may have an 85 percent lifetime risk of developing breast cancer. Screening for this gene can identify these women, but the clinical value of this information is unclear. Preventive management, for example with tamoxifen [6] or prophylactic bilateral mastectomy [7],[8],[9] is of uncertain benefit. There is no evidence that prophylactic mastectomy reduces breast cancer risk, but much of the public and the medical profession seems eager to accept its promise. In a recent survey, 90 of 700 Maryland surgeons reported performing at least one preventive mastectomy [10]. Moreover, most women who will eventually develop breast cancer do not carry a detectable mutation in this gene. Negative screening for BRCA1 mutations may create a false sense of security, leading women to forego conventional periodic breast examination or mammography [11]. Confusion is likely to increase among patients and clinicians as other genes related to breast cancer are uncovered.

Even for less significant disorders, genetic information may be more stressful than other types of medical information, and the communication of screening results and interpretation presents special challenges. Genetic traits are more likely to be a source of stigma and discrimination, perhaps because people often define personal identity in terms of one's genes. Genetic traits cluster along ethnic or racial lines, which also can be sources of discrimination, and they can create intense feelings of guilt and responsibility among relatives. Whatever the explanation, genetic testing often has more profound emotional consequences than other medical tests. At the same time, the test results can be difficult to contain, both because they have implications for family members, and because family members, for some tests, must also provide DNA samples to complete testing for the proband and may inadvertently receive information they wish to avoid [12].

Many social concerns of genetic testing resemble those of testing for infection with the human immunodeficiency virus. Both raise issues of privacy, stigma, discrimination, and voluntarism. Genetic disorders are not communicable, but they are transmitted vertically to future generations, and relatives in many ways resemble contacts. Individuals learning they carry a gene for Tay Sachs disease, for example, simultaneously learn that their siblings have a fifty percent chance of carrying that gene as well. Because this knowledge may be valuable to them, it may create obligations to disclose [13]. This conflict between the right of privacy and the duty to disclose is a perennial problem in public health that has no clear solution. Genetic testing provides another setting for this dilemma.

Another challenge is how to convey and interpret the results of genetic tests, which are often probabilistic in nature. One quarter of a sample of middle class pregnant women in one study equated chances of 1/1000 and 10%, showing the widespread difficulty that many people have in interpreting probabilistic information [14]. Many genetic tests have the potential for false positive and false negative results, which further complicates their interpretation.

As genetic tests become more widely available, we need to think about how the results will be used.

Thus genetic testing creates problems for informed consent and the need for more genetics education for the general public and health professionals. More testing will increase the need for genetic counseling. Counseling for cystic fibrosis carrier screening alone might require approximately one third of the currently board-certified genetic counselors [15].

Reproductive Planning

Genetic testing can provide couples with information useful in reproductive planning. Despite its obvious benefits, prenatal testing has been controversial because it is often linked with the option of abortion. Also, it may target certain genotypes as undesirable and promote the implicit goal of reducing their prevalence.

Opportunities for genetic testing in the reproductive setting are increasing rapidly. Approximately one in twenty five Caucasians in the United States carries a cystic fibrosis mutation; in approximately one in 625 couples both partners are carriers. For these couples, each pregnancy faces a one in four chance of being affected with cystic fibrosis. Carrier screening, now available, can help identify couples at risk, and prenatal diagnosis can help identify affected fetuses [16]. Prenatal diagnosis can help couples decide whether to continue a pregnancy, and if the pregnancy is not terminated, it can be valuable in planning the delivery and arranging future care.

Alternatively, couples can undergo carrier screening before a pregnancy and decide whether or not to conceive on the basis of test results, or whether to conceive with different partners through the donation of eggs or sperm from noncarriers.

For many, however, the important issue is not abortion, but whether genetic testing in a reproductive setting is a form of eugenics. The issue is what, if any, genetic characteristics are legitimate targets for screening. Nobody knows how many pregnancies are terminated for sex selection, although the practice certainly occurs. As we increase our ability to perform prenatal diagnostic tests earlier in pregnancy, less invasively, and for many new indications, more pregnancies will be terminated as a result of genetic tests. *In vitro* fertilization has further expanded prospects for prenatal diagnosis. A single cell can be removed from an embryo fertilized *in vitro* and new molecular techniques used to amplify and determine some of its genetic characteristics. From this information, one can decide whether to select that embryo, among the others similarly tested, for implantation [17]-[19]. The availability of such tests will lead to demand for using them, because couples may see the tests as answering personal needs, and because of external pressures to use them [20].

Physicians may be a source of such pressures. Obstetricians may fuel the demand for carrier screening and prenatal diagnosis if they fear future malpractice suits if a child is born with a condition that could have been identified in advance. Physicians may yield also to marketing pressures. Both of these factors have been cited in the rapid diffusion of alpha-fetoprotein screening for neural tube defects. Future genetic tests are likely to follow the same pattern.

Public opinion is a source of pressure as well. A recent survey found that 39% of Americans believe every pregnancy should be tested to determine if the fetus has any serious genetic defects; 22% believe that a pregnant woman should have an abortion if the fetus has a serious genetic defect; and 10% believe an indigent woman should be required by law to have an abortion rather than have the government help pay for the care of a child with a genetically based condition [21]. These public attitudes add to the stigma faced by parents of children with genetic conditions, and may increase demand for genetic tests. Women are particularly likely to be targets of such pressures, because they are often viewed as responsible for adverse outcomes of pregnancy, are more commonly the recipients of reproductive health care, and are more likely to become the subjects of screening programs.

Past experience with carrier screening has raised other concerns. Sickle cell carrier screening programs, begun in the 1970s, have been fraught with difficulties. In many states mandatory sickle cell screening raised concerns about racist and eugenic motivations. Patients and the public often equated carrier status with sickle cell disease, believed sickle cell disease to be communicable or sexually transmitted, believed that carriers should not have children, or believed that childbirth could be hazardous [22]. These misconceptions increased the stigma and discrimination faced by carriers and those affected with the disease.

These problems underscore the need for widespread efforts to educate medical professionals, the public, and screened individuals about the meaning of test results. Despite the obvious difficulties, it seems possible to design carrier screening programs that are sensitive to these concerns. Carrier screening for Tay-Sachs disease began at about the same time as screening for sickle cell trait, and has generally been regarded as a success [23]. Screening programs were designed with heavy participation from the Jewish communities at risk for the disease. Screening has been voluntary, although some have suggested that the community provides some coercive pressure [24]. One orthodox community in New York, where marriages are arranged, originated the program of Chevra Dor Yeshorim. In this program, young adults are screened for Tay Sachs disease and results are kept secret in a central registry. To minimize the chances of stigma and discrimination, no one is informed of his or her result until a marriage is proposed. Only if both prospective partners are carriers are they told their carrier status; they are then referred for counseling, and new matches may be sought [25]. These efforts suggest that screening programs can be designed to address cultural needs and concerns.

Screening programs like Chevra Dor Yeshorim illustrate the use of genetic screening tests by individuals in the selection of partners. Another example is from Cyprus, where the high prevalence of beta-thalassemia led to a national effort in the early 1970s to screen potential carriers and provide counseling and prenatal diagnosis to couples at risk. In 1983, the Greek Orthodox Church, which opposes abortion to prevent affected births, began requiring couples to show evidence of screening and counseling before blessing a marriage. This led to a 97% decrease in the number of children born with beta-thalassemia. Since 1985 there has been a 20% reduction in the number of couples in which both partners screen positive, suggesting that carriers are now less likely to marry other carriers [26].

Genetic Testing by Employers

Questions arise about the legitimacy of genetic screening in the employment setting for several reasons. First, employers might avoid hiring individuals with a genetic predisposition to medical illnesses that might increase their absenteeism, reduce their productivity, or increase the costs of recruitment and training new workers. A company might use a test to detect and exclude prospective workers who are likely to die of premature coronary artery disease, for example. Less likely, prospective employees might use genetic tests to choose career paths.

Second, employers might use genetic screening tests to avoid hiring individuals who are likely (or who have family members likely) to incur high health care or disability costs. Screening prospective employees for the purposes of identifying those likely to have high future health costs is prohibited under the Americans with Disabilities Act. Nevertheless, self-insured employers are permitted to limit coverage for specific conditions after employees are hired; the ADA provides only limited protections.

Third, employers might use genetic screening tests to avoid hiring workers who have genetic predispositions that might endanger public safety. Tremendous pressures might develop to use a test that could screen prospective airline pilots or bus drivers for a gene linked to alcoholism, for example, even if its predictive ability was limited [27].

Finally, employers might use genetic screening tests to identify workers especially vulnerable to workplace exposures. About fifty genetic conditions are known that enhance susceptibility to workplace exposures [28]. Tests for sickle cell trait, glucose-6-phosphate dehydrogenase deficiency, and alpha1-antitrypsin deficiency have been used at one time or another to exclude certain workers from jobs felt to present them with special risk [29], and some companies prohibit fertile women from working in settings that are potentially hazardous to fetal development [30].

Efforts to identify susceptible workers might benefit them, especially if they are offered alternative positions, but it is not clear that such screening has uncovered real risks which may be mitigated [31]. Testing in the workplace is less voluntary than testing in the physician's office, and the same safeguards for confidentiality and counseling do not exist. Patients can ignore the advice of their physicians, but employees, especially prospective employees, cannot similarly ignore the "advice" of employers.

Some critics suggest that screening workers for susceptibility to occupational hazards shifts the blame for occupational hazards from environmental exposures to genetic predisposition. Screening may substitute for efforts to clean up the workplace that might benefit others who are not hypersusceptible [32]. In a recent case involving the potential risk of lead exposure to fetuses carried by women involved in the production of batteries, the Supreme Court ruled that company fetal protection policies are unlawful if they ban fertile women from sites because of possible fetal injury [33].

With many of these concerns in mind, the American Medical Association (AMA) recently issued an opinion that identified no role for the screening of workers for non-occupationally related health risks, and only a very limited role for the screening of employees for susceptibility to workplace exposures. The AMA considered the latter to be appropriate only when the workplace hazard is so serious and develops so rapidly that monitoring the exposure would be ineffective in preventing the injury; when the test is highly predictive of a significantly increased susceptibility; when alternative means of lowering the exposure are extraordinarily expensive; and when the worker provides informed consent for the screening [34].

Despite the many ways employers *might* use genetic screening, it is unclear how prevalent the practice really is. A 1982 survey commissioned by the Office of Technology Assessment suggested that fewer than 2% of Fortune 500 companies engage in any form of genetic screening or monitoring. Another 1989 survey found that 5% of Fortune 500 companies engage in genetic screening or monitoring [35]. This suggests that the practice is neither widespread nor increasing rapidly. The potential savings of employee screening programs might not offset their costs. Nevertheless, the social hazards of workplace screening are real, and one reason for the expanded scope of the ADA is that clear motivations exist for engaging in more testing. Whether or not genetic testing by employers offers some benefits in limited settings, it clearly provides an opportunity for unfair employment discrimination. It is not yet clear how well the ADA or the opinions of organized medicine can steer the use of genetic tests by employers away from these discriminatory uses.

Genetic Testing by Insurers

Genetic testing by life and health insurers raises other concerns. Insurers rate individuals according to their perceived health risks and determine their insurability and the premium to charge. Genetic tests expand the ability of insurance companies to determine individual risks, and might change the practice of insurance risk rating.

Only about a third of health insurance policies and about half of life insurance policies in this country are individual policies. The rest are group insurance, typically obtained through employers and not subject to individual risk rating.

Because employers have parallel incentives to reduce their financial exposure, as discussed above, genetic information might be equally relevant to those covered by both group and individual policies.

Individuals might hesitate to use genetic tests if the results might later be used to deny insurance coverage. This is similar to "job lock," in which employees receiving employment-based health insurance are unable to change jobs because recently diagnosed conditions would represent "preexisting conditions" that would not be covered if health insurance policies are changed. Genetic tests are different from other medical tests in this setting because they change the very nature of what might be considered preexisting. Most health insurers that responded to a 1991 Office of Technology Assessment (OTA) survey considered genetic conditions, such as cystic fibrosis or Huntington's disease, to be "preexisting" conditions [36]. Predictive genetic tests blur the distinction between an increased risk of disease and a preexisting condition.

An example illustrates a related concern. A couple with a child affected with cystic fibrosis learned through prenatal diagnosis that another pregnancy was affected. When the couple decided to continue the pregnancy, their insurance company argued that it should not have to pay for health care of the child related to cystic fibrosis, since it had paid for the prenatal diagnosis ostensibly to avoid the birth of the affected child. Under pressure, the insurance company later reversed its position [37]. Other similar cases show that this was not an isolated instance [38]. This anecdotal evidence shows that insurance companies can use genetic information irresponsibly.

A related question is whether insurers would use genetic testing before issuing individual policies. OTA's 1991 survey of insurers found that most health insurers believe it is "fair for insurers to use genetic tests to identify individuals with increased risk of disease," and that they "should have the option of determining how to use genetic information in determining risks." Few insurers, however, predicted they would in fact be using genetic tests in this way in five or ten years [39]. This might change if accurate tests to predict breast cancer or low back pain, for example, became available.

Genetic screening will probably be more expensive than other methods of obtaining actuarially relevant medical information, such as questions about family medical history. Why then might insurers use such tests?

One reason is that individuals can undergo genetic testing on their own, and if they discover that they are at greater medical risk they might purchase more insurance. This phenomenon, known as adverse selection, puts insurance companies at a disadvantage if they lack access to the same information. To level the playing field, insurance companies might have a legitimate interest in the results of genetic tests already performed, and might have an incentive to initiate testing for selected conditions. When one insurer begins to use genetic information in its risk rating, others must follow or they will disproportionately attract individuals who are denied coverage elsewhere.

This suggests that insurance companies will inevitably begin to use genetic information in risk rating, despite the expense and complexities involved. Some analysts have argued that this inexorable trend contradicts the rationale for insurance in the first place. As genetic tests increase our ability to refine risk so that individuals are placed in ever smaller risk pools, administrative costs for insurance will soar as the ability to distribute financial risks vanishes [40]. These trends may constitute an independent argument for new systems of health care financing that eliminate individual insurance policies and risk rating.

Future Policy Options

Advances in molecular biology offer tremendous promise for diagnosing disease. But the lack of effective therapies following genetic diagnosis creates dilemmas. Genetic information that cannot be used for treatment might be used for undesirable social purposes. We need to gain the most good from our expanding knowledge and avoid the pitfalls. Clinicians, the public, and policy makers share this challenge.

Experience with past and present genetic testing programs provide some direction. Past problems with sickle cell screening programs vividly demonstrate the need to educate the public and health care professionals alike about the meaning and implications of genetic tests. We need to strengthen our genetic curriculum at all educational levels, and also train more health care providers who can counsel patients about the benefits and risks of these tests. The experience with sickle cell screening also illustrates the need for strict protection of confidentiality to avoid the stigma and discrimination that may occur in social, employment, and insurance settings. This may require the strengthening of the ADA and similar legislation.

We must identify the best balance in our *overall* approach to genetic disorders. Should we screen to avoid diseases, or devote those resources to helping those who are or will be affected?

Genetic testing raises broader health policy issues as well. These include: whether to foster the development of new diagnostic information when treatment is not immediately available, how to control the diffusion of new medical technologies in commercial environments or the role of social pressures and defensive medicine in the adoption of these technologies, and the role of health insurance in providing equitable access to care. These policy concerns, most of which are not unique to genetic testing, need to be addressed as the technology expands.

Perhaps most importantly, we need to match our increasing understanding of genetic diagnosis with efforts to increase our acceptance of those affected. As we learn to detect, prevent, and alter previously immutable characteristics, we risk reducing our tolerance for genetic variation. We must continue to view genetic diversity as both a biologic and a social good. The American values of diversity and individual rights will undoubtedly continue to inform U.S. policies, but we should recognize that other countries and other cultures may develop substantially different policies. Health policy for genetic testing must be formed through the exploration of human values.

References

[1] National Center for Human Genome Research, *Understanding our Genetic Inheritance: The U.S. Human Genome Project*, NIH Publication 90-1590. Springfield, VA: National Technical Information Service, 1990.

[2] House of Representatives Committee on Government Operations, "Designing genetic information policy: the need for an independent policy review of the ethical, legal, and social implications of the human genome project," *House of Representatives Rep. 102-478*, April 1, 1992, p. 25.

[3] G.J. Meissen, R.H. Myers, C.A. Mastromauro *et al.*, "Predictive testing for Huntington's disease with use of a linked DNA marker," *New Eng. J. Med.*, vol. 318, pp. 535-542, 1988.

[4] J.M. Hall, M.K. Lee, B. Newman, J.E. Morrow, L.A. Anderson, B. Huey and M.C. King, "Linkage of early onset breast cancer to chromosome 17q21," *Science*, vol. 250, pp. 1684-1689, 1990.

[5] Y. Miki, J. Swensen, S. Shattuck-Eidens *et al.*, "A strong candidate gene for the breast and ovarian cancer susceptibility gene BRCA 1," *Science*, vol. 266, pp. 66-71, 1994.

[6] S.G. Nayfield, J.E. Karp, L.G. Ford, P.A. Dorr and B.S. Kramer, "Potential role of tamoxifen in prevention of breast cancer," *J. Natl. Cancer Inst.*, vol. 83, pp. 1450-1459, 1991.

[7] L.D. Zeigler and S.S. Kroll, "Primary breast cancer after prophylactic mastectomy," *Amer. J. Clin. Oncol.*, vol. 14, pp. 451-454, 1991.

[8] H. Nelson, S.H. Miller, D. Buck, R.J. Demuth, W.S. Fletcher and P. Buehler, "Effectiveness of prophylactic mastectomy in the prevention of breast tumors in C3H mice, *Plast. Reconstr. Surg.*, vol. 83, pp. 662-669, 1989.

[9] J.E. Goodnight, J.M. Quagliana and L. Morton, "Failure of subcutaneous mastectomy to prevent the development of breast cancer," *J. Surg. Oncol.*, vol. 26, pp. 198-201, 1984.

[10] G. Cowley, "Family matters," *Newsweek*, pp. 46-52, December 6, 1993.

[11] B.B. Biesecker, M. Boehnke, K. Calzone, D.S. Markel, J.E. Garber, F.S. Collins and B.L. Weber, "Genetic counseling for families with inherited susceptibility to breast and ovarian cancer," *JAMA*, vol. 269, pp. 1970-1974, 1993.

[12] F.A. Millan, A. Curtis and M. Mennie, *et al.*, "Prenatal exclusion testing for Huntington's disease: A problem of too much information," *J. Med. Genetics*, vol. 26, pp. 83-85, 1989.

[13] M.Z. Pelias, "Duty to disclose in medical genetics: A legal perspective," *Amer. J. Med. Genetics*, vol. 39, pp. 347-354, 1991.

[14] G.A. Chase, R.R. Faden, N.A. Holtzman, *et al.*, "The assessment of genetic risk by pregnant women; implications for genetic counseling, *Soc. Biol.*, vol. 33, pp. 57-64, 1986.

[15] B.A. Wilfond and N. Fost, "The cystic fibrosis gene: medical and social implications for heterozygote detection," *JAMA*, vol. 263, pp. 2777-2783, 1990.

[16] D.A. Asch, J.P. Patton, J.C. Hershey and M.T. Mennuti, "Reporting the results of cystic fibrosis carrier screening," *Amer. J. Obstet. Gynecol.*, vol. 168, pp. 1-6, 1993.

[17] Y. Verlinsky, E. Pergament and C. Strom, "The preimplantation genetic diagnosis of genetic diseases," *J. In Vitro Fert. and Embryo Transfer*, vol. 7, pp. 1-5, 1990.

[18] A.H. Handyside, J.G. Lesko, J.J. Tar'n, M.L. Winston and M.R. Hughes, "Birth of a normal girl after in vitro fertilization and preimplantation diagnostic testing for cystic fibrosis," *New Eng. J. Med.*, vol. 327, pp. 905-909, 1993.

[19] J.L. Simpson and S.A. Carson, "Preimplantation genetic diagnosis," *New Eng. J. Med.*, vol. 327, pp. 951-953, 1992.

[20] A. Clarke, "Is non-directive genetic counseling possible?" *Lancet*, vol. 338, pp. 998-1002, 1991.

[21] E. Singer, "Public attitudes toward genetic testing," *Population Res. and Policy Rev.*, vol. 10, pp. 235-255, 1991.

[22] P. Reilly, *Genetics, Law and Social Policy*. Cambridge, MA: Harvard Univ. Press, 1977.

[23] L. Roberts, "One worked: The other didn't," *Science*, vol. 247, p. 18, 1990.

[24] N.A. Holtzman, *Proceed with Caution: Predicting Genetic Risks in the Recombinant DNA Era*. Baltimore, MD: Johns Hopkins Univ. Press, 1989, pp. 217-218.

[25] B. Metz, "Matchmaking scheme solves Tay-Sachs problem," *JAMA*, vol. 258, pp. 2636-2639, 1987.

[26] M. Angastiniotis, "Cyprus: Thalassaemia program," *Lancet*, pp. 119-129, 1990.

[27] American Medical Association Council on Ethical and Judicial Affairs, "Use of genetic testing by employers," *JAMA*, vol. 266, pp. 1827-1830, 1991.

[28] U.S. Congress, Office of Technology Assessment, *Genetic Monitoring and Screening in the Workplace, OTA-BA-455*. Washington DC: U.S. Government Printing Office, 1990, p. 83.

[29] C.F. Reinhardt, "Genetic hypersusceptibility," *J. Occup. Med.*, vol. 20, pp. 319-322, 1978.

[30] U.S. Congress, Office of Technology Assessment, *Genetic Monitoring and Screening in the Workplace, OTA-BA-455*. Washington DC: U.S. Government Printing Office, 1990, p. 148.

[31] American Medical Association Council on Ethical and Judicial Affairs, "Use of genetic testing by employers," *JAMA*, vol. 266, pp. 1827-1830, 1991.

[32] E. Draper, *Risky Business: Genetic Testing and Exclusionary Practices in the Hazardous Workplace*. Cambridge, MA: Cambridge Univ. Press, 1991.

[33] United Automobile Workers v. Johnson Controls, Inc. 111 SC 1196, 1991.

[34] American Medical Association Council on Ethical and Judicial Affairs, "Use of genetic testing by employers," *JAMA*, vol. 266, pp. 1827-1830, 1991.

[35] U.S. Congress, Office of Technology Assessment, *Genetic Monitoring and Screening in the Workplace, OTA-BA-455*. Washington DC: U.S. Government Printing Office, 1990.

[36] U.S. Congress, Office of Technology Assessment, *Cystic Fibrosis and DNA Tests: Implications for Carrier Screening, OTA-BA-532*. Washington DC: U.S. Government Printing Office, Aug. 1992, p. 195.

[37] C.T. Caskey, "New technology brings new ethical consideration," *Legal and Ethical Issues Raised by the Human Genome Project* (conference materials), Houston, TX, Mar. 7-9, 1991.

[38] P.R. Billings, M.A. Kohn, M. de Cuevas, J. Beckwith, J.S. Alper and M.R. Natowicz, "Discrimination as a consequence of genetic testings," *Amer. J. Human Genetics*, vol. 50, pp. 476-482, 1992.

[39] U.S. Congress, Office of Technology Assessment, *Cystic Fibrosis and DNA Tests: Implications for Carrier Screening, OTA-BA-532*. Washington DC: U.S. Government Printing Office, Aug. 1992, p. 33.

[40] D.W. Light, "The practice and ethics of risk-related health insurance," *JAMA*, vol. 267, pp. 2503-2508, 1992.

T&S

JOHN H. FIELDER

© The Image Bank/Chris Raschka/1999

Defects and Deceptions — The Bjork-Shiley Heart Valve

Despite extensive engineering analysis, bench and *in vitro* testing, and clinical trials, the behavior of medical devices cannot be entirely predicted under conditions of commercial use. While extensive testing greatly increases the probability that the device will perform safely and effectively, it cannot capture the range of conditions experienced in actual use. Instead, commercial use should be regarded as a second clinical trial, with patients having the right to be informed of any significant changes in performance results data.

As discussed in [1], the Safe Medical Devices Act of 1990 implicitly incorporates this concept, by establishing a registry and tracking provision for notification of patients with Class III medical devices, implants whose failure is likely to cause serious harm to the patient. The Act also authorizes post-marketing surveillance of devices about which questions have arisen concerning their safety.

Despite conscientious testing programs, new defects often appear, some serious enough to require withdrawal of the device. The Starr-Edwards caged-ball heart valve originally used a ball made of silicone elastomer, which absorbed lipids from the blood. This caused cracking, swelling, and abnormal wear, resulting in its withdrawal from the market. More suitable materials for the ball were found, and the valve is still in use. Other caged-ball valves featured a cloth cover for the cage struts to promote tissue

The author is a Professor of Philosophy at Villanova University, 800 Lancaster Ave., Villanova, PA 19085. Email: fielder@wild.ucis.vill.edu.

Reprinted from *IEEE Technology and Society Magazine,* Vol. 14, No. 3, pp. 17–22, Fall 1995.

ingrowth. However, the cloth abraded before this could occur, causing a variety of serious problems. These are only two of many examples of heart valves that were promising but were found to have unacceptable defects after being put on the market and implanted in patients.[1]

Heart valves like these, where new and substantial defects appear after being introduced into the market, are clearly defective, and there is little controversy about their withdrawal. Their performance falls substantially below the standard predicted from testing and clinical trials. Similarly, minor problems that do not significantly affect the valve's function do not result in it being designated as defective. For example, some patients complained that the Bjork-Shiley Convexo-Concave (C/C) heart valve made too much noise and caused them to be embarrassed in social situations.[2] But where is the line between these two poles, where devices cross over from being acceptably safe, to being defective and needing to be withdrawn? How serious does the defect have to be, and what characteristics of the defect are important, for deciding whether it should be withdrawn from the market?

It is tempting to set a threshold of failure — say 1% — to define "defective." Is any device that has a greater than 1% failure rate sufficiently defective to require discontinuation? Based on my conversations with individuals in government and industry who are familiar with the case, a majority would have said that a 1% failure rate would not disqualify a medical device. Yet the C/C valve was withdrawn despite having a failure rate that even its harshest critics did not place above half of that 1% level. Was the valve improperly withdrawn or are there other factors to consider? Are there other factors besides failure rates that should influence what counts as an unacceptable failure rate?

While I cannot answer this question in a general way, an examination of the Bjork-Shiley heart-valve case suggests that other factors were considered in the development of a consensus that the valve was defective and should be withdrawn. By examining a brief history of the Bjork-Shiley C/C valve and then discussing the factors that critically influenced the judgment that it was defective, I have reached the conclusion that the failure rate and the technical characteristics of the C/C valve alone were not responsible for its withdrawal. A number of critical ethical issues made the difference. I believe that if Shiley and its later owner, Pfizer, had responded to the problem in an ethically acceptable way, the valve could still be on the market.

[1]Both of these examples are discussed in [2].

[2]In a survey of 35 patients with C/C valve, about half reported that they were disturbed by the sound of the valve. See [3].

Mechanical Heart Valves[3]

The development of mechanical heart valves is one of the success stories of contemporary medicine. Many people with diseased heart valves become seriously disabled and soon die unless a prosthetic valve can be installed. The first mechanical heart valves were implanted in 1960, and since then their use has grown rapidly. Today over 40 000 Americans receive artificial heart valves each year [2, p. 410].

Shiley, Inc., later a subsidiary of Pfizer, was a pioneer in the development of mechanical heart valves. In 1974 the company developed a radial spherical (R/S) valve, consisting of a disk held in place by two wire struts, allowing it to swing open and closed in response to blood flow. The struts are welded to a metal ring, which is covered with a cloth sewing ring for attachment to the heart. In 1979 Shiley introduced a similar valve, the 60° Convexo-Concave (C/C), which it believed would improve blood flow through the valve. In this valve, the inlet strut was an integral part of the metal ring and only the outlet strut was welded. A C/C valve that opened to 70° was also manufactured, but it was not approved for sale in the U.S.

How serious does the defect have to be, and what characteristics of the defect are important, for deciding whether it should be withdrawn from the market?

Blood clots (thromboses) caused by the presence of artificial heart valves are a serious problem and are responsible for the greatest percentage of complications that occur. Because the movement of blood is obstructed by the valve, there are areas of relatively stagnant flow where clots can form. They may form on the valve, preventing it from fully opening or closing, or they may break free and cause strokes, heart attacks, and other serious complications. Drugs to "thin" the patient's blood and reduce clotting are an essential part of the treatment of

[3]Some of the material in this section is taken from other articles I have written on this topic [4], [5].

implantees, but there is a limit to their ability to reduce the incidence of thromboembolism (TE): blood clots that break free and lodge in an artery, cutting off blood flow to tissues or organs. Shiley's 60° C/C valve, which appeared to allow better blood flow and promised a lower incidence of TE, was regarded as a significant improvement in heart-valve technology.

Strut Fractures in the C/C Valve

A fracture of the outlet strut of a 60° C/C valve occurred during clinical trials in 1978, and as more of the valves were implanted, other reports of similar fractures began to reach Shiley. While the number was not large in relation to the number of valves implanted, the fact that all the fractures were similar —occurring at the point where the outlet strut was welded to the ring — strongly suggested that a design or manufacturing problem was responsible for the failures. There had been a few fractures of the inlet strut of the similar radial-spherical (R/S) valve, but a welding change was made in 1975 and no further fractures occurred.

When both legs of the outlet strut of the C/C valve fracture, the disk falls out of its ring, resulting in unrestricted blood flow through the heart. Without a valve to close one end of the chamber, contractions cannot force blood out of the heart. This is a form of heart failure, and it requires immediate open heart surgery. Depending upon the location of the valve in the heart, the person has anywhere from a few minutes — if it is an aortic valve — to one or two hours to live [2, p. 175]. About two-thirds of the persons who experience valve failure die, and many survivors have serious complications ([6]; also [2, appendix]).

The symptoms resulting from valve failure are similar to those of other forms of heart failure, and unless an autopsy is performed, it is usually not possible to determine that valve failure was the cause of death. Obviously, there are hidden or unreported fractures, and one of the controversial issues in the case is the estimated number of fractures. By 1993 the number of reported deaths by fractured heat valves was 386. However, the Public Citizen Health Research Group, an organization founded by Ralph Nader, estimated that the actual number of deaths was 50% higher than actually reported [6, p. 48].

Subsequent studies in the Netherlands added to this controversy. Twenty-two patients elected to have their Bjork-Shiley C/C valves explanted because epidemiological data (age, size of valve, opening angle, and position) indicated they were at risk for strut fracture [7].4 In some patients there were additional indications for reoperation, such as the presence of moderate cardiac impairment or other cardiac complaints unrelated to the valve. There was no diagnostic or clinical indication in any patient to suggest a valve defect.

The mean age of the 8 women and 14 men who chose valve explantation was 51 (26-68). Two patients had two valves replaced, making a total of 24 explanted valves. Ironically, one of the patients had previously survived a strut fracture of a 29-mm 60° mitral valve. Nondestructive examination of the valves was performed stereoscopically and with a scanning electron microscope to evaluate their structural integrity. Examination of the valve struts with a scanning electron microscope revealed no evidence of surgical mishandling in either implantation or explantation.

A total of eight 60° and sixteen 70° valves were explanted. The 70° valves have been shown to be more susceptible to strut fracture, particularly the larger sizes (29 mm). An earlier study showed mitral valves of this size to have the greatest risk of failure, a cumulative risk after eight years of 17.4% [8]. In contrast, the estimated failure rate for the 60° valves is less than 0.5% per year. Larger sizes of the 60° valve have a greater risk of fracture as well. The 70° valves were never approved for sale in the United States, and Shiley has been criticized for marketing them abroad when questions of their safety prevented them from being approved in the U.S.

Results of the Study

The results of the study of these twenty-four valves are breathtaking. *Seven of them (29%) had single-leg fractures (SLF): one leg of the outlet strut had already broken.* Two others showed characteristics of incipient fatigue failure (one was already cracked; the other showed intrusions and extrusions). *Counting these two and the seven instances of SLF, 37.5% of the valves were in various stages of strut fracture.* These fractures cannot be seen with the naked eye, but are easily detected at 25× magnification. Unless microscopic examination is undertaken, these fractures will not be noticed.

These new numbers place the risk of fractured valves orders of magnitude higher. They reinforce the position of those who have held that the actual number of fractures is far higher than what is reported. For example, 4 of 13 aortic valves had single leg fracture and one had a cracked strut. "This finding supports our hypothesis that, owing to the lethal character of the failing aortic valve strut, fractures remain under-reported; few such patients reach hospital and necropsy [autopsy] is rarely done" [7, p. 11]. As a result of this new

[4] For a discussion of [7], see [5].

study, suggesting that the fracture rate is far higher than originally expected, the Public Citizen Health Research Group increased its estimate of the number of deaths to "at least 1500" [9].

Shiley's Recalls

Soon after the C/C valve was introduced into commercial use in 1980, Shiley began to receive reports of fractures of the outlet strut. Between 1980 and 1986 Shiley/Pfizer initiated several voluntary recalls of the C/C valve and tried to identify the cause of the problem. This was difficult because of the small rate of failure. Although the failure rate was higher for large diameter valves, the reported failure rate for all sizes of valves was still much less than 1% per year. To reproduce failures in the lab would have required unrealistically long periods of testing. Consequently, scientists and engineers at Shiley had to construct hypotheses about the failures based upon the relatively meager data available and test them by reintroducing the valve back into the market [2, p. 374]. For example, one theory of failure was based on how the strut was bent before welding. New manufacturing techniques were instituted, and as new valves became available, Shiley recalled the old ones and replaced them, accompanied by "Dear Doctor" letters to surgeons that minimized the problem.[5]

Unfortunately, the failures continued despite several changes, and new explanations for the failures were put forward. Each one resulted in new methods of manufacturing or quality control and the replacement of "old" valves with new. The Food and Drug Administration (FDA) became increasingly concerned about Shiley's actions. Finally, Shiley withdrew the valve from the market voluntarily in the face of FDA movement toward an involuntary withdrawal. Ironically, Shiley may have finally hit upon the correct theory, for its last batch of valves appears to have had no known failures.[6]

Shiley's Combined Mortality Defense

When Shiley withdrew the valve from the market in 1986, the failure rate for its 60° C/C valve sold in the U.S. was estimated to be below 1%. Was this less than 1% failure rate enough to establish that the valve was defective and should be withdrawn from the market? Shiley argued that it was not, because the small failure rate is offset by the valve's superior performance in preventing TE. Most deaths (50%) of patients with valve implants are caused by the underlying heart disease of the patients. Another 30% of deaths are also unrelated to the artificial valve. The remaining 20% of deaths are caused by valve-related problems. TE is responsible for about half of valve-related deaths (10% of the total) and the rest are caused by anticoagulant hemorrhage, infection, leaks around the valve, and mechanical failure.

The fact that all the fractures were similar strongly suggested that a design or manufacturing problem caused the failures.

David Schoenfield prepared a report for Shiley in which he analyzed the performance of several artificial heart valves and concluded that "the overall combined mortality from thrombosis, embolism and strut fracture with the 60° C/C valve does not differ appreciably from other mechanical heart valves." Shiley claims that this reflects the improved design characteristics of the C/C valve in reducing the incidence of thrombosis and thromboembolism, and Schoenfield agrees [11]. Although Schoenfeld makes a strong case, his conclusion is not universally accepted. An important part of Shiley's conflict with the FDA concerned whether or not the data supplied by Shiley really established its claim of reduced TE. The issue is a statistical quagmire, and for my purposes it is more useful to assume that Shiley's claim was correct — that the C/C valve has a higher mechanical failure rate, which is offset by improved reduction of TE deaths — and examine what that means for the question of its defectiveness. If one cause of death is balanced by a roughly equal reduction in another cause of death, does this mean that the valve is not defective and should have remained on the market?

Ethical Dimensions of Defectiveness

If one looks at this defense from a utilitarian ethical perspective, it has a good deal of force. Utilitarianism, based on the work of Jeremy Bentham (1748-1832) and John Stuart Mill

[5] This issue is discussed in [10].

[6] At the hearing Shiley claimed that there had been no fractures in valves manufactured since April 1984; statistically, there would have been 11 fractures based on previous rates [2, p. 374].

(1806-1873), holds that the ethical choice is the one that results in the maximum amount of good consequences and minimizes the bad ones for all affected.[7] If the balance of good and bad consequences — utility — is the same for the C/C valve and other valves on the market, then the C/C valve is no more defective than any other. Forcing it off the market would be unfair and discriminatory, using a different standard for the C/C valves than the rest. If this were the only ethical dimension of the problem, Shiley would have a good case (assuming that the statistical analysis of TE reduction is correct).[8]

Utility is only one of several competing ethical traditions, however. Another major tradition, deontology, is derived from the thought of Immanuel Kant (1724-1804).[9] This tradition emphasizes respect for the individual as an autonomous person with the freedom to choose. Any ethically acceptable decision must take into account this concept of the individual. People should not be viewed as "means" to achieve a (good) end, but as "ends in themselves" (i.e., as autonomous moral agents). Whereas utilitarianism emphasizes the totality of good and bad consequences that accrue to individuals, deontology focuses on the intrinsic worth of individuals as free, rational beings. The language of rights is most at home in deontology, for rights protect individuals from intrusion by others. For Kant, individuals have a right to choose, as long as their choices treat others as free, autonomous choice-makers as well.

From a deontological standpoint, Shiley's defense is much less persuasive. While the balance of good and bad consequences may be the same, the effect on individual autonomy is not. Consider Shiley's program of making changes in their valve and testing whether those changes were effective by sending out modified valves to be implanted in patients. In effect, Shiley was conducting a clinical trial of their modified valve, but neither the surgeons nor the patients were afforded the ethical protections, particularly informed consent, associated with a clinical trial. Shiley's "Dear Doctor" letters to the surgeons minimized the problems with the valves and based the statistical evidence on reported fractures, ignoring the issue of hidden fractures. Most observers regard them as misleading. Patients were not informed of the valve's history or modifications; they were enlisted as research subjects without their knowledge or consent.

In a similar way, the Food and Drug Administration was also misled in order to prevent its taking action against Shiley. A 1990 FDA report concluded that there was "information that supports a belief that Shiley, Inc. has engaged in a continuing scheme to interrupt, deflect, and misdirect FDA's regulation of the Shiley Convexo-Concave heart valve." This was carried out by failing to reveal material information about the valve's performance, delaying reports of the defect to FDA, and making manufacturing and quality control changes (discussed above) without advising FDA and without their approval [15].

All of these practices are designed to interfere with the autonomous choices of surgeons, patients, and the FDA. Withholding information, providing misleading information, and preventing others from finding out what they need to know were direct attacks on the ability of these groups to make rational choices. Had they known the truth about the C/C valve, surgeons might have chosen another valve, patients might have refused to have a possibly defective valve implanted in them, and the FDA might have moved against Shiley more quickly. Shiley's efforts were designed to prevent surgeons, patients, and the FDA from making decisions that would have an adverse effect on the company's marketing of the C/C valve. As such, those efforts were clearly wrong from a deontological standpoint.

It is important to see that patients may have agreed with Shiley that the overall risk —fracture and TE — was the same for different valves, but that does not mean that each patient (or surgeon) would have given equal weight to the possibility of fracture and the reduction of TE. Many patients have chosen to have their C/C valves explanted because they did not want to live with the uncertainty of a possibly defective heart valve. For them, the reduction in TE is not an acceptable trade-off for the risk of fracture, and they would not choose it. In fact, it was important enough for them to take the risk of additional open-heart surgery with its 1-4% mortality rate and the pain and discomfort it brings.

Another factor contributed to the view that the Shiley valve was defective. The failures were all alike: they were not random component failures that can be expected with any device. There was clearly a mistake in the design and/or manufacturing of the valve, and therefore a presumption that it could be fixed. Shiley obviously believed this, too, but it did not want to discontinue marketing the valve while seeking a solution (hence the "earn as you learn" characterization). By working on changes while still marketing the valve, Shiley was treating surgeons and patients simply as a means to its marketing program and not acknowledging their right to make informed choices about participating in it.

Had Shiley not engaged in these deceptive

[7] There is another discussion of this approach in [12].

[8] For a critique of Shiley's statistical claim, see [6, p. 22-24].

[9] For an introductory discussion of these ethical traditions, see [13] and [14].

practices, it is possible that it could have kept its valve on the market. There is a risk that taking the valve off the market to determine what the problem is and how to fix it would result in not being able to successfully remarket the valve. But the Starr-Edwards caged-ball valve, mentioned earlier, which experienced early failures, was successfully redesigned and reintroduced into commercial use. Removing a product from the market to fix a problem is not an automatic kiss of death.

Defective Ethics

Was the C/C valve defective? Yes, but not simply because a small percentage fractured. What made the valve defective was the unethical practices that surrounded its marketing and development. A heart valve is a social as well as a material artifact, and the quality of its social relationships is as significant as its technical features. The valve cannot be considered apart from the company that makes it and determines what will be done when things go wrong. The same valve in the hands of a company that took its ethical obligations more seriously might not be defective, because what determines if it is defective is, in part, the quality of its relationships with patients, the medical community, and federal regulators. If the quality of those relationships at Shiley had been better, the valve might well have survived. Thus the valve must be regarded as a social artifact, not simply a material one. A heart valve with a small failure rate that is hidden from patients, surgeons, and the FDA is a defective device. Defects in both aspects of the C/C valve contributed to its being defective.

The concept of "defective device" is analogous to that of "approved device," in that both must meet socially defined criteria. A device that has been approved by FDA for commercial use is one that has certain technical features, as well as having passed all the regulatory requirements to secure the approval. A device that is technically capable of approval will not be approved if the manufacturer fails to meet the FDA requirements. Similarly, a heart valve whose small failure rate is surrounded by deceptive practices that substantially violate the rights of patients, surgeons, and the government agency whose job is to insure that medical devices are safe and effective, is a defective valve. For the purpose of determining defectiveness, the valve cannot be considered apart from its context,[10] the ethical relationships the manufacturer has with the major stakeholders — patients, surgeons, the FDA and, ultimately, Congress.

The Shiley heart-valve case is an instructive one. Because the defect in the valve was not so serious that it clearly had to be withdrawn, Shiley chose to continue to market the valve while trying to fix it. The company might have been successful had it not engaged in a pattern of deceptive practices. Perhaps if Shiley had realized that a medical device can be defective for ethical reasons, it might have chosen a different response to the small but persistent rate of fracture that developed in its heart valve.

[10]The analysis I am suggesting takes its inspiration from Arnold Pacey's *The Culture of Technology* [16], where he argues that an understanding of technology must include not only its technical features but also the cultural and organizational ones [16, p. 6].

References

[1] J.H. Fielder and J. Black, "But Doctor, it's my hip!: The fate of failed medical devices," *Kennedy Inst. Ethics J.*, to be published.
[2] Shiley presentation in *FDA and the Medical Device Industry*, Hearing Before the Subcommittee on Oversight and Investigations of the Committee on Energy and Commerce, House of Representatives, One Hundred First Congress, Second Session, Feb. 26, 1990, Ser. no. 101-127, p. 563.
[3] F. Schondube, H. Keusenm, B.J. Messmer, "Physical analysis of the Bjork-Shiley prosthetic valve sound," *J. Thoracic and Cardiovascular Surgery*, vol. 86, no. 1, pp. 136-141, 1983.
[4] "An ethical issue in the Bjork-Shiley artificial heart valve case," in *Proc. Twelfth Southern Biomedical Conf.*, Apr. 2-4, 1993.
[5] "More bad news about Bjork-Shiley C/C heart valves, *IEEE Eng. Med. Biology*, vol. 13, Apr. 1994.
[6] "The Bjork-Shiley heart valve: '"Earn as you learn,' Shiley Inc.'s breach of the honor system and FDA's failure in medical device regulation," staff rep. for The Subcommittee on Oversight and Investigations of the Committee on Energy and Commerce, U.S. House of Representatives, Feb. 1990.
[7] B.A. de Mol, M. Kallewaard, R.B. McLellan, J.J. van Herwerden, and Y. van der Graaf, "Single leg strut fractures in explanted Bjork-Shiley valves," *Lancet*, vol. 343, pp. 9-12, 1994.
[8] Yolanda van der Graaf, Frits de Waard, Lex A. van Herwerden, and Jo Defauw, "Risk of strut fracture of Bjork-Shiley valves," *Lancet*, vol. 339, pp. 257-261, 1992.
[9] Letter to Dr. David Kessler, Commissioner of the Food and Drug Administration, Jan. 13, 1994.
[10] J.H. Fielder, "Getting the bad news about your artificial heart valve," *Hastings Center Rep.*, pp. 22-28, Mar.-Apr. 1993.
[11] David Schoenfield, "Statistical analysis of the Bjork-Shiley 60^0 convexo-concave heart valve compared with other mechanical heart valves," in *FDA and the Medical Device Industry*, Hearing Before the Subcommittee on Oversight and Investigations of the Committee on Energy and Commerce, House of Representatives, One Hundred First Congress, Second Session, Feb. 26, 1990, Ser. no. 101-127, p. 442-443.
[12] J.H. Fielder and G. Rahmoeller, "Analyzing ethical problems in medical products: The role of conflicting ethical theories," *Clinical Res. Reg. Affairs*, vol. 11, no. 2, May 1994.
[13] James E. White, *Contemporary Moral Problems*. St. Paul, MN: West, 1994, ch. 1.
[14] Louis G. Lombardi, *Moral Analysis: Foundations, Guides, and Applications*. Albany, NY: State Univ. New York Press, 1988.
[15] *Report of the Shiley Task Group*, Dec. 1990, p. 2.
[16] Arnold Pacey, *The Culture of Technology*. Cambridge, MA: M.I.T. Press, 1984.

T&S

Chapter 10

Information Technology

OVER the past two decades, a number of social scientists and ethicists have focused attention on the social and ethical implications of information technology, a difficult task that is compounded by the rapid pace of technological development in the information technology field. These groups have struggled with such issues as human–machine interaction, privacy and security, equitable access, freedom of speech, and the impact of information technology on work, leisure, and education. It has often seemed that in their rush to develop a faster microprocessor or a "killer" application, the engineers and computer scientists responsible for developing information technology have had little or no concern for these matters.

Although this characterization may be true of most technical people, groups of information technology professionals concerned with the social and ethical implications of their work have also emerged over the past two decades (1). Indeed, of the topics covered in Part III of this book, information technology is perhaps the area in which engineers have shown the most interest in policy and ethical issues. The most visible of these groups are the Computer Professionals for Social Responsibility, the Special Interest Group on Computers and Society of the Association for Computing Machinery, and IEEE-SSIT (the latter of which is also concerned with other fields of electrotechnology that fall within the scope of IEEE). These groups sponsor publications, conferences, and electronic forums for the purpose of establishing dialogue with other information technology professionals, teachers, and scholars in the humanities and social sciences, policymakers, and the public on the social and ethical dimensions of existing and emerging information technology.

Through my involvement in IEEE-SSIT, my own research agenda has expanded to include the policy and ethical implications of information systems. For example, a colleague and I undertook a study of the social impacts of increased reliance by the IEEE on electronic communication and electronic dissemination of information. At the request of the Technical Activities Board, we formed an expert panel which we surveyed during multiple rounds of the study. The study technique, known as the Delphi method, is designed to identify areas of agreement and disagreement among expert participants. The study panel included a broad spectrum of IEEE members and customers, including diverse representation of regions, technical specialties, and information disseminators and receivers. When we took on this project, we expected that questions of access to information would be at the forefront. In addition, a number of other issues surfaced in the study:

- cultural differences in the perception of electronic information dissemination
- differing individual levels of experience with electronic communication (i.e., net-surfers versus more pedestrian users)
- standards of ethics and propriety on the Internet
- impacts of electronic publishing on the relative balance of industry and academic interests within IEEE
- questions relating to intellectual property rights

In all, we identified more than 70 obstacles to electronic communication and information dissemination by IEEE (2).

This study was followed up by a workshop we conducted with funding from the IEEE Foundation and the National Science Foundation on the socioeconomic dimensions of electronic publishing. The following recommendations, only part of those emerging from the workshop, illustrate the interplay between policy to promote electronic publishing and social and ethical considerations regarding its use (3):

- It is essential to address the critical issue of the integrity of the scientific intellectual enterprise in the electronic publishing world. The dynamic nature of the electronic product requires new processes be established for sustainability, certification, and ownership of scientific and technical information. A process of authentication of scholarly merit, equivalent in thoroughness and rigor to conventional peer review, must be established and maintained.
- Requirements and standards for output file formats must take into consideration the wide variety of workstation capabilities that are available globally. User tools should be affordable and user-friendly. Otherwise, access and participation will not be available to those less technologically advantaged.
- Care should be taken in recasting copyright standards for electronic media in order to preserve a balance between the interests of the copyright holder and the public good. Appropriate principles of "fair use" should be maintained in an electronic publishing environment, and technical innovations aimed at upholding these principles, such as "electronic reserves," should be implemented when possible.

Most recently, I conducted a study of the changing ethical responsibilities of editors, peer reviewers, and authors in a transition to electronic publishing (4). I also guest-edited a special issue of a scholarly journal focusing on the social implications of computing, which included papers from an annual conference on technology and society sponsored by IEEE-SSIT (5). That selection of papers illustrated that the social implications of information technology can be approached from a wide range of frames of reference, for example, organizations, the international policy arena, technological end-use sectors, and legislative bodies and the courts. Yet each paper, in its own way, argued for the need to ensure that social and ethical concerns are not left out in the rush to technological innovation.

One notable issue is the growing problem of computer misuse within organizations, including such issues as fraud, theft, and vandalism. Purely technical solutions to computer misuse (e.g., security controls) are unlikely to succeed; hence, there is a need to develop a culture of ethical practices within organizations. To be effective, such practices must account for three factors underlying computer misuse—personal factors, work situations, and opportunities for misuse—and should be enacted through a series of formal and informal practices designed to address the problem of computer misuse (6).

On a larger scale, it is possible that convergence of computer, mass media, and telecommunications technologies, along with other factors, including market liberalization and changes in international diplomacy and policymaking processes, are leading toward a situation in which the world will be split between the "information rich" and "information poor." This inequity in access to information results primarily from the failure of the United Nations agencies historically charged with information technology issues. The United Nations Education Science and Cultural Organization (UNESCO) and the International Telecommunications Union (ITU) have not successfully incorporated the expertise and concerns of private information providers, nongovernment organizations, end-users, and national governments (see the case study by Frempong in Chapter 8). On the other hand, the regional and private information technology initiatives that have emerged to fill this leadership void, which are largely based in the developed world, are incapable of incorporating public policy concerns such as universal access in developing countries. UNESCO and the ITU may be able to reassume their traditional role as leaders in development of international communication policy, but only if they are able to achieve meaningful cooperation with the other stakeholders (7).

The policy and ethical concerns regarding information technology even extend to the university community. For example, some recent issues of interest to colleges and universities include (1): censorship of indecent/inappropriate material versus freedom of speech; pirated software; educational value of Internet use, including its impact on the renewed push for "distance learning"; and academic honesty.

With information technology rapidly advancing, there is little room for error in formulating policies or dealing with ethical issues. As Brian O'Connell, a legal scholar who specializes in computer law and ethics, has noted (8), in dealing with information technology we are no longer afforded the luxury of isolated experimentation: "The chief laboratory of the digital age is now society as a whole" (p. 291). And more often than not, engineers are involved in conducting the experiments.

Background Reading

Michael C. McFarland contributed an essay on "Humanizing the Information Superhighway" in 1995 in which he discussed the emerging National Information Infrastructure (NII) which is resulting from the convergence of computers and communications technologies. McFarland's basic premise is that a gathering place would be a more appropriate metaphor for NII than the popular "information superhighway."

McFarland notes that the NII, though offering many potential benefits, also has the potential, like other technologies such as the automobile, to contribute to the continued deterioration of a sense of community in American society by feeding the notion of "utilitarian individualism," that is, the pursuit of one's own material self-interest. Not surprisingly, various private, government, and public interest groups have competing visions of how the NII will be implemented and how it will be controlled. McFarland argues that the vision of the NII implied by the superhighway metaphor is one of technical efficiency and tight control over its use—information is commodified, and social factors are ignored. The metaphor of a gathering place, such as the Greek agora, on the other hand, would signal that the purpose of the NII is to foster community, with social factors taking precedence over technical issues. Extending this metaphor implies that users of the network be regarded as citizens rather than consumers, with all the rights and responsibilities commonly associated with citizenship, including, for example, the right to privacy and the responsibility to respect the privacy of others.

Key to achieving McFarland's vision is a fair system of governance of the NII. Borrowing from the literature on urban planning, McFarland suggests that a well-functioning society needs two interdependent, but separate, role types: "guardians" who put the community first, and "commercial" entities that foster individualism. In the past, in the absence of a strong guardian role being filled by the government, two competing computer cultures that McFarland represents by IBM and the original (nonmalevolent) "hackers" were inhibited in following their primary commercial roles because they also had to fill the guardian void. As a result, a feeling of mistrust developed between the two cultures; the mistrust lingers to this day, as evidenced by their responses to problems posed by modern-day (malevolent) hackers. The corporate culture has enlisted government in rigorous prosecution of wrongdoers, while the old-guard hacker culture, though equally critical of grievous offenders, still celebrates the freedom of access symbolized by the original hacker ethic.

As a way out of this dilemma, McFarland proposes creating new guardian institutions based on the following principles: diversity of ownership of the network, with ownership resting in public hands or regulated private companies; user participation in system design and administration; control by independent institutions representative of all stakeholder groups; and government involvement in maintaining a competitive environment, protecting universal access, and protecting human rights.

Cases for Discussion

Richard O. Hundley and *Robert H. Anderson* present a case study of information technology in their 1995 article "Emerging Challenge: Security and Safety in Cyberspace." The authors are concerned both with "cyberspace security" against hostile actions ranging from hackers to agents of international espionage and with "cyberspace safety," which refers to protection against unforeseen failures in software and hardware. The consequences of cyberspace attacks may include annoyance, limited damage, major damage, or major disasters, where the damages in each case are not necessarily quantifiable in economic terms. Although major disasters have yet to occur, the authors assert that incidents in the other categories appear to be on the rise, and disasters such as failure in air traffic control systems are becoming more possible.

Threats to cyberspace security include hackers, criminals, terrorists, commercial organizations (e.g., in an effort to steal trade secrets), and nations. The nature of attacks on cyberspace security and safety may include some combination of attacks on insecure operations, user authentication, software, networks and hardware, as well as hardware failures. Other important factors involved in cyberspace security and safety include increasing transnationalism; limited use of security technology; lack of user acceptance of security and safety measures; growing inability to isolate systems; and blurring of the distinction between crime and warfare, thus creating ambiguity in government agency roles and missions as well as those of nongovernmental agencies.

Hundley and Anderson observe that the existing environment for computer security and safety can be described by various metaphors including "wild west," "medieval world," and "biological immune system," each of which suggests measures that would enhance system security. Drawing on the first two metaphors, the authors suggest a security strategy that entails "local enclaves protected by firewalls." They also recommend a number of protective procedures, including improved access controls, secure software, encryption, and active software defense agents (patterned after the immune system metaphor). Successful security strategies will also require motivated users, comprehensive national plans, and mechanisms for updating the security measures. The authors conclude by noting some unresolved issues for further study with respect to defining national interest areas, determining the robustness of existing infrastructures dependent on computer systems, and protecting against sabotage by persons inside an organization.

In their 1994 article "Hidden Costs and Benefits of Government Card Technologies," *Tom Hausken* and *Paula Bruening*, then staff members of the Office of Technology Assessment, present an interesting case study of an emerging information technology. They believe it is important to account for all costs and benefits of a new technology, including indirect costs and difficult-to-quantify costs. They illustrate the importance of this point by examining two proposed uses of card technologies by government agencies—food stamps and health care cards.

Card technologies range from simple, magnetic stripe cards to "smart cards" and other advanced technologies, including optical memory and hybrid cards. The technologies vary in their degree of sophistication, security, cost for the card, and cost for the card readers, and may be utilized in either off-line or on-line applications. The cards have a number of benefits, including elimination of paper transactions and large databases and ease of transactions.

Card technologies have been promoted as possible replacements for food stamps and other public assistance programs. Such technologies have the potential of reducing the significant administration costs of these programs. In principle, more than one program could be handled on a single card, although minor technical issues and significant institutional issues make such uses difficult. The direct costs of using card technologies include planning, implementing, and operating costs. On the other hand, recipients, retailers, banks, and government agencies stand to benefit from improved convenience or lower overall costs. Indirect costs and benefits include undermining of fraud, theft, and counterfeiting; improved self-esteem for users; inclusion of recipients in use of mainstream technologies; increased use of assistance programs; increased and decreased employment in the commercial and government sectors; and increased government liability in the event of card losses.

Cards used for medical records and other health care functions also have both direct and indirect costs and benefits. Applications include ordering prescription drugs, determining eligibility for assistance programs, and maintaining partial or comprehensive medical records. In the first two cases, the costs and benefits do not differ in quality from those discussed above, nor do the direct costs of the medical records cards. However, the authors argue that the medical records cards may have significant indirect costs with respect to accuracy and privacy of the patient's medical information. Protecting confidentiality may require the engineering of various security access zones into the design of the cards. In addition, the necessity of backup databases poses privacy questions. Legislative responses also carry costs in the form of enforcement and compliance with regulations designed to protect patient privacy, confidentiality, and access to their own information.

Hausken and Bruening conclude that it should not be assumed that such new technologies will automatically be beneficial without a thorough assessment of all the potential costs and benefits, including social costs such as threats to personal privacy.

References

[1] Herkert, J. R., and Cartwright, G. P. 1998. The conscience of computer science: Organizations that target the ethical & social implications of technology. *Change* (January–February): 61–63.

[2] Herkert, J. R., and Nielsen, C. (1998). Assessing the impact of shift to electronic communication and information dissemination by a professional organization: An analysis of the Institute of Electrical and Electronics Engineers (IEEE). *Technological Forecasting and Social Change* 57 (1): 75–103.

[3] Nielsen, C. S. and Herkert, J. R., eds. 1998. *Proceedings—socioeconomic dimensions of electronic publishing workshop: meeting the needs of the engineering and scientific communities.* Piscataway, N.J.: Institute of Electrical and Electronics Engineers.

[4] Herkert, J. R. 1998. Ethical issues in electronic publishing. Presented at 1998 International Symposium on Technology and Society, IEEE Society on Social Implications of Technology, South Bend, Indiana.

[5] Herkert, J. R., guest ed. 1998. Special Issue on "Computers and Society at a Time of Sweeping Change." *Social Science Computer Review* 16 (3).

[6] Kesar, S., and Rogerson, S. 1998. Developing ethical practices to minimise computer misuse. *Social Science Computer Review* 16 (3): 240–251.

[7] Ferguson, K. 1998. World information flows and the impact of new technology: Is there a need for international communication policy and regulation? *Social Science Computer Review* 16 (3): 252–267.

[8] O'Connell, B. M. 1998. Ethics, law and information technology: The transformative role of rhetoric. *Social Science Computer Review* 16 (3): 283–292.

MICHAEL C. MCFARLAND

©John Labbe/The Image Bank/1999

Humanizing the Information Superhighway

We are rapidly moving toward an integration of two of the most powerful technologies shaping modern life: computers and communications. The telephone and television systems are merging with the computer to form an even more powerful and all-encompassing network of communication and information services, known formally as the National Information Infrastructure (NII), and informally as the Information Superhighway.[1] The form this network finally takes will undoubtedly have an enormous impact on our individual lives, our businesses, and our society. The decisions that ultimately will determine it are even now being made. It is crucial that educators, engineers, business people, policy makers, and citizens participate in shaping the technology and the institutions surrounding it.

Some new ways of thinking about the NII are needed in order to achieve greater sensitivity to its social implications. Instead of regarding the network as a "superhighway," which emphasizes the technical infrastructure, it can be viewed as a gathering place, like a village square or marketplace, which supports a rich variety of human interactions. This perspective suggests a framework for understanding and analyzing

The author is with the Computer Science Department, Boston College, Chestnut Hill, MA 02167.

[1]For a description of the project and its current status, see [1].

Reprinted from *IEEE Technology and Society Magazine,* Vol. 14, No. 4, pp. 11–18, Winter 1995/1996.

public policy issues related to the NII and some principles for addressing those issues.

Technology as a Social Force

Robert N. Bellah and his coworkers, in *Habits of the Hearts*, their landmark study of moral discourse in the United States, have described how the prevailing ethos in the U.S. has developed from one that emphasized civic responsibility to one that is predominantly individualistic [2]. The study identified two strands of American individualism, both of which are very strong today. The first, which they call "utilitarian individualism," is the pursuit of one's material self-interest, as opposed to the common good. This is especially manifest in economic and commercial activity. The second, called "expressive individualism," focuses on personal development, self-realization, and self-expression, and therefore focuses more on the inner life of the person [2, pp. 32-35].

The authors do not suggest that individualism is bad in itself. It has been an important part of American culture from the beginning. However, they have observed that without a sense of community to balance the individualism, Americans are increasingly finding their lives unfulfilling. And the weakened sense of social obligation that has accompanied the growth of individualism has undermined the moral structures and moral action needed to protect both the individual and the common good.

Technology has certainly played a role in this transformation. Take for example the automobile. As one commentator has observed, "No other single innovation of the twentieth century, or possibly any century, has so profoundly influenced manners, customs, and living habits" [3]. The lifestyle and cultural changes that the automobile has helped bring about have further encouraged the drive toward individualism — and away from community — that Bellah *et al.* found to be so characteristic of Americans. The automobile has provided the freedom, the privacy, and the space for people to live their own lives. But because of that independence and the diminished need for cooperation, most people do not experience the same rich community life that existed before. Furthermore, the widespread commitment to the automobile led to population redistributions and radical changes in the nature of cities [5], which in turn facilitated changes in social relationships, habits, and structures. These changes further emphasized individualism and weakened the sense of coherence in society and the commitment to the common good. On the other hand, the establishment of the automobile as the dominant transportation technology in the U.S. was partly the result of the pursuit of individualism, through both the manipulation of public policy and economic structures by the auto industry for its own benefit (utilitarian individualism), and the preference of individuals for a mode of transportation that provides greater freedom and privacy (expressive individualism).

Global Information Network

The proposed information network, which is just now beginning to take shape organizationally and technologically, could in the long run have as great an influence on our society as the automobile. It will be a high-speed, high-bandwidth digital network, which means it will be capable of carrying not just text, but audio, video, graphics, animation and anything else that can be coded in digital form. Potentially it will be universal, with a connection to every home, school, and business in the country. And it will be interactive, offering the possibility of full two-way communication to all its users. This technology has tremendous possibilities, touching many aspects of our lives. It will certainly be used for entertainment, enhancing both the technical quality (e.g., the resolution) and the variety of TV programs, and providing movies on demand. It will offer multi-way, multimedia communications. It will give users access to all kinds of information services, including stock quotes, airline schedules, weather maps, calendars of events, government documents, and so on. It will be used in education, not just for delivering lectures remotely, but for interactive tutorials, for simulations, for research, and so on. It will carry an endless array of interest group discussions, where anyone can enter a discussion of just about any topic at any time — talk radio gone wild. And of course it will be used for a wide variety of commercial transactions, including home shopping and banking.

The potential commercial and political value of the network has not gone unnoticed. Numerous interest groups have been maneuvering for influence over its design and ownership. Chief among these have been various commercial interests, especially those that already control the technological base on which it will be built. This includes the local operating companies of the telephone system, such as NYNEX, Bell Atlantic, and so on, and the cable companies, both of whom already own wires into most homes and, for the phone companies, businesses. Other potential players are the entertainment industry and news and information outlets, who will provide much of the content on the network, and computer, electronics and communications companies, who will build the switches, communications gear and home terminals to make it all work. All of these have been trying to position themselves, through mergers, acquisitions and strategic alliances, to guarantee that they will have a part in, or even ownership and control of, the system.

Other interested groups, more representative of potential users, have begun to take positions on what the network should be and how it should be run. The Federal Government, especially through the Clinton-Gore proposal for a National Information Infrastructure, is one. Another is Computer Professionals for Social Responsibility (CPSR), a group of computer professionals involved in education and advocacy on a variety of social issues related to computing. CPSR has its own plan for the NII [7]. There are many other potential users who have an interest in how the network will develop. These include current users of the Internet, a network of computer networks that is in some ways a precursor of the NII.

Corbis/Bettman Archive

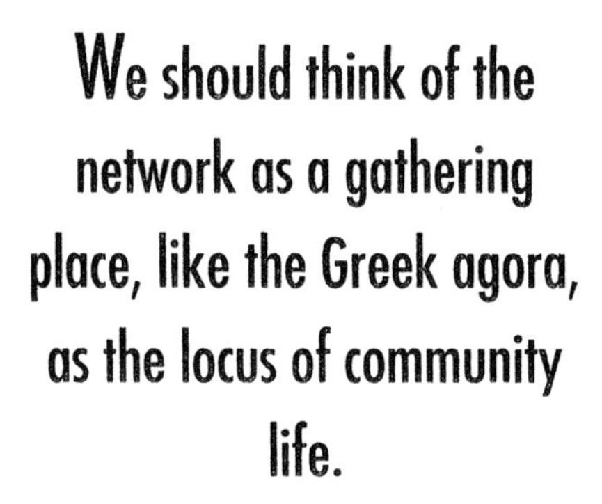
We should think of the network as a gathering place, like the Greek agora, as the locus of community life.

Each of these groups has its own distinctive vision for the NII, including who will have access, what it will cost, what services will be available, what equipment will be needed to access it, how it will be governed and how it will work. These visions have very different implications for society. The struggle that is unfolding now is over which vision will prevail.

Imaging the Network

The winner will probably be the one that best "sells" its image of the network to the financial community, policy makers, and the public. One technique that has always been very effective in selling an image is the use of metaphor. Most of us tend to think concretely, so we respond strongly to a good, concrete image. An effective metaphor can not only shape our conceptualization of an issue, it can also capture our emotions, thus controlling what we value, and ultimately our actions and policies.

Political operatives have known this for a long time. For example, in the 1988 Presidential election, the Republicans used the image of Willie Horton, the convict who, after being furloughed by Michael Dukakis, the Democratic candidate and then-governor of Massachusetts, committed a brutal rape. This became a metaphor for all the fears shared by a sizable segment of voters about crime, race, and liberal policies, and associated them with Dukakis. It was not the only reason he lost the election, but it helped.

Advertisers too are masters at using metaphor to influence people's thinking and valuing. Consider, for example, the image of the Marlboro man, the metaphor for the strong, independent, freedom-loving smoker. The cartoon character Joe Camel performs a similar function for the MTV generation. When the tobacco industry decided to go after the female market, they created the image of the Virginia Slims woman, independent, free, and fun-loving. All of these suggested that smoking made one independent and attractive.

What metaphor is being used to sell the global network? It is the "information superhighway." What is the message here? The purpose of a superhighway is to get the vehicles where they are going as fast and efficiently as possible. There are people involved, but we don't see them. They are just part of the cargo to be moved. The vehicles do not interact; at least we hope they don't. If they do, it means trouble. Furthermore, access is tightly controlled. If you don't have a car, or you can't pay the toll, you are excluded. The important considerations in designing and operating a superhighway, therefore, are efficiency, security and protection, order and stability. These are promoted by maintaining very tight control over the system and who uses it.

When we view the network as an information superhighway, then, we are making certain assumptions about its nature and purpose. In this view its purpose is to move around packets of information as efficiently as possible, where information is viewed as a commodity. The focus is on technical issues, such as how to route the packets, how to move them faster, and how to control and measure use. Access should be controlled and limited to those who are viewed as safe and can pay. Social factors are not important. In fact interaction among users is not to be encouraged. The important thing is to maintain a stable, predictable, secure environment.

This view has important implications for how the network would be used. For example, business on the network is viewed as the delivery of services, especially entertainment and information, by those who own a piece of the system or are sanctioned by the owners. It ignores many other possibilities, such as using the network to develop new ideas and new modes for interactive use, or using the network as a way for businesses to restructure themselves, improve communications, share ideas more effectively, and get closer to their customers. In education it would encourage a view of teaching as the transmittal of information rather than as a way of encouraging exploration, interaction, discovery, and growth.

Individualists, utilitarian especially, but also expressive, find the superhighway metaphor congenial. Once they are on, they can pursue their own ends free of interference. There is no need to think about, or take responsibility for, the needs or wants of others. The users can go where they want and get there quickly. It is ironic, though, that this freedom for the users requires a rigid system of control of the medium and the outside environment.

The superhighway metaphor does address some important technical and organizational issues. But it is still an impoverished way to look at the network. It does not regard the network as a human institution or consider the possibilities for human action and community. Therefore I would like to suggest an alternative metaphor. We should think of the network as a gathering place, like the classical Greek αγορα (agora) or the Roman forum (cf., [8]). Agora is usually translated as "marketplace," but that should not be understood in a purely commercial sense. The agora was the center of the city, the place where people gathered. It was the locus of community life, social, commercial, and political. Not only did people come together to buy and sell there, but it was where most of the socializing took place as well. Moreover, it was a place of political and moral debate, the place where the political community shaped its ideas and formed its policies.

The agora is a dynamic environment. It thrives on flexibility and change. It brings people together, encourages participation, and supports creativity. By its nature it is open and inclusive. Anyone can join the conversation. It tolerates, indeed relishes, the messiness and inefficiency that marks such an environment.

If we see the network as a meeting place, an agora, we adopt a very different set of assumptions. The purpose of the network is to bring people together, to foster community. Access is open to everyone and interaction is encouraged. The important issues are social, not technical: how do people interact; what kind of facilities are needed to support cooperation and group work; how can people with similar interests find one another; how are the benefits and costs of the system shared equitably? The environment is seen as always growing, adapting, and changing in response to new ideas and initiatives. Governance is not so much imposing rigid control as building an environment that fosters cooperation and trust.

Citizenship on the Network

The network as agora, the place where one participates in the life of society, suggests another metaphor: the user as citizen, rather than consumer. The network, like citizenship, should be open to all, and should encourage all to participate fully, not just to be passive observers. And like citizenship, this participation is defined by the rights and responsibilities of each member.

The rights of a network user are similar to the rights of every citizen. Preeminent among these is of course freedom. This includes the freedom to pursue any legitimate interests, whether personal, social, or commercial. It also includes freedom from prejudice and discrimination, from manipulation and control by outside forces, and from unwanted intrusion. Essential to protecting these freedoms is the right to privacy, including privacy of personal data, of creative products, and of opinions. There also needs to be a respect for property rights, including the control of one's intellectual property and the security of one's computer and communications resources.

Citizenship is not a one-way street. With the rights of citizenship come certain responsibilities. Chief among these is the responsibility to respect the rights of others. It is also essential that everyone accept the responsibility for maintaining a supportive and cooperative environment. That means maintaining order and decorum, being honest and trustworthy, and participating in the common life. It is only when members accept these responsibilities that it is possible to maintain an open and flexible environment without falling into the chaos that makes rigid, centralized control inevitable.

Governance Issues

It is not enough, however, to determine that rights and responsibilities of individuals within the system. As with any complex social system, it is important to build an environment that guarantees those rights and facilitates those responsibilities. Freedom, equality, justice, and human respect are not attainable without supporting social structures. It is therefore imperative that those who participate in the system and those who are responsible in any way for shaping it, work together to build a just, fair, and humane system of governance. In what follows we will discuss what that involves.

Analytical Framework

As noted above, the proposed "information infrastructure" will play the role that the city has traditionally played as a "space" for commercial, social, and civic life. Therefore, it makes sense to consult our experience with urban planning to understand how the network ought and ought not to be formed and governed. In this section, I will use the work of Jane Jacobs, the well-known social critic whose book *The Death and Life of Great American Cities* [9] has been described by *The New York Times Book Review* as "perhaps the most influential single work in the history of town planning."

In her most recent book, *Systems of Survival* [10], Ms. Jacobs draws on her long experience studying urban systems to give her reflections on what kind of structures are needed for a complex social system to function successfully. She first notes that there are two fundamental roles that people take on in working life. The first involves *taking*, that is, acquiring and holding the property and goods necessary to sustain one's life. The other involves *trading*, meaning commerce and producing what is needed for commerce. Then she makes the startling observation that each role has its own set of values and moral imperatives, which she calls a *moral syndrome*; and that while each syndrome contains a consistent set of tenets, the two systems are quite different from each other. It may be too much to say that they are contradictory ethical systems, for they ultimately recognize the same moral principles. But the way they prioritize and apply them are so different that they lead to quite different norms and behavior.

The first syndrome, the one connected with taking, Ms. Jacobs calls the *guardian* syndrome, after the guardian class introduced in Book III of Plato's *Republic*. The role of the guardian is to fight and win battles for territory and to provide stability and protection. Government, religion, and the military are traditional guardian functions. Therefore, the moral qualities that are most highly valued for the guardian are those that promote coherence, strength, and stability. According to Ms. Jacobs, the following precepts define the guardian syndrome [10, p. 24]:

Shun trading
Make rich use of leisure
Exert prowess
Be ostentatious
Be obedient and disciplined
Dispense largesse
Adhere to tradition
Be exclusive
Respect hierarchy
Show fortitude
Be loyal
Be fatalistic
Take vengeance
Treasure honor
Deceive for the sake of the task

The other syndrome is called the *commercial* syndrome. It is oriented toward economic activity: invention, production, buying, and selling. Therefore it values openness, innovation, pragmatism, and fair and honest dealing. Thus the following precepts define the commercial syndrome, according to Ms. Jacobs [10, p. 23]:

Shun force
Be efficient
Come to voluntary agreements
Promote comfort and convenience
Be honest
Dissent for the sake of the task
Collaborate easily with strangers and aliens
Invest for productive purposes
Compete
Be industrious
Respect contracts
Be thrifty
Use initiative and enterprise
Be optimistic
Be open to inventiveness and novelty

The commercial syndrome encourages individualism. It is an ethic of self-improvement, personal integrity, creativity, and personal profit. Therefore it is easy for those of us who have been formed by the individualism of American culture to appreciate it.

The guardian syndrome, on the other hand, puts the community first, often sacrificing individual welfare, achievement, and freedom for it. To most of us that is foreign, even threatening. That is why some of the precepts of the guardian syndrome seem strange, almost to the point of being immoral. We do not have the right context to understand them. For example, it is necessary for the guardian to "take vengeance," not for personal protection or satisfaction, but because the guardian is the symbol and guarantor of the stability and welfare of the community, and a threat to the community cannot be tolerated. A guardian should "be ostentatious," not for personal glorification, but to celebrate the value of the community.

Instead of regarding the network as a "superhighway," some new ways of thinking are needed to achieve greater sensitivity to its social implications.

The important point is that any society needs both syndromes to be successful. The commercial syndrome is what makes the society grow and flourish, to rise above subsistence. But the commercial cannot function without the stability, protection, and regulation provided by the guardian syndrome. Yet even though the two syndromes are highly interdependent, *the two cannot be mixed.* To do so leads to inefficiency and corruption, what Ms. Jacobs refers to as "monstrous hybrids" [10, pp. 92-111]. That is Ms. Jacobs' key ethical insight. For example, when the government tries to run a private business, it ends up being clumsy, inefficient, and wasteful. When the police start cutting deals for themselves, there is corruption. And when commercial interests become more worried about acquiring and holding turf than about production and trading, we get monopolies, fraud, rigidity, and injustice. The key to a properly functioning social system, therefore, is for members of each syndrome to be clear on their roles and true to their own ethic, and not to try to take on the functions and characteristics of their opposites.

Computing and Communications Environment

Unfortunately the situation now in computing and communications is very confused, because there is no clear delineation of roles. In Jane Jacobs' terms, there is some serious syndrome-mixing going on, with all the possibilities for corruption that it implies. To see how this came to be we have to look back a bit.

In the early days of computing, in the fifties, there were two cultures that grew up around computing, and these defined two different sets of attitudes toward the technology.

The first was that of the large corporate interests, symbolized of course by IBM. IBM was tremendously successful commercially, relying on hard work, initiative and aggressive marketing, all good commercial virtues. But there was another side to IBM and firms like it. They emphasized loyalty, obedience, and discipline, had a very rigid hierarchy, and were closed to outsiders. The effects could be seen in IBM's corporate behavior. While IBM sponsored leading-edge research, their products were very conservative technically. For example, when in the late 1980s RISC computers became the hottest new development in computer design, IBM revealed that they had actually pioneered the concept ten years earlier with their 801 computer project. The trouble is, they had not told anyone, and management had killed the project. As a result, IBM lost a potential ten-year head start on the next generation of desktop computers.

In fact, so much did IBM reflect the guardian syndrome that the journalist Steven Levy, in his book *Hackers* [15], described them as follows:

> "All these people in charge of punching cards, feeding them into readers, and pressing buttons and switches on the machine were what was commonly called a Priesthood, and those privileged enough to submit data to those most holy priests were the official acolytes."

The other pole is represented by so-called hackers, a term that was not at the time pejorative. These mostly young, noncommercial computer enthusiasts were characterized by what Levy termed the Hacker ethic [11, pp. 40-49]. Here are some of its tenets, which are similar to those found in Jane Jacobs' syndromes:

- Access to computers — and everything which might teach you something about the way the world works — should be unlimited and total. (Use initiative and enterprise. Be open to inventiveness and novelty.)
- Mistrust authority — promote decentralization. (Dissent for the sake of the task. Come to voluntary agreements.)
- You can create art and beauty on a computer. Computers can change your life for

the better. (Promote comfort and convenience. Be optimistic.)

So far the Hacker Ethic seems identical to the commercial syndrome. But, there were some guardian elements:

- ▾ All information should be free. This does not just mean the free exchange of information. It means *free*: you should never pay for it. (Shun trading.)
- ▾ Hackers should be judged by their hacking, not bogus criteria such as degrees, age, race, or position. This certainly rejects caste, but there was a culture of Winners and Losers, an elitist culture. (Exert prowess. Be exclusive. Treasure honor.)

In fact some of the hackers were almost monastic in their insistence on self-abnegation and purity for the sake of their ideals. They were not traders.

The government, which might be expected to function as a guardian, was instead a benign presence in the background, dispensing largesse and exercising very little control. The government paid for much of the research that fueled the computer revolution, but left the technical people free to innovate and explore. Ironically, many of the hackers, who were so distrustful of authority, could afford their high-minded and pure lifestyle because they were support by grants from the Advanced Research Projects Administration (ARPA) of the Department of Defense.

IBM also took from the government, but resisted its control. For example, it fought off an attempt by the government to break the company up because of its near-monopoly of the computer industry. Most analysts now agree that IBM would be much better off now if the government had succeeded.

In essence, IBM wanted to pursue utilitarian individualism and the hackers their expressive individualism, and both should have been free to do so. Therefore, both should have adhered to the commercial syndrome. But because there were no separate guardian institutions to provide stability and security and to promote the common good, both groups ended up taking on guardian functions, which confused their efforts and distracted them from their proper purpose.

Where are we today? The two groups may have mellowed some, but they still show some of the same characteristics, and some of the same confusion of syndromes. The large commercial interests still seem more interested in taking and defending "turf," whatever turf may mean in such an abstract environment; and while the hacker culture may have lost some of its elitism, many computer enthusiasts and innovators are still suspicious of commercial activity.

The conflict between the two groups is being fought out symbolically in the controversy over computer security and break-ins by modern-day "hackers," where that name now has a more sinister connotation. There are other important issues in computer ethics, but this is the one that has captured the popular imagination, so it must have some deeper significance. The IBMs, AT&Ts, and the Baby Bells have fought hard, especially in the legal arena, to punish the hackers for their break-ins. And they have coopted the government into helping them. Of course the government has a legitimate interest in deterring lawbreakers, but they do not seem too sure of what is going on. John Perry Barlow, cofounder of the Electronic Frontier Foundation, computer enthusiast, Montana rancher, and former Grateful Dead lyricist, tells of a visit by an FBI agent who was investigating him because he had once participated in a roundtable with some hackers. Barlow had to sit the agent down and explain to him what he was investigating [12]. This would be funny except that there have been several instances in which the unsuspecting parent of a sixteen-year-old with a modem has answered a knock at the door at 6 a.m. only to have a shotgun shoved in his face and twenty agents swarm over his home, grabbing everything with a plug on it, plus disks, tapes, and CDs, never to be returned [12, p. 10].

There are some malicious lawbreakers who break into computer systems to steal and destroy. But there are others who are not interested either in destruction or in personal gain. They just want to breach the walls that they feel have no right to exist. The rest of the user community, while deploring the rudeness and irresponsibility of these young hackers, takes some satisfaction in the discomfiture they cause. They resent the large companies because they are taking when they should be trading. The hackers, although they do nothing beneficial in practice, are seen in a way as Robin Hoods, taking back the control or power that the big companies are seen to have taken illegitimately. It is not a healthy situation.

These attitudes are likely to continue in the future as the new, all-encompassing, interactive, multimedia Net takes shape. Already we see powerful companies forming cartels in hopes of controlling the medium and therefore what goes on it, like the great transportation monopolies of the early twentieth century. At the same time user groups are clamoring for access and for democratic control.

So what is the outcome going to be? Surely the network has the potential to enable increased communication, participation, innovation, and economic growth. But only if the users put aside their guardian impulses and behave according to the commercial syndrome. But where are the guardians that will free, or compel, them to do so? Without legitimate guardians, all we will have are a group of commercial warlords vying for control, with a few hacker commandos making annoying but ineffective strikes at their outposts.

The fundamental ethical principles according to which the network should operate are clear enough and were stated earlier in the discussion on network citizenship. There should be universal access, free flow of information, and plenty of room for diversity and competition. The rights of users must be protected, including the rights to privacy, autonomy, and participation in the opportunities offered by the network.

But we also have to think about the institutions and practices that will guarantee these principles and inspire the trust necessary for the agents to choose trading over taking. This is a critical issue. For, as Jane Jacobs points out, *without strong guardian institutions, commercial interests will arrogate guardian functions to themselves, and that way leads to "monstrous hybrids" and corruption.*

As a start, here are some principles that I think are important in shaping those institutions:

- The medium should not be owned by any one commercial interest. Because of the expense of building the connectivity, it is likely that there will only be one wire into each home or business, not four or five, so there will be no competition. Giving one company or alliance control of the medium is giving them too much power. The medium should be publicly held, or, if it is privately owned, the owner should be treated like a utility and subject to careful regulation.
- Users should have an opportunity to participate in the design and administration of the system. That is the only way to guarantee that their interests are respected and promoted. Furthermore their experience and expertise are needed to make the system flexible, innovative and workable.
- There should be independent institutions dedicated solely to administering the network and enforcing the principles mentioned above. These institutions can and should include representative users, but they should be separate from both commercial concerns and user interest groups. This is the minimum needed to separate commercial activities from guardian activities. It is especially important when those whose activities and interests are, and should be, commercial participate in guardian functions. The informal arrangements that have worked fairly well on the Internet will not be adequate for the new information infrastructure because of the dominance of commercial interests in the latter.
- The government, particularly the federal government, must be involved. It is the only guardian institution strong enough to counterbalance the powerful commercial interests. There are a number of responsibilities the government should assume:
  - Maintain an environment that supports competition and innovation.
  - Guarantee universal access and participation, acting especially on behalf of those on the margins of society.
  - Protect human rights.

On the other hand, the government should not favor one commercial interest or technology over another or be involved in commercial activity itself. It is probably not a good idea for the government to administer the entire system. But it should be involved as the ultimate guarantor of the integrity and independence of the administrative structures.

What shape these new institutions will take is not an easy question. Such institutions cannot just be imposed, nor do they spring up overnight. We have to grow them, partly from democratic participation, partly from government policy. This will be difficult, especially where there has been such a long history of mistrust. But to fail to do so would be to condemn ourselves to repeat the inequities, distrust, violence, and corruption that accompanied the industrialization of the United States in the first part of this century.

References

[1] A. Reinhardt, "Building the data highway," *Byte*, vol. 19, pp. 46-74, Mar. 1994.

[2] R.N. Bellah *et al.*, *Habits of the Heart: Individualism and Commitment in American Life*. New York: Harper & Row, 1985.

[3] J.B. Rae, *The Road and the Car in American Life*. Cambridge, MA: M.I.T. Press, 1971, p. 133.

[4] J.R. Meyer, J.F. Kain, and M. Wohl, *The Urban Transportation Problem*. Cambridge, MA: Harvard Univ. Press, 1966.

[5] B.C. Snell, "American ground transport," Rep. of the Subcommittee on Antitrust and Monopoly, U.S. Senate, Feb. 26, 1974.

[6] J. Kwitny, "The great transportation conspiracy," *Harper's*, pp. 14-21, Feb. 1981.

[7] Computer Professionals for Social Responsibility, "Serving the community: A public-interest vision of the national information infrastructure," Oct. 1993.

[8] S. Steinberg, "Travels on the net," *Technol. Rev.*, vol. 97, pp. 20-31, July 1994.

[9] J. Jacobs, *The Death and Life of Great American Cities*. New York: Random House, 1961.

[10] J. Jacobs, *Systems of Survival: A Dialogue on the Moral Foundations of Commerce and Politics*. New York: Random House, 1992.

[11] S. Levy, *Hackers: Heroes of the Computer Revolution*. Garden City, NY : Anchor, 1984, p. 19.

[12] J.P. Barlow, "Crime and puzzlement," *CPSR Newsl.*, vol. 8, pp. 1-15, Fall 1990.

T&S

RICHARD O. HUNDLEY AND ROBERT H. ANDERSON

Emerging Challenge: Security and Safety in Cyberspace

With more and more of the activities of individuals, organizations, and nations being conducted in cyberspace,[1] the security of those activities is an emerging challenge for society. The medium has thus created new potentials for criminal or hostile actions, "bad actors" in cyberspace carrying out these hostile actions, and threats to societal interests as a result of these hostile actions.

Potential Hostile Actions

Security holes in current computer and telecommunications systems allow these systems to be subject to a broad spectrum of adverse or hostile actions. The spectrum includes: inserting false data or harmful programs into information systems; stealing valuable data or programs from a system, or even taking over control of its operation; manipulating the performance of a system, by changing data or programs, introducing communications delays, etc.; and disrupting the performance of a system, by causing erratic behavior or destroying data or programs, or by denying access to the system. Taken together, the surreptitious and remote nature of these actions can make their detection difficult and the identification of the perpetrator even more difficult. Furthermore, new possibilities for hostile actions arise every day as a result of new developments and applications of information technology.

The bad actors who might perpetrate these actions include: hackers, zealots or disgruntled insiders, to satisfy personal agendas; criminals, for personal financial gain, etc.; terrorists or other malevolent groups, to advance their cause; commercial organizations, for industrial espionage or to disrupt competitors; nations, for espionage or economic advantage or as a tool of warfare. Cyberspace attacks mounted by these different types of actors are indistinguishable from each other, insofar as the perceptions of the target personnel are concerned. In this cyberspace world, the distinction between "crime" and "warfare" in cyberspace also blurs the distinction between police responsibilities, to protect societal interests from criminal acts in cyberspace, and military responsibilities, to protect societal interests from acts of war in cyberspace.

We call protecting targets in cyberspace, such as government, business, individuals, and society as a whole, against these actions by bad actors in cyberspace, "cyberspace security." In addition to deliberate threats, information systems operating in cyberspace can also cause unforeseen actions or events — without the intervention of any bad actors — that create unintended (potentially or actually) dangerous

The authors are with the Rand Corporation, Santa Monica, CA. Email:Richard_Hundley@rand.org and Robert_Anderson@rand.org. *This work was partly supported by the Office of the Secretary of Defense, and by the Advanced Research Projects Agency.*

[1] As one consequence of the electronic digitization of information and the worldwide internetting of computer systems, more and more activities throughout the world are mediated and controlled by information systems. The global world of internetted computers and communications systems in which these activities are being carried out has come to be called "cyberspace," a term originated by William Gibson in his novel "Neuromancer."

Reprinted from *IEEE Technology and Society Magazine,* Vol. 14, No. 4, pp. 19–28, Winter 1995/1996.

situations for themselves or for the physical and human environments in which they are embedded. Such safety hazards can result from both software errors and hardware failures. We call protection against this additional set of cyberspace hazards "cyberspace safety." In the new cyberspace world, government, business, individuals, and society as a whole require a comprehensive program of cyberspace security and safety (CSS) [1]-[5].[2]

Cyberspace information systems are subject to a broad spectrum of adverse or hostile actions.

Consequence Categories

We have used four categories to define the consequences of cyberspace attacks, categories based on the degree of economic, human, or societal damage caused. From the least to the most consequential, they are:

1) *minor annoyance or inconvenience*, which causes no important damage or loss, and is generally self-healing, with no significant recovery efforts being required;

2) *limited misfortune*, which causes limited economic or human or societal damage, relative to the resources of the individuals, organizations, or societal elements involved, and for which the recovery is straightforward, with the recovery efforts being well within the recuperative resources of those affected, organizations, or societal elements;

3) *major or widespread loss*, which causes significant economic or human or societal damage, relative to the resources of those involved, and/or which may affect, or threaten to affect, a major portion of society, and for which recovery is possible but difficult, and strains the recuperative resources of the affected individuals, organizations, or societal elements; and

4) *major disaster*, which causes great damage or loss to affected individuals or organizations, and for which recovery is extremely difficult, if not impossible, and puts an enormous, if not overwhelming, load on the recuperative resources of those affected.

We assert that it is not always possible to measure human or societal damage in purely economic terms.

Past Incidents

CSS incidents constituting a minor annoyance or inconvenience have been a frequent occurrence across the entire spectrum of target categories. For some targets (e.g., the AT&T Bell Labs computer network or the unclassified Pentagon network) such minor annoyances can occur one or more times every day. For many computer installations, such incidents have become so commonplace that they are no longer reported.

CSS incidents constituting a limited misfortune —e.g., computer installations disrupted for limited periods of time, or limited financial losses (relative to the resources of the target) — have occurred less frequently, but nevertheless numerous examples exist across the entire spectrum of targets. A number of these are reported in [1] and [4].

There have even been a few cases of incidents which many observers would class as major or widespread loss to the target(s) involved. Examples include the "AIDS Trojan" attack in December 1989, which caused (among many other things) an AIDS research center at the University of Bologna in Italy to lose 10 years of irreplaceable data [4]; the AT&T network failure on January 15, 1990, due to a software error, which disrupted and virtually shut down a major portion of the U.S. nationwide long-distance network for a period of about nine hours [1], [4]; the almost total disruption of the computers and computer networks at the Rome (NY) Air Force Base for a period of 18 days in early 1994, during which time most (if not all) of the information systems at Rome were "disconnected from the Net" [6]; and the MCI calling-card scam during 1992-1994, in which malicious software was installed on MCI switching equipment to record and steal about 100 000 calling card numbers and personal identification codes that were then sold to hackers throughout the U.S. and Europe and posted on bulletin boards, resulting in an estimated $50 million in unauthorized long-distance calls [7].

[2] In addressing questions of cyberspace security and safety, we have relied on a variety of anecdotal information obtained from a number of sources. The anecdotal data by no means constitute a comprehensive, statistically valid sample. In principle, one could develop such a sample from databases from the various computer emergency response teams (CERTs), law enforcement databases, and private sector incident data. However, we have yet to find anyone who has done so.

There are a number of reasons for this. One is that many if not most cyberspace security incidents apparently go unreported to authorities, particularly in the financial community. It is therefore unclear if the incidents that are reported are "the tip of the iceberg," or all there is to the problem.

Lacking a comprehensive sample, the total quantitative dimensions of the cyberspace security problem are unclear. Therefore, we present here our qualitative impressions of the problem.

Table I
Internet Penetration Incidents Reported to Carnegie-Mellon CERT

Year	Incidents Reported
1988	6
1989	132
1990	252
1991	406
1992	773
1993	1334
1994	2241

We know of no clear examples to date of a CSS incident constituting a major disaster.

Potential Future Incidents

Whatever may have happened in the past, we expect cyberspace security and safety incidents to become much more prevalent in the future, due to the facts that more and more people are becoming "computer smart" all over the world; bad actors of many different types are becoming more and more aware of opportunities in cyberspace; connectivity is becoming more widespread and universal; more and more systems and infrastructures are shifting from mechanical/electrical control to electronic/software control; and human activities in cyberspace are expanding much faster than security efforts.

Recent data support this expectation. For example, the number of Internet penetration incidents reported to the computer emergency response team (CERT) at Carnegie-Mellon University each year since 1988 are shown in Table I [8].

Accordingly, we expect that, in the future, CSS incidents constituting a minor annoyance or inconvenience will become commonplace across the entire spectrum of targets; incidents constituting a limited misfortune could also become a common occurrence; CSS incidents constituting a major or widespread loss are quite possible for all targets in cyberspace; and CSS incidents constituting a major disaster are definitely possible for some targets in special cases.

Some examples of special cases in which major disasters may be possible include the following:

- *Physical and functional infrastructures*, such as the air traffic control system, possibly leading to the crashes of one or more aircraft.
- *Military and national security.* For example, if a cyberspace-based attack were to bring down an essential military command and control system at a critical moment in a battle, it might lead to the loss of the battle. If the battle were pivotal, or the stakes otherwise high enough, this could ultimately lead to military disaster.
- *Other societal organizations and activities.* With medical care becoming increasingly dependent on information systems, many of them internetted, a perpetrator could make changes to data or software, possibly resulting in the loss of life.

Other examples of possible cases leading to major disasters may occur to the reader. Today these examples are all hypothetical. Tomorrow one or more of them could well be real. Our impression is that CSS incidents will become much more prevalent; they will impact almost every corner of society in the developed nations of the world; and the consequences could become much greater.

Infrastructure Fragility

There are many uncertainties associated with this projection of future cyberspace security and safety incidents. Attacks on vital infrastructures are one of the things most likely to cause widespread repercussions for society. Accordingly, one of the most important uncertainties has to do with the degree of robustness of current and future infrastructures: Are the key physical and functional infrastructures in various nations highly robust, due to built-in redundancies and self-healing capabilities? Or do some infrastructures have hidden fragilities that could lead to failures having important consequences?

Conventional wisdom regarding these questions is not always correct. For example, prior to 1990, the AT&T long distance network in the U.S. was usually thought to be very robust, with many alternative paths for long distance calls to take, going through different switching centers. But all of these switching centers use the same software, and when new software was introduced in 1990, every long-distance switch had the same bad line of code. So at the software level, there was no redundancy at all, but rather a fragility that brought a large part of the AT&T long-distance network down [1], [4].

The message is clear: many infrastructures may not be as robust as they seem; a detailed look at vulnerabilities of specific infrastructures is needed.

Actors Responsible for Incidents

By far the greatest portion of past cyberspace security incidents have been perpetrated by "hackers": individuals satisfying a variety of personal agendas, which in their view do not include criminal motives [9], [10]. This continues to be the case regarding current incidents.

In recent years, the role of criminals in cyberspace incidents has increased. According to law enforcement professionals consulted by the authors, this has come about not as a result of the criminal element becoming more aware of opportunities in cyberspace, but rather primarily as a result of computer hackers "growing up" and some (small) fraction of them realizing and exploiting the financial opportunities open to them via criminal acts.

There are no known cases in the open literature of cyberspace security incidents perpetrated by terrorists or other malevolent groups, commercial organizations, or nations. However, there are plenty of rumors of business organizations and intelligence agencies outside the U.S. that have mounted cyberspace-based attacks against companies in other nations as a means of industrial or economic espionage.

New possibilities for hostile actions arise every day.

In addition, police authorities in Europe have recently begun to discern a number of potentially more dangerous actors manipulating and guiding some malicious hacker activity. This appears to include professional hackers, who are often the source of the penetration tools used by the "ordinary" hackers; information brokers, who frequently post notices on European hacker bulletin boards offering various forms of "payment" for specific information; private detectives, who also often use the European hacker bulletin boards as a means of obtaining information regarding targeted individuals or organizations; foreign embassies, who appear to have been behind the bulletin board activities of least some European private detectives and information brokers; and organized crime.

Whatever may have happened in the past, in the future we expect all five of our classes of bad actors to continue participating in cyberspace security incidents.

Mechanisms: Past and Future

A number of mechanisms have been prevalent in past cyberspace security and safety incidents and are likely to be prevalent in future incidents as well. Many incidents involve more than one of these mechanisms, which include:

- *Operations-based attacks*, taking advantage of inadequate or lax security environments. Exploitation of deficient security environments has been a feature of many/most past successful cyberspace penetrations and is likely to continue to be prevalent in the future — as long as lax security continues to be commonplace.
- *User authentication-based attacks*, which bypass or penetrate login and password protections. Such attacks are a common feature of many/most past cyberspace security incidents and are also likely to be prevalent in the future.
- *Software-based attacks*, exploiting software features (e.g., maintenance backdoors), programmatic flaws, and logical errors or misjudgments in software implementation, as well as the insertion of malicious software.
- *Network-based attacks*, which take advantage of network design, protocol, or topology in order to gather data, gain unauthorized system access, or disrupt network connectivity. This can include alterations of routing tables, password sniffing, and the spoofing of TCP/IP packet addresses. Attacks of this type have not been common in the past. However, beginning in 1994 hackers have been detected penetrating Internet routers to install password sniffers, etc.; TCP/IP packet address spoofing was first detected in early 1995. Such attacks — including attempts to disrupt Internet connectivity — could become much more common in the future, unless Internet security is markedly improved.
- *Hardware-based attacks or failures*, exploiting programmatic or logical flaws in hardware design and implementation, or component failures. These have not been a feature of past cyberspace security incidents (i.e., deliberately perpetrated incidents), but have played a role in occasional safety hazards (i.e., accidental incidents). This is likely to continue in the future.

Additional Key Factors

There are a number of additional factors impacting on the cyberspace security problem and of necessity shaping any effective protective strategies.

Increasing Transnationalism

As is well known, cyberspace does not respect national boundaries. In recent years more and more nations throughout the world have become "connected" to the world network, and within those nations connectivity has become more and more universal.

Every year greater numbers of individuals and organizations in the U.S. are taking advantage of this increasing worldwide connectivity to become involved, via cyberspace, in economic or social activities with individuals and organizations in other nations. These transnational activities are becoming increasingly important to the U.S. individuals and organizations involved; they will not willingly give them up.

Since threats in cyberspace pay no regard to regional or national boundaries, knowledge of computer hacking techniques has spread around the globe, and the perpetrator of a security incident can just as well be on the other side of the world as across the street.

For both of these reasons—the nature of activities in cyberspace and the nature of threats—cyberspace has become effectively transnational. No nation has effective sovereignty over cyberspace. Any effective cyberspace protective strategy must take this into account.

▼ *Current Security Inadequate*

The information processing systems and telecommunications systems currently in use throughout the world are full of security flaws, and new security flaws are being uncovered almost every day, usually as a result of hacker activity. As new developments and applications of information technology become available and as human activities in cyberspace continually expand, security efforts appear to be lagging behind. There is currently no effective way to police cyberspace. Considering the rapid increase in the number of reported security incidents in recent years, along with the apparent increase in the severity of these incidents, it does not appear that the "good guys" are winning; they may not even be holding their own.

Current security operations in cyberspace are inadequate. This is not the result of a lack of security technology. Rather, it reflects a very limited application of available technology; most of the available computer security technology is not used in most of the computers in the world.

▼ *Acceptance Lacking*

The U.S. has had a computer security program since the 1960s. In spite of these efforts, the U.S. is full of insecure computers today. There are several reasons for this. A primary reason is that user acceptance and utilization of available computer security safeguards has been reluctant and limited. There are several causes of this lack of user acceptance.

- ▼ Typically, user interfaces accompanying security features are awkward. As a result, the secure systems are more difficult to use than the nonsecure systems. Many users are not motivated to take the extra effort.
- ▼ Users have not considered security features as adding value, and therefore are reluctant to pay extra for such features.
- ▼ Computer hardware and software manufactures have not perceived the security market as being attractive. Rather, it has usually been considered a limited, niche market. Therefore the largest commercial manufacturers (Microsoft, Apple, etc.) have not included many security features in their primary product lines.
- ▼ Many individual users do not understand the need for a communal role in cyberspace security and do not accept responsibility for such a role.
- ▼ Most users don't take computer security seriously until something bad has happened to them or to their immediate organization.

For reasons such as these, most of the computer security technology currently available is not used on most of the computers in the world. A typical computer on the Internet uses a garden variety Unix operating system with few additional security safeguards. Similarly, a typical desktop computer uses the MS-DOS, MS-DOS plus Windows, or Macintosh operating systems, once again with few additional security safeguards. The various secure operating systems, multilevel security systems, and Orange Book[3] compliant software systems that have been developed are primarily used in restricted, niche applications.

▼ *Isolation Disappearing as Option*

Twenty or thirty years ago there was a simple solution to this problem: the physical isolation of computer systems, what is now called an "air gap." This is no longer a viable option. As more and more human activities move into cyberspace to take advantage of the efficiencies provided by interconnection, organizations and individuals who fail or refuse to connect will increasingly fall behind the pace of economic and social activity, will become increasingly noncompetitive in their area of activity, and will have difficulty accomplishing their missions. This idea is stated succinctly in a report of the Joint Security Commission appointed by the U.S. Secretary of Defense and the Director of Central Intelligence to develop a new approach to security to meet the challenges facing the Department of Defense and the Intelligence Community in the post-Cold War era [13]:

"Those who steadfastly resist connectivity will be perceived as unresponsive and will ultimately be considered as offering little value to

[3]The "Orange Book" is a common term for the DOD Trusted Computer System Evaluation Criteria (TCSEC) [12].

their customers. ... The defense and intelligence communities share this imperative to connect."

▾ *Roles and Missions Blurred*

By their nature, developments in cyberspace blur the distinction between crime and warfare, thereby also blurring the distinction between police responsibilities to protect U.S. interests from criminal acts in cyberspace, and military responsibilities to protect U.S. interests from acts of war in cyberspace.

In addition, providing protection against transnational threats in cyberspace, and apprehending their perpetrators, frequently goes well beyond the reach and resources of local and regional authorities.

These two characteristics of security in cyberspace — the blurring of the distinction between crime and warfare, and the transnational nature of many security incidents — raise new questions regarding the proper roles and missions in cyberspace security and safety. Some of the agencies, organizations, and institutions that have essential roles to play, from the viewpoint of one living in the U.S., include:

- *U.S. federal government,* including intelligence agencies, the Department of Defense, federal law enforcement agencies; civilian regulatory agencies; and other civilian agencies;
- *U.S. State and local governments*, including law enforcement agencies and regulatory agencies;
- *Nongovernmental organizations* such as CERTs, business and professional associations, vendors, industry standard-setting bodies, and private businesses;
- *Governments of other nations*, including intelligence agencies, ministries of defense, and law enforcement agencies;
- *International organizations* such as the United Nations, supranational governing bodies, Interpol, and international standards bodies.

Today this is "everybody's" problem, and therefore "nobody's" problem. It falls into all of the cracks.

Useful Metaphors

These various characteristics of the current security situation in cyberspace suggest three metaphors which may stimulate thinking about protective strategies.

▾ *"Wild West" World*

Cyberspace has many similarities to a Wild West world.

- In the Wild West almost anything could occur. There was no one to enforce overall law and order, only isolated packets of local law. The same is true in cyberspace.
- There were both "good guys" and "outlaws" in the Wild West, often very difficult to tell apart. "Friends" were the only ones a person could trust, even though he or she would frequently have to deal with "strangers." This is also true in cyberspace.
- Outside of the occasional local enclaves of law and order, everyone in the Wild West was primarily dependent for security on their own resources and those of their trusted friends . This is also true in cyberspace.

The message of this metaphor for cyberspace security is clear: If there is no way to enforce law and order throughout all of cyberspace, which appears to be the case, one must rely on local enclaves of law and order, and trusted friends.

▾ *Medieval World*

The medieval world depended on local enclaves for security: castles and fortified cities, protected by a variety of fortifications — moats, walls, and drawbridges. Communication and commerce between these fortified enclaves was carried out and/or protected by groups of armored individuals.

This metaphor also suggests a message for cyberspace security: cyberspace fortifications (i.e., firewalls) can protect the local enclaves in cyberspace, just as moats and walls protected the castles in the medieval world.

We have found the security concepts suggested by these two metaphors — local enclaves and firewalls — to be very compelling, and usable as part of a basic paradigm for cyberspace security.

▾ *Biological Immune System*

The problems faced by biological immune systems have a number of similarities to the challenges confronting cyberspace security. This suggests that the "security" solutions employed by immune systems could serve as another useful model for cyberspace security. For example:

- Higher-level biological organisms are comprised of a large number of diverse, complex, highly interdependent components. So is cyberspace.
- Biological organisms face diverse dangers (from microbes) that cannot always be described in detail before an individual attack occurs, and which evolve over time. Organisms cannot defend against these dangers by "disconnecting" from their environment. The same is true of information systems exposed to threats in cyberspace.

- Biological organisms employ a variety of complementary defense mechanisms, including both barrier defense strategies involving the skin and cell membranes, and active defense strategies that sense the presence of outsiders (i.e., antigens) and respond with circulating killers (i.e., antibodies). The cyberspace firewalls are an obvious analogue to the biological barrier defenses. But what about the active defenses? Perhaps software agents could be created providing a cyberspace active defense analogue to biological antibodies.

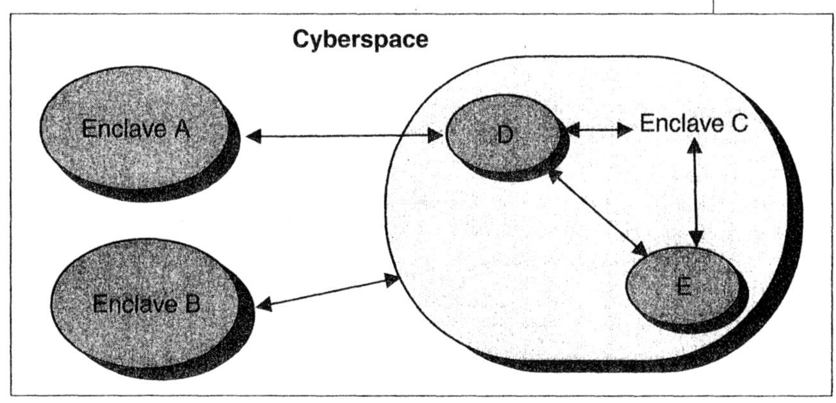

Fig. 1. An architectural concept and basic paradigm for cyberspace security: local enclaves, of various sizes, some of them nested, protected by firewalls.

The biological agents providing the active defense portion of the immune system employ certain critical capabilities: the ability to distinguish "self" from "nonself"; the ability to create and transmit recognition templates and killer mechanisms throughout the organism; and the ability to evolve defenses as the "threat" changes.

Software agents providing a cyberspace active defense analogue to these biological antibodies would need the same capabilities.[4]

The message of this metaphor is clear: Cyberspace security would be enhanced by active defenses capable of evolving over time.

We find this third metaphor as compelling as the first two; however, we are not as far along in exploiting it in our analysis.

Security Strategy

Using the concepts suggested by the Wild West and Medieval metaphors, Fig. 1 depicts the basic paradigm and overarching architectural concept we suggest for cyberspace security: local enclaves protected by firewalls. These enclaves can be of various sizes, some of them can be nested, and the firewalls can be of various permeabilities. The enclaves have protected connections to other trusted enclaves, and limited connections to the rest of cyberspace.

In this architectural concept, no attempt is made to maintain centralized law and order throughout all of cyberspace. Each authority maintains local law and order in its own enclave. Everything outside of the enclaves is left to the "wild west."

These enclaves can come in a variety of sizes, ranging from an individual computer to a complete network. The firewalls protecting these various size enclaves come in several different types, with different degrees of permeability.[5]

In the most extreme case, one can have an air gap, i.e., the absence of any electronic connection between the interior of the enclave and the outside world. Within this overall category, there can be various degrees of permeability, depending upon what software and/or data are allowed in and out, on diskettes, tapes, etc., and how rigorously this software and data are checked.

When electronic connections are allowed, a firewall computer stands between the world outside the enclave and the internal machines. Two main categories of variations are possible:

1) Different services can be allowed to come in or to go out, depending on the permeability desired of the firewall. Typical service categories include electronic mail, file transfer (e.g., FTP), information servers (e.g., World Wide Web browsers), and remote execution (e.g., Telnet). Of these four categories, electronic mail is the safest to interchange with the outside world and remote execution is the most dangerous — in the sense of providing opportunities that hackers can exploit to penetrate the firewall barrier and gain control of internal machines. Accordingly, even the tightest firewalls usually allow the passage of electronic mail in both directions, whereas only the loosest firewalls allow the passage of remote execution services, particularly in the inward direction.

2) Some allowed services can terminate (or originate) at the firewall machine, while others can go right through the firewall to the internal machines (incoming services) or to the outside world (outgoing services). The fewer services that pass through the firewall, the tighter it is.

These variations in the permeability of electronic firewalls can be tuned to the circumstances of the particular enclave.

[4]We are not the first to be intrigued by this metaphor. Forrest *et al.* [14] and Kephart [15] discuss software implementations of certain aspects of the biological immune system metaphor.

[5]We are certainly not the first to suggest firewalls as a protective technique or as a central element of a protective strategy. See [16]-[18].

▾ *Protective Techniques and Procedures*

In addition to firewalls, there are a number of other protective techniques and procedures which have important roles to play in our strawman protective strategy. These include:

- Improved access controls, including one-time passwords, smart cards, and shadow passwords.
- More secure software. This could include expanded use of software independent verification and validation (IV&V) techniques, to find and eliminate software bugs and security holes in widely used software, as well as more secure operating systems.
- Encrypted communications, both between and within protected enclaves.
- Encrypted files, for data that is particularly sensitive.
- Improved capabilities to detect penetrations, including user and file-access profiling.
- Active counteractions, to harass and suppress bad actors. This is something that is woefully lacking today; almost all current computer security measures are either passive or counteractive, leaving the initiative to the perpetrator.
- Software agents, perhaps acting in a manner similar to a biological immune system.

▾ *Motivating Users*

The best protective strategy in the world and the best set of protective techniques and procedures will be ineffective if users do not employ them. Necessary (and hopefully sufficient) ways to motivate users include:

1) A vigorous program of education and training, of both users and managers concerned with information systems in potential target organizations — education, so that people will understand the magnitude of the risk to their interests and the importance of cyberspace security, and training, so that people will know how to protect themselves.

2) Proactive programs to demonstrate vulnerabilities — sometimes called "red teams" — and thereby to increase organizational and individual awareness of cyberspace vulnerabilities. The Vulnerability Analysis and Assistance Program (VAAP) of the U.S. Center for Information Systems Security (CISS) is a good example of such a proactive program [20].

3) Mandates, tailored to different societal elements. These can include mandatory security procedures established by an organization for all of its employees or members to follow, mandatory security standards that a computer host must meet in order to be permitted to connect to a network, security standards and procedures that organizations and individuals must adhere to in order not to incur legal liability, and even possibly laws mandating certain minimum levels of security standards for information systems engaged in certain types of public activity.

4) Sanctions, to enforce the mandates.

▾ *Complete Protective Strategy*

In addition to the elements we have discussed thus far, a complete cyberspace protective strategy needs at least two additional elements.

1) A set of prescriptions governing the application of the basic security paradigm and the set of protective techniques and procedures to different security situations: for protecting different elements of society; for countering different actors; and for determining what role various agencies and organizations should play in cyberspace security, in which situations. These prescriptions — in particular those associated with the assignment of roles and missions in cyberspace security — may well differ from nation to nation.

2) A built-in mechanism or mechanisms to continually update the protective techniques and procedures, and the overall strategy, as information technology continues to evolve and its applications to expand, and as new threats emerge.

These elements remain to be developed.

Open Questions, Key Issues

A number of open questions and key issues should be resolved in the process of proceeding further. These include:

- *What specific organizations and activities comprise what we will call the "National Interest Element" in the U.S. or any other nation?* That is, what organizations, information systems, and activities play such vital roles in society that their disruption due to cyberspace attacks would have national consequences, and their protection should therefore be of national concern?
- *Which organizations (in each nation) should play what roles in the protection of the National Interest Element?*
- *How robust or fragile are essential infrastructures contained in the National Interest Element of each nation?* This is one of the key uncertainties in our current understanding of the cyberspace security situation. A detailed look at the vulnerabilities of specific infrastructures in various nations is needed to resolve this issue.
- *How does one protect against the trusted insider?* Our basic security paradigm of local enclaves protected by firewalls protects against malicious outsiders, but not necessarily against malicious insiders, individuals inside the firewall with all of the

access privileges of a trusted member of the enclave. As knowledge of hacker techniques spreads throughout the population, adverse actions by malicious insiders is becoming more and more of a problem. We have not discussed this here, but it is an important threat with which any complete cyberspace security strategy should deal. It becomes particularly important for very large protected enclaves, encompassing large numbers of individuals; the more people within an enclave, the greater the probability that at least one of them might be a bad actor.

Increasingly Complex World, Expanding Security Concerns

A number of points are worth emphasizing:

Fifty years after ENIAC, the network has become the computer (paraphrasing the Sun Microsystems slogan "The Network Is the Computer").

In the future, cyberspace security and safety incidents in this networked environment will become much more prevalent; cyberspace security and safety incidents will impact almost every corner of society; and the consequences of cyberspace security and safety incidents could become much greater.

Local enclaves protected by firewalls appear promising as a basic cyberspace security paradigm, applicable to a wide range of security situations.

We're all in this together; weak links in the net created by any of us (software developers, end users, network providers, etc.) increase the problem for all of us.

Much more attention must be paid to user motivation, for all classes of users, with different approaches required for each class. Inadequate user acceptance and utilization of security techniques and procedures has been the bane of most previous attempts at cyberspace security.

No one's in charge; the problem transcends all usual categories. The question of "roles and missions" is an important one, both philosophically (e.g., do we need more centralized control, or are there decentralized effective solutions) and pragmatically (what roles do we give DoD versus FBI versus CIA; UN versus U.S.; Interpol versus whom?).

The world has become much more complex. It is a useful complexity, but with this complexity has come security and safety problems that we are only beginning to understand and appreciate.

Acknowledgment

The authors thank their RAND colleagues Steven C. Bankes, Mark Gabriele, James J. Gillogly, Anthony C. Hearn, and Willis H. Ware for numerous suggestions and insights that have contributed very significantly to the course of this research and to the results presented here.

The authors are indebted to Special Agent Jim Christy and his team at the Computer Crime Unit, Air Force Office of Special Investigations, Bolling Air Force Base; Supervisory Special Agent Harold Hendershot, Economic Crimes Unit, Federal Bureau of Investigation; Jack Lewis, Electronic Crimes Branch, Financial Crimes Division, U.S. Secret Service; Detective Inspector John Austen, Computer Crime Unit, Metropolitan Police, New Scotland Yard, London; and Harry Onderwater, Computer Crime Unit, National Criminal Intelligence Division, National Police Agency, The Netherlands, for sharing their perceptions of computer crime, its current perpetrators, and likely future trends.

The authors are also indebted for stimulating discussions to Michael Higgins, Defense Information Systems Agency; Professor Roger Needham and Ross Anderson, University of Cambridge Computer Laboratory, Cambridge, England; Dennis Jackson, JANET-CERT Coordinator, UKERNA, Chilton, England; Christoph Fischer, Director, Micro-BIT Virus Center/CERT, University of Karlsruhe, Germany; Professor Dr. Ulrich Sieber, Chair for Criminal Law, Criminal Procedural Law & Legal Philosophy, University of Würzburg, Germany; Klaus-Peter Kossakowski, Deutsches Forschungnetz Computer Emergency Response Team (DFN-CERT), University of Hamburg, Germany; Dr. Wietse Venema, Department of Mathematics and Computing Science, Eindhoven University of Technology, The Netherlands; and Dr. Giampiero E.G. Beroggi, Delft University of Technology, The Netherlands.

References

[1] P. Neumann, *Computer Related Risks.* Reading, MA: Addison-Wesley, 1994.

[2] P.J. Denning, *Computers Under Attack: Intruders, Worms, and Viruses.* Reading, MA: Addison-Wesley, 1990.

[3] K. Hafner and J. Markoff, *Cyberpunk: Outlaws and Hackers on the Computer Frontier.* New York, NY: Simon & Schuster, 1991.

[4] P. Mungo and B. Clough, *Approaching Zero: The Extraordinary Underworld of Hackers, Phreakers, Virus Writers, and Keyboard Criminals.* New York, NY: Random House, 1992.

[5] P. Wallach, "Wire pirates," *Sci. Amer.*, vol. 270, pp. 90-101, Mar. 1994.

[6] Presentation by Air Force Computer Emergency Response Team (AFCERT), Kelly AFB, at Sixth Ann. Computer Security Incident Handling Wkshp., hosted by the Forum of Incident Response and Security Teams (FIRST), Boston, MA, July 25-29, 1994.

[7] R.E. Yates, "Hackers stole phone card numbers in $50 million scam," *Chicago Trib.*, pp. 1, 6, Nov. 2, 1994.

[8] Data presented by Computer Emergency Response Team (CERT), Carnegie Mellon University, at Sixth Ann. Computer Security Incident Handling Wkshp., hosted by the Forum of Incident Response and Security Teams (FIRST), Boston, MA, July 25-29, 1994 — supplemented by CERT 1994 Ann. Rep. web homepage (http://www.sei.cmu.edu/SEI/ programs/cert/1994_CERT_Summary.html)

[9] S. Levy, *Hackers, Heroes of the Computer Revolution.* Anchor, 1984.
[10] D. G. Johnson, *Computer Ethics,* 2nd Ed. Englewood Cliffs, NJ: Prentice Hall, 1994.
[11] B. Hoffman, "Responding to terrorism across the technological spectrum," RAND, Rep. P-7874, 1994.
[12] *DOD Trusted Computer System Evaluation Criteria (TCSEC),* DoD 5200.28-STD. Washington, DC: U.S. Government Printing Office, Dec. 1985.
[13] "Redefining security," report by the Joint Security Commission, Washington, DC 20505, Feb. 28, 1994.
[14] S. Forrest, A.S. Perelson, L. Allen and R. Cherukuri, "Self-nonself discrimination in a computer," in *Proc. 1994 IEEE Symp. Res. in Security and Privacy,* 1994.
[15] J.O. Kephart, "A biologically inspired immune system for computers," in *Artificial Life IV, Proc. Fourth Int. Wkshp Systhesis and Simulation of Living Systems,* R.A. Brooks and P. Maes, Eds. Cambridge, MA: M.I.T. Press, 1994, pp. 130-139.
[16] W.R. Cheswick and S.M. Bellovin, *Firewalls and Internet Security: Repelling the Wily Hacker.* Reading, MA: Addison-Wesley, 1994.
[17] S. Garfinkel and G. Spafford, *Practical UNIX Security.* Sebastopol, CA: O'Reilly & Associates, 1991.
[18] *Proc. 17th Nat. Computer Security Conf.,* vols. 1 and 2, National Inst. of Standards and Technology/National Computer Security Center, Oct. 11-14, 1994.
[19] M.R. Higgins, "Threats to DoD unclassified systems," DoD Center for Information Systems Security, Automated Systems Incident Support Team (ASSIST), 1994.
[20] R.L. Ayers, "Center for Information Systems Security, Functions and Services," Center for Information Systems Security, Defense Information Systems Agency, 1994. T&S

TOM HAUSKEN AND PAULA BRUENING

Hidden Costs and Benefits of Government Card Technologies

The current health-care debate and the ongoing budget debates highlight the widely varying cost estimates for government programs, many of which use information technology. Information technology alone does not necessarily improve productivity or reduce costs, however, and it is often oversold as a panacea for solving all of what ails an organization or a society. The infusion of information technology often may change the balance of costs divided among various providers and recipients of government programs. Information technology also can bring with it risks or enhancements to personal privacy or other social costs and benefits (depending upon how it is applied), the importance of which goes well beyond any immediate cost-effectiveness calculations.

The Federal government, nevertheless, has a responsibility to citizens for prudence in its expenditures, suggesting a need for precise accounting and cost-effectiveness in government programs. Some government programs are managed from a business perspective, sometimes even privatizing services that can be handled "more efficiently" by the private sector. The government has many responsibilities that private enterprise does not, however. The government is expected to procure its goods and services fairly, for example, and make information about its decisions freely available. The public has no marketplace alternative if the government should fail in these responsibilities.

The government also is expected to oversee the overall health of the nation and not only its own immediate or short term interests. A comprehensive cost and benefit perspective suggests that future and less quantifiable social costs and benefits be included in any cost-effectiveness evaluation of a single government program. Some of these indirect and less tangible costs and benefits are difficult to estimate, but are nevertheless real. When changes are made to existing programs, some costs and benefits might have to be reallocated to balance the overall accounts of the various stakeholders.

We explore here two specific examples of this cost/benefit evaluation applied to card technologies: food stamps and other assistance cards, and health-care cards. The U.S. Federal government is currently promoting both of these applications in conjunction with State governments and providers. The applications are promoted as modernizing and cost-saving measures in a period of tight budgets, but without necessarily considering the full costs or benefits. The cases presented in this article clearly

---

An earlier version of this article was presented at the IEEE-SSIT conference "Technology: Whose Costs? Whose Benefits?," held October 22-23, 1993, in Washington, DC. The authors are members of the analytical staff at the Office of Technology Assessment, U.S. Congress, Washington, DC 20510-8025. Email: thausken@ota.gov. The opinions expressed in this article are solely those of the authors and not necessarily those of the Office of Technology Assessment or of the Technology Assessment Board.

Reprinted from *IEEE Technology and Society Magazine,* Vol. 13, No. 2, pp. 24–32, Summer 1994.

do not represent all Federal programs, but the comprehensive approach discussed here is appropriate to others.

Card Technologies

Card technologies are widely used for such applications as credit and bank cards, subway tokens, and hotel keys. In some government programs, card applications have advanced beyond pilot testing to full operation. Promoters note that cards can eliminate printing and mailing checks and coupons, and save in direct paper-handling costs. Cards also reduce errors and fraud through the use of passwords and encryption. Some cards can reduce or eliminate the need for a large central on-line database, simplifying and speeding many transactions [1].

Smart cards include a microprocessor and substantial memory embedded in the card that enhance security and can manage complicated transactions without the need for on-line processing. The smart card is literally a portable computer, but an external point-of-sale terminal essentially provides the power supply, keyboard, and display for the card. Even if an unauthorized user could read the data in a smart card memory, the data are encrypted and the computer chip itself is virtually impossible to duplicate. Smart cards generally contain 2 to 8 kbits of rewriteable EEPROM data, and cost $5 to $25 per card. The point-of-sale terminals, or readers, cost about $500 apiece.

The smart card can be programmed so that different parts of its memory can be accessed by different people. For example, it could be programmed so that the issuer can access only specific records, the cardholder can monitor only account balances, store personnel can monitor only transaction data, emergency room physicians can access only information on allergies, pharmacists can access only prescriptions, and so on.

Optical memory cards, on the other hand, hold as much information as a book on a wallet-sized card. Such a card can carry biometric information used to identify the cardholder, store medical information, or contain reference data that might be used for a portable laptop computer. Optical memory cards use technology similar to CD-ROMs and are not rewriteable. They can store up to 30 Mbits of data and cost $5 to $20 per card. The optical card readers are relatively expensive at about $1500 to $4000 apiece.

Magnetic stripe cards are most common and least expensive, but have the least memory capacity. The format used for these cards is relatively standardized, although advanced proprietary versions also exist. Magnetic stripe cards can require the use of a password, but the stored data are not encrypted and can be read by any card reader. The card can also be easily duplicated using inexpensive parts. Magnetic stripe cards can store 1-7 kbits of data, and cost $0.20 to $1 per card. The card readers are also relatively inexpensive, costing $50 apiece and up.

Card technologies can bring risks or enhancements to personal privacy, depending on how they are applied.

Hybrid cards combine the features of smart or optical memory cards with a magnetic stripe to gain the advantages of each, except for added cost. Integrated circuit (IC) memory cards function like magnetic stripe cards, but look like smart cards. These cards have an integrated circuit like a smart card but without a microprocessor, and the memory can be rewriteable EEPROM or non-rewriteable EPROM. They have neither the security nor the computing functions of the smart cards, nor the ubiquitous card readers like magnetic stripe cards. IC memory cards contain 100 bits to 64 kbits of memory, and cost from $1 to $6 per card. The readers are equivalent in cost to magnetic stripe card readers.

Card systems can be either off-line (subway fare cards and hotel keys, for example) or on-line (automatic teller machines, for example). Many credit card transactions are fully off-line; no verification is made, and a record is kept only on paper. Such a system does not take advantage of electronic record-keeping and is not practical for dispensing public-assistance benefits.

Other off-line systems complete the transaction at the point of sale using a local computer (in the case of some magnetic stripe cards), or a microprocessor on the card itself (in the case of some smart cards) that stores the transaction on a local computer. These individual transactions are then bundled together and sent periodically from the local computer to a central computer in a single telephone connection. These bundled transactions are usually sent at night when telecommunication charges are low. Magnetic stripe cards can be used off-line when a personal identification number (PIN) is not necessary, or in combination with a driver's license or other identification. Smart cards can be used off-line when PIN verification is necessary, since the computer on the card can verify the PIN given by the cardholder.

On-line applications use a central computer to check the card information and an entered PIN against a stored database. Account balances and all other information are stored in the central computer. On-line systems can use inexpensive and simple magnetic stripe cards, and the central database can be updated quickly. On the other hand, fully on-line configurations require a telephone connection for every transaction, leading to delays and telecommunication costs. If the central computer fails, store transactions may be impossible or may require considerably more effort.

Public-Assistance Cards

The present food stamp program of the Department of Agriculture issues denominations of paper coupons that are accepted in lieu of paper currency in stores. Food stamps currently provide about $22 billion of benefits every year, reaching about 1 in 10 Americans. The Federal and State administrative costs for food stamps are about $1.8 billion and $1.4 billion, respectively. Clearly, any savings in administering programs such as food stamps could be used to reach more eligible citizens. The most obvious improvement is to convert the existing paper-based coupon system to an electronic system to eliminate paperwork.

Some current public-assistance programs operate as follows [2]-[5]:

The Treasury Department prints the coupons. The coupons are distributed to States, which distribute them to citizens, who can redeem the coupons in stores for food items. Store clerks must make sure that the coupons are used only for food items. Separate transactions, counting, and stamping the coupons attract the attention of other customers and create delays. The store counts and recounts the coupons again at day's end, and passes them with paperwork to its bank. The bank must recount the coupons, and then pass them on with paperwork to the Federal Reserve Bank. The Federal Reserve must count the coupons again, manage more paperwork, and settle accounts with the Treasury Department. Printing, distributing the coupons to States, and reconciling the coupons costs over $40 million per year for the Federal government, and over $20 million per year for State governments.[1]

Cards eliminate the paper transactions entirely. The recipient must separate items as before, but can use the card for food stamp purchases like mainstream customers. From the point-of-sale terminal, the retailer electronically records a debit transaction in the store computer. The store computer passes this debit to any intermediaries, and the intermediaries pass it to a bank. The bank passes it electronically to the Federal Reserve Bank, which debits the Treasury Department. Existing operational "food card" applications include the on-line use of magnetic stripe cards in Reading, PA (since 1984), and the off-line use of smart cards in Dayton, OH (since 1992).

The Special Supplemental Food Program for Women, Infants, and Children (WIC) of the Department of Health and Human Services (HHS) provides checks instead of coupons. WIC assistance reaches about one in seven children in the United States, and is intended to be preventative. The WIC program claims, for example, that every dollar spent for WIC assistance saves four dollars in health-care costs later in the child's life, about one-half of this in the first 90 days after birth.

The WIC checks can be applied only for certain foods prescribed by a public health nurse specifically for child care. Recipients must periodically return to a WIC office to renew the prescription and to receive counseling. Grocery store clerks handling WIC checks must be attentive to verify that the exact item to be purchased is listed on the check. One check is assigned per child, and therefore, a mother may have several checks. This process delays waiting customers. Card technologies can make this verification electronic by comparing the bar code on the product with the code registered in a computer. The computer can be centralized (on-line), or internal to the smart card (off-line). An off-line system using a smart card for WIC assistance has been tested in Casper, WY, since 1991.

Aid to Families with Dependent Children (AFDC), Social Security, Supplemental Security Income (SSI), and state and local general assistance are less restrictive. Checks are cashed and the recipient has total control over how the money is spent. Over 14 million Americans receive AFDC assistance, 46 million receive social security, and 300 000 receive SSI. Nationally, 54% of social security recipients receive their assistance by direct deposit to banks. Recipients can withdraw these funds with their bank cards. AFDC assistance is currently provided by an on-line system in Linn County, IA (since 1989); SSI assistance is provided in a demonstration on-line project in Houston, TX (since 1992).

One card could be used for all the items in a purchase if different programs were combined

[1] These figures represent the costs of actually handling the paper coupons. Other costs included in the $1.8 billion and $1.4 billion administrative costs include eligibility determination, training programs, matching funds to states, and other management costs not necessarily associated with the paper-based system

on one card, since many recipients independently receive assistance from more than one program. The Department of Labor estimates that four-fifths of AFDC and general assistance recipients, and 43% of SSI recipients, also receive food stamps. Conversely, 52% of food stamps recipients also receive AFDC and general assistance, and 21% also receive SSI. There are minor engineering problems regarding the sharing of memory and functions on the microprocessor chip. Much more important, however, government agencies are not accustomed to working together and may have widely different objectives and requirements that first must be accommodated. For example, in existing programs, recipients who live near state borders cannot shop in bordering states, and neighboring states do not share costs.

Cards have the potential to reduce errors and fraud through the use of passwords and encryption.

To accommodate these different programs, the computer storing the account information (whether stored on the card or in a central database) could be programmed in a number of ways. For example, the WIC account could be drawn first for its prescribed products, and then other accounts could be drawn according to a pre-arrangement between the different participating programs. Alternatively, funds could be drawn according to a percentage of each program's total contribution, or sequentially by the order that funds were deposited, or through other arrangements.

Multi-program systems are operational in the State of Maryland (since 1993), Ramsey County, MN (1988), and Albuquerque, NM (1990), with plans in New Jersey, Delaware, Wyoming, and other states (expected 1994-1995).

Direct Costs and Benefits of "Food Cards"

The various direct costs of using "food cards" to replace coupons can be divided into several categories: planning costs, implementation costs, and operational costs. Planning costs include plans prepared by the State agencies, preparing training materials, and so forth.

Implementation costs for the food stamps and WIC programs are mainly for installing point-of-sale (POS) terminals. Even if stores can use existing terminals, perhaps 300 000 more POS terminals may have to be installed nationwide to accommodate food stamp and WIC recipients. Some remote or small "mom and pop" stores, in particular, might need help to ensure that their stores could continue to provide assistance to eligible recipients. This cost could be reduced significantly by installing fewer terminals in large stores, or by waiting for stores, banks, or intermediaries to install terminals on their own. Existing automated teller machines (ATMs) could be used for AFDC, SSI, and general assistance through arrangements with ATM providers, therefore avoiding much of the installation costs for these programs.

Operating costs are mainly for transaction fees, and for the amortized cost of terminals distributed over 5 years (in the case of food stamps and WIC). Unless a single system operator owns the entire system, an intermediary generally charges from 50 cents to $1 to handle each transaction, and there may be several transactions per month. In programs that use checks, on the other hand, the checks are cashed only once per month. There may also be communication charges for dedicated or long-distance telephone lines.

Recipients save money due to improved convenience, particularly with programs that presently issue checks. Some check cashing services charge 2-5% or more of the face value of the check. On the other hand, the transaction and communication costs can be reduced by limiting the number of free transactions per month, increasing costs and inconvenience to the recipient.

Combining programs with existing point-of-sale systems may reduce overall costs for retailers and banks. In Albuquerque, NM, and Ramsey County, MN, pilot tests showed that retailers' costs dropped 13% and 22%, respectively, and the financial institutions' costs were virtually eliminated or showed a small profit.

Each stakeholder may also gain or lose interest income due to changes in float as a result of the new system. For example, if retailers and banks settle their accounts more quickly using electronic transactions as opposed to paper, the government loses (and banks gain) interest income for that difference in time. In the case of SSI assistance, funds are deposited initially in banks, and the banks lose (and recipients collectively gain) some interest income as the float time decreases.

Combining programs to use one card could reduce operational costs for the Federal government, compared to managing cards for each program, but the overall Federal and State costs may increase compared to paper coupons unless some of the retailer and bank savings are redistributed. The government could charge retailers, for example, for the use of the POS terminals in the same way that credit card issuers currently charge stores. Retailers, on the other hand, may wish to charge the government or the recipient for convenience, in the same way that ATM service providers currently charge cardholders.

The law, however, discourages redistributions of costs and benefits among programs, among State and Federal governments, and among retailers and the government. The law enacting the food stamps program requires that new electronic systems be "cost-neutral." Cost-neutrality means that the operational cost, including startup costs, of any new electronic benefits issuance program must "not exceed, in any one year, the operational cost of issuance systems in use prior to the implementation of the on-line electronic benefit transfer system" [6]. That is, the conversion is required to be not simply "cost-effective," but cost-neutral, since it only compares annual operational cost and investment, and not future savings or social benefits. Furthermore, tangible losses are passed to the State participating in the program, deterring States from innovation. Finally, public law stipulates that "the cost of documents or systems that may be required [for alternative issuance systems] may not be imposed upon a retail food store participating in the food stamp program" [7]. A 1987 Department of Agriculture report estimated that small, single-program food stamp systems would be unlikely to meet these cost neutrality provisions, but larger programs and programs that span multiple agencies might be cost-neutral [5].

Indirect Costs and Benefits of "Food Cards"

Aside from the direct costs, there are significant costs and benefits that are indirect or cannot be easily measured. These costs and benefits are real but cannot be included legally in the cost-neutrality calculation to determine if cards can be used in specific food stamp programs (as described above), although recent legislation seeks to change the current policy [8]. The indirect costs and benefits are numerous:

First, cards can greatly reduce various diversions of funds and counterfeiting. An estimated $1 billion per year is lost in the food stamps program through losses such as errors, overissuances, counterfeiting, and fraud committed by retailers and others [9]. Of this, an estimated $100 million per year is lost through trafficking in stolen coupons. These coupons are stolen and sold at a discount for cash. An estimated $1 million in counterfeit coupons have been discovered, but the actual amount of counterfeit coupons is unknown.

The Office of Management and Budget considers these diversions as a leak in the system, even though a reduction in diversions would not necessarily change the funds allocated to these programs.

Cards, on the other hand, can include a PIN and/or photo identification to deter theft or sale of the card. Smart cards are also particularly difficult to counterfeit. Tighter security would help assure that the funds were used by those for whom they were intended, and might therefore reach more people.

Second, cards can improve self-esteem on the part of assistance recipients. Recipients are often enthusiastic about cards: cards look like the credit or bank debit cards that are used more and more by mainstream customers. The cards simplify the transaction and therefore reduce the waiting time for other customers. The recipients also can carry over balances; an effect which is not possible with food stamp coupons or WIC checks. Store clerks welcome the cards because they simplify their task. Taxpayers may also view such assistance programs more positively if they perceive that the government is keeping up with technology and trying to improve services.

Third, cards could help to "mainstream" many recipients who are less accustomed to electronic funds and bank accounts. General assistance benefits could be disbursed via local banks and automatic teller machines, for example. Recipients would save time and some would also be less likely to be left behind in a cash-only economy, as society turns more and more to electronic transactions. Some savings may accrue by removing recipients sooner from assistance rolls. Other recipients, however, might resist changes or have difficulty with an electronic system.

A more attractive system could bring a larger number of applicants, possibly resulting in increased direct expenditures for taxpayers while serving new recipients. Many eligible citizens forego benefits because of the stigma that the current system attaches to the coupons and checks. Only about 60% of those qualified for food stamps actually apply and receive them [10]. The Wyoming WIC program estimates that only one-half of eligible citizens actually apply and receive WIC benefits. These citizens are accustomed to being self-sufficient. Many know the customers or clerks in the store.

Fourth, manufacturing and banking jobs may be created to fabricate the cards and to manage the intermediary functions. On the other hand, jobs may also be lost in the government, stores, and banks because of the elimination of paperwork and printing — an effect often not mentioned. In effect, unskilled workers may be replaced, in part, by fewer and higher-skilled workers.

Fifth, the extension of the so-called "Regulation E" by the Federal Reserve Board of Governors may also force Federal and State governments to absorb some losses that were previously shouldered by recipients. With a paper-based system, the recipients carry all liability if the coupons are lost or stolen. Regulation E establishes the rights and responsibilities of card providers to cardholders, including a limit on debit card liability for cardholders of $50 if the lost card is reported promptly [11]. Federal and State governments would be forced to accept the rest of any liability. By using passwords and quickly closing bad accounts, however, the providers should not expect any additional liability from lost or stolen cards. Meanwhile, the recipients benefit from added security of their benefits.

Finally, the funds used to finance the programs can be leveraged to add a stimulus to the economy much larger than the actual funds — perhaps as much as $1 billion annually. That is, if food stamp, WIC, and AFDC funds were deposited initially in banks like SSI funds, banks could reinvest the funds during the float period, but at the expense of interest income to Federal and State governments.

It may be that some costs and benefits must be excluded from consideration, simply because they are too imprecise, or because there are too many other variables. The decision to include some of these other costs and benefits still takes a more comprehensive perspective, however, than the present one of simply minimizing government costs. Long-term costs and benefits are difficult and imprecise to calculate, but are nevertheless important in order to understand the overall success of such programs.

ZONE 1 - Card holder's identifying information

This usually involves the full name, sex, date of birth, next of kin, and administrative numbers. It may also include access and PIN codes.

ZONE 2 - Emergency information

Information usually considered important in the first few minutes of an emergency.

ZONE 3 - Vaccination history

Information on vaccinations including travel immunizations.

ZONE 4 - Pharmaceutical and medications

Prescription drugs and over-the-counter drugs taken on a regular basis; allergies and intolerance to specific drugs. This zone could include such specifics as drug name, quantity, renewal schedule, and duration of treatment.

ZONE 5 - Medicine history

Details relating to medical history of family members, personal history, current care, preventive care, data justifying specific follow-up procedures.

Fig. 1. Possible applications of smart-card memory zones for medical information: illustration shows how the health care information contained on the smart card may be accessed and used. Zone 1: Identification information. All care providers would have access to this level. Only physicians, pharmacists, and the issuing organization would be permitted to make entries. Zone 2: Emergency information. All care providers would be authorized to read this zone. Only physicians would be authorized to make entries. Zone 3: Vaccination information. All providers with the exception of ambulance personnel would be authorized to read this zone. Zone 4: Medication information. Only physicians and pharmacists would be permitted to read or write in this zone. Zone 5: Medicine history. Only physicians would be permitted to read or write in this zone. Source: [17].

Medical-Record and Other Health-Care Cards

Another proposed card application is for a national or several regional health-care cards, or "health passports," which could also overlap with assistance cards. Both the Bush and Clinton Administrations have put forth health card proposals, and some states are moving ahead on their own. These health cards would be carried by citizens to interoperate with an information network, either centralized or distributed.

Several applications are possible. First, a health-care card could directly pay for certain services, such as prescription drugs. This might be similar to the Wyoming WIC card, but the health-care card could carry prescriptions for medicines or services instead of food, and could be redeemable at a medical clinic or a pharmacy instead of a grocery store.

Estimation of the associated planning, implementation, and operational costs might be relatively straightforward and would follow the analysis given above for public-assistance cards. Costs might include point-of-sale equipment and transaction fees, but possibly shared with other stakeholders. Cost savings could include eliminating reimbursement checks, postal fees, paperwork, and reduction of errors.

In a second application, the health-care card might act as a common interface to improve processing among many incompatible systems, but without completely eliminating paper or other existing on-line systems. The State of Arkansas, for example, is using a magnetic stripe card to speed Medicare eligibility verification when a patient walks into a doctor's office. The recipient is recognized or must provide an identification card such as a driver's license along with the Medicare card. The mag-

Secret zone
Unreadable For storage of passwords and cryptographic keys
Confidential zone
Read-only, with Password For storage of an audit trail of card transactions
Usage zone
Read/Write Access, with Password For storage of information actively used in applications
Public zone
Read-Only, without Password For storage of nonsensitive information, such as the issuer's name and address

Fig. 2. This figure illustrates a possible smart card memory divided into four zones: a secret zone, a confidential zone, a usage zone, and a public zone. A secret zone could be used for storage of information that can be used only by the microprocessor itself. Passwords, cryptographic keys, the card-bearer's digitized fingerprint, or any other information wich should never be readable outside of the smart card could be stored in this zone. Source: [13].

netic stripe card replaces a paper card and makes the processing automatic, eliminating forms, delays, and errors. Another example is the proposed "Health Passport" to be developed jointly by Wyoming, North Dakota, Montana, Idaho, and Nevada as a regional health-care card. This card would combine services such as Medicaid, Head Start, Indian Health Services, and many other medical programs onto one eligibility card.

The cost calculation in this second application might also be relatively straightforward and follow the same analysis as the public-assistance cards. The widespread use of electronic transactions in the health-care industry (not limited to the use of cards) may save $4 billion to $10 billion per year in administration and billing costs alone, according to one estimate [12]. Many physicians — perhaps even 40% — may already be capable of making such transactions. Only POS terminals would need to be added to make such systems fully electronic.

In a third application, a "medical-record" card might contain medical information immediately accessible in case of emergency, as a check against errors when prescribing medicines, or to reduce the need for redundant medical tests. Such a card might include only information about allergies or medications that a patient is currently taking, or the entire medical history of the citizen. The medical center at Lackland Air Force Base in Texas, for example, is using optical cards to store such personal medical information of local military personnel. Medical-record cards are being planned or implemented in France, Australia, Germany, Sweden, Italy, and the United Kingdom.

The direct cost of such medical-record cards does not appear significantly different from other health-care cards, except that higher performance and more expensive smart or optical cards would be necessary. The indirect costs are considerably different, however.

Indirect Costs of Medical-Record Cards

The indirect cost-benefit calculation for the medical-record card application is more complicated since, unlike other card applications, it contains personal information. Medical-record cards would not merely replace paper with electronic records, but they also would change the way that medical information is stored. Among the claimed indirect benefits derived from use of a card system are improved patient care, reduced health-care costs, improved record keeping for research on epidemiology, and enhanced education of health-care professionals. On the other hand, these changes in information storage could shift costs considerably from one player to another — to the patient in particular.

The information contained on such a health-care card, i.e., health information and the medical record, includes sensitive personal data revealing some of the most intimate aspects of an individual's life. In addition to diagnostic and testing information, the medical record includes the details of a person's family history, genetic testing, history of diseases and treatment, history of drug use, sexual orientation and practices, and testing for sexually transmitted diseases. Subjective remarks about a patient's demeanor, character, and mental state are sometimes a part of the record. The medical record is the primary source for much of the health-care information sought by parties outside the direct health-care delivery relationship. These data are important from a privacy standpoint, because health-care information can influence decisions about an individual's access to credit, admission to educational institutions, and his or her ability to secure employment and obtain insurance. Inaccuracies in the information, or its improper use, can affect an individual's access to these necessities of life. Yet accurate and comprehensive health-care information is critical to the physician-patient relationship, and as a result, the quality of health-care delivery.

Thus, hidden costs of a medical-record card include any necessary modifications to the technology to assure confidentiality in the data. The hidden costs also include broad policy innovations to clearly delineate the individual's right to this information, and approved uses of the information outside the health-care relationship.

With respect to the technology, it may be necessary to make use of the ability to store and to access the card's expansive memory to ensure that the information stored in it is secure. The memory of a smart card can be divided into several zones, each with different levels of security and requirements for access, as required for each specific application. The smart card microprocessor and its associated operating system can keep track of which memory addresses belong to which zones, and the conditions under which each zone can be accessed. (Figs. 1 and 2 illustrate the possible design of the card described below.)

An indirect benefit of card use may be improved self-esteem on the part of assistance recipients.

A confidential zone could be used to store an audit trail listing all transactions, or attempted transactions, made with the card. The confidential zone could have a password known only to the card issuer, who could examine the history of the card for evidence of misuses of the system. To prevent any attempts to modify the card's audit trail, the confidential zone could have a read-only access restriction, so that the system could write to the zone, but information could not be changed from the outside.

A usage zone could be used for storage of information that is specific to the smart card application and that requires periodic updates and modification. For example, the date of the card bearer's last access to the host computer or the amount of computer time used could be stored in the usage zone. Depending upon the sensitivity of the data, the usage zone could have both read and write access protected by a password.

A public zone could hold nonsensitive information, such as the card issuer's name and address. The public zone could have read-only access, without a password.

Crucial secret information can be maintained in separate protected memory locations. Passwords, cryptographic keys, and other information which should never be readable outside of the smart card could be located here. It may also be possible to produce a smart card that would ensure that the entire secret zone will be destroyed if any attempt is made to access the data in that zone [13].

Care providers would be equipped with a reader, microcomputer, and necessary software. Each provider would be given an accreditation card to gain access to the patient's smart card, which defines the accessible zones. The provider would also have to enter a PIN before the smart card could be accessed.

Concerns have been raised about patient compliance with carrying the card, which is useless if lost, forgotten, or stolen. For this reason, none of the current proposals for use of the card suggest that the medical data reside solely on the card for that reason. The proposed solution, creation of a back-up database containing the patient information,[2] presents its own information privacy problems. These concerns include the security of the information at the host, the patient's lack of awareness of the nature of the information he or she is required to carry, and the dependence of the patient on the health-care administrators to determine what information should be accessed by which health-care provider, insurer, or other third party. In addition, there is some concern that patients will not want information about certain highly sensitive conditions recorded on the card. As a result, some people question whether a smart card would contain a comprehensive medical record or an abbreviated version with its attendant limitations.

The proposed legislative solutions to these problems carry with them inherent costs. Because existing laws to ensure privacy of this information are not consistent throughout the country and do not necessarily address information in a computerized environment, proposed solutions include enacting a comprehensive health-care information privacy law. Such a law would be based upon longstanding principles of fair information practices, which are also reflected in the Privacy Act of 1974 [14]:

1. There must be no secret personal-data record-keeping system.
2. There must be a way for individuals to discover what personal information is recorded and how it is used.
3. There must be a way for individuals to prevent information obtained for one purpose from being used or made available for other purposes without their consent.
4. There must be a way for individuals to correct or amend a record of information about themselves.
5. An organization creating, maintaining, using or disseminating records of identifiable

[2]Each of the current proposals for implementation of an electronic card system also calls for one or more databases on the other end of the medical/insurance transaction, keeping track of every claim filed and every medical treatment administered.

personal data must assure the reliability of the data for its intended use and must take reasonable precautions to prevent misuses of the data [15].

Such a law might address the potential for abuse of information by authorized parties with appropriate access to a computer system, create criminal and civil recourse for illegally obtaining or disclosing records, and provide appropriate sanctions for such activities. Rules might be established for patient education about collection and handling of health-care information, including access to information; amendment, correction and deletion of information; and creation of databases. Some also suggest that requirements for consent by patients to disclose information might be established through legislation, in addition to protocols for access to health-care information by secondary users, and a clear determination of the rights and responsibilities in the information they access.[3]

The cost of these policies would be borne by a variety of parties, health-care professionals, administrators, and support staff, as well as by individual patients who would be required to take more responsibility for their own health data. Computer system and software developers would likely be required to create security and application systems that would facilitate compliance with the new law.

New Technologies, Old Policies

Integrated circuit "smart" cards and magnetic stripe cards are being increasingly promoted by federal and state agencies to disburse food stamps and other public assistance, and to assist in the administration of health care. These new technologies are invariably presented as cost-effective improvements over existing paper-based systems, although social costs and benefits are often not included in the analysis.

In the case of food stamps, current law restricts the inclusion of important indirect costs and benefits in cost-effectiveness calculations. Current policy also restricts redistribution of costs and benefits that might be necessary to find an equitable solution among providers.

In the health-care industry, cards and centralized databases are also being promoted to streamline processing and reduce future health-care costs. However, a national medical-record card or a centralized medical-records database would greatly change individuals' sense and expectation of privacy. For such applications, enactment of a broad privacy law would be necessary to protect the individual and to shift some costs from individuals to the health-care system. Despite its difficulty, such comprehensive cost/benefit evaluations should be made when making policy decisions, to account for all the costs and redistributions of costs among the players that may be necessary.

[3]For a detailed discussion of proposals for health-care information privacy legislation, see [16].

References

[1] Jerome Svigals, *Smart Cards: The New Bank Cards*. New York, NY: Macmillan Pub. Co., 1987.

[2] U.S. Congress, Office of Technology Assessment, *Making Government Work: Electronic Delivery of Federal Services*. Washington, DC: U.S. Government Printing Office, Sept. 1993, Pub. OTA-TCT-578, ch. 4.

[3] Gary L. Glickman, Phoenix Planning & Evaluation, Ltd., "Multi-program cards for the delivery of social services," contractor report prepared for the Office of Technology Assessment, Pub. PB94-113354. Springfield, VA: National Technical Information Service, Jan. 1993.

[4] John Harris, Alan F. Westin, and Anne L. Finger, Reference Point Foundation, "Innovations for federal service: A study of innovative technologies for federal government services to older Americans and consumers," contractor report prepared for the Office of Technology Assessment, Pub. PB94-113297. Springfield, VA: National Technical Information Service, Feb. 1993.

[5] U.S. Department of Agriculture, Food and Nutrition Service, Office of Analysis and Evaluation, "Electronic benefit transfer in the food stamp program: The first decade," Mar. 1992.

[6] The Mickey Leland Memorial Domestic Hunger Relief Act, Public Law 101-624, Section 1729 (a), 104 Stat. 3789, codified as 7 U.S.C. 2016 (i).

[7] The Omnibus Budget Reconciliation Act of 1982, Public Law 97-253, Section 162, 96 Stat. 778, codified as 7 U.S.C. 2016 (g).

[8] S. 1646, "The Food Stamp Fraud Reduction Act of 1993," Congressional Record, Nov. 8, 1993, S15336.

[9] U.S. Senate, Committee on Agriculture, Nutrition, and Forestry, hearing on "Fraud in Federal Nutrition Programs," Feb. 2, 1994.

[10] Jason DeParle, "Food stamp users up sharply in sign of weak recovery," *New York Times*, Mar. 2, 1993, p. A18.

[11] The Electronic Funds Transfer Act, Public Law 95-630, Section 909, 92 Stat. 3734 codified as 15 U.S.C. 1693g.

[12] Workgroup for Electronic Data Interchange (WEDI), report to the Secretary of the U.S. Department of Health and Human Services, July, 1992.

[13] Martha E. Haykin and Robert B. J. Warnar, U.S. Department of Commerce, National Institute of Standards and Technology, "Smart card technology: New methods for computer access control," NIST Spec. Pub. 500-147, Sept. 1988, pp. 13-26.

[14] The Privacy Act of 1974, Public Law 579, 88 Stat. 1896, codified as 5 U.S.C. Sec. 552a.

[15] U.S. Department of Health, Education and Welfare, *The Secretary's Advisory Committee on Automated Personal Data Systems, Computers and the Rights of Citizens*. Washington D.C.: U.S. Government Printing Office, 1973.

[16] U.S. Congress, Office of Technology Assessment, *Protecting Privacy in Computerized Medical Information*, Rep. OTA-TCT-576. Washington, DC: U.S. Government Printing Office, Sept. 1993.

[17] Simon Davies, *Big Brother: Australia's Growing Web of Surveillance*. Australia: Simon and Schuster, 1992; and Office of Technology Assessment 1993.

T&S

Appendix

IEEE Code of Ethics

WE, the members of the IEEE, in recognition of the importance of our technologies in affecting the quality of life throughout the world, and in accepting a personal obligation to our profession, its members and the communities we serve, do hereby commit ourselves to the highest ethical and professional conduct and agree:

1. to accept responsibility in making engineering decisions consistent with the safety, health and welfare of the public, and to disclose promptly factors that might endanger the public or the environment;
2. to avoid real or perceived conflicts of interest whenever possible, and to disclose them to affected parties when they do exist;
3. to be honest and realistic in stating claims or estimates based on available data;
4. to reject bribery in all its forms;
5. to improve the understanding of technology, its appropriate application, and potential consequences;
6. to maintain and improve our technical competence and to undertake technological tasks for others only if qualified by training or experience, or after full disclosure of pertinent limitations;
7. to seek, accept, and offer honest criticism of technical work, to acknowledge and correct errors, and to credit properly the contributions of others;
8. to treat fairly all persons regardless of such factors as race, religion, gender, disability, age, or national origin;
9. to avoid injuring others, their property, reputation, or employment by false or malicious action;
10. to assist colleagues and co-workers in their professional development and to support them in following this code of ethics.

Approved by the IEEE Board of Directors, August 1990.

NSPE Code of Ethics for Engineers

Preamble

Engineering is an important and learned profession. As members of this profession, engineers are expected to exhibit the highest standards of honesty and integrity. Engineering has a direct and vital impact on the quality of life for all people. Accordingly, the services provided by engineers require honesty, impartiality, fairness and equity, and must be dedicated to the protection of the public health, safety, and welfare. Engineers must perform under a standard of professional behavior that requires adherence to the highest principles of ethical conduct.

I. Fundamental Canons

Engineers, in the fulfillment of their professional duties, shall:

1. Hold paramount the safety, health and welfare of the public.
2. Perform services only in areas of their competence.
3. Issue public statements only in an objective and truthful manner.
4. Act for each employer or client as faithful agents or trustees.
5. Avoid deceptive acts.
6. Conduct themselves honorably, responsibly, ethically, and lawfully so as to enhance the honor, reputation, and usefulness of the profession.

II. Rules of Practice

1. Engineers shall hold paramount the safety, health, and welfare of the public.
   (a) If engineers' judgment is overruled under circumstances that endanger life or property, they shall

notify their employer or client and such other authority as may be appropriate.

(b) Engineers shall approve only those engineering documents that are in conformity with applicable standards.

(c) Engineers shall not reveal facts, data or information without the prior consent of the client or employer except as authorized or required by law or this Code.

(d) Engineers shall not permit the use of their name or associate in business ventures with any person or firm that they believe are engaged in fraudulent or dishonest enterprise.

(e) Engineers having knowledge of any alleged violation of this Code shall report thereon to appropriate professional bodies and, when relevant, also to public authorities, and cooperate with the proper authorities in furnishing such information or assistance as may be required.

2. Engineers shall perform services only in the areas of their competence.

(a) Engineers shall undertake assignments only when qualified by education or experience in the specific technical fields involved.

(b) Engineers shall not affix their signatures to any plans or documents dealing with subject matter in which they lack competence, nor to any plan or document not prepared under their direction and control.

(c) Engineers may accept assignments and assume responsibility for coordination of an entire project and sign and seal the engineering documents for the entire project, provided that each technical segment is signed and sealed only by the qualified engineers who prepared the segment.

3. Engineers shall issue public statements only in an objective and truthful manner.

(a) Engineers shall be objective and truthful in professional reports, statements, or testimony. They shall include all relevant and pertinent information in such reports, statements, or testimony, which should bear the date indicating when it was current.

(b) Engineers may express publicly technical opinions that are founded upon knowledge of the facts and competence in the subject matter.

(c) Engineers shall issue no statements, criticisms, or arguments on technical matters that are inspired or paid for by interested parties, unless they have prefaced their comments by explicitly identifying the interested parties on whose behalf they are speaking, and by revealing the existence of any interest the engineers may have in the matters.

4. Engineers shall act for each employer or client as faithful agents or trustees.

(a) Engineers shall disclose all known or potential conflicts of interest that could influence or appear to influence their judgment or the quality of their services.

(b) Engineers shall not accept compensation, financial or otherwise, from more than one party for services on the same project, or for services pertaining to the same project, unless the circumstances are fully disclosed and agreed to by all interested parties.

(c) Engineers shall not solicit or accept financial or other valuable consideration, directly or indirectly, from outside agents in connection with the work for which they are responsible.

(d) Engineers in public service as members, advisors, or employees of a governmental or quasi-governmental body or department shall not participate in decisions with respect to services solicited or provided by them or their organizations in private or public engineering practice.

(e) Engineers shall not solicit or accept a contract from a governmental body on which a principal or officer of their organization serves as a member.

5. Engineers shall avoid deceptive acts.

(a) Engineers shall not falsify their qualifications or permit misrepresentation of their or their associates' qualifications. They shall not misrepresent or exaggerate their responsibility in or for the subject matter of prior assignments. Brochures or other presentations incident to the solicitation of employment shall not misrepresent pertinent facts concerning employers, employees, associates, joint venturers, or past accomplishments.

(b) Engineers shall not offer, give, solicit or receive, either directly or indirectly, any contribution to influence the award of a contract by public authority, or which may be reasonably construed by the public as having the effect of intent to influencing the awarding of a contract. They shall not offer any gift or other valuable consideration in order to secure work. They shall not pay a commission, percentage, or brokerage fee in order to secure work, except to a bona fide employee or bona fide established commercial or marketing agencies retained by them.

III. Professional Obligations

1. Engineers shall be guided in all their relations by the highest standards of honesty and integrity.

(a) Engineers shall acknowledge their errors and shall not distort or alter the facts.

(b) Engineers shall advise their clients or employers when they believe a project will not be successful.

(c) Engineers shall not accept outside employment to the detriment of their regular work or interest. Before accepting any outside engineering employment they will notify their employers.

(d) Engineers shall not attempt to attract an engineer from another employer by false or misleading pretenses.
(e) Engineers shall not actively participate in strikes, picket lines, or other collective coercive action.
(f) Engineers shall not promote their own interest at the expense of the dignity and integrity of the profession.

2. Engineers shall at all times strive to serve the public interest.
   (a) Engineers shall seek opportunities to participate in civic affairs; career guidance for youths; and work for the advancement of the safety, health and wellbeing of their community.
   (b) Engineers shall not complete, sign, or seal plans and/or specifications that are not in conformity with applicable engineering standards. If the client or employer insists on such unprofessional conduct, they shall notify the proper authorities and withdraw from further service on the project.
   (c) Engineers shall endeavor to extend public knowledge and appreciation of engineering and its achievements.
3. Engineers shall avoid all conduct or practice that deceives the public.
   (a) Engineers shall avoid the use of statements containing a material misrepresentation of fact or omitting a material fact.
   (b) Consistent with the foregoing, Engineers may advertise for recruitment of personnel.
   (c) Consistent with the foregoing, Engineers may prepare articles for the lay or technical press, but such articles shall not imply credit to the author for work performed by others.
4. Engineers shall not disclose, without consent, confidential information concerning the business affairs or technical processes of any present or former client or employer, or public body on which they serve.
   (a) Engineers shall not, without the consent of all interested parties, promote or arrange for new employment or practice in connection with a specific project for which the Engineer has gained particular and specialized knowledge.
   (b) Engineers shall not, without the consent of all interested parties, participate in or represent an adversary interest in connection with a specific project or proceeding in which the Engineer has gained particular specialized knowledge on behalf of a former client or employer.
5. Engineers shall not be influenced in their professional duties by conflicting interests.
   (a) Engineers shall not accept financial or other considerations, including free engineering designs, from material or equipment suppliers for specifying their product.
   (b) Engineers shall not accept commissions or allowances, directly or indirectly, from contractors or other parties dealing with clients or employers of the Engineer in connection with work for which the Engineer is responsible.
6. Engineers shall not attempt to obtain employment or advancement or professional engagements by untruthfully criticizing other engineers, or by other improper or questionable methods.
   (a) Engineers shall not request, propose, or accept a commission on a contingent basis under circumstances in which their judgment may be compromised.
   (b) Engineers in salaried positions shall accept part-time engineering work only to the extent consistent with policies of the employer and in accordance with ethical considerations.
   (c) Engineers shall not, without consent, use equipment, supplies, laboratory, or office facilities of an employer to carry on outside private practice.
7. Engineers shall not attempt to injure, maliciously or falsely, directly or indirectly, the professional reputation, prospects, practice, or employment of other engineers. Engineers who believe others are guilty of unethical or illegal practice shall present such information to the proper authority for action.
   (a) Engineers in private practice shall not review the work of another engineer for the same client, except with the knowledge of such engineer, or unless the connection of such engineer with the work has been terminated.
   (b) Engineers in governmental, industrial, or educational employ are entitled to review and evaluate the work of other engineers when so required by their employment duties.
   (c) Engineers in sales or industrial employ are entitled to make engineering comparisons of represented products with products of other suppliers.
8. Engineers shall accept personal responsibility for their professional activities, provided, however, that Engineers may seek indemnification for services arising out of their practice for other than gross negligence, where the Engineer's interests cannot otherwise be protected.
   (a) Engineers shall conform with state registration laws in the practice of engineering.
   (b) Engineers shall not use association with a nonengineer, a corporation, or partnership as a "cloak" for unethical acts.
9. Engineers shall give credit for engineering work to those to whom credit is due and will recognize the proprietary interests of others.
   (a) Engineers shall, whenever possible, name the person or persons who may be individually responsible for designs, inventions, writings, or other accomplishments.
   (b) Engineers using designs supplied by a client recognize that the designs remain the property of the

client and may not be duplicated by the Engineer for others without express permission.

(c) Engineers, before undertaking work for others in connection with which the Engineer may make improvements, plans, designs, inventions, or other records that may justify copyrights or patents, should enter into a positive agreement regarding ownership.

(d) Engineers' designs, data, records, and notes referring exclusively to an employer's work are the employer's property. Employer should indemnify the Engineer for use of the information for any purpose other than the original purpose.

As Revised July 1996

"By order of the United States District Court for the District of Columbia, former Section 11(c) of the NSPE Code of Ethics prohibiting competitive bidding, and all policy statements, opinions, rulings or other guidelines interpreting its scope, have been rescinded as unlawfully interfering with the legal right of engineers, protected under the antitrust laws, to provide price information to prospective clients; accordingly, nothing contained in the NSPE Code of Ethics, policy statements, opinions, rulings or other guidelines prohibits the submission of price quotations or competitive bids for engineering services at any time or in any amount."

Statement by NSPE Executive Committee

In order to correct misunderstandings which have been indicated in some instances since the issuance of the Supreme Court decision and the entry of the Final Judgment, it is noted that in its decision of April 25, 1978, the Supreme Court of the United States declared: "The Sherman Act does not require competitive bidding."

It is further noted that as made clear in the Supreme Court decision:

1. Engineers and firms may individually refuse to bid for engineering services.
2. Clients are not required to seek bids for engineering services.
3. Federal, state, and local laws governing procedures to procure engineering services are not affected, and remain in full force and effect.
4. State societies and local chapters are free to actively and aggressively seek legislation for professional selection and negotiation procedures by public agencies.
5. State registration board rules of professional conduct, including rules prohibiting competitive bidding for engineering services, are not affected and remain in full force and effect. State registration boards with authority to adopt rules of professional conduct may adopt rules governing procedures to obtain engineering services.
6. As noted by the Supreme Court, "nothing in the judgment prevents NSPE and its members from attempting to influence governmental action . . ."

NOTE: In regard to the question of application of the Code to corporations vis-à-vis real persons, business form or type should not negate nor influence conformance of individuals to the Code. The Code deals with professional services, which services must be performed by real persons. Real persons in turn establish and implement policies within business structures. The Code is clearly written to apply to the Engineer and items incumbent on members of NSPE to endeavor to live up to its provisions. This applies to all pertinent sections of the Code.

Author Index

Subject Index

About the Editor

Joseph R. Herkert is assistant professor of Multidisciplinary Studies at North Carolina State University where he teaches in the Science, Technology and Society Program and is director of the Benjamin Franklin Scholars Program, a dual-degree program in engineering and humanities/social sciences. Dr. Herkert also holds an Associate Appointment in the Department of Electrical and Computer Engineering. The courses Dr. Herkert offers include: Engineering Ethics; Contemporary Science, Technology and Values; Technological Catastrophes; Technology Assessment and Policy; and graduate seminars in Technology and Society and Energy for Sustainable Development.

Dr. Herkert's research interests include engineering ethics, social impacts of electronic media, and energy/environmental policy. His work has appeared in such journals as *IEEE Technology and Society Magazine, IEEE Transactions on Education, Technological Forecasting and Social Change, Science and Engineering Ethics,* and *The Journal of Professional Issues in Engineering Education and Practice.* He recently served as guest-editor of special issues of *Social Science Computer Review* and *The Journal of Electronic Publishing* (http://www.press.umich.edu/jep/04-02/index.html).

Dr. Herkert received his doctorate in Engineering and Policy in 1987 from Washington University in St. Louis, Missouri, and his B.S. in Electrical Engineering in 1970 from Southern Methodist University. He is a registered professional engineer with nearly six years experience as a consultant in the electric utility industry, including planning and financial feasibility studies of major power generation facilities. Dr. Herkert is a senior member of IEEE and a past-president of the IEEE Society on Social Implications of Technology (SSIT). He currently serves as Publications Chair of SSIT and Editorial Board Chair of *IEEE Technology and Society Magazine.*

Website: http://www4.ncsu.edu/unity/users/j/jherkert/jrh.html